Introduction to Statistical Machine Learning

Introduction to Statistical Machine Learning

Masashi Sugiyama

AMSTERDAM • BOSTON • HEIDELBERG • LONDON
NEW YORK • OXFORD • PARIS • SAN DIEGO
SAN FRANCISCO • SINGAPORE • SYDNEY • TOKYO
Morgan Kaufmann Publishers is an Imprint of Elsevier

Acquiring Editor: Todd Green
Editorial Project Manager: Amy Invernizzi
Project Manager: Mohanambal Natarajan
Designer: Maria Ines Cruz

Morgan Kaufmann is an imprint of Elsevier
225 Wyman Street, Waltham, MA 02451 USA

Library of Congress Cataloging-in-Publication Data
A catalog record for this book is available from the Library of Congress

British Library Cataloging-in-Publication Data
A catalogue record for this book is available from the British Library.

ISBN: 978-0-12-802121-7

Contents

List of Figures

List of Tables

Biography

MASASHI SUGIYAMA

Masashi Sugiyama received the degrees of Bachelor of Engineering, Master of Engineering, and Doctor of Engineering in Computer Science from Tokyo Institute of Technology, Japan in 1997, 1999, and 2001, respectively. In 2001, he was appointed Assistant Professor in the same institute, and he was promoted to Associate Professor in 2003. He moved to the University of Tokyo as Professor in 2014. He received an Alexander von Humboldt Foundation Research Fellowship and researched at Fraunhofer Institute, Berlin, Germany, from 2003 to 2004. In 2006, he received a European Commission Program Erasmus Mundus Scholarship and researched at the University of Edinburgh, Edinburgh, UK. He received the Faculty Award from IBM in 2007 for his contribution to machine learning under nonstationarity, the Nagao Special Researcher Award from the Information Processing Society of Japan in 2011 and the Young Scientists' Prize from the Commendation for Science and Technology by the Minister of Education, Culture, Sports, Science and Technology Japan for his contribution to the density-ratio paradigm of machine learning. His research interests include theories and algorithms of machine learning and data mining, and a wide range of applications such as signal processing, image processing, and robot control.

Preface

Machine learning is a subject in computer science, aimed at studying theories, algorithms, and applications of systems that learn like humans. Recent development of computers and sensors allows us to access a huge amount of data in diverse domains such as documents, audio, images, movies, e-commerce, electric power, medicine, and biology. Machine learning plays a central role in analyzing and benefiting from such *big data*.

This textbook is devoted to presenting mathematical backgrounds and practical algorithms of various machine learning techniques, targeting undergraduate and graduate students in computer science and related fields. Engineers who are applying machine learning techniques in their business and scientists who are analyzing their data can also benefit from this book.

A distinctive feature of this book is that each chapter concisely summarizes the main idea and mathematical derivation of particular machine learning techniques, followed by compact MATLAB programs. Thus, readers can study both mathematical concepts and practical values of various machine learning techniques simultaneously. All MATLAB programs are available from

"`http://www.ms.k.u-tokyo.ac.jp/software/SMLbook.zip`".

This book begins by giving a brief overview of the field of machine learning in Part 1. Then Part 2 introduces fundamental concepts of *probability and statistics*, which form the mathematical basis of statistical machine learning. Part 2 was written based on

> Sugiyama, M.
> Probability and Statistics for Machine Learning,
> Kodansha, Tokyo, Japan, 2015. (in Japanese).

Part 3 and Part 4 present a variety of practical machine learning algorithms in the *generative* and *discriminative* frameworks, respectively. Then Part 5 covers various advanced topics for tackling more challenging machine learning tasks. Part 3 was written based on

> Sugiyama, M.
> Statistical Pattern Recognition: Pattern Recognition Based on Generative Models,
> Ohmsha, Tokyo, Japan, 2009. (in Japanese),

and Part 4 and Part 5 were written based on

> Sugiyama, M.
> An Illustrated Guide to Machine Learning,
> Kodansha, Tokyo, Japan, 2013. (in Japanese).

The author would like to thank researchers and students in his groups at the University of Tokyo and Tokyo Institute of Technology for their valuable feedback on earlier manuscripts.

Masashi Sugiyama
The University of Tokyo

PART

INTRODUCTION

1

CHAPTER

STATISTICAL MACHINE LEARNING

1

CHAPTER CONTENTS

Recent development of computers and the Internet allows us to immediately access a vast amount of information such as texts, sounds, images, and movies. Furthermore, a wide range of personal data such as search logs, purchase records, and diagnosis history are accumulated everyday. Such a huge amount of data is called *big data*, and there is a growing tendency to create new values and business opportunities by extracting useful knowledge from data. This process is often called *data mining*, and *machine learning* is the key technology for extracting useful knowledge. In this chapter, an overview of the field of machine learning is provided.

1.1 TYPES OF LEARNING

Depending on the type of available data, machine learning can be categorized into *supervised learning*, *unsupervised learning*, and *reinforcement learning*.

Supervised learning would be the most fundamental type of machine learning, which considers a student learning from a supervisor through questioning and answering. In the context of machine learning, a student corresponds to a computer and a supervisor corresponds to a user of the computer, and the computer learns a mapping from a question to its answer from paired samples of questions and answers. The objective of supervised learning is to acquire the *generalization ability*, which refers to the capability that an appropriate answer can be guessed for unlearned questions. Thus, the user does not have to teach everything to the computer, but the computer can automatically cope with unknown situations by learning only a fraction of knowledge. Supervised learning has been successfully applied to a wide range of real-world problems, such as hand-written letter recognition, speech recognition, image recognition, spam filtering, information retrieval, online advertisement, recommendation, brain signal analysis, gene analysis, stock price prediction, weather forecasting, and astronomy data analysis. The supervised learning problem is particularly called *regression* if the answer is a real value (such as the temperature), *classification* if the answer is a categorical value (such as "yes" or "no"), and *ranking* if the answer is an ordinal value (such as "good," "normal," or "poor").

Unsupervised learning considers the situation where no supervisor exists and a student learns by himself/herself. In the context of machine learning, the computer autonomously collects data through the Internet and tries to extract useful knowledge without any guidance from the user. Thus, unsupervised learning is more automatic than supervised learning, although its objective is not necessarily specified clearly. Typical tasks of unsupervised learning include *data clustering* and *outlier detection*, and these unsupervised learning techniques have achieved great success in a wide range of real-world problems, such as system diagnosis, security, event detection, and social network analysis. Unsupervised learning is also often used as a preprocessing step of supervised learning.

Reinforcement learning is aimed at acquiring the generalization ability in the same way as supervised learning, but the supervisor does not directly give answers to the student's questions. Instead, the supervisor *evaluates* the student's behavior and gives feedback about it. The objective of reinforcement learning is, based on the feedback from the supervisor, to let the student improve his/her behavior to maximize the supervisor's evaluation. Reinforcement learning is an important model of the behavior of humans and robots, and it has been applied to various areas such as autonomous robot control, computer games, and marketing strategy optimization. Behind reinforcement learning, supervised and unsupervised learning methods such as regression, classification, and clustering are often utilized.

The focus on this textbook is supervised learning and unsupervised learning. For reinforcement learning, see references [99, 105]

1.2 EXAMPLES OF MACHINE LEARNING TASKS

In this section, various supervised and unsupervised learning tasks are introduced in more detail.

1.2.1 SUPERVISED LEARNING

The objective of *regression* is to approximate a real-valued function from its samples (Fig. 1.1). Let us denote the input by d-dimensional real vector $\boldsymbol{x}$, the output by a real scalar y, and the learning target function by $y = f(\boldsymbol{x})$. The learning target function f is assumed to be unknown, but its input-output paired samples $\{(\boldsymbol{x}_i, y_i)\}_{i=1}^{n}$ are observed. In practice, the observed output value y_i may be corrupted by some noise ϵ_i, i.e., $y_i = f(\boldsymbol{x}_i) + \epsilon_i$. In this setup, $\boldsymbol{x}_i$ corresponds to a question that a student asks the supervisor, and y_i corresponds to the answer that the supervisor gives to the student. Noise ϵ_i may correspond to the supervisor's mistake or

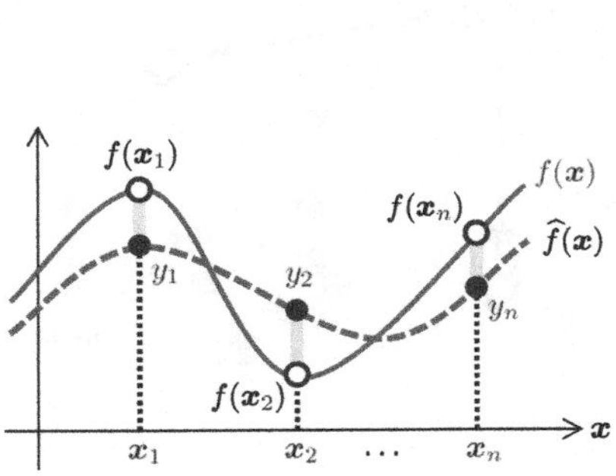

FIGURE 1.1

Regression.

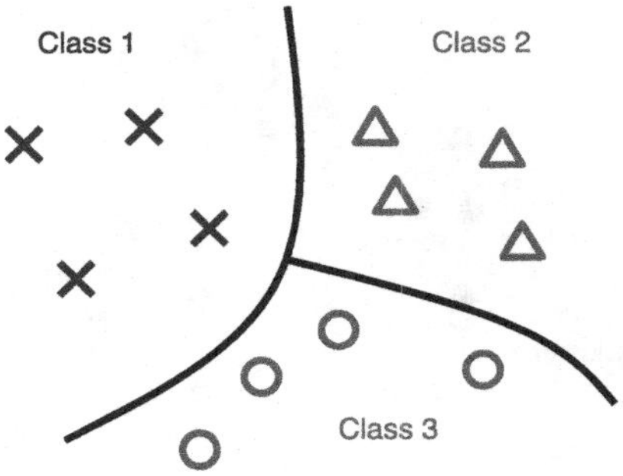

FIGURE 1.2

Classification.

the student's misunderstanding. The learning target function f corresponds to the supervisor's knowledge, which allows him/her to answer any questions. The objective of regression is to let the student learn this function, by which he/she can also answer any questions. The level of generalization can be measured by the closeness between the true function f and its approximation $\widehat{f}$.

On the other hand, *classification* is a *pattern recognition* problem in a supervised manner (Fig. 1.2). Let us denote the input pattern by d-dimensional vector $\boldsymbol{x}$ and its class by a scalar $y \in \{1,\ldots,c\}$, where c denotes the number of classes. For training a classifier, input-output paired samples $\{(\boldsymbol{x}_i, y_i)\}_{i=1}^n$ are provided in the same way as regression. If the true classification rule is denoted by $y = f(\boldsymbol{x})$, classification can also be regarded as a function approximation problem. However, an essential difference is that there is no notion of closeness in y: $y = 2$ is closer to $y = 1$ than $y = 3$ in the case of regression, but whether y and y' are the same is the only concern in classification.

The problem of *ranking* in supervised learning is to learn the rank y of a sample $\boldsymbol{x}$. Since the rank has the order, such as $1 < 2 < 3$, ranking would be more similar to regression than classification. For this reason, the problem of ranking is also referred to as *ordinal regression*. However, different from regression, exact output value y is not necessary to be predicted, but only its relative value is needed. For example, suppose that "values" of three instances are 1, 2, and 3. Then, since only the ordinal relation $1 < 2 < 3$ is important in the ranking problem, predicting the values as $2 < 4 < 9$ is still a perfect solution.

1.2.2 UNSUPERVISED LEARNING

Clustering is an unsupervised counter part of classification (Fig. 1.3), and its objective is to categorize input samples $\{\boldsymbol{x}_i\}_{i=1}^n$ into clusters $1,2,\ldots,c$ without any supervision $\{y_i\}_{i=1}^n$. Usually, similar samples are supposed to belong to the same cluster, and

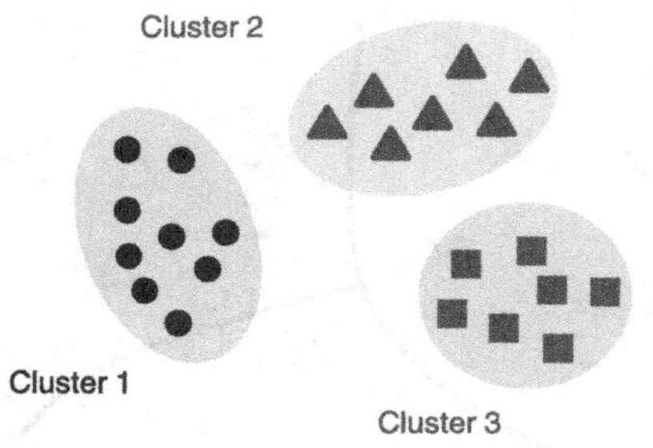

FIGURE 1.3

Clustering.

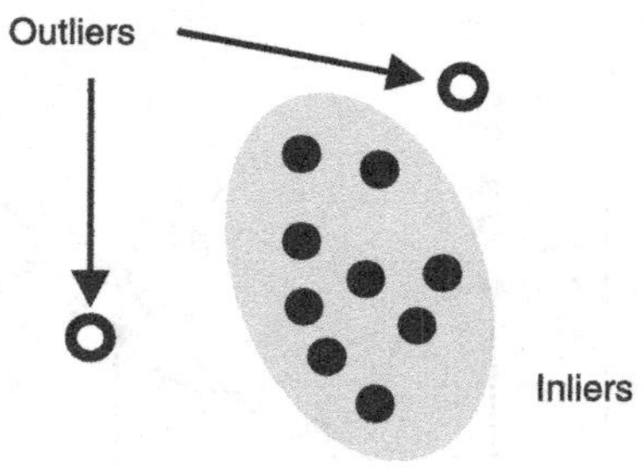

FIGURE 1.4

Outlier detection.

dissimilar samples are supposed to belong to different clusters. Thus, how to measure the *similarity* between samples is the key issue in clustering.

Outlier detection, which is also referred to as *anomaly detection*, is aimed at finding irregular samples in a given data set $\{\boldsymbol{x}_i\}_{i=1}^n$. In the same way as clustering, the definition of similarity between samples plays a central role in outlier detection, because samples that are dissimilar from others are usually regarded as outliers (Fig. 1.4).

The objective of *change detection*, which is also referred to as *novelty detection*, is to judge whether a newly given data set $\{\boldsymbol{x}'_{i'}\}_{i'=1}^{n'}$ has the same property as the original data set $\{\boldsymbol{x}_i\}_{i=1}^n$. Similarity between samples is utilized in outlier detection, while similarity between data sets is needed in change detection. If $n' = 1$, i.e., only a single point is provided for detecting change, the problem of change detection may be reduced to an outlier problem.

1.2.3 FURTHER TOPICS

In addition to supervised and unsupervised learnings, various useful techniques are available in machine learning.

Input-output paired samples $\{(\boldsymbol{x}_i, y_i)\}_{i=1}^n$ are used for training in supervised learning, while input-only samples $\{\boldsymbol{x}_i\}_{i=1}^n$ are utilized in unsupervised learning. In many supervised learning techniques, collecting input-only samples $\{\boldsymbol{x}_i\}_{i=1}^n$ may be easy, but obtaining output samples $\{y_i\}_{i=1}^n$ for $\{\boldsymbol{x}_i\}_{i=1}^n$ is laborious. In such a case, output samples may be collected only for $m \ll n$ input samples, and the remaining $n - m$ samples are input-only. *Semisupervised learning* is aimed at learning from both input-output paired samples $\{(\boldsymbol{x}_i, y_i)\}_{i=1}^m$ and input-only samples $\{\boldsymbol{x}_i\}_{i=m+1}^n$. Typically, semisupervised learning methods extract distributional information such as cluster structure from the input-only samples $\{\boldsymbol{x}_i\}_{i=m+1}^n$ and utilize that information for improving supervised learning from input-output paired samples $\{(\boldsymbol{x}_i, y_i)\}_{i=1}^m$.

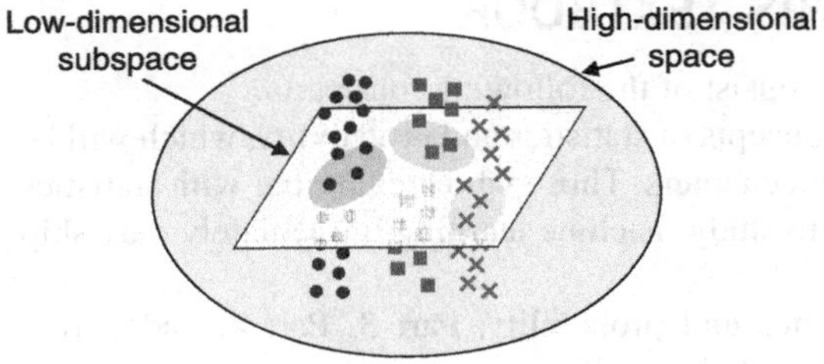

FIGURE 1.5

Dimensionality reduction.

Given weak learning algorithms that perform only slightly better than a random guess, *ensemble learning* is aimed at constructing a strong learning algorithm by combining such weak learning algorithms. One of the most popular approaches is voting by the weak learning algorithms, which may complement the weakness of each other.

Standard learning algorithms consider vectorial data $\boldsymbol{x}$. However, if data have a two-dimensional structure such as an image, directly learning from matrix data would be more promising than vectorizing the matrix data. Studies of *matrix learning* or *tensor learning* are directed to handle such higher-order data.

When data samples are provided sequentially one by one, updating the learning results to incorporate new data would be more natural than re-learning all data from scratch. *Online learning* is aimed at efficiently handling such sequentially given data.

When solving a learning task, transferring knowledge from other similar learning tasks would be helpful. Such a paradigm is called *transfer learning* or *domain adaptation*. If multiple related learning tasks need to be solved, solving them simultaneously would be more promising than solving them individually. This idea is called *multitask learning* and can be regarded as a bidirectional variant of transfer learning.

Learning from high-dimensional data is challenging, which is often referred to as the *curse of dimensionality*. The objective of *dimensionality reduction* is to extract essential information from high-dimensional data samples $\{\boldsymbol{x}_i\}_{i=1}^n$ and obtain their low-dimensional expressions $\{\boldsymbol{z}_i\}_{i=1}^n$ (Fig. 1.5). In linear dimensionality reduction, the low-dimensional expressions $\{\boldsymbol{z}_i\}_{i=1}^n$ are obtained as $\boldsymbol{z}_i = \boldsymbol{T}\boldsymbol{x}_i$ using a fat matrix $\boldsymbol{T}$. Supervised dimensionality reduction tries to find low-dimensional expressions $\{\boldsymbol{z}_i\}_{i=1}^n$ as preprocessing, so that the subsequent supervised learning tasks can be solved easily. On the other hand, unsupervised dimensionality reduction tries to find low-dimensional expressions $\{\boldsymbol{z}_i\}_{i=1}^n$ such that certain structure of original data is maintained, for example, for visualization purposes. *Metric learning* is similar to dimensionality reduction, but it has more emphasis on learning the metric in the original high-dimensional space rather than reducing the dimensionality of data.

1.3 STRUCTURE OF THIS TEXTBOOK

The main contents of this textbook consist of the following four parts.

Part 2 introduces fundamental concepts of statistics and probability, which will be extensively utilized in the subsequent chapters. Those who are familiar with statistics and probability or those who want to study machine learning immediately may skip Part 2.

Based on the concept of statistics and probability, Part 3, Part 4, and Part 5 introduce various machine learning techniques. These parts are rather independent, and therefore readers may start from any part based on their interests.

Part 3 targets the *generative approach* to statistical pattern recognition. The basic idea of generative methods is that the probability distribution of data is modeled to perform pattern recognition. When prior knowledge on data generation mechanisms is available, the generative approach is highly useful. Various classification and clustering techniques based on generative model estimation in the *frequentist* and *Bayesian* frameworks will be introduced.

Part 4 focuses on the *discriminative approach* to statistical machine learning. The basic idea of discriminative methods is to solve target machine learning tasks such as regression and classification directly without modeling data-generating probability distributions. If no prior knowledge is available for data generation mechanisms, the discriminative approach would be more promising than the generative approach. In addition to statistics and probability, knowledge of *optimization theory* also plays an important role in the discriminative approach, which will also be covered.

Part 5 is devoted to introducing various advanced issues in machine learning, including *ensemble learning*, *online learning*, *confidence of prediction*, *semisupervised learning*, *transfer learning*, *multitask learning*, *dimensionality reduction*, *clustering*, *outlier detection*, and *change detection*.

Compact MATLAB codes are provided for the methods introduced in Part 3, Part 4, and Part 5. So readers can immediately test the algorithms and learn their numerical behaviors.

PART

STATISTICS AND PROBABILITY

2

Probability and statistics are important mathematical tools used in the state-of-the-art machine learning methods, and are becoming indispensable subjects of science in the era of big data. Part 2 of this book is devoted to providing fundamentals of probability and statistics.

Chapter 2 overviews the basic notions of random variables and probability distributions. Chapter 3 and Chapter 4 illustrate examples of discrete and continuous probability distributions, respectively. Chapter 5 introduces concepts used in multidimensional data analysis, and Chapter 6 give examples of multidimensional probability distributions. Chapter 7 discusses the asymptotic behavior of the sum of independent random variables, and Chapter 8 shows various inequalities related to random variables. Finally, Chapter 9 and Chapter 10 cover fundamentals of statistical estimation and hypothesis testing, respectively.

CHAPTER

RANDOM VARIABLES AND PROBABILITY DISTRIBUTIONS

2

CHAPTER CONTENTS

In this chapter, the notions of random variables and probability distributions are introduced, which form the basis of probability and statistics. Then simple statistics that summarize probability distributions are discussed.

2.1 MATHEMATICAL PRELIMINARIES

When throwing a six-sided die, the possible outcomes are only 1, 2, 3, 4, 5, 6, and no others. Such possible outcomes are called *sample points* and the set of all sample points is called the *sample space*.

An *event* is defined as a subset of the sample space. For example, event A that any odd number appears is expressed as

$$A = \{1,3,5\}.$$

The event with no sample point is called the *empty event* and denoted by $\emptyset$. An event consisting only of a single sample point is called an *elementary event*, while an event consisting of multiple sample points is called a *composite event*. An event that includes all possible sample points is called the *whole event*. Below, the notion of combining events is explained using Fig. 2.1.

The event that at least one of the events A and B occurs is called the *union of events* and denoted by $A \cup B$. For example, the union of event A that an odd number appears and event B that a number less than or equal to three appears is expressed as

$$A \cup B = \{1,3,5\} \cup \{1,2,3\} = \{1,2,3,5\}.$$

(a) Event A (b) Event B (c) Complementary event A^c

$A \cup B$
(d) Union of events

$A \cap B$
(e) Intersection of events

$A \cap B = \emptyset$
(f) Disjoint events

$(A \cup B) \cap C = (A \cap C) \cup (B \cap C)$ $(A \cap B) \cup C = (A \cup C) \cap (B \cup C)$
(g) Distributive law 1   (h) Distributive law 2

$(A \cup B)^c = A^c \cap B^c$ $(A \cap B)^c = A^c \cup B^c$
(i) De Morgan's law 1   (j) De Morgan's law 2

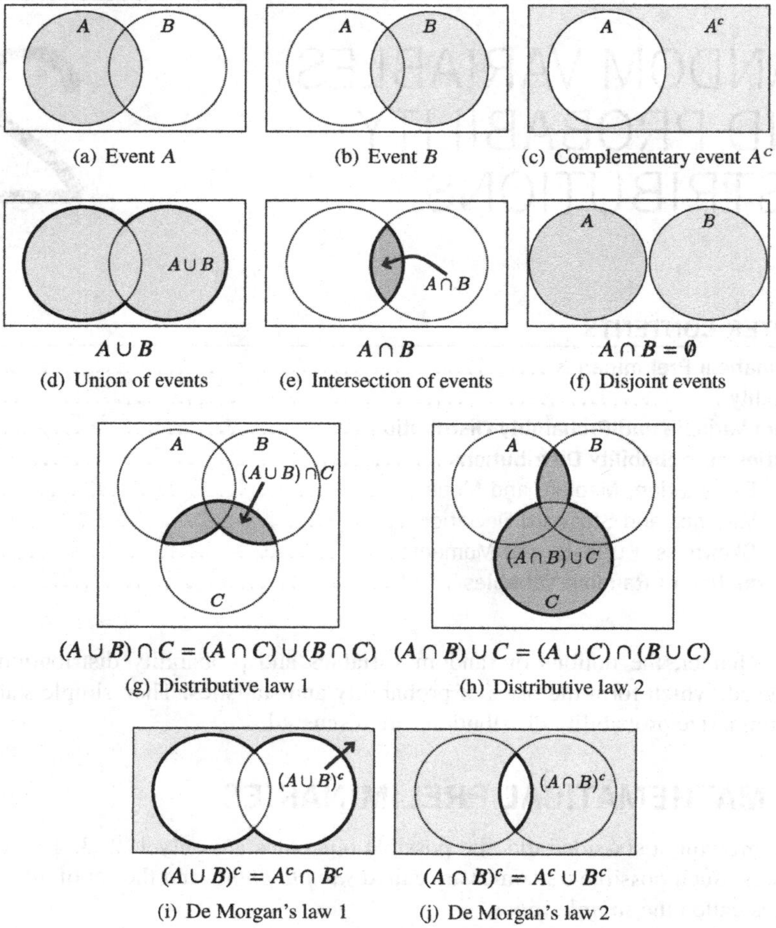

FIGURE 2.1

Combination of events.

On the other hand, the event that both events A and B occur simultaneously is called the *intersection of events* and denoted by $A \cap B$. The intersection of the above events A and B is given by

$$A \cap B = \{1,3,5\} \cap \{1,2,3\} = \{1,3\}.$$

If events A and B never occur at the same time, i.e.,

$$A \cap B = \emptyset,$$

events A and B are called *disjoint events*. The event that an odd number appears and the event that an even number appears cannot occur simultaneously and thus are disjoint. For events A, B, and C, the following *distributive laws* hold:

$$(A \cup B) \cap C = (A \cap C) \cup (B \cap C),$$
$$(A \cap B) \cup C = (A \cup C) \cap (B \cup C).$$

The event that event A does not occur is called the *complementary event* of A and denoted by A^c. The complementary event of the event that an odd number appears is that an odd number does not appear, i.e., an even number appears. For the union and intersection of events A and B, the following *De Morgan's laws* hold:

$$(A \cup B)^c = A^c \cap B^c,$$
$$(A \cap B)^c = A^c \cup B^c.$$

2.2 PROBABILITY

Probability is a measure of likeliness that an event will occur and the probability that event A occurs is denoted by $\Pr(A)$. A Russian mathematician, *Kolmogorov*, defined the probability by the following three axioms as abstraction of the evident properties that the probability should satisfy.

1. **Non-negativity:** For any event A_i,

$$0 \leq \Pr(A_i) \leq 1.$$

2. **Unitarity:** For entire sample space Ω,

$$\Pr(\Omega) = 1.$$

3. **Additivity:** For any countable sequence of disjoint events $A_1, A_2, \ldots$,

$$\Pr(A_1 \cup A_2 \cup \cdots) = \Pr(A_1) + \Pr(A_2) + \cdots.$$

From the above axioms, events A and B are shown to satisfy the following *additive law*:

$$\Pr(A \cup B) = \Pr(A) + \Pr(B) - \Pr(A \cap B).$$

This can be extended to more than two events: for events A, B, and C,

$$\begin{aligned}\Pr(A \cup B \cup C) = {} & \Pr(A) + \Pr(B) + \Pr(C) \\ & - \Pr(A \cap B) - \Pr(A \cap C) - \Pr(B \cap C) \\ & + \Pr(A \cap B \cap C).\end{aligned}$$

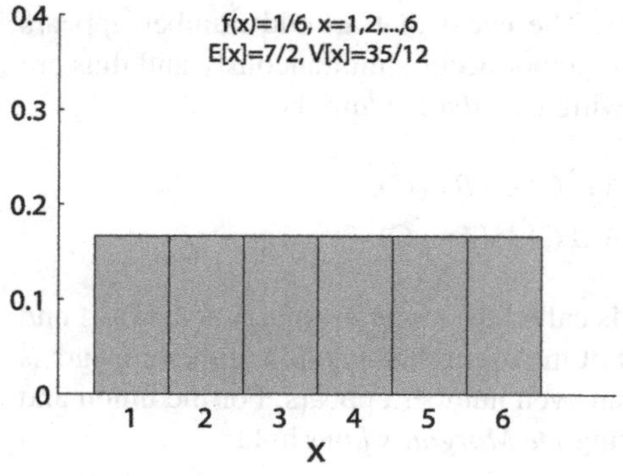

FIGURE 2.2

Examples of probability mass function. Outcome of throwing a fair six-sided dice (discrete uniform distribution $U\{1,2,\ldots,6\}$).

2.3 RANDOM VARIABLE AND PROBABILITY DISTRIBUTION

A variable is called a *random variable* if probability is assigned to each *realization* of the variable. A *probability distribution* is the function that describes the mapping from any realized value of the random variable to probability.

A *countable set* is a set whose elements can be enumerated as $1,2,3,\ldots$. A random variable that takes a value in a countable set is called a *discrete random variable*. Note that the size of a countable set does not have to be finite but can be infinite such as the set of all natural numbers. If probability for each value of discrete random variable x is given by

$$\Pr(x) = f(x),$$

$f(x)$ is called the *probability mass function*. Note that $f(x)$ should satisfy

$$\forall x,\ \ f(x) \geq 0, \ \ \text{and} \ \ \sum_x f(x) = 1.$$

The outcome of throwing a fair six-sided die, $x \in \{1,2,3,4,5,6\}$, is a discrete random variable, and its probability mass function is given by $f(x) = 1/6$ (Fig. 2.2).

A random variable that takes a continuous value is called a *continuous random variable*. If probability that continuous random variable x takes a value in $[a,b]$ is given by

$$\Pr(a \leq x \leq b) = \int_a^b f(x)\mathrm{d}x, \tag{2.1}$$

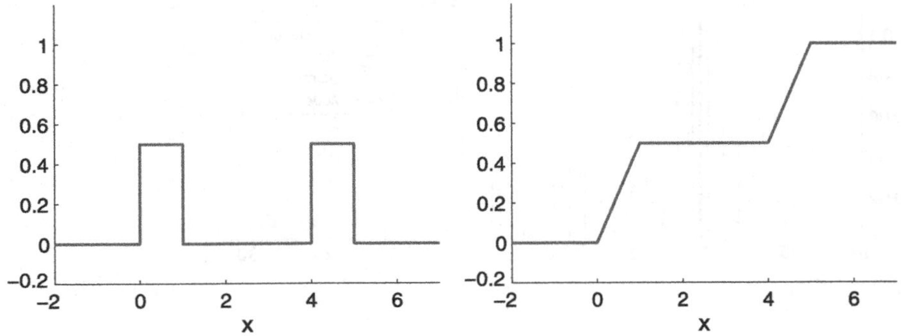

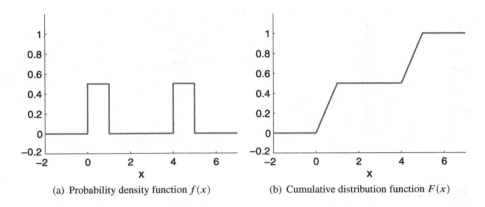

(a) Probability density function $f(x)$ (b) Cumulative distribution function $F(x)$

FIGURE 2.3

Example of probability density function and its cumulative distribution function.

$f(x)$ is called a *probability density function* (Fig. 2.3(a)). Note that $f(x)$ should satisfy

$$\forall x, \ f(x) \geq 0, \quad \text{and} \quad \int f(x)\mathrm{d}x = 1.$$

For example, the outcome of spinning a roulette, $x \in [0, 2\pi)$, is a continuous random variable, and its probability density function is given by $f(x) = 1/(2\pi)$. Note that Eq. (2.1) also has an important implication, i.e., the probability that continuous random variable x exactly takes value b is actually zero:

$$\Pr(b \leq x \leq b) = \int_b^b f(x)\mathrm{d}x = 0.$$

Thus, the probability that the outcome of spinning a roulette is exactly a particular angle is zero.

The probability that continuous random variable x takes a value less than or equal to b,

$$F(b) = \Pr(x \leq b) = \int_{-\infty}^{b} f(x)\mathrm{d}x,$$

is called the *cumulative distribution function* (Fig. 2.3(b)). The cumulative distribution function F satisfies the following properties:

- **Monotone nondecreasing:** $x < x'$ implies $F(x) \leq F(x')$.
- **Left limit:** $\lim_{x \to -\infty} F(x) = 0$.
- **Right limit:** $\lim_{x \to +\infty} F(x) = 1$.

If the derivative of a cumulative distribution function exists, it agrees with the probability density function:

$$F'(x) = f(x).$$

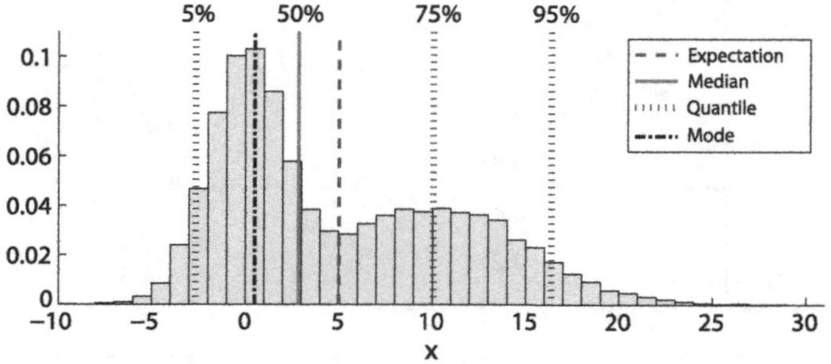

FIGURE 2.4

Expectation is the average of x weighted according to $f(x)$, and median is the 50% point both from the left-hand and right-hand sides. α-quantile for $0 \le \alpha \le 1$ is a generalization of the median that gives the $100\alpha\%$ point from the left-hand side. Mode is the maximizer of $f(x)$.

$\Pr(a \le x)$ is called the *upper-tail probability* or the *right-tail probability*, while $\Pr(x \le b)$ is called the *lower-tail probability* or the *left-tail probability*. The upper-tail and lower-tail probabilities together are called the *two-sided probability*, and either of them is called a *one-sided probability*.

2.4 PROPERTIES OF PROBABILITY DISTRIBUTIONS

When discussing properties of probability distributions, it is convenient to have simple statistics that summarize probability mass/density functions. In this section, such statistics are introduced.

2.4.1 EXPECTATION, MEDIAN, AND MODE

The *expectation* is the value that a random variable is expected to take (Fig. 2.4). The expectation of random variable x, denoted by $E[x]$, is defined as the average of x weighted according to probability mass/density function $f(x)$:

$$\text{Discrete: } E[x] = \sum_x x f(x),$$

$$\text{Continuous: } E[x] = \int x f(x)\mathrm{d}x.$$

Note that, as explained in Section 4.5, there are probability distributions such as the Cauchy distribution that the expectation does not exist (diverges to infinity).

The expectation can be defined for any function ξ of x similarly:

$$\text{Discrete: } E[\xi(x)] = \sum_x \xi(x) f(x),$$

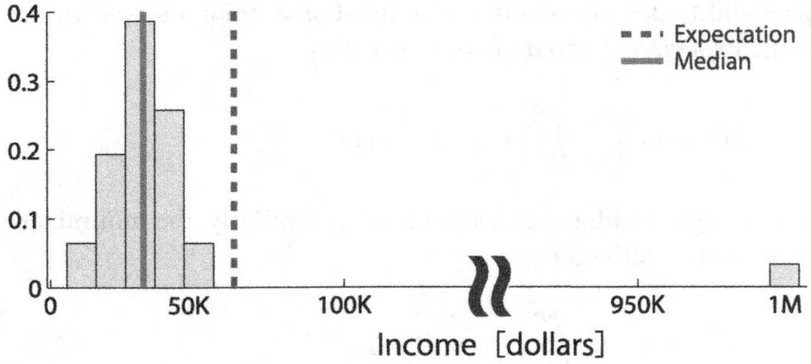

FIGURE 2.5

Income distribution. The expectation is 62.1 thousand dollars, while the median is 31.3 thousand dollars.

$$\text{Continuous: } E[\xi(x)] = \int \xi(x)f(x)\mathrm{d}x.$$

For constant c, the expectation operator E satisfies the following properties:

$$E[c] = c,$$
$$E[x + c] = E[x] + c,$$
$$E[cx] = cE[x].$$

Although the expectation represents the "center" of a probability distribution, it can be quite different from what is intuitively expected in the presence of *outliers*. For example, in the income distribution illustrated in Fig. 2.5, because one person earns 1 million dollars, all other people are below the expectation, 62.1 thousand dollars. In such a situation, the *median* is more appropriate than the expectation, which is defined as b such that

$$\Pr(x \leq b) = 1/2.$$

That is, the median is the "center" of a probability distribution in the sense that it is the 50% point both from the left-hand and right-hand sides. In the example of Fig. 2.5, the median is 31.3 thousand dollars and it is indeed in the middle of everybody.

The *α-quantile* for $0 \leq \alpha \leq 1$ is a generalization of the median that gives b such that

$$\Pr(x \leq b) = \alpha.$$

That is, the α-quantile gives the $100\alpha\%$ point from the left-hand side (Fig. 2.4) and is reduced to the median when $\alpha = 0.5$.

Let us consider a probability density function f defined on a finite interval $[a,b]$. Then the minimizer of the *expected squared error*, defined by

$$E\left[(x-y)^2\right] = \int_a^b (x-y)^2 f(x)\mathrm{d}x,$$

with respect to y is shown to agree with the expectation of x. Similarly, the minimizer y of the *expected absolute error*, defined by

$$E\left[|x-y|\right] = \int_a^b |x-y| f(x)\mathrm{d}x, \tag{2.2}$$

with respect to y is shown to agree with the expectation of x. Furthermore, a weighted variant of Eq. (2.2),

$$\int_a^b |x-y|_\alpha f(x)\mathrm{d}x, \quad |x-y|_\alpha = \begin{cases} (1-\alpha)(x-y) & (x > y), \\ \alpha(y-x) & (x \leq y), \end{cases}$$

is minimized with respect to y by the α-quantile of x.

Another popular statistic is the *mode*, which is defined as the maximizer of $f(x)$ (Fig. 2.4).

2.4.2 VARIANCE AND STANDARD DEVIATION

Although the expectation is a useful statistic to characterize probability distributions, probability distributions can be different even when they share the same expectation. Here, another statistic called the *variance* is introduced to represent the spread of the probability distribution.

The variance of random variable x, denoted by $V[x]$, is defined as

$$V[x] = E\left[(x - E[x])^2\right].$$

In practice, expanding the above expression as

$$V[x] = E\left[x^2 - 2xE[x] + (E[x])^2\right] = E[x^2] - (E[x])^2$$

often makes the computation easier. For constant c, variance operator V satisfies the following properties:

$$V[c] = 0,$$
$$V[x+c] = V[x],$$
$$V[cx] = c^2 V[x].$$

Note that these properties are quite different from those of the expectation.

The square root of the variance is called the *standard deviation* and is denoted by $D[x]$:

$$D[x] = \sqrt{V[x]}.$$

Conventionally, the variance and the standard deviation are denoted by σ^2 and σ, respectively.

2.4.3 SKEWNESS, KURTOSIS, AND MOMENTS

In addition to the expectation and variance, higher-order statistics such as the *skewness* and *kurtosis* are also often used. The skewness and kurtosis represent asymmetry and sharpness of probability distributions, respectively, and defined as

$$\text{Skewness: } \frac{E\left[(x - E[x])^3\right]}{(D[x])^3},$$

$$\text{Kurtosis: } \frac{E\left[(x - E[x])^4\right]}{(D[x])^4} - 3.$$

$(D[x])^3$ and $(D[x])^4$ in the denominators are for normalization purposes and -3 included in the definition of the kurtosis is to zero the kurtosis of the normal distribution (see Section 4.2). As illustrated in Fig. 2.6, the right tail is longer than the left tail if the skewness is positive, while the left tail is longer than the right tail if the skewness is negative. The distribution is perfectly symmetric if the skewness is zero. As illustrated in Fig. 2.7, the probability distribution is sharper than the normal distribution if the kurtosis is positive, while the probability distribution is duller than the normal distribution if the kurtosis is positive.

The above discussions imply that the statistic,

$$\nu_k = E\left[(x - E[x])^k\right],$$

plays an important role in characterizing probability distributions. ν_k is called the kth *moment* about the expectation, while

$$\mu_k = E[x^k]$$

is called the kth *moment* about the origin. The expectation, variance, skewness, and kurtosis can be expressed by using μ_k as

$$\text{Expectation: } \mu_1,$$

$$\text{Variance: } \mu_2 - \mu_1^2,$$

$$\text{Skewness: } \frac{\mu_3 - 3\mu_2\mu_1 + 2\mu_1^3}{(\mu_2 - \mu_1^2)^{\frac{3}{2}}},$$

$$\text{Kurtosis: } \frac{\mu_4 - 4\mu_3\mu_1 + 6\mu_2\mu_1^2 - 3\mu_1^4}{(\mu_2 - \mu_1^2)^2} - 3.$$

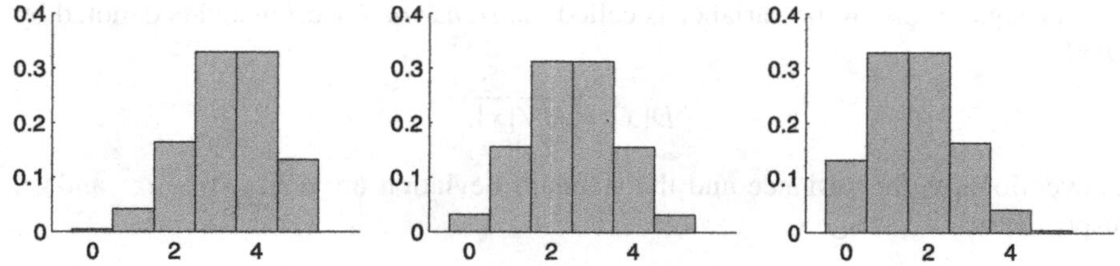

(a) Skewness: −0.32 (b) Skewness: 0 (c) Skewness: 0.32

FIGURE 2.6

Skewness.

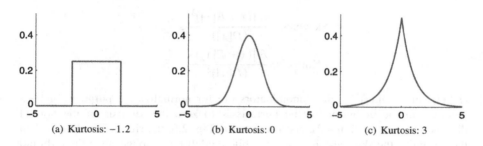

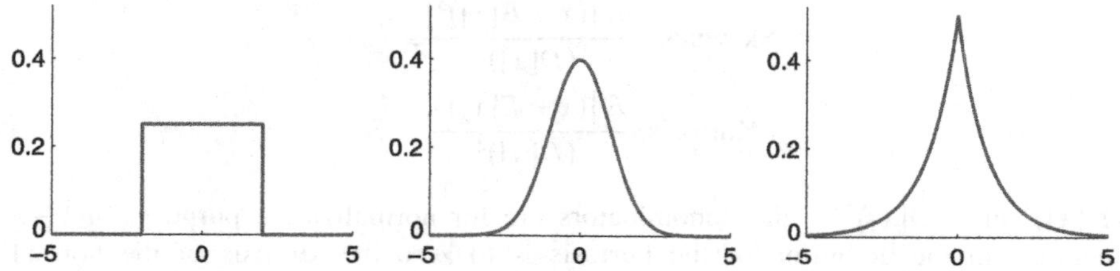

(a) Kurtosis: −1.2 (b) Kurtosis: 0 (c) Kurtosis: 3

FIGURE 2.7

Kurtosis.

Probability distributions will be more constrained if the expectation, variance, skewness, and kurtosis are specified. As the limit, if the moments of all orders are specified, the probability distribution is uniquely determined. The *moment-generating function* allows us to handle the moments of all orders in a systematic way:

$$M_x(t) = E[e^{tx}] = \begin{cases} \displaystyle\sum_x e^{tx} f(x) & \text{(Discrete)}, \\ \displaystyle\int e^{tx} f(x)\mathrm{d}x & \text{(Continuous)}. \end{cases}$$

Indeed, substituting zero to the kth derivative of the moment-generating function with respect to t, $M_x^{(k)}(t)$, gives the kth moment:

$$M_x^{(k)}(0) = \mu_k.$$

Below, this fact is proved.

The value of function g at point t can be expressed as

$$g(t) = g(0) + t\frac{g'(0)}{1!} + t^2\frac{g''(0)}{2!} + \cdots .$$

If higher-order terms in the right-hand side are ignored and the infinite sum is approximated by a finite sum, an approximation to $g(t)$ can be obtained. When only the first-order term $g(0)$ is used, $g(t)$ is simply approximated by $g(0)$, which is too rough. However, when the second-order term $tg'(0)$ is included, the approximation gets better, as illustrated below. By further including higher-order terms, the approximation gets more accurate and converges to $g(t)$ if all terms are included.

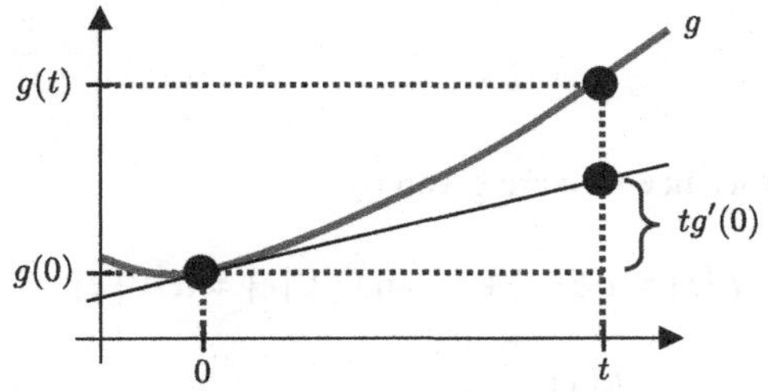

FIGURE 2.8

Taylor series expansion at the origin.

Given that the kth derivative of function e^{tx} with respect to t is $x^k e^{tx}$, the *Taylor series expansion* (Fig. 2.8) of function e^{tx} at the origin with respect to t yields

$$e^{tx} = 1 + (tx) + \frac{(tx)^2}{2!} + \frac{(tx)^3}{3!} + \cdots .$$

Taking the expectation of both sides gives

$$E[e^{tx}] = M_x(t) = 1 + t\mu_1 + t^2\frac{\mu_2}{2!} + t^3\frac{\mu_3}{3!} + \cdots .$$

Taking the derivative of both sides yields

$$M'_x(t) = \mu_1 + \mu_2 t + \frac{\mu_3}{2!}t^2 + \frac{\mu_4}{3!}t^3 + \cdots ,$$

$$M''_x(t) = \mu_2 + \mu_3 t + \frac{\mu_4}{2!}t^2 + \frac{\mu_5}{3!}t^3 + \cdots ,$$

$$\vdots$$

$$M_x^{(k)}(t) = \mu_k + \mu_{k+1} t + \frac{\mu_{k+2}}{2!}t^2 + \frac{\mu_{k+3}}{3!}t^3 + \cdots .$$

Substituting zero into this gives $M_x^{(k)}(0) = \mu_k$.

Depending on probability distributions, the moment-generating function does not exist (diverges to infinity). On the other hand, its sibling called the *characteristic function* always exists:

$$\varphi_x(t) = M_{ix}(t) = M_x(it),$$

where i denotes the *imaginary unit* such that $i^2 = -1$. The characteristic function corresponds to the *Fourier transform* of a probability density function.

2.5 TRANSFORMATION OF RANDOM VARIABLES

If random variable x is transformed as

$$r = ax + b,$$

the expectation and variance of r are given by

$$E[r] = aE[x] + b \quad \text{and} \quad V[r] = a^2 V[x].$$

Setting $a = \dfrac{1}{D[x]}$ and $b = -\dfrac{E[x]}{D[x]}$ yields

$$z = \frac{x}{D[x]} - \frac{E[x]}{D[x]} = \frac{x - E[x]}{D[x]},$$

which has expectation 0 and variance 1. This transformation from x to z is called *standardization*.

Suppose that random variable x that has probability density $f(x)$ defined on $\mathcal{X}$ is obtained by using transformation ξ as

$$x = \xi(r).$$

Then the probability density function of z is not simply given by $f(\xi(r))$, because $f(\xi(r))$ is not integrated to 1 in general. For example, when x is the height of a person in centimeter and r is its transformation in meter, $f(\xi(r))$ should be divided by 100 to be integrated to 1.

More generally, as explained in Fig. 2.9, if the *Jacobian* $\frac{\mathrm{d}x}{\mathrm{d}r}$ is not zero, the scale should be adjusted by multiplying the absolute Jacobian as

$$g(r) = f(\xi(r)) \left| \frac{\mathrm{d}x}{\mathrm{d}r} \right|.$$

$g(r)$ is integrated to 1 for any transform $x = \xi(r)$ such that $\frac{\mathrm{d}x}{\mathrm{d}r} \neq 0$.

Integration of function $f(x)$ over $\mathcal{X}$ can be expressed by using function $g(r)$ on $\mathcal{R}$ such that

$$x = g(r) \quad \text{and} \quad \mathcal{X} = g(\mathcal{R})$$

as

$$\int_{\mathcal{X}} f(x)\mathrm{d}x = \int_{\mathcal{R}} f(g(r)) \left| \frac{\mathrm{d}x}{\mathrm{d}r} \right| \mathrm{d}r.$$

This allows us to change variables of integration from x to r. $|\frac{\mathrm{d}x}{\mathrm{d}r}|$ in the right-hand side corresponds to the ratio of lengths when variables of integration are changed from x to r. For example, for

$$f(x) = x \quad \text{and} \quad \mathcal{X} = [2, 3],$$

integration of function $f(x)$ over $\mathcal{X}$ is computed as

$$\int_{\mathcal{X}} f(x)\mathrm{d}x = \int_2^3 x\mathrm{d}x = \left[\frac{1}{2}x^2\right]_2^3 = \frac{5}{2}.$$

On the other hand, $g(r) = r^2$ yields

$$\mathcal{R} = [\sqrt{2}, \sqrt{3}], \quad f(g(r)) = r^2, \quad \text{and} \quad \frac{\mathrm{d}x}{\mathrm{d}r} = 2r.$$

This results in

$$\int_{\mathcal{R}} f(g(r)) \left| \frac{\mathrm{d}x}{\mathrm{d}r} \right| \mathrm{d}r = \int_{\sqrt{2}}^{\sqrt{3}} r^2 \cdot 2r\mathrm{d}r = \left[\frac{1}{2}r^4\right]_{\sqrt{2}}^{\sqrt{3}} = \frac{5}{2}.$$

FIGURE 2.9

One-dimensional change of variables in integration. For multidimensional cases, see Fig. 4.2.

For linear transformation

$$r = ax + b \quad \text{and} \quad a \neq 0,$$

$x = \frac{r-b}{a}$ yields $\frac{\mathrm{d}x}{\mathrm{d}r} = \frac{1}{a}$, and thus

$$g(r) = \frac{1}{|a|} f\left(\frac{r-b}{a}\right)$$

is obtained.

CHAPTER

EXAMPLES OF DISCRETE PROBABILITY DISTRIBUTIONS

3

CHAPTER CONTENTS

In this chapter, popular discrete probability distributions and their properties such as the expectation, the variance, and the moment-generating functions are illustrated.

3.1 DISCRETE UNIFORM DISTRIBUTION

The *discrete uniform distribution* is the probability distribution for N events $\{1,\ldots,N\}$ that occur with equal probability. It is denoted by $U\{1,\ldots,N\}$, and its probability mass function is given by

$$f(x) = \frac{1}{N} \quad \text{for } x = 1,\ldots,N.$$

From the series formulas,

$$\sum_{x=1}^{N} x = \frac{N(N+1)}{2} \quad \text{and} \quad \sum_{x=1}^{N} x^2 = \frac{N(N+1)(2N+1)}{6},$$

the expectation and variance of $U\{1,\ldots,N\}$ can be computed as

$$E[x] = \frac{N+1}{2} \quad \text{and} \quad V[x] = \frac{N^2-1}{12}.$$

The probability mass function of $U\{1,\ldots,6\}$ is plotted in Fig. 2.2.

The probability mass function of the discrete uniform distribution $U\{a,a+1,\ldots,b\}$ for finite $a < b$ is given by

$$f(x) = \frac{1}{b-a+1} \quad \text{for } x = a, a+1,\ldots,b.$$

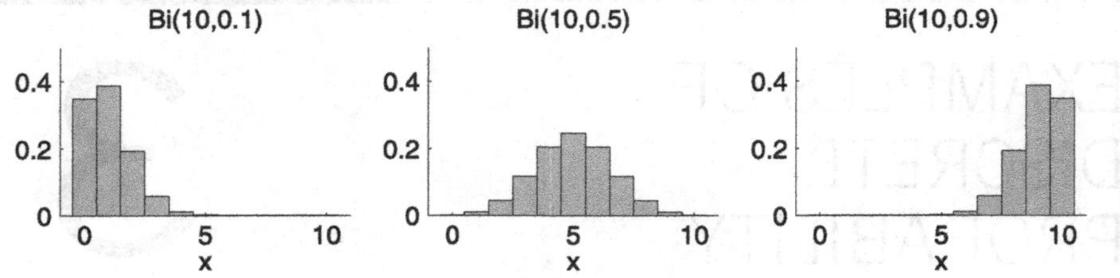

FIGURE 3.1

Probability mass functions of binomial distribution Bi(n, p).

Its expectation and variance are given by

$$E[x] = \frac{a+b}{2} \quad \text{and} \quad V[x] = \frac{(b-a+1)^2 - 1}{12}.$$

3.2 BINOMIAL DISTRIBUTION

Bernoulli trials are independent repeated trials of an experiment with two possible outcomes, say success and failure. Repeated independent tosses of the same coin are typical Bernoulli trials. Let p be the probability of success (getting heads in the coin toss) and q $(= 1 - p)$ be the probability of failure (getting tails in the coin toss). The *binomial distribution* is the probability distribution of the number x of successful trials in n Bernoulli trials and is denoted by Bi(n, p).

The probability of having x successful trials is given by p^x, while the probability of having $n - x$ unsuccessful trials is given by q^{n-x}. The number of combinations of x successful trials and $n - x$ unsuccessful trials is given by $\frac{n!}{x!(n-x)!}$, where

$$n! = n \times (n-1) \times \cdots \times 2 \times 1$$

denotes the *factorial*. $\frac{n!}{x!(n-x)!}$ is called the *binomial coefficient* and is denoted by $\binom{n}{x}$:

$$\binom{n}{x} = \frac{n!}{x!(n-x)!}.$$

Putting them together based on the axioms of the probability provided in Section 2.2, the probability mass function of Bi(n, p) is given by

$$f(x) = p^x q^{n-x} \binom{n}{x} \quad \text{for } x = 0, 1, \ldots, N.$$

Probability mass functions of the binomial distribution for $n = 10$ are illustrated in Fig. 3.1.

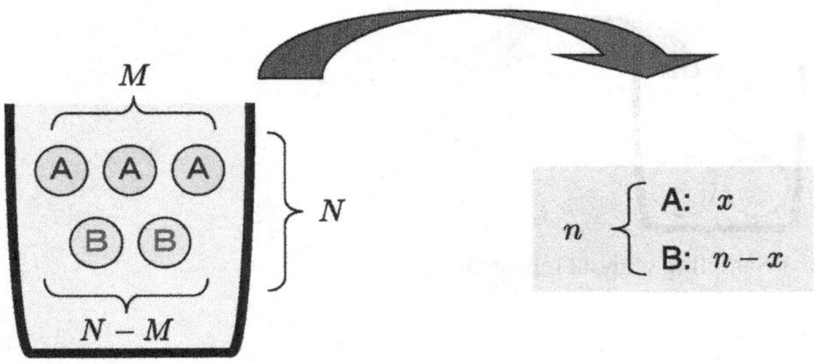

FIGURE 3.2

Sampling from a bag. The bag contains N balls which consist of $M < N$ balls labeled as "A" and $N - M$ balls labeled as "B." n balls are sampled from the bag, which consists of x balls labeled as "A" and $n - x$ balls labeled as "B."

The moment-generating function of $\mathrm{Bi}(n,p)$ can be obtained by using the *binomial theorem*,

$$(p+q)^n = \sum_{x=0}^{n} \binom{n}{x} p^x q^{n-x},$$

as

$$\begin{aligned} M_x(t) &= \sum_{x=0}^{n} e^{tx} \binom{n}{x} p^x q^{n-x} = \sum_{x=0}^{n} \binom{n}{x} (pe^t)^x q^{n-x} \\ &= (pe^t + q)^n. \end{aligned}$$

From this, the expectation and variance of $\mathrm{Bi}(n,p)$ can be computed as

$$E[x] = np \quad \text{and} \quad V[x] = npq.$$

The expectation np would be intuitive because a Bernoulli trial with probability of success p is repeated n times. The variance npq is maximized when $p = 0.5$, while it is minimized when $p = 0$ or $p = 1$. This is also intuitive because it is difficult to predict the success or failure of a trial when the probability of success is 0.5.

The binomial distribution with $n = 1$, $\mathrm{Bi}(1,p)$, is specifically called the *Bernoulli distribution.*

3.3 HYPERGEOMETRIC DISTRIBUTION

Let us consider the situation illustrated in Fig. 3.2: n balls are sampled from the bag containing N balls, where M balls are labeled as "A" and $N - M$ balls are labeled

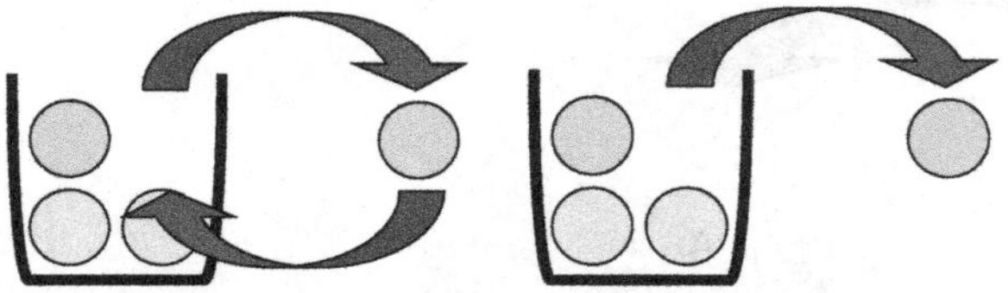

(a) Sampling with replacement (b) Sampling without replacement

FIGURE 3.3

Sampling with and without replacement. The sampled ball is returned to the bag before the next ball is sampled in sampling with replacement, while the next ball is sampled without returning the previously sampled ball in sampling without replacement.

as "B." In this situation, there are two sampling schemes, as illustrated in Fig. 3.3: The first scheme is called *sampling with replacement*, which requires to return the sampled ball to the bag before the next ball is sampled. The other scheme is called *sampling without replacement*, where the next ball is sampled without returning the previously sampled ball to the bag.

In sampling with replacement, a ball is always sampled from the bag containing all N balls. This sampling process corresponds to the Bernoulli trials, and thus the probability distribution of obtaining x balls labeled as "A" when n balls are sampled with replacement is given by $\mathrm{Bi}(n, M/N)$.

On the other hand, in sampling without replacement, the number of balls contained in the bag is decreasing as the sampling process is progressed. Thus, the ratio of balls labeled as "A" and "B" in the bag depends on the history of sampling. The probability distribution of obtaining x balls labeled as "A" when n balls are sampled without replacement is called the *hypergeometric distribution* and denoted by $\mathrm{HG}(N, M, n)$.

The number of combinations of sampling x balls labeled as "A" from M balls is given by $\binom{M}{x}$, the number of combinations of sampling $n-x$ balls labeled as "B" from $N-M$ balls is given by $\binom{N-M}{n-x}$, and the number of combinations of sampling n balls from N balls is given by $\binom{N}{n}$. Putting them together, the probability mass function of $\mathrm{HG}(N, M, n)$ is given by

$$f(x) = \frac{\binom{M}{x}\binom{N-M}{n-x}}{\binom{N}{n}} \quad \text{for } x = 0, 1, \ldots, n.$$

Although the domain of x is $\{0, 1, \ldots, n\}$, the actual range of x is given by

$$\{\max(0, n-(N-M)), \ldots, \min(n, M)\},$$

because the number of balls in the bag is limited.

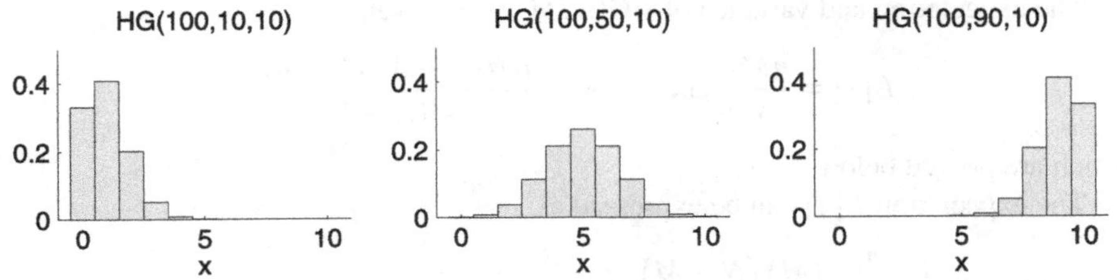

FIGURE 3.4

Probability mass functions of hypergeometric distribution HG(N, M, n).

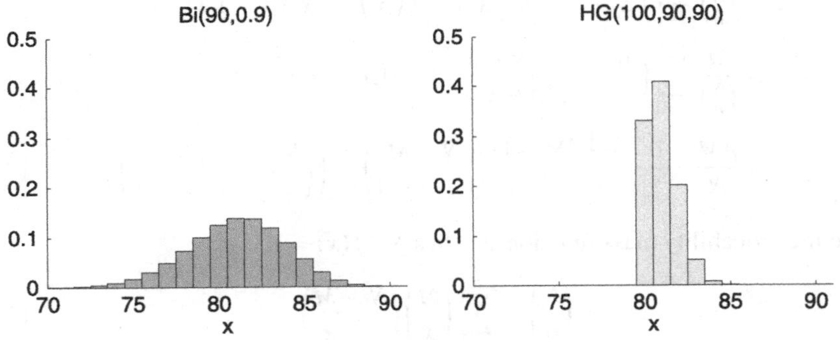

FIGURE 3.5

Probability mass functions of Bi$(n, M/N)$ and HG(N, M, n) for $N = 100$, $M = 90$, and $n = 90$.

For $N = 100$ and $n = 10$, the probability mass functions of HG(N, M, n) and Bi$(n, M/N)$ are plotted in Fig. 3.4 and Fig. 3.1, respectively. These graphs show that the probability mass functions are quite similar to each other, meaning that sampling with or without replacement does not affect that much when only $n = 10$ balls are sampled from $N = 100$ balls. Indeed, as N and M tend to infinity under n and the ratio N/M fixed, HG(N, M, n) is shown to agree with Bi$(n, M/N)$.

Next, let us sample $n = 90$ from the bag of $N = 100$ balls where $M = 90$ balls are labeled as "A." In this situation, even though the probability is very low, sampling with replacement can result in selecting one of the 10 balls labeled as "B" in all 90 trials. On the other hand, sampling without replacement results in selecting balls labeled as "B" at most 10 times and balls labeled as "A" at least 80 times. Thus, in this situation, the probability mass functions of the hypergeometric distribution and the binomial distribution are quite different, as illustrated in Fig. 3.5.

The expectation and variance of HG(N, M, n) are given by

$$E[x] = \frac{nM}{N} \quad \text{and} \quad V[x] = \frac{nM(N-M)(N-n)}{N^2(N-1)},$$

which are proved below.

The expectation $E[x]$ can be expressed as

$$\begin{aligned}
E[x] &= \frac{1}{\binom{N}{n}} \sum_{x=0}^{n} x \binom{M}{x}\binom{N-M}{n-x} \\
&= \frac{1}{\binom{N}{n}} \sum_{x=1}^{n} x \binom{M}{x}\binom{N-M}{n-x} \quad \text{(the term with } x = 0 \text{ is zero)} \\
&= \frac{M}{\binom{N}{n}} \sum_{x=1}^{n} \binom{M-1}{x-1}\binom{N-M}{n-x} \quad \left(\binom{M}{x} = \frac{M}{x}\binom{M-1}{x-1}\right) \\
&= \frac{M}{\binom{N}{n}} \sum_{x=0}^{n-1} \binom{M-1}{x}\binom{N-M}{n-x-1} \quad \text{(let } x \leftarrow x-1\text{)} \\
&= \frac{nM}{N} \frac{1}{\binom{N-1}{n-1}} \sum_{x=0}^{n-1} \binom{M-1}{x}\binom{N-M}{n-x-1} \quad \left(\binom{N}{n} = \frac{N}{n}\binom{N-1}{n-1}\right).
\end{aligned} \tag{3.1}$$

Since the probability mass function satisfies $\sum_x f(x) = 1$,

$$\binom{N}{n} = \sum_{x=0}^{n} \binom{M}{x}\binom{N-M}{n-x}. \tag{3.2}$$

Letting $M \leftarrow M-1$, $N \leftarrow N-1$, and $n \leftarrow n-1$ in Eq. (3.2) yields

$$\binom{N-1}{n-1} = \sum_{x=0}^{n-1} \binom{M-1}{x}\binom{N-M}{n-x-1},$$

and substituting this into Eq. (3.1) gives $E[x] = \frac{nM}{N}$.

The variance $V[x]$ can be expressed as

$$V[x] = E[x(x-1)] + E[x] - (E[x])^2. \tag{3.3}$$

Similar derivation to the expectation and using Eq. (3.2) yield

$$E[x(x-1)] = \frac{n(n-1)M(M-1)}{N(N-1)},$$

and substituting this into Eq. (3.3) gives $V[x] = \frac{nM(N-M)(N-n)}{N^2(N-1)}$.

The moment-generating function of HG(N, M, n) is given by

$$M_x(t) = E[e^{tx}] = \frac{\binom{N-M}{n}}{\binom{N}{n}} F(-n, -M, N-M-n+1, e^t),$$

where

$$F(a,b,c,d) = \sum_{x=0}^{\infty} \frac{(a)_x (b)_x}{(c)_x} \frac{d^x}{x!},$$

$$(a)_x = \begin{cases} a(a+1)\cdots(a+x-1) & (x > 0) \\ 1 & (x = 0) \end{cases}$$

is the *hypergeometric series*. The name, the hypergeometric distribution, stems from the fact that its moment-generating function can be expressed using the hypergeometric series.

3.4 POISSON DISTRIBUTION

If the probability of success in Bernoulli trials is very small, every trial is almost always failed. However, even if the probability of success, p, is extremely small, the Bernoulli trials will succeed a small number of times, as long as the number of trials, n, is large enough. Indeed, given that the expectation of binomial distribution $\mathrm{Bi}(n,p)$ is np, repeating the Bernoulli trials $n = 10000000$ times with probability of success $p = 0.0000003$ yields three successful trials on average:

$$np = 10000000 \times 0.0000003 = 3.$$

This implies that the probability that the number of successful trials x is nonzero may not be that small, if the number of trials n is large enough. More precisely, given that the probability mass function of $\mathrm{Bi}(n,p)$ is

$$f(x) = \binom{n}{x} p^x (1-p)^{n-x},$$

the probability for $x = 5$ is

$$\binom{10000000}{5} (0.0000003)^5 (0.9999997)^{9999995}.$$

However, calculating $(0.9999997)^{9999995}$ requires 9999995-time multiplications of 0.9999997, which is computationally expensive. At a glance, the use of approximation $0.9999997 \approx 1$ works fine:

$$(0.9999997)^{9999995} \approx 1^{9999995} = 1.$$

However, the correct value is

$$(0.9999997)^{9999995} \approx 0.0498 \ll 1,$$

and thus the above approximation is very poor.

This problem can be overcome by applying *Poisson's law of small numbers*: for $p = \lambda/n$,

$$\lim_{n\to\infty}\binom{n}{x}p^x(1-p)^{n-x} = \frac{e^{-\lambda}\lambda^x}{x!}.$$

Let us prove this. First, the left-hand side of the above equation can be expressed as

$$\begin{aligned}
&\lim_{n\to\infty}\binom{n}{x}\left(\frac{\lambda}{n}\right)^x\left(1-\frac{\lambda}{n}\right)^{n-x}\\
&=\lim_{n\to\infty}\frac{n!}{x!(n-x)!}\left(\frac{\lambda}{n}\right)^x\left(1-\frac{\lambda}{n}\right)^{n-x}\\
&=\frac{\lambda^x}{x!}\lim_{n\to\infty}\frac{n!}{(n-x)!n^x}\left(1-\frac{\lambda}{n}\right)^n\left(1-\frac{\lambda}{n}\right)^{-x}.
\end{aligned} \tag{3.4}$$

Here, it holds that

$$\begin{aligned}
\lim_{n\to\infty}\frac{n!}{(n-x)!n^x} &= \lim_{n\to\infty}\frac{n}{n}\times\frac{n-1}{n}\times\cdots\times\frac{n-x+1}{n}\\
&=\lim_{n\to\infty}1\times\frac{1-\frac{1}{n}}{1}\times\cdots\times\frac{1-\frac{x}{n}+\frac{1}{n}}{1}=1.
\end{aligned}$$

Also, setting $t = -\frac{\lambda}{n}$ in the definition of the *Euler number* e,

$$e = \lim_{t\to 0}(1+t)^{\frac{1}{t}}, \tag{3.5}$$

yields

$$\lim_{n\to\infty}\left(1-\frac{\lambda}{n}\right)^n = e^{-\lambda}.$$

Furthermore,

$$\lim_{n\to\infty}\left(1-\frac{\lambda}{n}\right)^{-x} = 1$$

holds. Putting them together yields

$$\lim_{n\to\infty}\binom{n}{x}\left(\frac{\lambda}{n}\right)^x\left(1-\frac{\lambda}{n}\right)^{n-x} = \frac{e^{-\lambda}\lambda^x}{x!}.$$

The *Poisson distribution*, denoted by $\mathrm{Po}(\lambda)$, is the probability distribution whose probability mass function is given by

$$f(x) = \frac{e^{-\lambda}\lambda^x}{x!}. \tag{3.6}$$

Since this corresponds to the binomial distribution with $p = \lambda/n$, if $1/n$ is regarded as time and an event occurs λ times on average in unit time, $f(x)$ corresponds to the probability that the event occurs x times in unit time.

Eq. (3.6) is non-negative and the Taylor series expansion of the exponential function at the origin,

$$e^{\lambda} = 1 + \frac{\lambda^1}{1!} + \frac{\lambda^2}{2!} + \cdots = \sum_{x=0}^{\infty} \frac{\lambda^x}{x!}, \tag{3.7}$$

yields

$$\sum_{x=0}^{\infty} f(x) = \sum_{x=0}^{\infty} \frac{e^{-\lambda}\lambda^x}{x!} = e^{-\lambda} \sum_{x=0}^{\infty} \frac{\lambda^x}{x!} = e^{-\lambda}e^{\lambda} = 1.$$

Thus, Eq. (3.6) is shown to be the probability mass function.

The moment-generating function of $\mathrm{Po}(\lambda)$ is given by

$$M_x(t) = E[e^{tx}] = \sum_{x=0}^{\infty} \frac{e^{tx}e^{-\lambda}\lambda^x}{x!} = \exp\left(\lambda(e^t - 1)\right),$$

where Eq. (3.7) is used for the derivation. From this, the expectation and variance of $\mathrm{Po}(\lambda)$ are obtained as

$$E[x] = \lambda \quad \text{and} \quad V[x] = \lambda.$$

Interestingly, the expectation and variance of the Poisson distribution are equal.

It was shown in Section 3.2 that the expectation and variance of the binomial distribution (i.e., without applying Poisson's law of small numbers) are given by

$$E[x] = np \quad \text{and} \quad V[x] = np(1-p).$$

Setting $p = \lambda/n$ yields

$$\lim_{n\to\infty} np = \lambda \quad \text{and} \quad \lim_{n\to\infty} np(1-p) = \lambda,$$

which imply that application of Poisson's law of small numbers does not essentially change the expectation and variance.

Probability mass functions of $\mathrm{Po}(\lambda)$ are illustrated in Fig. 3.6, showing that the expectation and variance grow as λ is increased.

3.5 NEGATIVE BINOMIAL DISTRIBUTION

Let us consider Bernoulli trials with probability of success p. Then the number of unsuccessful trials x until the kth success is obtained follows the *negative binomial distribution*, which is denoted by $\mathrm{NB}(k,p)$.

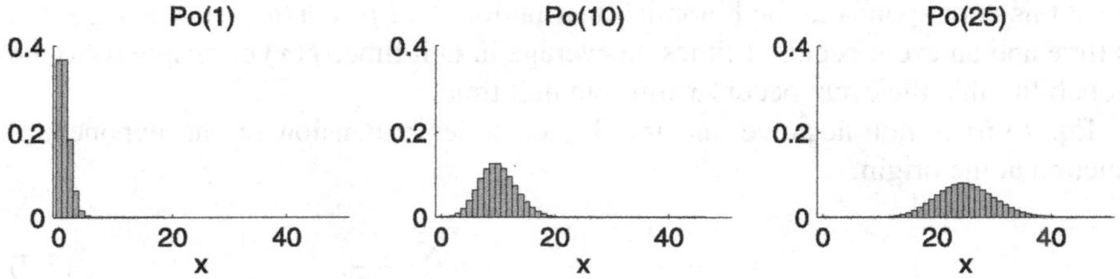

FIGURE 3.6

Probability mass functions of Poisson distribution Po(λ).

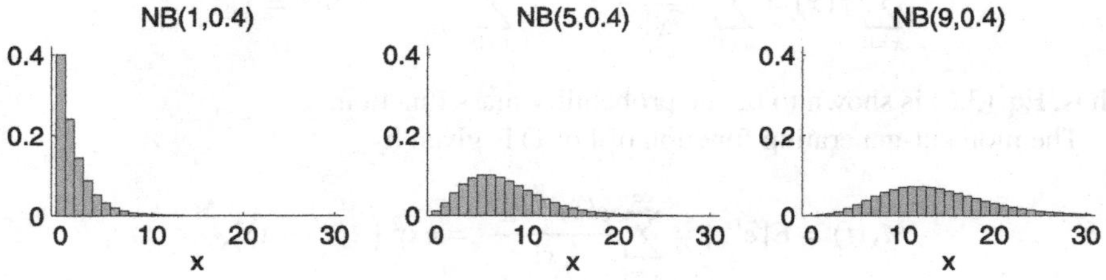

FIGURE 3.7

Probability mass functions of negative binomial distribution NB(k,p).

Since the kth success is obtained at the $(k + x)$th trial, the $(k + x)$th trial is always successful. On the other hand, the number of combinations of x unsuccessful trials in the first $(k + x - 1)$ trials is given by $\binom{k+x-1}{x}$. Putting them together, the probability mass function of NB(k,p) is given by

$$f(x) = \binom{k+x-1}{x} p^k (1-p)^x. \tag{3.8}$$

Probability mass functions of NB(k,p) for $p = 0.4$ are illustrated in Fig. 3.7.

The binomial coefficient $\binom{r}{x}$ can be extended to negative number $r = -k < 0$:

$$\binom{-k}{x} = \frac{(-k-x+1)(-k-x+2)\cdots(-k-1)(-k)}{x(x-1)\cdots 2\cdot 1}.$$

With this *negative binomial coefficient*, the probability mass function defined in (3.8) can be expressed as

$$\begin{aligned} f(x) &= \frac{(k+x-1)(k+x-2)\cdots(k+1)k}{x(x-1)\cdots 2\cdot 1} p^k (1-p)^x \\ &= (-1)^x \binom{-k}{x} p^k (1-p)^x. \end{aligned} \tag{3.9}$$

The name, the negative binomial distribution, stems from this fact.

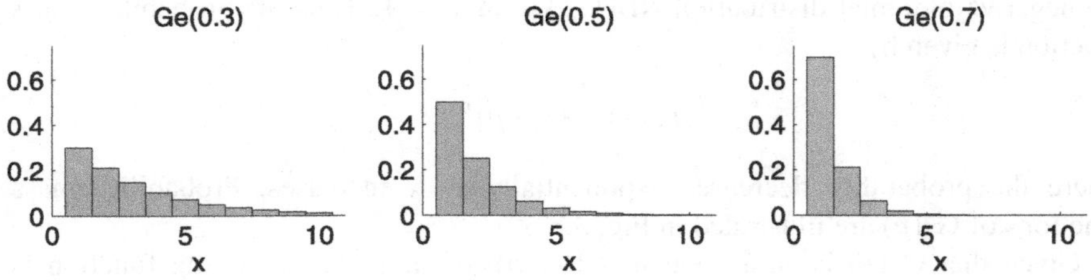

FIGURE 3.8

Probability mass functions of geometric distribution Ge(p).

The *binomial theorem* explained in Section 3.2 can also be generalized to negative numbers as

$$\sum_{x=0}^{\infty} \binom{-k}{x} t^x = (1+t)^{-k}.$$

Setting $t = p - 1$ in the above equation yields that Eq. (3.9) satisfies

$$\sum_{x=0}^{\infty} f(x) = p^k \sum_{x=0}^{\infty} \binom{-k}{x} (p-1)^x = 1.$$

This generalized binomial theorem allows us to obtain the moment-generating function of NB(k, p) as

$$\begin{aligned} M_x(t) = E[e^{tx}] &= \sum_{x=0}^{\infty} e^{tx} \binom{-k}{x} p^k (p-1)^x \\ &= p^k \sum_{x=0}^{\infty} \binom{-k}{x} \{(p-1)e^t\}^x = \left(\frac{p}{1-(1-p)e^t}\right)^k. \end{aligned}$$

From this, the expectation and variance of NB(k, p) are obtained as

$$E[x] = \frac{k(1-p)}{p} \quad \text{and} \quad V[x] = \frac{k(1-p)}{p^2}.$$

The negative binomial distribution is also referred to as the *Pascal distribution*.

3.6 GEOMETRIC DISTRIBUTION

Let us consider Bernoulli trials with probability of success p. Then the number of unsuccessful trials x until the first success is obtained follows the *geometric distribution*, which is denoted by Ge(p). The geometric distribution Ge(p) is equivalent to the negative binomial distribution NB(k,p) with $k = 1$. Thus, its probability mass function is given by

$$f(x) = p(1-p)^x,$$

where the probability decreases exponentially as x increases. Probability mass functions of Ge(p) are illustrated in Fig. 3.8.

Given that Ge(p) is equivalent to NB$(1,p)$, its moment-generating function is given by

$$M_x(t) = \frac{p}{1-(1-p)e^t},$$

and the expectation and variance are given by

$$E[x] = \frac{1-p}{p} \quad \text{and} \quad V[x] = \frac{1-p}{p^2}.$$

CHAPTER

EXAMPLES OF CONTINUOUS PROBABILITY DISTRIBUTIONS

4

CHAPTER CONTENTS

In this chapter, popular continuous probability distributions and their properties such as the expectation, the variance, and the moment-generating functions are illustrated.

4.1 CONTINUOUS UNIFORM DISTRIBUTION

The *continuous uniform distribution* has a constant probability density over a finite interval $[a, b]$:

$$f(x) = \begin{cases} \frac{1}{b-a} & (a \leq x \leq b), \\ 0 & (\text{otherwise}). \end{cases}$$

The continuous uniform distribution is denoted by $U(a, b)$, and its expectation and variance are given by

$$E[x] = \frac{a+b}{2} \quad \text{and} \quad V[x] = \frac{(b-a)^2}{12}.$$

4.2 NORMAL DISTRIBUTION

The *normal distribution*, also known as the *Gaussian distribution*, would be the most important continuous distribution. For $-\infty < \mu < \infty$ and $\sigma > 0$, the normal distribution is denoted by $N(\mu,\sigma^2)$, and its probability density is given by

$$f(x) = \frac{1}{\sigma\sqrt{2\pi}} \exp\left(-\frac{(x-\mu)^2}{2\sigma^2}\right).$$

The fact that the above $f(x)$ is integrated to one can be proven by the *Gaussian integral* shown in Fig. 4.1 as

$$\begin{aligned}\int_{-\infty}^{\infty} f(x)\mathrm{d}x &= \frac{1}{\sigma\sqrt{2\pi}}\int_{-\infty}^{\infty}\exp\left(-\frac{(x-\mu)^2}{2\sigma^2}\right)\mathrm{d}x\\ &= \frac{\sigma\sqrt{2}}{\sigma\sqrt{2\pi}}\int_{-\infty}^{\infty}\exp\left(-r^2\right)\mathrm{d}r = 1.\end{aligned}$$

Here, variables of integration are changed from x to $r = \frac{x-\mu}{\sigma\sqrt{2}}$ (i.e., $\frac{\mathrm{d}x}{\mathrm{d}r} = \sigma\sqrt{2}$), as explained in Fig. 2.9.

The constants μ and σ included in normal distribution $N(\mu,\sigma^2)$ correspond to the expectation and standard deviation:

$$E[x] = \mu \quad\text{and}\quad V[x] = \sigma^2.$$

As explained in Section 2.4.3, the expectation and variance can be obtained through the moment-generating function: Indeed, the moment-generating function of $N(\mu,\sigma^2)$ is given by

$$\begin{aligned}M_x(t) &= \int_{-\infty}^{\infty} e^{tx} f(x)\mathrm{d}x\\ &= \frac{1}{\sigma\sqrt{2\pi}}\int_{-\infty}^{\infty}\exp\left(-\frac{(x-\mu)^2}{2\sigma^2}+tx\right)\mathrm{d}x\\ &= \frac{1}{\sigma\sqrt{2\pi}}\int_{-\infty}^{\infty}\exp\left(-\frac{x^2-2(\mu+\sigma^2 t)x+\mu^2}{2\sigma^2}\right)\mathrm{d}x\\ &= \frac{1}{\sigma\sqrt{2\pi}}\int_{-\infty}^{\infty}\exp\left(-\frac{\left(x-(\mu+\sigma^2 t)\right)^2}{2\sigma^2}+\mu t+\frac{\sigma^2 t^2}{2}\right)\mathrm{d}x\\ &= \exp\left(\mu t+\frac{\sigma^2 t^2}{2}\right)\int_{-\infty}^{\infty}\frac{1}{\sigma\sqrt{2\pi}}\exp\left(-\frac{\left(x-(\mu+\sigma^2 t)\right)^2}{2\sigma^2}\right)\mathrm{d}x\\ &= \exp\left(\mu t+\frac{\sigma^2 t^2}{2}\right).\end{aligned} \tag{4.1}$$

Note that, in the above derivation, *completing the square*,

$$x^2 + 2ax + b = 0 \iff (x+a)^2 - a^2 + b = 0, \tag{4.2}$$

is used to have the probability density function of $N(\mu+\sigma^2 t,\sigma^2)$, which is integrated to 1.

Probability density functions of $N(\mu,\sigma^2)$ for $\mu = 0$ are illustrated in Fig. 4.3, showing that the normal density is symmetric and bell-shaped.

The Gaussian integral bridges the two well-known *irrational numbers*, $e = 2.71828\cdots$ and $\pi = 3.14159\cdots$, as

$$\int_{-\infty}^{\infty} e^{-x^2}\mathrm{d}x = \sqrt{\pi}.$$

To prove this, let us consider change of variables in integration (see Fig. 4.2) for $f(x,y) = e^{-(x^2+y^2)}$ and $\mathcal{X} = \mathcal{Y} = [-\infty,\infty]$. Let $g(r,\theta) = r\cos\theta$ and $h(r,\theta) = r\sin\theta$. Then

$$\mathcal{R} = [0,\infty],\ \ \Theta = [0,2\pi],\ \ \boldsymbol{J} = \begin{pmatrix} \cos\theta & -r\sin\theta \\ \sin\theta & r\cos\theta \end{pmatrix},\ \ \text{and}\ \ \det(\boldsymbol{J}) = r,$$

which yields

$$\int_{-\infty}^{\infty}\int_{-\infty}^{\infty} e^{-(x^2+y^2)}\mathrm{d}x\mathrm{d}y = \int_{0}^{2\pi}\int_{0}^{\infty} re^{-r^2}\mathrm{d}r\mathrm{d}\theta = \int_{0}^{2\pi}\mathrm{d}\theta\int_{0}^{\infty} re^{-r^2}\mathrm{d}r$$

$$= \int_{0}^{2\pi}\mathrm{d}\theta\int_{0}^{\infty} re^{-r^2}\mathrm{d}r = 2\pi\left[-\frac{1}{2}e^{-r^2}\right]_0^{\infty} = \pi.$$

Consequently,

$$\int_{-\infty}^{\infty} e^{-x^2}\mathrm{d}x = \sqrt{\left(\int_{-\infty}^{\infty} e^{-x^2}\mathrm{d}x\right)^2} = \sqrt{\left(\int_{-\infty}^{\infty} e^{-x^2}\mathrm{d}x\right)\left(\int_{-\infty}^{\infty} e^{-y^2}\mathrm{d}y\right)}$$

$$= \sqrt{\int_{-\infty}^{\infty}\int_{-\infty}^{\infty} e^{-(x^2+y^2)}\mathrm{d}x\mathrm{d}y} = \sqrt{\pi}.$$

FIGURE 4.1

Gaussian integral.

If random variable x follows $N(\mu,\sigma^2)$, its affine transformation

$$r = ax + b$$

follows $N(a\mu + b, a^2\sigma^2)$. This can be proved by the fact that the probability density function of r is given as follows (see Section 2.5):

$$g(r) = \frac{1}{|a|}f\left(\frac{r-b}{a}\right).$$

Let us extend change of variables in integration from one dimension (see Fig. 2.9) to two dimensions. Integration of function $f(x,y)$ over $\mathcal{X} \times \mathcal{Y}$ can be computed with $x = g(r,\theta)$, $y = h(r,\theta)$, $\mathcal{X} = g(\mathcal{R},\Theta)$, and $\mathcal{Y} = h(\mathcal{R},\Theta)$ as

$$\int_{\mathcal{X}}\int_{\mathcal{Y}} f(x,y)\mathrm{d}y\mathrm{d}x = \int_{\mathcal{R}}\int_{\Theta} f(g(r,\theta),h(r,\theta))|\det(\boldsymbol{J})|\mathrm{d}\theta\mathrm{d}r,$$

where $\boldsymbol{J}$ is called the *Jacobian matrix* and $\det(\boldsymbol{J})$ denotes the *determinant* of $\boldsymbol{J}$:

$$\boldsymbol{J} = \begin{pmatrix} \frac{\partial x}{\partial r} & \frac{\partial x}{\partial \theta} \\ \frac{\partial y}{\partial r} & \frac{\partial y}{\partial \theta} \end{pmatrix} \quad \text{and} \quad \det(\boldsymbol{J}) = \frac{\partial x}{\partial r}\frac{\partial y}{\partial \theta} - \frac{\partial x}{\partial \theta}\frac{\partial y}{\partial r}.$$

The determinant is the product of all eigenvalues and it corresponds to the ratio of volumes when (x,y) is changed to (r,θ). $\det(\boldsymbol{J})$ is often called the *Jacobian*. The above formula can be extended to more than two dimensions.

FIGURE 4.2

Two-dimensional change of variables in integration.

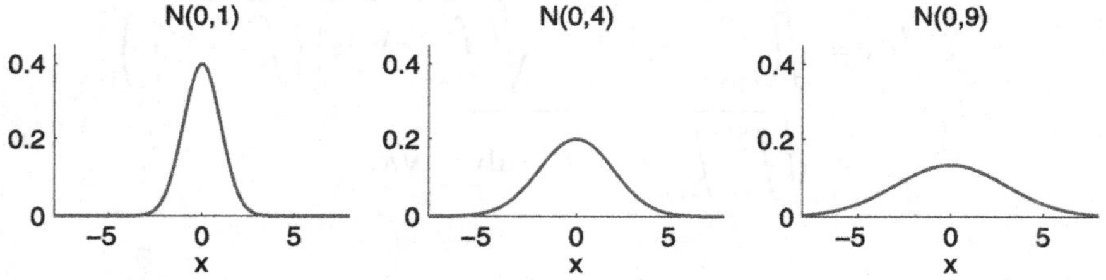

FIGURE 4.3

Probability density functions of normal density $N(\mu,\sigma^2)$.

Setting $a = \frac{1}{D[x]}$ and $b = -\frac{E[x]}{D[x]}$ yields

$$z = \frac{x}{D[x]} - \frac{E[x]}{D[x]} = \frac{x - E[x]}{D[x]},$$

which follows *standard normal distribution* $N(0,1)$ (Fig. 4.4).

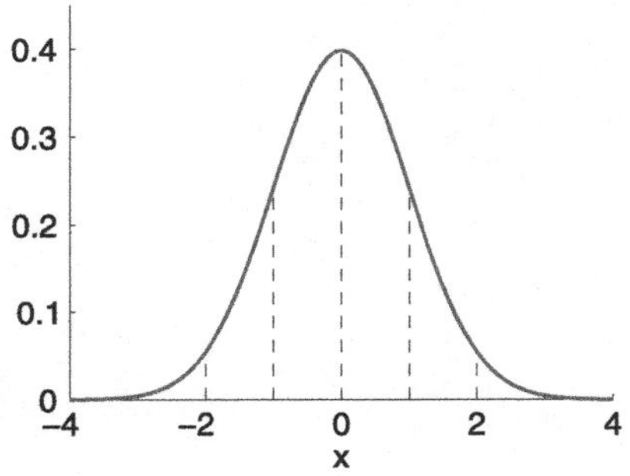

FIGURE 4.4

Standard normal distribution $N(0,1)$. A random variable following $N(0,1)$ is included in $[-1,1]$ with probability 68.27%, in $[-2,2]$ with probability 95.45%, and in $[-3,3]$ with probability 99.73%.

4.3 GAMMA DISTRIBUTION, EXPONENTIAL DISTRIBUTION, AND CHI-SQUARED DISTRIBUTION

As explained in Section 3.4, the Poisson distribution represents the probability that the event occurring λ times on average in unit time occurs x times in unit time. On the other hand, the elapsed time x that the event occurring λ times on average in unit time occurs α times follows the *gamma distribution*. The gamma distribution for positive real constants α and λ is denoted by $\mathrm{Ga}(\alpha,\lambda)$.

The probability density function of $\mathrm{Ga}(\alpha,\lambda)$ is given by

$$f(x) = \frac{\lambda^\alpha}{\Gamma(\alpha)} x^{\alpha-1} e^{-\lambda x} \quad \text{for } x \geq 0,$$

where

$$\Gamma(\alpha) = \int_0^\infty x^{\alpha-1} e^{-x} \mathrm{d}x \tag{4.3}$$

is called the *gamma function*. $\int_0^\infty f(x)\mathrm{d}x = 1$ can be proved by changing variables of integration as $y = \lambda x$ (see Fig. 2.9):

$$\begin{aligned}\int_{-\infty}^\infty f(x)\mathrm{d}x &= \frac{\lambda^\alpha}{\Gamma(\alpha)} \int_0^\infty x^{\alpha-1} e^{-\lambda x} \mathrm{d}x = \frac{\lambda^\alpha}{\Gamma(\alpha)} \int_0^\infty \left(\frac{y}{\lambda}\right)^{\alpha-1} e^{-y} \frac{1}{\lambda} \mathrm{d}y \\ &= \frac{1}{\Gamma(\alpha)} \int_0^\infty y^{\alpha-1} e^{-y} \mathrm{d}y = 1.\end{aligned}$$

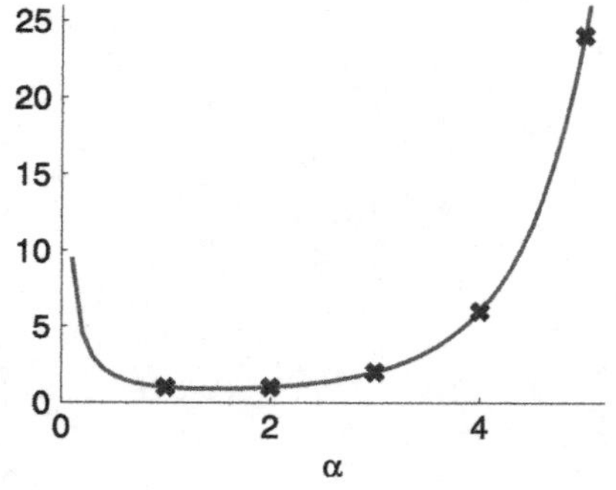

FIGURE 4.5

Gamma function. $\Gamma(\alpha + 1) = \alpha!$ holds for non-negative integer α, and the gamma function smoothly interpolates the factorials.

The gamma function fulfills

$$\Gamma(1) = \int_0^\infty e^{-x}\mathrm{d}x = [-e^{-x}]_0^\infty = 1.$$

Furthermore, *integration by parts* for functions $u(x)$ and $v(x)$ and their derivatives $u'(x)$ and $v'(x)$ given by

$$\int_a^b u(x)v'(x)\mathrm{d}x = \Big[u(x)v(x)\Big]_a^b - \int_a^b u'(x)v(x)\mathrm{d}x \tag{4.4}$$

yields

$$\begin{aligned}\Gamma(\alpha) &= \int_0^\infty e^{-x}x^{\alpha-1}\mathrm{d}x = \left[e^{-x}\frac{x^\alpha}{\alpha}\right]_0^\infty - \int_0^\infty (-e^{-x})\frac{x^\alpha}{\alpha}\mathrm{d}x \\ &= \frac{1}{\alpha}\int_0^\infty e^{-x}x^{(\alpha+1)-1}\mathrm{d}x = \frac{\Gamma(\alpha+1)}{\alpha}.\end{aligned}$$

Putting them together, the gamma function for non-negative integer α is shown to fulfill

$$\Gamma(\alpha+1) = \alpha\Gamma(\alpha) = \alpha(\alpha-1)\Gamma(\alpha-1) = \cdots = \alpha!\Gamma(1) = \alpha!.$$

Thus, the gamma function can be regarded as generalization of the factorial to real numbers (see Fig. 4.5). Furthermore, change of variables $x = y^2$ in integration yields

$$\Gamma(\alpha) = \int_0^\infty y^{2(\alpha-1)}e^{-y^2}\frac{\mathrm{d}x}{\mathrm{d}y}\mathrm{d}y = 2\int_0^\infty y^{2\alpha-1}e^{-y^2}\mathrm{d}y, \tag{4.5}$$

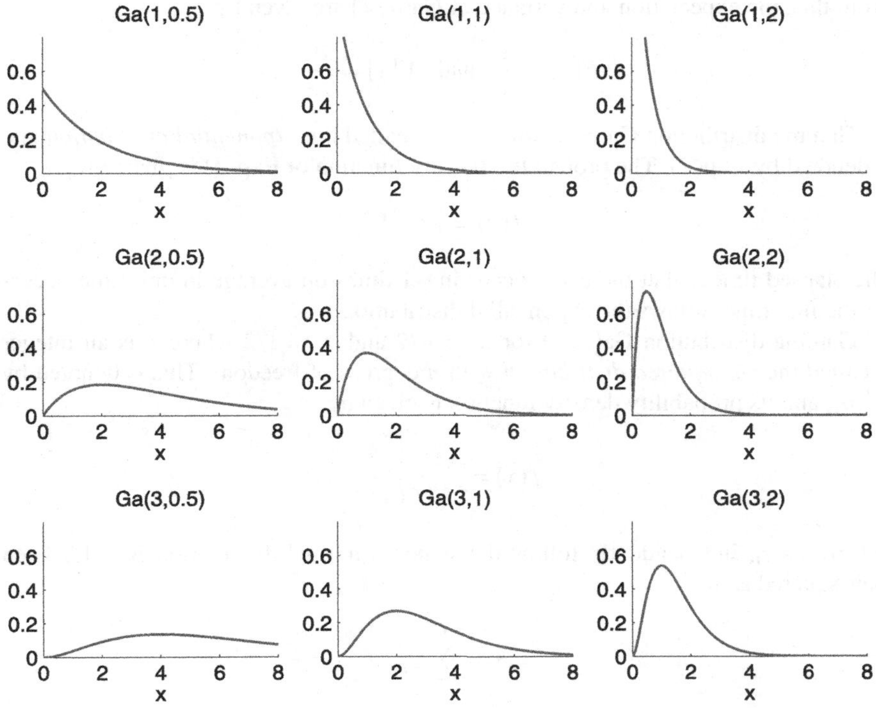

FIGURE 4.6

Probability density functions of gamma distribution Ga(α, λ).

and applying the *Gaussian integral* (Fig. 4.1) to this results in

$$\Gamma\left(\frac{1}{2}\right) = 2\int_0^{\infty} e^{-y^2}\mathrm{d}y = \int_{-\infty}^{\infty} e^{-y^2}\mathrm{d}y = \sqrt{\pi}.$$

Probability density functions of Ga(α, λ) are illustrated in Fig. 4.6. The probability density is monotone decreasing as x increases when $\alpha \leq 1$, while the probability density increases and then decreases as x increases when $\alpha > 1$.

By changing variables of integration as $y = (\lambda - t)x$, the moment-generating function of Ga(α, λ) can be expressed as

$$\begin{aligned}
M_x(t) = E[e^{tx}] &= \frac{\lambda^\alpha}{\Gamma(\alpha)}\int_0^{\infty} x^{\alpha-1}e^{-(\lambda-t)x}\mathrm{d}x \\
&= \frac{\lambda^\alpha}{\Gamma(\alpha)}\int_0^{\infty}\left(\frac{y}{\lambda-t}\right)^{\alpha-1} e^{-y}\frac{1}{\lambda-t}\mathrm{d}y = \frac{\lambda^\alpha}{\Gamma(\alpha)}\frac{\Gamma(\alpha)}{(\lambda-t)^\alpha} = \left(\frac{\lambda}{\lambda-t}\right)^\alpha.
\end{aligned}$$

From this, the expectation and variance of $\mathrm{Ga}(\alpha, \lambda)$ are given by

$$E[x] = \frac{\alpha}{\lambda} \quad \text{and} \quad V[x] = \frac{\alpha}{\lambda^2}.$$

Gamma distribution $\mathrm{Ga}(\alpha, \lambda)$ for $\alpha = 1$ is called the *exponential distribution* and is denoted by $\mathrm{Exp}(\lambda)$. The probability density function of $\mathrm{Exp}(\lambda)$ is given by

$$f(x) = \lambda e^{-\lambda x}.$$

The elapsed time x that the event occurring λ times on average in unit time occurs for the first time follows the exponential distribution.

Gamma distribution $\mathrm{Ga}(\alpha, \lambda)$ for $\alpha = n/2$ and $\lambda = 1/2$ where n is an integer is called the *chi-squared distribution* with n degrees of freedom. This is denoted by $\chi^2(n)$, and its probability density function is given by

$$f(x) = \frac{x^{\frac{n}{2}-1} e^{-\frac{x}{2}}}{2^{\frac{n}{2}} \Gamma(\frac{n}{2})}.$$

Let $z_1, \ldots, z_n$ independently follow the standard normal distribution $N(0, 1)$. Then their squared sum,

$$x = \sum_{i=1}^{n} z_i^2,$$

follows $\chi^2(n)$. The chi-squared distribution plays an important role in hypothesis testing explained in Chapter 10.

4.4 BETA DISTRIBUTION

For positive real scalars α and β, the *beta distribution*, denoted by $\mathrm{Be}(\alpha, \beta)$, is the probability distribution whose probability density is given by

$$f(x) = \frac{x^{\alpha-1}(1-x)^{\beta-1}}{B(\alpha, \beta)} \quad \text{for } 0 \le x \le 1,$$

where

$$B(\alpha, \beta) = \int_0^1 x^{\alpha-1}(1-x)^{\beta-1} \mathrm{d}x$$

is the *beta function*. When α and β are positive integers, αth smallest value (or equivalently βth largest value) among the $\alpha + \beta - 1$ random variables independently following the continuous uniform distribution $U(0, 1)$ follows the beta distribution.

Change of variables in integration as $x = (\sin\theta)^2$ allows us to express the beta function using sinusoidal functions as

$$B(\alpha,\beta) = \int_0^{\frac{\pi}{2}} (\sin\theta)^{2(\alpha-1)} \left(1 - (\sin\theta)^2\right)^{\beta-1} \frac{\mathrm{d}x}{\mathrm{d}\theta}\mathrm{d}\theta$$
$$= \int_0^{\frac{\pi}{2}} (\sin\theta)^{2(\alpha-1)}(\cos\theta)^{2(\beta-1)} \cdot 2\sin\theta\cos\theta\mathrm{d}\theta$$
$$= 2\int_0^{\frac{\pi}{2}} (\sin\theta)^{2\alpha-1}(\cos\theta)^{2\beta-1}\mathrm{d}\theta. \tag{4.6}$$

Furthermore, from Eq. (4.5), the gamma product $\Gamma(\alpha)\Gamma(\beta)$ can be expressed as

$$\Gamma(\alpha)\Gamma(\beta) = \left(2\int_0^{\infty} u^{2\alpha-1}e^{-u^2}\mathrm{d}u\right)\left(2\int_0^{\infty} v^{2\beta-1}e^{-v^2}\mathrm{d}v\right)$$
$$= 4\int_0^{\infty}\int_0^{\infty} u^{2\alpha-1}v^{2\beta-1}e^{-(u^2+v^2)}\mathrm{d}u\mathrm{d}v.$$

If variables of integrations are changed as $u = r\sin\theta$ and $v = r\cos\theta$ (see Fig. 4.2) in the above equation, then

$$\Gamma(\alpha)\Gamma(\beta) = 4\int_0^{\frac{\pi}{2}}\int_0^{\infty} r^{2(\alpha+\beta)-2}e^{-r^2}(\sin\theta)^{2\alpha-1}(\cos\theta)^{2\beta-1}r\mathrm{d}r\mathrm{d}\theta$$
$$= \left(2\int_0^{\infty} r^{2(\alpha+\beta)-1}e^{-r^2}\mathrm{d}r\right)\left(2\int_0^{\frac{\pi}{2}} (\sin\theta)^{2\alpha-1}(\cos\theta)^{2\beta-1}\mathrm{d}\theta\right)$$
$$= \Gamma(\alpha+\beta)B(\alpha,\beta),$$

where Eq. (4.6) was used. Thus, the beta function can be expressed by using the gamma function as

$$B(\alpha,\beta) = \frac{\Gamma(\alpha)\Gamma(\beta)}{\Gamma(\alpha+\beta)}. \tag{4.7}$$

This allows us to compute, e.g., the following integrals:

$$\int_0^{\frac{\pi}{2}} (\sin\theta)^{2n}\mathrm{d}\theta = \int_0^{\frac{\pi}{2}} (\sin\theta)^{2(n+\frac{1}{2})-1}(\cos\theta)^{2\frac{1}{2}-1}\mathrm{d}\theta = B\left(n+\frac{1}{2},\frac{1}{2}\right),$$
$$\int_0^{\frac{\pi}{2}} (\sin\theta)^{2n+1}\mathrm{d}\theta = \int_0^{\frac{\pi}{2}} (\sin\theta)^{2(n+1)-1}(\cos\theta)^{2\frac{1}{2}-1}\mathrm{d}\theta = B\left(n+1,\frac{1}{2}\right).$$

Probability density functions of $\mathrm{Be}(\alpha,\beta)$ are illustrated in Fig. 4.7. This shows that the profile of the beta density drastically changes depending on the values of α and β, and the beta distribution is reduced to the continuous uniform distribution when $\alpha = \beta = 1$.

The expectation and variance of $\mathrm{Be}(\alpha,\beta)$ are given by

$$E[x] = \frac{\alpha}{\alpha+\beta} \quad \text{and} \quad V[x] = \frac{\alpha\beta}{(\alpha+\beta)^2(\alpha+\beta+1)}.$$

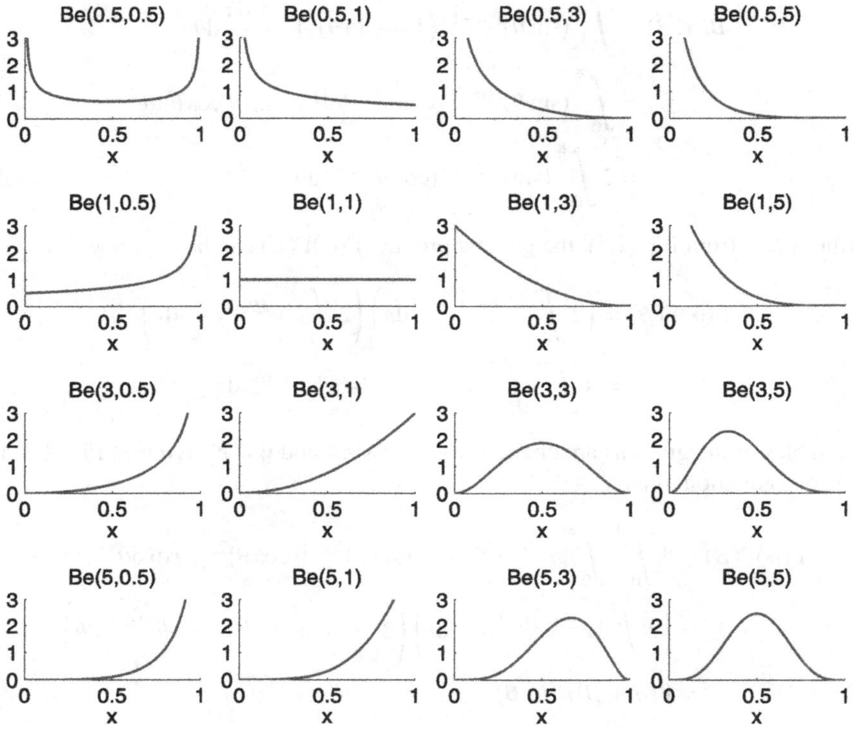

FIGURE 4.7

Probability density functions of beta distribution $\mathrm{Be}(\alpha, \beta)$.

The expectation can be proved by applying integration by parts as

$$\begin{aligned}
E[x] &= \frac{1}{B(\alpha,\beta)}\int_0^1 x x^{\alpha-1}(1-x)^{\beta-1}\mathrm{d}x \\
&= \frac{1}{B(\alpha,\beta)}\int_0^1 x^{\alpha}(1-x)^{\beta-1}\mathrm{d}x \\
&= \frac{1}{B(\alpha,\beta)}\left\{\left[x^{\alpha}\left(-\frac{(1-x)^{\beta}}{\beta}\right)\right]_0^1 - \int_0^1 \alpha x^{\alpha-1}\left(-\frac{(1-x)^{\beta}}{\beta}\right)\mathrm{d}x\right\} \\
&= \frac{\alpha}{\beta}\frac{1}{B(\alpha,\beta)}\int_0^1 x^{\alpha-1}(1-x)^{\beta-1}(1-x)\mathrm{d}x \\
&= \frac{\alpha}{\beta}\frac{1}{B(\alpha,\beta)}\left\{\int_0^1 x^{\alpha-1}(1-x)^{\beta-1}\mathrm{d}x - \int_0^1 x x^{\alpha-1}(1-x)^{\beta-1}\mathrm{d}x\right\} \\
&= \frac{\alpha}{\beta}(1-E[x]).
\end{aligned}$$

Similar computation applied to $E[x^2]$ gives

$$E[x^2] = \frac{\alpha+1}{\beta}\big(E[x] - E[x^2]\big),$$

which yields

$$E[x^2] = \frac{\alpha(\alpha+1)}{(\alpha+\beta)(\alpha+\beta+1)}.$$

Plugging this into

$$V[x] = E[x^2] - (E[x])^2$$

gives $V[x] = \frac{\alpha\beta}{(\alpha+\beta)^2(\alpha+\beta+1)}$.

In Section 6.3, the beta distribution will be extended to multiple dimensions.

4.5 CAUCHY DISTRIBUTION AND LAPLACE DISTRIBUTION

Let z and z' be the random variables independently following the standard normal distribution $N(0,1)$. Then their ratio,

$$x = \frac{z}{z'},$$

follows the standard Cauchy distribution, whose probability density function is given by $f(x) = \frac{1}{\pi(x^2+1)}$. Its generalization using a real scalar a and a positive real b is given by

$$f(x) = \frac{b}{\pi((x-a)^2+b^2)}.$$

This is called the *Cauchy distribution* and is denoted by $\mathrm{Ca}(a,b)$.

Computing the expectation of the standard Cauchy distribution yields

$$\begin{aligned} E[x] &= \int_{-\infty}^{+\infty} x f(x)\mathrm{d}x = \int_{-\infty}^{+\infty} \frac{x}{\pi(x^2+1)}\mathrm{d}x \\ &= \frac{1}{2\pi}\Big[\log(1+x^2)\Big]_{-\infty}^{+\infty} = \frac{1}{2\pi}\lim_{\alpha\to+\infty,\beta\to-\infty}\log\frac{1+\alpha^2}{1+\beta^2}, \end{aligned}$$

which means that the value depends on how fast α approaches $+\infty$ and β approaches $-\infty$. For this reason, $\mathrm{Ca}(a,b)$ does not have the expectation. The limiting value when $\alpha = -\beta$ is called the *principal value*, which is given by a and represents the "location" of the Cauchy distribution. Since the expectation does not exist, the Cauchy distribution does not have the variance and all higher moments, either. The positive scalar b represents the "scale" of the Cauchy distribution.

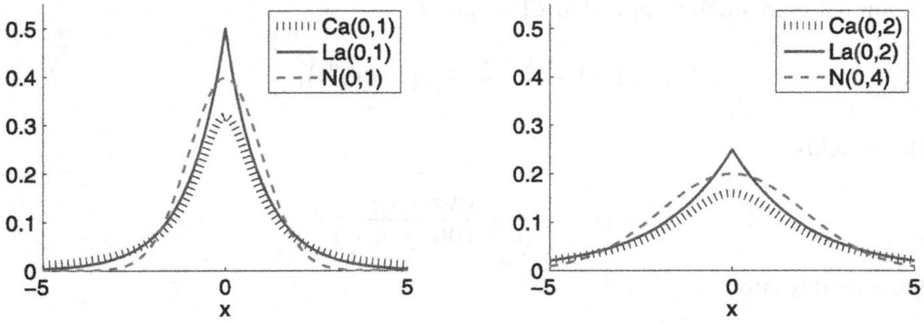

FIGURE 4.8

Probability density functions of Cauchy distribution $\mathrm{Ca}(a, b)$, Laplace distribution $\mathrm{La}(a, b)$, and normal distribution $N(a, b^2)$.

Let y and y' be random variables independently following the exponential distribution $\mathrm{Exp}(1)$. Then their difference,

$$x = y - y',$$

follows the standard Laplace distribution, whose probability density function is given by $f(x) = \frac{1}{2}\exp(-|x|)$. Its generalization using a real scalar a and a positive real b is given by

$$f(x) = \frac{1}{2b}\exp\left(-\frac{|x-a|}{b}\right).$$

This is called the *Laplace distribution* and is denoted by $\mathrm{La}(a, b)$. Since the Laplace distribution can be regarded as extending the exponential distribution to the negative domain, it is also referred to as the *double exponential distribution*.

When $|t| < 1/b$, the moment-generating function of $\mathrm{La}(a, b)$ is given by

$$\begin{aligned} M_x(t) &= \frac{1}{2b}\int_{-\infty}^{a} \exp\left(xt + \frac{x}{b} - \frac{a}{b}\right)\mathrm{d}x + \frac{1}{2b}\int_{a}^{+\infty} \exp\left(xt - \frac{x}{b} + \frac{a}{b}\right)\mathrm{d}x \\ &= \frac{1}{2}\left[\frac{1}{1+bt}\exp\left(xt + \frac{x}{b} - \frac{a}{b}\right)\right]_{-\infty}^{a} - \frac{1}{2}\left[\frac{1}{1-bt}\exp\left(xt - \frac{x}{b} + \frac{a}{b}\right)\right]_{-\infty}^{a} \\ &= \frac{\exp(at)}{1-b^2t^2}. \end{aligned}$$

From this, the expectation and variance of $\mathrm{La}(a, b)$ are given by

$$E[x] = a \quad \text{and} \quad V[x] = 2b^2.$$

Probability density functions of Cauchy distribution $\mathrm{Ca}(a, b)$, Laplace distribution $\mathrm{La}(a, b)$, and normal distribution $N(a, b^2)$ are illustrated in Fig. 4.8. Since the Cauchy

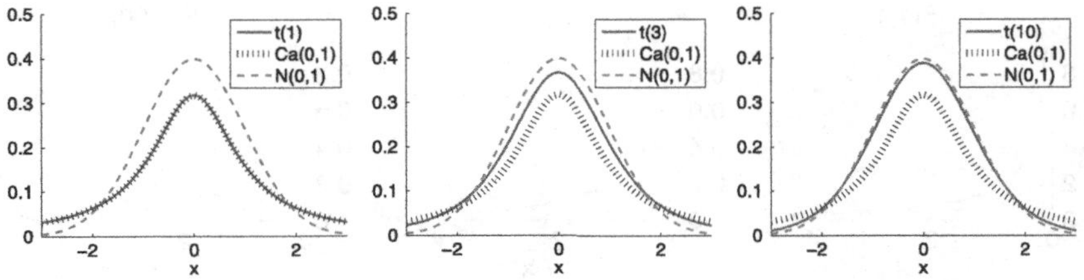

FIGURE 4.9

Probability density functions of t-distribution $t(d)$, Cauchy distribution Ca(0, 1), and normal distribution $N(0, 1)$.

and Laplace distributions have heavier tails than the normal distribution, realized values can be quite far from the origin. For this reason, the Cauchy and Laplace distributions are often used for modeling data with *outliers*. Note that the Laplace density is not differentiable at the origin.

4.6 t-DISTRIBUTION AND F-DISTRIBUTION

Let z be a random variable following the standard normal distribution $N(0, 1)$ and y be a random variable following the chi-squared distribution with d degrees of freedom $\chi^2(d)$. Then their ratio,

$$x = \frac{z}{\sqrt{y/d}},$$

follows the *t-distribution* denoted by $t(d)$. Following its inventor's pseudonym, the t-distribution is also referred to as *Student's t-distribution.*

The probability density function of t-distribution $t(d)$ is given as follows (Fig. 4.9):

$$f(x) = \frac{1}{B(\frac{d}{2}, \frac{1}{2})\sqrt{d}} \left(1 + \frac{x^2}{d}\right)^{-\frac{d+1}{2}},$$

where B is the beta function explained in Section 4.4. The t-distribution agrees with the Cauchy distribution when the degree of freedom is $d = 1$, and it is reduced to the normal distribution when the degree of freedom tends to ∞. The expectation exists when $d \geq 2$, and the variance exists when $d \geq 3$, which are given by

$$E[x] = 0 \quad \text{and} \quad V[x] = \frac{d}{d-2}.$$

Let y and y' be random variables following the chi-squared distributions with d and d' degrees of freedom, respectively. Then their ratio,

$$x = \frac{y/d}{y'/d'},$$

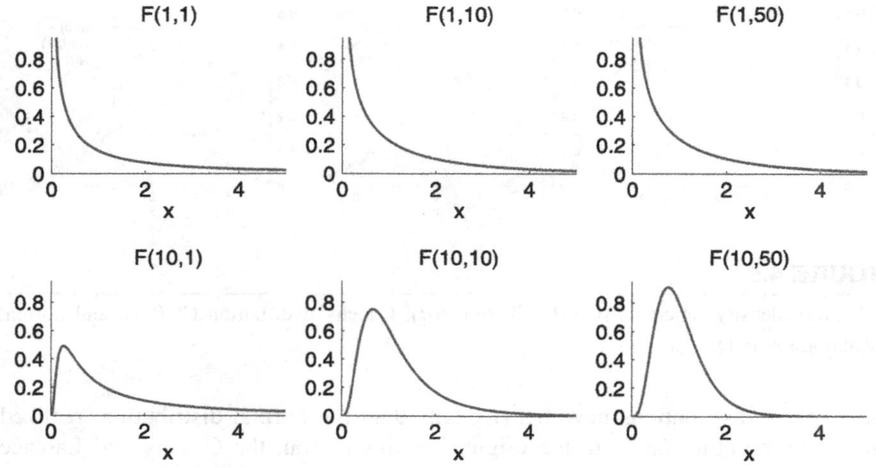

FIGURE 4.10

Probability density functions of F-distribution $F(d,d')$.

follows the *F-distribution* denoted by $F(d,d')$. Following its inventor's name, the F-distribution is also referred to as *Snedecor's F-distribution.*

The probability density function of F-distribution $F(d,d')$ is given as follows (Fig. 4.10):

$$f(x) = \frac{1}{B(d/2,d'/2)}\left(\frac{d}{d'}\right)^{\frac{d}{2}} x^{\frac{d}{2}-1}\left(1+\frac{d}{d'}x\right)^{-\frac{d+d'}{2}} \quad \text{for } x \geq 0.$$

The expectation exists when $d' \geq 3$, and the variance exists when $d' \geq 5$, which are given by

$$E[x] = \frac{d'}{d'-2} \quad \text{and} \quad V[x] = \frac{2d'^2(d+d'-2)}{d(d'-2)^2(d'-4)}.$$

If y follows the t-distribution $t(d)$, then y^2 follows the F-distribution $F(1,d)$.

The t-distribution and the F-distribution play important roles in hypothesis testing explained in Chapter 10. The t-distribution is also utilized for deriving the confidence interval in Section 9.3.1.

CHAPTER

MULTIDIMENSIONAL PROBABILITY DISTRIBUTIONS

5

CHAPTER CONTENTS

So far, properties of one-dimensional random variable x are discussed. When multiple random variables are available, we may be interested in knowing the dependency of one variable on another, which will give us more information. In this chapter, the relation between two random variables x and y is discussed.

5.1 JOINT PROBABILITY DISTRIBUTION

The probability that discrete random variables x and y take a value in a countable set is denoted by $\Pr(x, y)$. The function that describes the mapping from any realized value of the random variables to probability is called the *joint probability distribution*, and its probability mass function $f(x, y)$ is called the *joint probability mass function*:

$$\Pr(x, y) = f(x, y).$$

Similarly to the one-dimensional case, $f(x, y)$ should satisfy

$$f(x, y) \geq 0 \quad \text{and} \quad \sum_{x,y} f(x, y) = 1.$$

The probability mass functions of x or y alone, $g(x)$ and $h(y)$, can be expressed by using $f(x, y)$ as

$$g(x) = \sum_{y} f(x, y) \quad \text{and} \quad h(y) = \sum_{x} f(x, y).$$

These are called the *marginal probability mass functions* and the corresponding probability distributions are called the *marginal probability distributions*. The process of computing a marginal probability distribution from the joint probability distribution is called *marginalization.*

When x and y are continuous random variables, the *joint probability density function* $f(x, y)$ is defined as

$$\Pr(a \leq x \leq b,\ c \leq y \leq d) = \int_c^d \int_a^b f(x, y) \mathrm{d}x \mathrm{d}y.$$

Similarly to the one-dimensional case, $f(x, y)$ should satisfy

$$f(x, y) \geq 0 \quad \text{and} \quad \iint f(x, y) \mathrm{d}x \mathrm{d}y = 1.$$

The probability density functions of x or y alone, $g(x)$ and $h(y)$, can be expressed by using $f(x, y)$ as

$$\Pr(a \leq x \leq b) = \int_a^b \int f(x, y) \mathrm{d}y \mathrm{d}x = \int_a^b g(x) \mathrm{d}x,$$
$$\Pr(c \leq y \leq d) = \int_c^d \int f(x, y) \mathrm{d}x \mathrm{d}y = \int_c^d h(y) \mathrm{d}y,$$

where

$$g(x) = \int f(x, y) \mathrm{d}y \quad \text{and} \quad h(y) = \int f(x, y) \mathrm{d}x$$

are the *marginal probability density functions.*

5.2 CONDITIONAL PROBABILITY DISTRIBUTION

For discrete random variables x and y, the probability of x given y is denoted by $\Pr(x|y)$ and called the *conditional probability distribution.* Since $\Pr(x|y)$ is the probability that x occurs after y occurs, it is given by

$$\Pr(x|y) = \frac{\Pr(x, y)}{\Pr(y)}. \tag{5.1}$$

Based on this, the *conditional probability mass function* is given by

$$g(x|y) = \frac{f(x, y)}{h(y)}. \tag{5.2}$$

Since the conditional probability distribution is a probability distribution, its expectation and variance can also be defined, which are called the *conditional expectation* and *conditional variance*, respectively,

$$E[x|y] = \sum_x x\, g(x|y) \quad \text{and} \quad V[x|y] = E\left[(x - E[x|y])^2 |y\right].$$

Table 5.1 Example of Contingency Table

$x \setminus y$	Sleepy during the Lecture	Not Sleepy during the Lecture	Total
Like statistics and probability	20	40	60
Dislike statistics and probability	20	20	40
Total	40	60	100

When x and y are continuous random variables, $\Pr(y) = 0$ and thus the conditional probability cannot be defined by Eq. (5.1). However, the *conditional probability density function* can be defined in the same way as Eq. (5.2), and the conditional expectation is given by

$$E[x|y] = \int x\, g(x|y)\mathrm{d}x.$$

5.3 CONTINGENCY TABLE

A *contingency table* summarizes information of multiple discrete random variables. An example of the contingency table is given in Table 5.1: x is the random variable representing students' likes and dislikes of probability and statistics, while y is the random variable representing their drowsiness during the lecture.

A contingency table corresponds to a probability mass function. The right-most column is called the *row marginal total*, the bottom row is called the *column marginal total*, and the right-bottom cell is called the *grand total*. The row marginal total and the column marginal total correspond to marginal probability mass functions, while each row and each column correspond to conditional probability mass functions. Hypothesis testing using the contingency table will be explained in Section 10.4.

5.4 BAYES' THEOREM

When probability $\Pr(y|x)$ of effect y given cause x is known, *Bayes' theorem* allows us to compute the probability $\Pr(x|y)$ of cause x given effect y as

$$\Pr(x|y) = \frac{\Pr(y|x)\Pr(x)}{\Pr(y)}.$$

Since $\Pr(x)$ is the probability of cause x before effect y is known, it is called the *prior probability* of x. On the other hand, $\Pr(x|y)$ is the probability of cause x after effect y is known, and it is called the *posterior probability* of x. Bayes' theorem can be immediately proved by the definition of joint probability distributions:

$$\Pr(x|y)\Pr(y) = \Pr(x, y) = \Pr(y|x)\Pr(x).$$

Bayes' theorem also holds for continuous random variables x and y, using probability density functions as

$$g(x|y) = \frac{h(y|x)g(x)}{h(y)}.$$

Let us illustrate the usefulness of Bayes' theorem through an example of a *polygraph*. Let x be a random variable representing whether a subject's word is true or false, and let y be its prediction by a polygraph. The polygraph has excellent performance, such that

$$\Pr(y = \text{false}|\ x = \text{false}) = 0.99,$$
$$\Pr(y = \text{true}|\ x = \text{true}) = 0.95.$$

Suppose that the prior probability is

$$\Pr(x = \text{false}) = 0.001.$$

If the polygraph says that the subject's word is false, can we believe its decision? The reliability of the polygraph can be evaluated by comparing

$$\Pr(x = \text{false}|\ y = \text{false}) \quad \text{and} \quad \Pr(x = \text{true}|\ y = \text{false}).$$

Since marginalization of x yields

$$\begin{aligned}
\Pr(y = \text{false}) &= \Pr(y = \text{false}|\ x = \text{false})\Pr(x = \text{false}) \\
&\quad + \Pr(y = \text{false}|\ x = \text{true})\Pr(x = \text{true}) \\
&= \Pr(y = \text{false}|\ x = \text{false})\Pr(x = \text{false}) \\
&\quad + \big(1 - \Pr(y = \text{true}|\ x = \text{true})\big)\big(1 - \Pr(x = \text{false})\big) \\
&= 0.99 \times 0.001 + (1 - 0.95) \times (1 - 0.001) \\
&\approx 0.051,
\end{aligned}$$

Bayes' theorem results in

$$\begin{aligned}
\Pr(x = \text{false}|\ y = \text{false}) &= \frac{\Pr(y = \text{false}|\ x = \text{false})\Pr(x = \text{false})}{\Pr(y = \text{false})} \\
&\approx \frac{0.99 \times 0.001}{0.051} \approx 0.019.
\end{aligned}$$

Thus,

$$\begin{aligned}
\Pr(x = \text{true}|\ y = \text{false}) &= 1 - \Pr(x = \text{false}|\ y = \text{false}) \\
&\approx 0.981
\end{aligned}$$

holds and therefore

$$\Pr(x = \text{false}|\ y = \text{false}) \ll \Pr(x = \text{true}|\ y = \text{false}).$$

Consequently, we conclude that the output of the polygraph, y = false, is not reliable.

The above analysis shows that, if the prior probability that the subject tells a lie, $\Pr(x = \text{false})$, is small, the output of the polygraph, $y = \text{false}$, is not reliable. If

$$\Pr(x = \text{false}) > 0.048,$$

it follows that

$$\Pr(x = \text{false} \mid y = \text{false}) > \Pr(x = \text{true} \mid y = \text{false}),$$

showing that the output of the polygraph, $y = \text{false}$, becomes reliable.

5.5 COVARIANCE AND CORRELATION

If random variables x and y are related to each other, change in one variable may affect the other one.

The variance of random variables x and y, $V[x+y]$, does not generally agree with the sum of each variance, $V[x] + V[y]$. Indeed, $V[x + y]$ and $V[x] + V[y]$ are related as

$$V[x + y] = V[x] + V[y] + 2\mathrm{Cov}[x, y],$$

where $\mathrm{Cov}[x, y]$ is the *covariance* of x and y defined by

$$\mathrm{Cov}[x, y] = E\left[(x - E[x])(y - E[y])\right].$$

Increasing x tends to increase y if $\mathrm{Cov}[x, y] > 0$, while increasing x tends to decrease y if $\mathrm{Cov}[x, y] < 0$. If $\mathrm{Cov}[x, y] \approx 0$, x and y are unrelated to each other.

The covariance is useful to create a *portfolio* of stocks. Let x and y be the stock prices of companies A and B, respectively. If $\mathrm{Cov}[x, y] > 0$, buying the stocks of both companies increases the variance. Therefore, the property tends to fluctuate and there is chance to gain a big profit (and at the same time, there is the possibility to lose a lot). On the other hand, if $\mathrm{Cov}[x, y] < 0$, buying the stocks of both companies decreases the variance. Therefore, the risk is hedged and the property is more stabilized (and at the same time, there is less opportunities to gain a big profit).

The matrix that summarizes the variance and covariance of x and y is called the *variance–covariance matrix*:

$$\begin{aligned}\boldsymbol{\Sigma} &= E\left[\left\{\begin{pmatrix} x \\ y \end{pmatrix} - E\begin{pmatrix} x \\ y \end{pmatrix}\right\}\left\{\begin{pmatrix} x \\ y \end{pmatrix} - E\begin{pmatrix} x \\ y \end{pmatrix}\right\}^{\top}\right] \\ &= \begin{pmatrix} V[x] & \mathrm{Cov}[x, y] \\ \mathrm{Cov}[y, x] & V[y] \end{pmatrix},\end{aligned}$$

where $^{\top}$ denotes the transposes. Since $\mathrm{Cov}[y, x] = \mathrm{Cov}[x, y]$, the variance–covariance matrix is symmetric.

The *correlation coefficient* between x and y, denoted by $\rho_{x,y}$, is defined as $\mathrm{Cov}[x, y]$ normalized by the product of standard deviations $\sqrt{V[x]}$ and $\sqrt{V[y]}$,

$$\rho_{x,y} = \frac{\mathrm{Cov}[x,y]}{\sqrt{V[x]}\sqrt{V[y]}}.$$

Since the correlation coefficient is a normalized variant of the covariance, it essentially plays the same role as the covariance. However, the correlation coefficient is bounded as

$$-1 \le \rho_{x,y} \le 1, \tag{5.3}$$

and therefore the absolute strength of the relation between x and y can be known. Eq. (5.3) can be proved by the generic inequality

$$|E[x]| \le E[|x|]$$

and Schwarz's inequality explained in Section 8.3.2,

$$|\mathrm{Cov}[x,y]| \le E[|(x - E[x])(y - E[y])|] \le \sqrt{V[x]}\sqrt{V[y]}.$$

Examples of the correlation coefficient are illustrated in Fig. 5.1. If $\rho_{x,y} > 0$, x and y have the same tendency and x and y are said to be *positively correlated.* On the other hand, if $\rho_{x,y} < 0$, x and y have the opposite tendency and x and y are said to be *negatively correlated.* x and y are deterministically proportional to each other if $\rho_{x,y} = \pm 1$, and x and y are unrelated if $\rho_{x,y} \approx 0$. If $\rho_{x,y} = 0$, x and y are said to be *uncorrelated* to each other. Thus, the correlation coefficient allows us to capture the relatedness between random variables. However, if x and y have nonlinear relation, the correlation coefficient can be close to zero, as illustrated in Fig. 5.2.

5.6 INDEPENDENCE

x and y are said to be statistically *independent* if

$$f(x,y) = g(x)h(y) \quad \text{for all } x \text{ and } y.$$

Conversely, if x and y are not independent, they are said to be *dependent*. If x and y are independent, the following properties hold:

- Conditional probability is independent of the condition,

$$g(x|y) = g(x) \quad \text{and} \quad h(y|x) = h(y).$$

- The expectation of the product agrees with the product of the expectations,

$$E[xy] = E[x]E[y].$$

- The moment-generating function of the sum agrees with the product of the moment-generating functions,

$$M_{x+y}(t) = M_x(t)M_y(t).$$

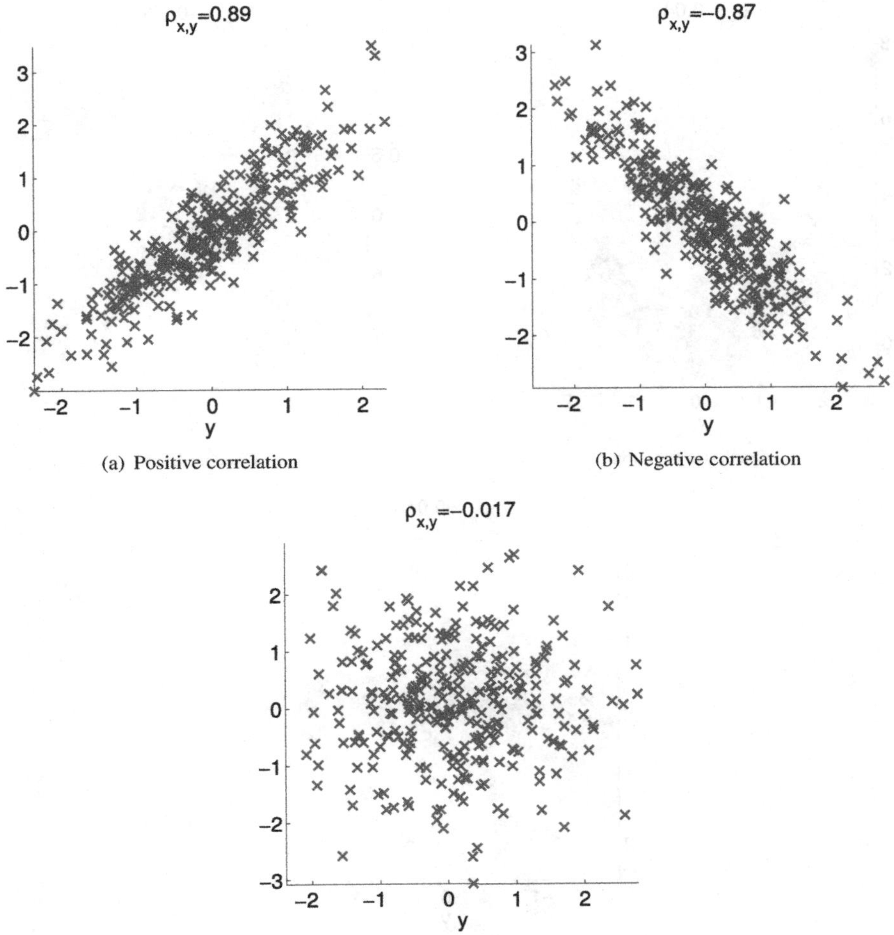

(a) Positive correlation

(b) Negative correlation

(c) No correlation

FIGURE 5.1

Correlation coefficient $\rho_{x,y}$. Linear relation between x and y can be captured.

- Uncorrelated,

$$\mathrm{Cov}[x, y] = 0.$$

Although independence and no correlation both mean that x and y are "unrelated," independence is stronger than no correlation. Indeed, independence implies no correlation but not *vice versa*. For example, random variables x and y whose joint

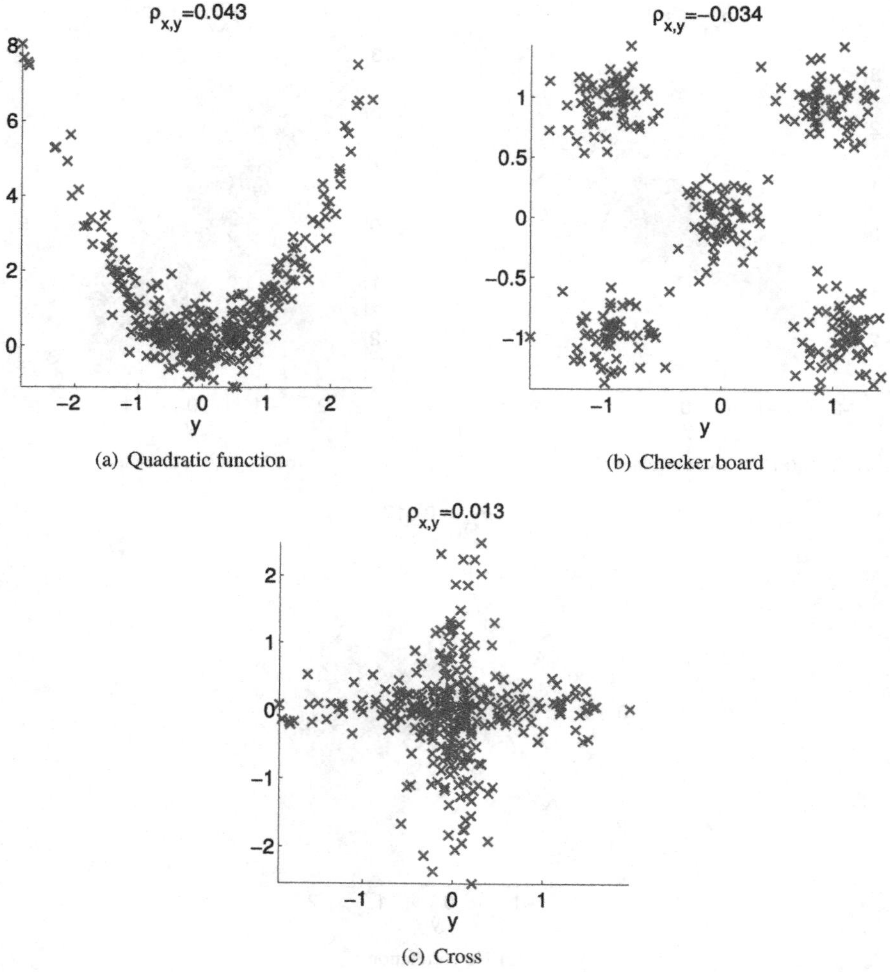

FIGURE 5.2

Correlation coefficient for nonlinear relations. Even when there is a nonlinear relation between x and y, the correlation coefficient can be close to zero if the probability distribution is symmetric.

probability density function is given by

$$f(x,y) = \begin{cases} 1 & (|x| + |y| \leq \frac{1}{\sqrt{2}}) \\ 0 & (\text{otherwise}) \end{cases}$$

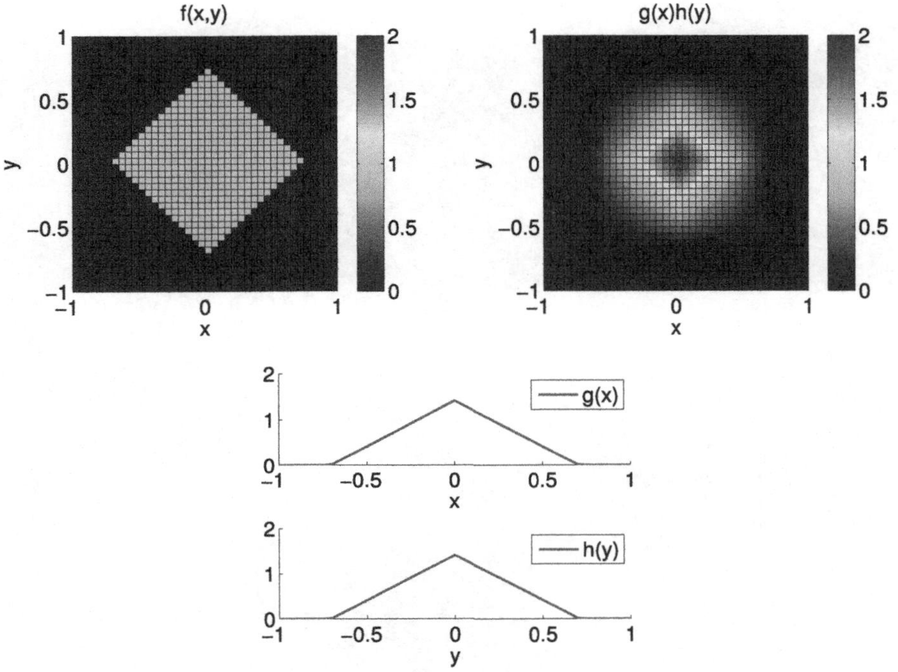

FIGURE 5.3

Example of x and y which are uncorrelated but dependent.

have no correlation but dependent (see Fig. 5.3). More precisely, the uncorrelatedness of x and y can be confirmed by

$$\begin{aligned}\mathrm{Cov}[x,y] &= E\Big[(x-E[x])\,(y-E[y])\Big] = E[xy]\\ &= \int_{-\frac{1}{\sqrt{2}}}^{\frac{1}{\sqrt{2}}} x\left(\int_{-\frac{1}{\sqrt{2}}+|x|}^{\frac{1}{\sqrt{2}}-|x|} y\mathrm{d}y\right)\mathrm{d}x = -\int_{-\frac{1}{\sqrt{2}}}^{\frac{1}{\sqrt{2}}} \sqrt{2}x|x|\mathrm{d}x = 0,\end{aligned}$$

while dependence of x and y, $f(x,y) \neq g(x)h(y)$, can be confirmed by

$$\begin{aligned} g(x) &= \max\left(0,\ \sqrt{2}-2|x|\right),\\ h(y) &= \max\left(0,\ \sqrt{2}-2|y|\right).\end{aligned}$$

CHAPTER

EXAMPLES OF MULTIDIMENSIONAL PROBABILITY DISTRIBUTIONS

6

CHAPTER CONTENTS

In this chapter, popular multidimensional probability distributions and their properties such as the expectation, the variance, and the moment-generating functions are illustrated.

6.1 MULTINOMIAL DISTRIBUTION

The binomial distribution explained in Section 3.2 is the probability distribution of the number x of successful trials in n Bernoulli trials with the probability of success p. The *multinomial distribution* is an extension of the binomial distribution to multidimensional cases.

Let us consider a d-sided dice with the probability of obtaining each side $\boldsymbol{p} = (p_1, \ldots, p_d)^\top$, where

$$p_1, \ldots, p_d \geq 0 \quad \text{and} \quad \sum_{j=1}^{d} p_j = 1.$$

Let $\boldsymbol{x} = (x^{(1)}, \ldots, x^{(d)})^\top$ be the number of times each side appears when the dice is thrown n times, where

$$\boldsymbol{x} \in \Delta_{d,n} = \left\{ \boldsymbol{x} \;\middle|\; x^{(1)}, \ldots, x^{(d)} \geq 0,\ x^{(1)} + \cdots + x^{(d)} = n \right\}.$$

The probability distribution that $\boldsymbol{x}$ follows is the multinomial distribution and is denoted by $\text{Mult}(n, \boldsymbol{p})$.

The probability that the jth side appears $x^{(j)}$ times is $(p_j)^{x^{(j)}}$, and the number of combinations each of the d sides appears $x^{(1)}, \ldots, x^{(d)}$ times for n trials is given by

$\frac{n!}{x^{(1)}!\cdots x^{(d)}!}$. Putting them together, the probability mass function of Mult$(n, \boldsymbol{p})$ is given by

$$f(\boldsymbol{x}) = \frac{n!}{x^{(1)}! \cdots x^{(d)}!}(p_1)^{x^{(1)}} \cdots (p_d)^{x^{(d)}}.$$

When $d = 2$, Mult$(n, \boldsymbol{p})$ is reduced to Bi(n, p_1).

The binomial theorem can also be extended to *multinomial theorem*:

$$(p_1 + \cdots + p_d)^n = \sum_{\boldsymbol{x} \in \Delta_{d,n}} \frac{n!}{x^{(1)}! \cdots x^{(d)}!}(p_1)^{x^{(1)}} \cdots (p_d)^{x^{(d)}},$$

with which the moment-generating function of Mult$(n, \boldsymbol{p})$ can be computed as

$$\begin{aligned} M_{\boldsymbol{x}}(\boldsymbol{t}) = E[e^{\boldsymbol{t}^\top \boldsymbol{x}}] &= \sum_{\boldsymbol{x} \in \Delta_{d,n}} e^{t_1 x^{(1)}} \cdots e^{t_d x^{(d)}} \frac{n!}{x^{(1)}! \cdots x^{(d)}!}(p_1)^{x^{(1)}} \cdots (p_d)^{x^{(d)}} \\ &= \sum_{\boldsymbol{x} \in \Delta_{d,n}} \frac{n!}{x^{(1)}! \cdots x^{(d)}!}(p_1 e^{t_1})^{x^{(1)}} \cdots (p_d e^{t_d})^{x^{(d)}} \\ &= (p_1 e^{t_1} + \cdots + p_d e^{t_d})^n. \end{aligned}$$

From this, the expectation of and (co)variance of Mult$(n, \boldsymbol{p})$ can be computed as

$$E[x^{(j)}] = np_j \quad \text{and} \quad \mathrm{Cov}[x^{(j)}, x^{(j')}] = \begin{cases} np_j(1 - p_j) & (j = j'), \\ -np_j p_{j'} & (j \neq j'). \end{cases}$$

6.2 MULTIVARIATE NORMAL DISTRIBUTION

When random variables $y^{(1)}, \ldots, y^{(d)}$ independently follow the standard normal distribution $N(0, 1)$, the joint probability density function of $\boldsymbol{y} = (y^{(1)}, \ldots, y^{(d)})^\top$ is given by

$$\begin{aligned} g(\boldsymbol{y}) &= \prod_{j=1}^{d} \frac{1}{\sqrt{2\pi}} \exp\left(-\frac{(y^{(j)})^2}{2}\right) \\ &= \frac{1}{(2\pi)^{d/2}} \exp\left(-\frac{1}{2}\boldsymbol{y}^\top \boldsymbol{y}\right). \end{aligned}$$

The expectation and variance-covariance matrices of $\boldsymbol{y}$ are given by

$$E[\boldsymbol{y}] = \boldsymbol{0} \quad \text{and} \quad V[\boldsymbol{y}] = \boldsymbol{I},$$

where $\boldsymbol{0}$ denotes the zero vector and $\boldsymbol{I}$ denotes the identity matrix.

Let us transform $\boldsymbol{y}$ by $d \times d$ *invertible matrix* $\boldsymbol{T}$ and d-dimensional vector $\boldsymbol{\mu}$ as

$$\boldsymbol{x} = \boldsymbol{T}\boldsymbol{y} + \boldsymbol{\mu}.$$

Then the joint probability density function of $\boldsymbol{x}$ is given by

$$\begin{aligned} f(\boldsymbol{x}) &= g(\boldsymbol{y})|\det(\boldsymbol{T})|^{-1} \\ &= \frac{1}{(2\pi)^{d/2}\det(\boldsymbol{\Sigma})^{1/2}} \exp\left(-\frac{1}{2}(\boldsymbol{x}-\boldsymbol{\mu})^\top \boldsymbol{\Sigma}^{-1}(\boldsymbol{x}-\boldsymbol{\mu})\right), \end{aligned}$$

where $\det(\boldsymbol{T})$ is the Jacobian (Fig. 4.2), and

$$\boldsymbol{\Sigma} = \boldsymbol{T}\boldsymbol{T}^\top.$$

This is the general form of the *multivariate normal distribution* and is denoted by $N(\boldsymbol{\mu}, \boldsymbol{\Sigma})$. The expectation and variance-covariance matrices of $N(\boldsymbol{\mu}, \boldsymbol{\Sigma})$ are given by

$$\begin{aligned} E[\boldsymbol{x}] &= \boldsymbol{T}E[\boldsymbol{y}] + \boldsymbol{\mu} = \boldsymbol{\mu}, \\ V[\boldsymbol{x}] &= V[\boldsymbol{T}\boldsymbol{y} + \boldsymbol{\mu}] = \boldsymbol{T}V[\boldsymbol{y}]\boldsymbol{T}^\top = \boldsymbol{\Sigma}. \end{aligned}$$

Probability density functions of two-dimensional normal distribution $N(\boldsymbol{\mu}, \boldsymbol{\Sigma})$ are illustrated in Fig. 6.1. The contour lines of two-dimensional normal distributions are *elliptic*, and their *principal axes* agree with the coordinate axes if the variance-covariance matrix is diagonal. Furthermore, the elliptic contour lines become spherical if all diagonal elements are equal (i.e., the variance-covariance matrix is proportional to the identity matrix).

Eigendecomposition of variance-covariance matrix $\boldsymbol{\Sigma}$ (see Fig. 6.2) shows that the principal axes of the ellipse are parallel to the eigenvectors of $\boldsymbol{\Sigma}$, and their length is proportional to the square root of the eigenvalues (Fig. 6.3).

6.3 DIRICHLET DISTRIBUTION

Let $\boldsymbol{\alpha} = (\alpha_1, \ldots, \alpha_d)^\top$ be a d-dimensional vector with positive entries and let $y^{(1)}, \ldots, y^{(d)}$ be random variables that independently follow the gamma distribution $\mathrm{Ga}(\alpha_j, \lambda)$. For $V = \sum_{j=1}^{d} y^{(j)}$, let

$$\boldsymbol{x} = (x^{(1)}, \ldots, x^{(d)})^\top = \left(\frac{y^{(1)}}{V}, \ldots, \frac{y^{(d)}}{V}\right)^\top.$$

Then the distribution that the above d-dimensional vector $\boldsymbol{x}$ follows is called the *Dirichlet distribution* and is denoted by $\mathrm{Dir}(\boldsymbol{\alpha})$. The domain of $\boldsymbol{x}$ is given by

$$\Delta_d = \left\{\boldsymbol{x} \;\middle|\; x^{(1)}, \ldots, x^{(d)} \geq 0,\ x^{(1)} + \cdots + x^{(d)} = 1\right\}.$$

Drawing a value from a Dirichlet distribution corresponds to generating a (unfair) d-sided die.

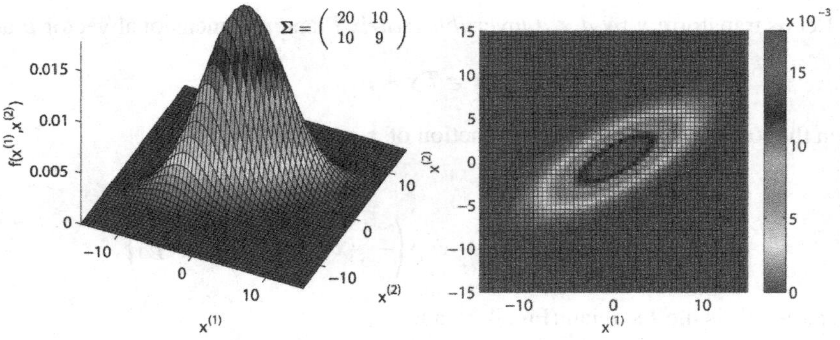

(a) When $\boldsymbol{\Sigma}$ is generic, the contour lines are elliptic

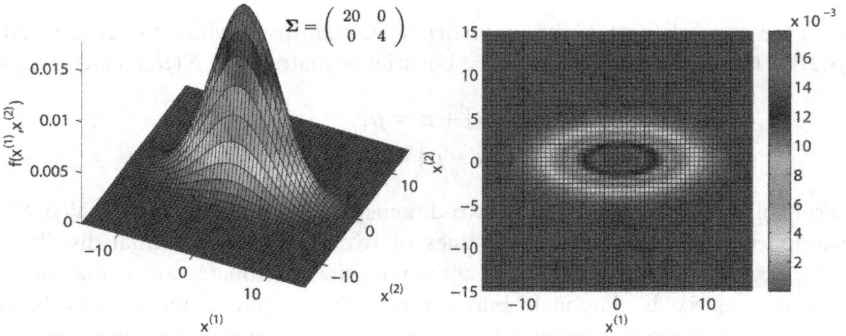

(b) When $\boldsymbol{\Sigma}$ is diagonal, the principal axes of the elliptic contour lines agree with the coordinate axes

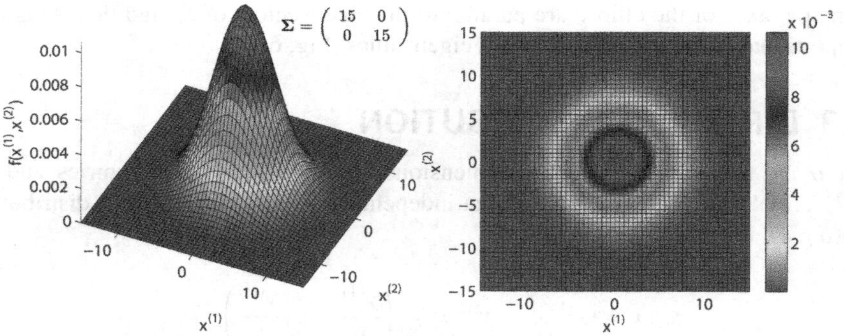

(c) When $\boldsymbol{\Sigma}$ is proportional to identity, the contour lines are spherical

FIGURE 6.1

Probability density functions of two-dimensional normal distribution $N(\boldsymbol{\mu},\boldsymbol{\Sigma})$ with $\boldsymbol{\mu} = (0,0)^\top$.

For $d \times d$ matrix $\boldsymbol{A}$, a nonzero vector $\boldsymbol{\phi}$ and a scalar λ that satisfy

$$\boldsymbol{A}\boldsymbol{\phi} = \lambda\boldsymbol{\phi}$$

are called an *eigenvector* and an *eigenvalue* of $\boldsymbol{A}$, respectively. Generally, there exist d eigenvalues $\lambda_1, \ldots, \lambda_d$ and they are all real when $\boldsymbol{A}$ is symmetric. A matrix whose eigenvalues are all positive is called a *positive definite matrix*, and a matrix whose eigenvalues are all non-negative is called a *positive semidefinite matrix*. Eigenvectors $\boldsymbol{\phi}_1, \ldots, \boldsymbol{\phi}_d$ corresponding to eigenvalues $\lambda_1, \ldots, \lambda_d$ are orthogonal and are usually normalized, i.e., they are *orthonormal* as

$$\boldsymbol{\phi}_j^\top \boldsymbol{\phi}_{j'} = \begin{cases} 1 & (j = j'), \\ 0 & (j \neq j'). \end{cases}$$

A matrix $\boldsymbol{A}$ can be expressed by using its eigenvectors and eigenvalues as

$$\boldsymbol{A} = \sum_{j=1}^{d} \lambda_j \boldsymbol{\phi}_j \boldsymbol{\phi}_j^\top = \boldsymbol{\Phi}\boldsymbol{\Lambda}\boldsymbol{\Phi}^\top,$$

where $\boldsymbol{\Phi} = (\boldsymbol{\phi}_1, \ldots, \boldsymbol{\phi}_d)$ and $\boldsymbol{\Lambda}$ is the diagonal matrix with diagonal elements $\lambda_1, \ldots, \lambda_d$. This is called *eigenvalue decomposition* of $\boldsymbol{A}$. In MATLAB, eigenvalue decomposition can be performed by the `eig` function. When all eigenvalues are nonzero, the inverse of $\boldsymbol{A}$ can be expressed as

$$\boldsymbol{A}^{-1} = \sum_{j=1}^{d} \lambda_j^{-1} \boldsymbol{\phi}_j \boldsymbol{\phi}_j^\top = \boldsymbol{\Phi}\boldsymbol{\Lambda}^{-1}\boldsymbol{\Phi}^\top.$$

For $d \times d$ matrix $\boldsymbol{A}$ and $d \times d$ positive symmetric matrix $\boldsymbol{B}$, a nonzero vector $\boldsymbol{\phi}$ and a scalar λ that satisfy

$$\boldsymbol{A}\boldsymbol{\phi} = \lambda\boldsymbol{B}\boldsymbol{\phi}$$

are called a *generalized eigenvector* and a *generalized eigenvalue* of $\boldsymbol{A}$, respectively.

FIGURE 6.2

Eigenvalue decomposition.

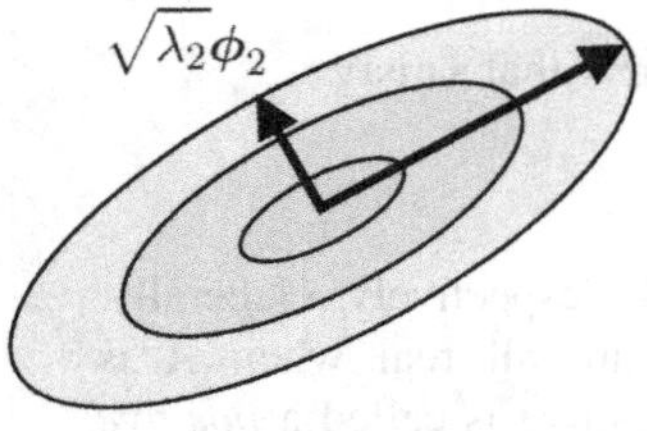

FIGURE 6.3

Contour lines of the normal density. The principal axes of the ellipse are parallel to the eigenvectors of variance-covariance matrix $\boldsymbol{\Sigma}$, and their length is proportional to the square root of the eigenvalues.

The probability density function of $\mathrm{Dir}(\boldsymbol{\alpha})$ is given by

$$f(\boldsymbol{x}) = \frac{\prod_{j=1}^{d}(x^{(j)})^{\alpha_j - 1}}{B_d(\boldsymbol{\alpha})},$$

where

$$B_d(\boldsymbol{\alpha}) = \int_{\Delta_d} \prod_{j=1}^{d} (x^{(j)})^{\alpha_j - 1} \mathrm{d}\boldsymbol{x} \tag{6.1}$$

is the d-dimensional *beta function.*

For $x = (\sqrt{p}\sin\theta)^2$ with $0 \le p \le 1$, Eq. (4.6) implies

$$\begin{aligned}
&\int_0^p x^{\alpha-1}(p-x)^{\beta-1}\mathrm{d}x \\
&= \int_0^{\frac{\pi}{2}} (\sqrt{p}\sin\theta)^{2(\alpha-1)} \left(p - (\sqrt{p}\sin\theta)^2\right)^{\beta-1} \frac{\mathrm{d}x}{\mathrm{d}\theta}\mathrm{d}\theta \\
&= \int_0^{\frac{\pi}{2}} p^{\alpha-1}(\sin\theta)^{2(\alpha-1)} p^{\beta-1}(\cos\theta)^{2(\beta-1)} \cdot 2p\sin\theta\cos\theta\mathrm{d}\theta \\
&= p^{\alpha+\beta-1}\left(2\int_0^{\frac{\pi}{2}} (\sin\theta)^{2\alpha-1}(\cos\theta)^{2\beta-1}\mathrm{d}\theta\right) \\
&= p^{\alpha+\beta-1}B(\alpha,\beta).
\end{aligned}$$

Letting $p = 1 - \sum_{j=1}^{d-2} x^{(j)}$, the integration with respect to $x^{(d-1)}$ in

$$B_d(\alpha) = \int_0^1 (x^{(1)})^{\alpha_1 - 1} \times \int_0^{1-x^{(1)}} (x^{(2)})^{\alpha_2 - 1} \times \cdots \times \int_0^{1-\sum_{j=1}^{d-2} x^{(j)}} (x^{(d-1)})^{\alpha_{d-1}-1} \times \left(1 - \sum_{j=1}^{d-1} x^{(j)}\right)^{\alpha_d - 1} \mathrm{d}x^{(1)}\mathrm{d}x^{(2)} \cdots \mathrm{d}x^{(d-1)}$$

can be computed as

$$\int_0^{1-\sum_{j=1}^{d-2} x^{(j)}} (x^{(d-1)})^{\alpha_{d-1}-1} \left(1 - \sum_{j=1}^{d-1} x^{(j)}\right)^{\alpha_d - 1} \mathrm{d}x^{(d-1)} = \left(1 - \sum_{j=1}^{d-2} x^{(j)}\right)^{\sum_{j=d-1}^{d} \alpha_j - 1} B(\alpha_{d-1}, \alpha_d).$$

Similarly, letting $p = 1 - \sum_{j=1}^{d-3} x^{(j)}$, the integration with respect to $x^{(d-2)}$ in the above equation can be computed as

$$\int_0^{1-\sum_{j=1}^{d-3} x^{(j)}} (x^{(d-2)})^{\alpha_{d-2}-1} \left(1 - \sum_{j=1}^{d-2} x^{(j)}\right)^{\sum_{j=d-1}^{d} \alpha_j - 1} \mathrm{d}x^{(d-2)} = \left(1 - \sum_{j=1}^{d-3} x^{(j)}\right)^{\sum_{j=d-2}^{d} \alpha_j - 1} B\left(\alpha_{d-2}, \sum_{j=d-1}^{d} \alpha_j\right).$$

Repeating this computation yields

$$B_d(\alpha) = \prod_{j=1}^{d-1} B\left(\alpha_j, \sum_{j'=j+1}^{d} \alpha_{j'}\right).$$

Applying

$$B(\alpha, \beta) = \frac{\Gamma(\alpha)\Gamma(\beta)}{\Gamma(\alpha + \beta)}$$

given in Eq. (4.7) to the above equation, the d-dimensional beta function can be expressed by using the gamma function as

$$B_d(\alpha) = \prod_{j=1}^{d-1} \frac{\Gamma(\alpha_j)\Gamma\left(\sum_{j'=j+1}^{d} \alpha_{j'}\right)}{\Gamma\left(\sum_{j'=j}^{d} \alpha_{j'}\right)} = \frac{\prod_{j=1}^{d} \Gamma(\alpha_j)}{\Gamma(\alpha_0)},$$

where

$$\alpha_0 = \sum_{j=1}^{d} \alpha_j.$$

When $d = 2$, $x^{(2)} = 1 - x^{(1)}$ holds and thus the Dirichlet distribution is reduced to the beta distribution shown in Section 4.4:

$$f(x) = \frac{x^{\alpha_1-1}(1-x)^{\alpha_2-1}}{B(\alpha_1,\alpha_2)}.$$

Thus, the Dirichlet distribution can be regarded as a multidimensional extension of the beta distribution.

The expectation of $\mathrm{Dir}(\boldsymbol{\alpha})$ is given by

$$\begin{aligned}
E[x^{(j)}] &= \frac{\int x^{(j)} \prod_{j'=1}^{d} (x^{(j')})^{\alpha_{j'}-1} \mathrm{d}x^{(j)}}{B_d(\boldsymbol{\alpha})} \\
&= \frac{B_d(\alpha_1,\ldots,\alpha_{j-1},\alpha_j+1,\alpha_{j+1},\ldots,\alpha_d)}{B_d(\alpha_1,\ldots,\alpha_{j-1},\alpha_j,\alpha_{j+1},\ldots,\alpha_d)} \\
&= \frac{\Gamma(\alpha_1)\cdots\Gamma(\alpha_{j-1})\Gamma(\alpha_j+1)\Gamma(\alpha_{j+1})\cdots\Gamma(\alpha_d)\times\Gamma(\alpha_0)}{\Gamma(\alpha_0+1)\times\Gamma(\alpha_1)\cdots\Gamma(\alpha_{j-1})\Gamma(\alpha_j)\Gamma(\alpha_{j+1})\cdots\Gamma(\alpha_d)} \\
&= \frac{\Gamma(\alpha_j+1)\Gamma(\alpha_0)}{\Gamma(\alpha_0+1)\Gamma(\alpha_j)} = \frac{\alpha_j\Gamma(\alpha_j)\Gamma(\alpha_0)}{\alpha_0\Gamma(\alpha_0)\Gamma(\alpha_j)} \\
&= \frac{\alpha_j}{\alpha_0}.
\end{aligned}$$

The variance and covariance of $\mathrm{Dir}(\boldsymbol{\alpha})$ can also be obtained similarly as

$$\mathrm{Cov}[x^{(j)}, x^{(j')}] = \begin{cases} \dfrac{\alpha_j(\alpha_0-\alpha_j)}{\alpha_0^2(\alpha_0+1)} & (j = j'), \\ -\dfrac{\alpha_j\alpha_{j'}}{\alpha_0^2(\alpha_0+1)} & (j \neq j'). \end{cases}$$

Probability density functions of $\mathrm{Dir}(\boldsymbol{\alpha})$ when $d = 3$ are illustrated in Fig. 6.4. When $\alpha_1 = \alpha_2 = \alpha_3 = \alpha$, the probability density takes larger values around the corners of the triangle if each $\alpha < 1$, the probability density is uniform if each $\alpha = 1$, and the probability density takes larger values around the center if each $\alpha > 1$. When $\alpha_1, \alpha_2, \alpha_3$ are nonuniform, the probability density tends to take large values around the corners of the triangle with large α_j.

When the Dirichlet parameters $\boldsymbol{\alpha}$ are all equal, i.e.,

$$\alpha_1 = \cdots = \alpha_d = \alpha,$$

(a) $\alpha = (0.99, 0.99, 0.99)^\top$

(b) $\alpha = (1, 1, 1)^\top$

(c) $\alpha = (3, 3, 3)^\top$

(d) $\alpha = (1.05, 0.95, 0.95)^\top$

(e) $\alpha = (4, 4, 1)^\top$

(f) $\alpha = (16, 4, 4)^\top$

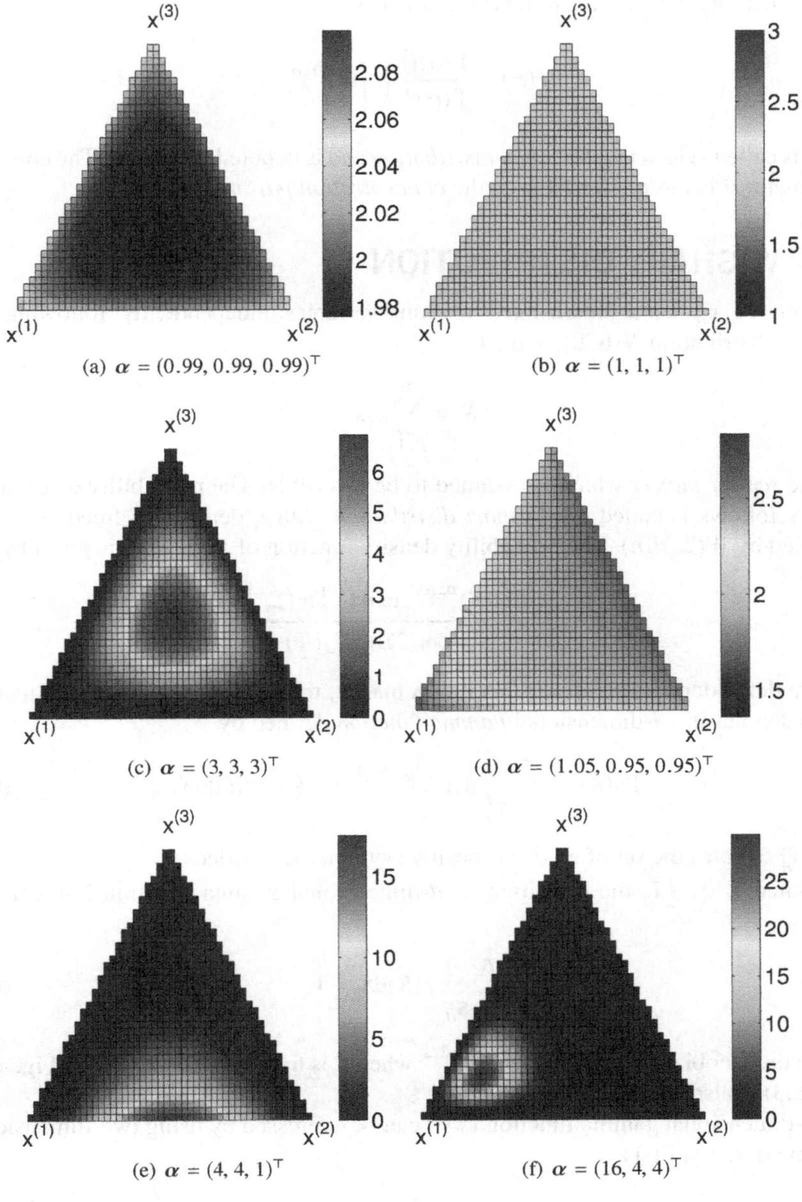

FIGURE 6.4

Probability density functions of Dirichlet distribution $\mathrm{Dir}(\alpha)$. The center of gravity of the triangle corresponds to $x^{(1)} = x^{(2)} = x^{(3)} = 1/3$, and each vertex represents the point that the corresponding variable takes one and the others take zeros.

the probability density function is simplified as

$$f(\boldsymbol{x}) = \frac{\Gamma(\alpha d)}{\Gamma(\alpha)^d} \prod_{j=1}^{d} (x^{(j)})^{\alpha-1}.$$

This is called *symmetric Dirichlet distribution* and is denoted by $\mathrm{Dir}(\alpha)$. The common parameter α is often referred to as the *concentration parameter*.

6.4 WISHART DISTRIBUTION

Let $\boldsymbol{x}_1, \ldots, \boldsymbol{x}_n$ be d-dimensional random variables independently following the normal distribution $N(\boldsymbol{0}, \boldsymbol{\Sigma})$, and let

$$\boldsymbol{S} = \sum_{i=1}^{n} \boldsymbol{x}_i \boldsymbol{x}_i^\top$$

be the *scatter matrix* which is assumed to be invertible. The probability distribution that $\boldsymbol{S}$ follows is called the *Wishart distribution* with n degrees of freedom and is denoted by $W(\boldsymbol{\Sigma}, d, n)$. The probability density function of $W(\boldsymbol{\Sigma}, d, n)$ is given by

$$f(\boldsymbol{S}) = \frac{\det(\boldsymbol{S})^{\frac{n-d-1}{2}} \exp\left(-\frac{1}{2}\mathrm{tr}\left(\boldsymbol{\Sigma}^{-1}\boldsymbol{S}\right)\right)}{\det(2\boldsymbol{\Sigma})^{\frac{n}{2}} \Gamma_d(\frac{n}{2})},$$

where $\det(\cdot)$ denotes the determinant of a matrix, $\mathrm{tr}(\cdot)$ denotes the trace of a matrix, $\Gamma_d(\cdot)$ denotes the d-dimensional *gamma function* defined by

$$\Gamma_d(a) = \int_{\boldsymbol{S} \in \mathbb{S}_d^+} \det(\boldsymbol{S})^{a-\frac{d+1}{2}} \exp\left(-\mathrm{tr}(\boldsymbol{S})\right) \mathrm{d}\boldsymbol{S}, \tag{6.2}$$

and $\mathbb{S}_d^+$ denotes the set of all $d \times d$ positive symmetric matrices.

When $\boldsymbol{\Sigma} = \frac{1}{2}\boldsymbol{I}$, the definition of d-dimensional gamma function immediately yields

$$\int_{\boldsymbol{S} \in \mathbb{S}_d^+} f(\boldsymbol{S}) \mathrm{d}\boldsymbol{S} = 1. \tag{6.3}$$

Since the Jacobian is given by $\det(2\boldsymbol{\Sigma})^{\frac{d+1}{2}}$ when $\boldsymbol{S}$ is transformed to $\frac{1}{2}\boldsymbol{\Sigma}^{-1}\boldsymbol{S}$ (Fig. 4.2), Eq. (6.3) is also satisfied for generic $\boldsymbol{\Sigma} \in \mathbb{S}_d^+$.

d-dimensional gamma function $\Gamma_d(\cdot)$ can be expressed by using two-dimensional gamma function $\Gamma(\cdot)$ as

$$\begin{aligned}\Gamma_d(a) &= \pi^{\frac{d(d-1)}{4}} \prod_{j=1}^{d} \Gamma\left(a + \frac{1-j}{2}\right) \\ &= \pi^{\frac{d-1}{2}} \Gamma_{d-1}(a) \Gamma\left(a + \frac{1-d}{2}\right).\end{aligned}$$

An operator that transforms an $m \times n$ matrix $\boldsymbol{A} = (\boldsymbol{a}_1, \ldots, \boldsymbol{a}_n)$ to the mn-dimensional vector,

$$\mathrm{vec}(\boldsymbol{A}) = (\boldsymbol{a}_1^\top, \ldots, \boldsymbol{a}_n^\top)^\top,$$

is called the *vectorization operator*. The operator that transforms an $m \times n$ matrix $\boldsymbol{A}$ and a $p \times q$ matrix $\boldsymbol{B}$ to an $mp \times nq$ matrix as

$$\boldsymbol{A} \otimes \boldsymbol{B} = \begin{pmatrix} a_{1,1}\boldsymbol{B} & \cdots & a_{1,n}\boldsymbol{B} \\ \vdots & \ddots & \vdots \\ a_{m,1}\boldsymbol{B} & \cdots & a_{m,n}\boldsymbol{B} \end{pmatrix}$$

is called the *Kronecker product*. The vectorization operator and the Kronecker product satisfy the following properties:

$$\begin{aligned} \mathrm{vec}(\boldsymbol{ABC}) &= (\boldsymbol{C}^\top \otimes \boldsymbol{A})\mathrm{vec}(\boldsymbol{B}) \\ &= (\boldsymbol{I} \otimes \boldsymbol{AB})\mathrm{vec}(\boldsymbol{C}) \\ &= (\boldsymbol{C}^\top \boldsymbol{B}^\top \otimes \boldsymbol{I})\mathrm{vec}(\boldsymbol{A}), \\ (\boldsymbol{A} \otimes \boldsymbol{B})^{-1} &= \boldsymbol{A}^{-1} \otimes \boldsymbol{B}^{-1}, \\ (\boldsymbol{A} \otimes \boldsymbol{B})(\boldsymbol{C} \otimes \boldsymbol{D}) &= (\boldsymbol{AC} \otimes \boldsymbol{BD}), \\ \mathrm{tr}(\boldsymbol{A} \otimes \boldsymbol{B}) &= \mathrm{tr}(\boldsymbol{A})\mathrm{tr}(\boldsymbol{B}), \\ \mathrm{tr}(\boldsymbol{AB}) &= \mathrm{vec}(\boldsymbol{A}^\top)^\top \mathrm{vec}(\boldsymbol{B}), \\ \mathrm{tr}(\boldsymbol{ABCD}) &= \mathrm{vec}(\boldsymbol{A}^\top)^\top (\boldsymbol{D}^\top \otimes \boldsymbol{B})\mathrm{vec}(\boldsymbol{C}). \end{aligned}$$

These formulas allow us to compute the product of big matrices efficiently.

FIGURE 6.5

Vectorization operator and Kronecker product.

Thus, the Wishart distribution can be regarded as a multidimensional extension of the gamma distribution. When $d = 1$, $\boldsymbol{S}$ and $\boldsymbol{\Sigma}$ become scalars, and letting $S = x$ and $\Sigma = 1$ yields

$$f(x) = \frac{x^{\frac{n}{2}-1}\exp\left(-\frac{x}{2}\right)}{2^{\frac{n}{2}}\Gamma\left(\frac{n}{2}\right)}.$$

This is equivalent to the probability density function of the chi-squared distribution with n degrees of freedom explained in Section 4.3.

The moment-generating function of $W(\mathbf{\Sigma}, d, n)$ can be obtained by transforming $\boldsymbol{S}$ to $(\frac{1}{2}\mathbf{\Sigma}^{-1} - \boldsymbol{T})\boldsymbol{S}$ as

$$\begin{aligned} M_{\boldsymbol{S}}(\boldsymbol{T}) &= E[e^{\mathrm{tr}(\boldsymbol{TS})}] \\ &= \int_{\boldsymbol{S}\in\mathbb{S}_d^+} \frac{\det(\boldsymbol{S})^{\frac{n-d-1}{2}} \exp\left(-\mathrm{tr}\left((\frac{1}{2}\mathbf{\Sigma}^{-1} - \boldsymbol{T})\boldsymbol{S}\right)\right)}{2^{\frac{dn}{2}} \det(\mathbf{\Sigma})^{\frac{n}{2}} \Gamma_d(\frac{n}{2})} \mathrm{d}\boldsymbol{S} \\ &= \det(\boldsymbol{I} - 2\boldsymbol{T}\mathbf{\Sigma})^{-\frac{n}{2}}. \end{aligned}$$

The expectation and variance-covariance matrices of $\boldsymbol{S}$ that follows $W(\mathbf{\Sigma}, d, n)$ are given by

$$E[\boldsymbol{S}] = n\mathbf{\Sigma} \quad \text{and} \quad V[\mathrm{vec}(\boldsymbol{S})] = 2n\mathbf{\Sigma} \otimes \mathbf{\Sigma},$$

where $\mathrm{vec}(\cdot)$ is the vectorization operator and $\otimes$ denotes the *Kronecker product* (see Fig. 6.5).

CHAPTER

SUM OF INDEPENDENT RANDOM VARIABLES

7

CHAPTER CONTENTS

In this chapter, the behavior of the sum of independent random variables is first investigated. Then the limiting behavior of the mean of independent and identically distributed (i.i.d.) samples when the number of samples tends to infinity is discussed.

7.1 CONVOLUTION

Let x and y be independent discrete variables, and z be their sum:

$$z = x + y.$$

Since $x + y = z$ is satisfied when $y = z - x$, the probability of z can be computed by summing the probability of x and $z - x$ over all x. For example, let z be the sum of the outcomes of two 6-sided dice, x and y. When $z = 7$, these dice take

$$(x, y) = (1,6), (2,5), (3,4), (4,3), (5,2), (6,1),$$

and summing up the probabilities of occurring these combinations gives the probability of $z = 7$.

The probability mass function of z, denoted by $k(z)$, can be expressed as

$$k(z) = \sum_x g(x)h(z - x),$$

where $g(x)$ and $h(y)$ are the probability mass functions of x and y, respectively. This operation is called the *convolution* of x and y and denoted by $x * y$. When x and y are continuous, the probability density function of $z = x + y$, denoted by $k(z)$, is given similarly as

$$k(z) = \int g(x)h(z - x)\mathrm{d}x,$$

where $g(x)$ and $h(y)$ are the probability density functions of x and y, respectively.

7.2 REPRODUCTIVE PROPERTY

When the convolution of two probability distributions in the same family again yields a probability distribution in the same family, that family of probability distributions is said to be *reproductive*. For example, the normal distribution is reproductive, i.e., the convolution of normal distributions $N(\mu_x, \sigma_x^2)$ and $N(\mu_y, \sigma_y^2)$ yields $N(\mu_x + \mu_y, \sigma_x^2 + \sigma_y^2)$.

When x and y are independent, the moment-generating function of their sum, $x + y$, agrees with the product of their moment-generating functions:

$$M_{x+y}(t) = M_x(t)M_y(t).$$

Let x and y follow $N(\mu_x, \sigma_x^2)$ and $N(\mu_y, \sigma_y^2)$, respectively. As shown in Eq. (4.1), the moment-generating function of normal distribution $N(\mu_x, \sigma_x^2)$ is given by

$$M_x(t) = \exp\left(\mu_x t + \frac{\sigma_x^2 t^2}{2}\right).$$

Thus, the moment-generating function of the sum, $M_{x+y}(t)$, is given by

$$\begin{aligned} M_{x+y}(t) &= M_x(t)M_y(t) \\ &= \exp\left(\mu_x t + \frac{\sigma_x^2 t^2}{2}\right) \exp\left(\mu_y t + \frac{\sigma_y^2 t^2}{2}\right) \\ &= \exp\left((\mu_x + \mu_y)t + \frac{(\sigma_x^2 + \sigma_y^2)t^2}{2}\right). \end{aligned}$$

Since this is the moment-generating function of $N(\mu_x+\mu_y, \sigma_x^2+\sigma_y^2)$, the reproductive property of normal distributions is proved.

Similarly, computation of the moment-generating function of $M_{x+y}(t)$ for independent random variables x and y proves the reproductive properties for the binomial, Poisson, negative binomial, gamma, and chi-squared distributions (see Table 7.1). The Cauchy distribution does not have the moment-generating function, but the computation of the characteristic function $\varphi_x(t) = M_{ix}(t)$ (see Section 2.4.3) shows that the convolution of $\mathrm{Ca}(a_x, b_x)$ and $\mathrm{Ca}(a_y, b_y)$ yields $\mathrm{Ca}(a_x + a_y, b_x + b_y)$.

On the other hand, the geometric distribution $\mathrm{Ge}(p)$ (which is equivalent to the binomial distribution $\mathrm{NB}(1, p)$) and the exponential distribution $\mathrm{Exp}(\lambda)$ (which is equivalent to the gamma distribution $\mathrm{Ga}(1, \lambda)$) do not have the reproductive properties for p and λ.

7.3 LAW OF LARGE NUMBERS

Let $x_1, \ldots, x_n$ be random variables and $f(x_1, \ldots, x_n)$ be their joint probability mass/density function. If $f(x_1, \ldots, x_n)$ can be represented by using a probability mass/density function $g(x)$ as

$$f(x_1, \ldots, x_n) = g(x_1) \times \cdots \times g(x_n),$$

Table 7.1 Convolution

Distribution	x	y	$x+y$
Normal	$N(\mu_x, \sigma_x^2)$	$N(\mu_y, \sigma_y^2)$	$N(\mu_x + \mu_y, \sigma_x^2 + \sigma_y^2)$
Binomial	$\mathrm{Bi}(n_x, p)$	$\mathrm{Bi}(n_y, p)$	$\mathrm{Bi}(n_x + n_y, p)$
Poisson	$\mathrm{Po}(\lambda_x)$	$\mathrm{Po}(\lambda_y)$	$\mathrm{Po}(\lambda_x + \lambda_y)$
Negative binomial	$\mathrm{NB}(k_x, p)$	$\mathrm{NB}(k_y, p)$	$\mathrm{NB}(k_x + k_y, p)$
Gamma	$\mathrm{Ga}(\alpha_x, \lambda)$	$\mathrm{Ga}(\alpha_y, \lambda)$	$\mathrm{Ga}(\alpha_x + \alpha_y, \lambda)$
Chi-squared	$\chi^2(n_x)$	$\chi^2(n_y)$	$\chi^2(n_x + n_y)$
Cauchy	$\mathrm{Ca}(a_x, b_x)$	$\mathrm{Ca}(a_y, b_y)$	$\mathrm{Ca}(a_x + a_y, b_x + b_y)$

$x_1, \ldots, x_n$ are mutually independent and follow the same probability distribution. Such $x_1, \ldots, x_n$ are said to be i.i.d. with probability density/mass function $g(x)$ and denoted by

$$x_1, \ldots, x_n \overset{\text{i.i.d.}}{\sim} g(x).$$

When $x_1, \ldots, x_n$ are i.i.d. random variables having expectation μ and variance σ^2, the sample mean (Fig. 7.1),

$$\overline{x} = \frac{1}{n} \sum_{i=1}^{n} x_i,$$

satisfies

$$E[\overline{x}] = \frac{1}{n} \sum_{i=1}^{n} E[x_i] = \mu,$$

$$V[\overline{x}] = \frac{1}{n^2} \sum_{i=1}^{n} V[x_i] = \frac{\sigma^2}{n}.$$

This means that the average of n samples has the same expectation as the original single sample, while the variance is reduced by factor $1/n$. Thus, if the number of samples tends to infinity, the variance vanishes and thus the sample average $\overline{x}$ converges to the true expectation μ.

The *weak law of large numbers* asserts this fact more precisely. When the original distribution has expectation μ, the characteristic function $\varphi_{\overline{x}}(t)$ of the average of independent samples can be expressed by using the characteristic function $\varphi_x(t)$ of a single sample x as

$$\varphi_{\overline{x}}(t) = \left[\varphi_x\left(\frac{t}{n}\right)\right]^n = \left[1 + i\mu\frac{t}{n} + \cdots\right]^n.$$

The mean of samples $x_1, \ldots, x_n$ usually refers to the *arithmetic mean*, but other means such as the *geometric mean* and the *harmonic mean* are also often used:

$$\text{Arithmetic mean: } \frac{1}{n}\sum_{i=1}^{n} x_i,$$

$$\text{Geometric mean: } \left(\prod_{i=1}^{n} x_i\right)^{\frac{1}{n}},$$

$$\text{Harmonic mean: } \frac{1}{\frac{1}{n}\sum_{i=1}^{n}\frac{1}{x_i}}.$$

For example, suppose that the weight increased by the factors 2%, 12%, and 4% in the last three years, respectively. Then the average increase rate is not given by the arithmetic mean $(0.02 + 0.12 + 0.04)/3 = 0.06$, but the geometric mean $(1.02 \times 1.12 \times 1.04)^{\frac{1}{3}} \approx 1.0591$. When climbing up a mountain at 2 kilometer per hour and going back at 6 kilometer per hour, the mean velocity is not given by the arithmetic mean $(2+6)/2 = 4$ but by the harmonic mean $2d/(\frac{d}{2} + \frac{d}{6}) = 3$ for distance d, according to the formula "velocity = distance/time." When $x_1, \ldots, x_n > 0$, the arithmetic, geometric, and harmonic means satisfy

$$\frac{1}{n}\sum_{i=1}^{n} x_i \geq \left(\prod_{i=1}^{n} x_i\right)^{\frac{1}{n}} \geq \frac{1}{\frac{1}{n}\sum_{i=1}^{n}\frac{1}{x_i}},$$

and the equality holds if and only if $x_1 = \cdots = x_n$. The *generalized mean* is defined for $p \neq 0$ as

$$\left(\frac{1}{n}\sum_{i=1}^{n} x_i^p\right)^{\frac{1}{p}}.$$

The generalized mean is reduced to the arithmetic mean when $p = 1$, the geometric mean when $p \to 0$, and the harmonic mean when $p = -1$. The maximum of $x_1, \ldots, x_n$ is given when $p \to +\infty$, and the minimum of $x_1, \ldots, x_n$ is given when $p \to -\infty$. When $p = 2$, it is called the *root mean square*.

FIGURE 7.1

Arithmetic mean, geometric mean, and harmonic mean.

Then Eq. (3.5) shows that the limit $n \to \infty$ of the above equation yields

$$\lim_{n\to\infty} \varphi_{\overline{x}}(t) = e^{it\mu}.$$

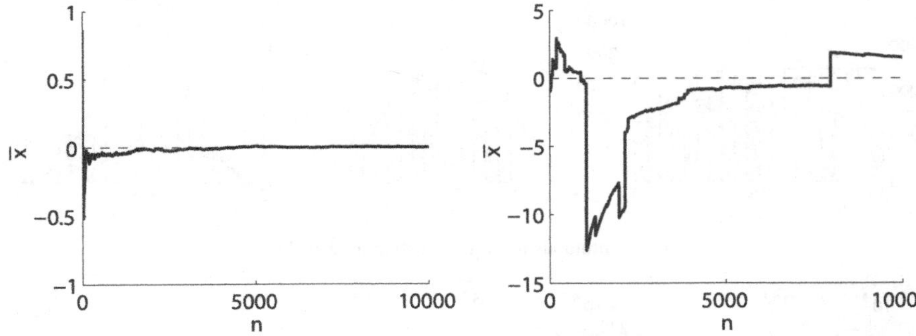

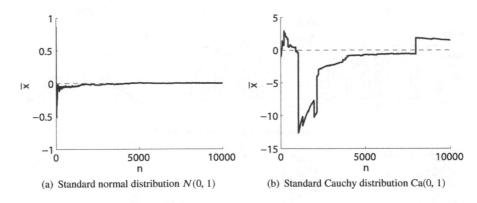

(a) Standard normal distribution $N(0, 1)$ (b) Standard Cauchy distribution $\mathrm{Ca}(0, 1)$

FIGURE 7.2

Law of large numbers.

Since $e^{it\mu}$ is the characteristic function of a constant μ,

$$\lim_{n\to\infty} \Pr(|\overline{x} - \mu| < \varepsilon) = 1$$

holds for any $\varepsilon > 0$. This is the weak law of large numbers and $\overline{x}$ is said to *converge in probability* to μ. If the original distribution has the variance, its proof is straightforward by considering the limit $n \to \infty$ of Chebyshev's inequality (8.4) (see Section 8.2.2).

On the other hand, the *strong law of large numbers* asserts

$$\Pr\left(\lim_{n\to\infty} \overline{x} = \mu\right) = 1,$$

and $\overline{x}$ is said to *almost surely converge* to μ. The almost sure convergence is a more direct and stronger concept than the convergence in probability.

Fig. 7.2 exhibits the behavior of the sample average $\overline{x} = \frac{1}{n}\sum_{i=1}^{n} x_i$ when $x_1, \ldots, x_n$ are i.i.d. with the standard normal distribution $N(0, 1)$ or the standard Cauchy distribution $\mathrm{Ca}(0, 1)$. The graphs show that, for the normal distribution which possesses the expectation, the increase of n yields the convergence of the sample average $\overline{x}$ to the true expectation 0. On the other hand, for the Cauchy distribution which does not have the expectation, the sample average $\overline{x}$ does not converge even if n is increased.

7.4 CENTRAL LIMIT THEOREM

As explained in Section 7.2, the average of independent normal samples follows the normal distribution. If the samples follow other distributions, which distribution does

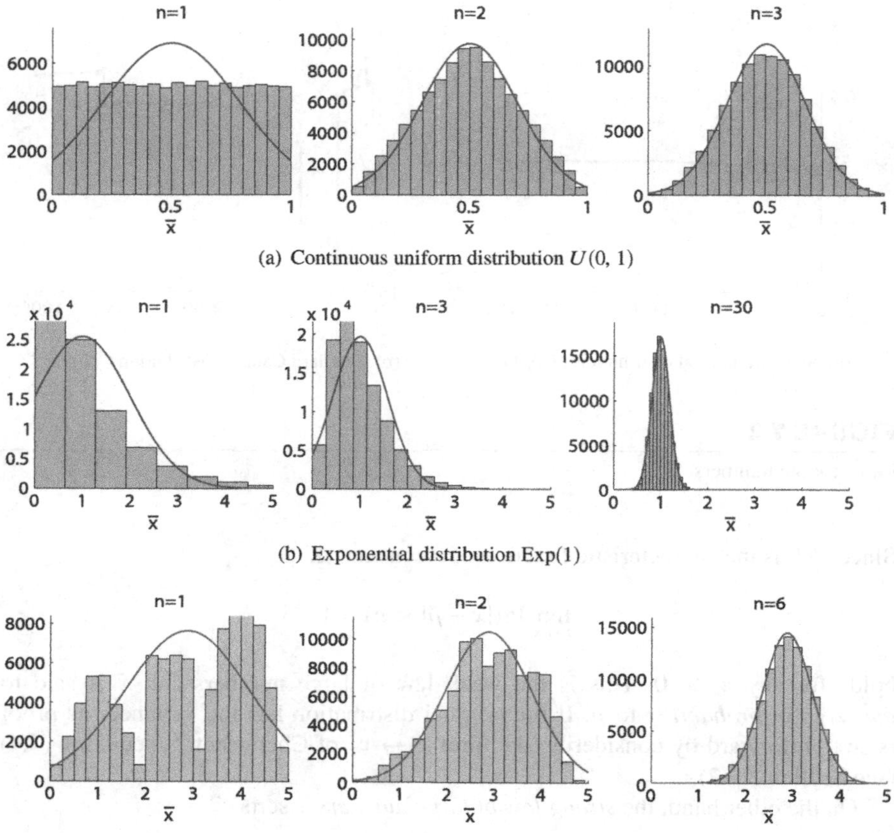

(a) Continuous uniform distribution $U(0, 1)$

(b) Exponential distribution Exp(1)

(c) Distribution used in Fig. 19.11

FIGURE 7.3

Central limit theorem. The solid lines denote the normal densities.

the sample average follow? Fig. 7.3 exhibits the histograms of the sample averages for the continuous uniform distribution $U(0, 1)$, the exponential distribution Exp(1), and the probability distribution used in Fig. 19.11, together with the normal densities with the same expectation and variance. This shows that the histogram of the sample average approaches the normal density as the number of samples, n, increases.

The *central limit theorem* asserts this fact more precisely: for standardized random variable

$$z = \frac{\overline{x} - \mu}{\sigma / \sqrt{n}},$$

the following property holds:

$$\lim_{n\to\infty} \Pr(a \le z \le b) = \int_a^b \frac{1}{\sqrt{2\pi}} e^{-x^2/2} \mathrm{d}x.$$

Since the right-hand side is the probability density function of the standard normal distribution integrated from a to b, z is shown to follow the standard normal distribution in the limit $n \to \infty$. In this case, z is said to *converge in law* or *converge in distribution* to the standard normal distribution. More informally, z is said to *asymptotically* follow the normal distribution or z has *asymptotic normality*. Intuitively, the central limit theorem shows that, for any distribution, as long as it has the expectation μ and variance σ^2, the sample average $\overline{x}$ approximately follows the normal distribution with expectation μ and variance σ^2/n when n is large.

Let us prove the central limit theorem by showing that the moment-generating function of

$$z = \frac{\overline{x} - \mu}{\sigma / \sqrt{n}}$$

is given by the moment-generating function of the standard normal distribution, $e^{t^2/2}$. Let

$$y_i = \frac{x_i - \mu}{\sigma}$$

and express z as

$$z = \frac{1}{\sqrt{n}} \sum_{i=1}^{n} \frac{x_i - \mu}{\sigma} = \frac{1}{\sqrt{n}} \sum_{i=1}^{n} y_i.$$

Since y_i has expectation 0 and variance 1, the moment-generating function of y_i is given by

$$M_{y_i}(t) = 1 + \frac{1}{2}t^2 + \cdots.$$

This implies that the moment-generating function of z is given by

$$M_z(t) = \left[M_{y_i/\sqrt{n}}(t)\right]^n = \left[M_{y_i}\left(\frac{t}{\sqrt{n}}\right)\right]^n = \left[1 + \frac{t^2}{2n} + \cdots\right]^n.$$

If the limit $n \to \infty$ of the above equation is considered, Eq. (3.5) yields

$$\lim_{n\to\infty} M_z(t) = e^{t^2/2},$$

which means that z follows the standard normal distribution.

CHAPTER

PROBABILITY INEQUALITIES

8

CHAPTER CONTENTS

If probability mass/density function $f(x)$ is given explicitly, the values of the probability (density) can be computed. However, in reality, $f(x)$ itself may not be given explicitly, but only partial information such as the expectation $E[x]$ or the variance $V[x]$ is given. In this chapter, various inequalities are introduced that can be used for evaluating the probability only from partial information. See [19] for more details.

8.1 UNION BOUND

Let us recall the additive law of probabilities shown in Section 2.2:

$$\Pr(A \cup B) = \Pr(A) + \Pr(B) - \Pr(A \cap B).$$

Since $\Pr(A \cap B)$ is non-negative, the following inequality is immediately obtained:

$$\Pr(A \cup B) \leq \Pr(A) + \Pr(B),$$

which is called the *union bound*. Even if $\Pr(A \cup B)$ is difficult to obtain explicitly, the union bound gives its upper bound from the probabilities of each event. The union bound can be extended to multiple events: for $A_1, \ldots, A_N$,

$$\Pr(A_1 \cup \cdots \cup A_N) \leq \Pr(A_1) + \cdots + \Pr(A_N).$$

8.2 INEQUALITIES FOR PROBABILITIES

In this section, inequalities for probabilities based on the expectation and variance are introduced.

8.2.1 MARKOV'S INEQUALITY AND CHERNOFF'S INEQUALITY

For *non-negative* random variable x having expectation $E[x]$,

$$\Pr(x \geq a) \leq \frac{E[x]}{a} \tag{8.1}$$

holds for any positive scalar a (Fig. 8.1). This is called *Markov's inequality*, which allows us to know the upper bound of the probability only from the expectation. Since $\Pr(x < a) = 1 - \Pr(x \geq a)$, a lower bound can also be obtained similarly:

$$\Pr(x < a) \geq 1 - \frac{E[x]}{a}.$$

Markov's inequality can be proved by the fact that the function

$$g(x) = \begin{cases} a & (x \geq a), \\ 0 & (0 \leq x < a), \end{cases}$$

defined for $x \geq 0$ satisfies $x \geq g(x)$:

$$E[x] \geq E[g(x)] = a\Pr(x \geq a).$$

For arbitrary non-negative and monotone increasing function $\phi(x)$, Markov's inequality can be generalized as

$$\Pr(x \geq a) = \Pr\big(\phi(x) \geq \phi(a)\big) \leq \frac{E[\phi(x)]}{\phi(a)}. \tag{8.2}$$

Setting $\phi(x) = e^{tx}$ for $t > 0$ in Eq. (8.2) yields

$$\Pr(x \geq a) = \Pr\big(e^{tx} \geq e^{ta}\big) \leq \frac{E[e^{tx}]}{e^{ta}}, \tag{8.3}$$

which is called *Chernoff's inequality*. Minimizing the right-hand side of (8.3) with respect to t yields a tighter upper bound.

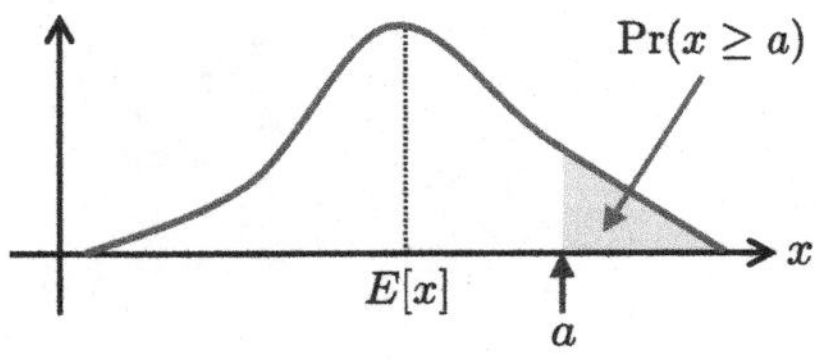

FIGURE 8.1

Markov's inequality.

8.2.2 CANTELLI'S INEQUALITY AND CHEBYSHEV'S INEQUALITY

Markov's inequality upper-bounds the probability based on the expectation $E[x]$. Here, upper bounds of the probability based on the variance $V[x]$ in addition to the expectation $E[x]$ are introduced.

When a random variable x possesses the expectation $E[x]$ and variance $V[x]$, the generic inequality (coming from $a \geq b \Longrightarrow a^2 \geq b^2$)

$$\Pr(a \geq b) \leq \Pr(a^2 \geq b^2)$$

and Markov's inequality (8.1) yield that for a positive scalar ε,

$$\begin{aligned}\Pr\Big(x - E[x] \geq \varepsilon\Big) &= \Pr\Big(\varepsilon(x - E[x]) + V[x] \geq V[x] + \varepsilon^2\Big)\\ &\leq \Pr\Big(\{\varepsilon(x - E[x]) + V[x]\}^2 \geq \{V[x] + \varepsilon^2\}^2\Big)\\ &\leq \frac{E[\{\varepsilon(x - E[x]) + V[x]\}^2]}{\{V[x] + \varepsilon^2\}^2} = \frac{V[x]}{V[x] + \varepsilon^2}.\end{aligned}$$

This is called *Cantelli's inequality* or *one-sided Chebyshev's inequality*. Similarly, the following inequality also holds:

$$\Pr\Big(x - E[x] \leq -\varepsilon\Big) \leq \frac{V[x]}{V[x] + \varepsilon^2}.$$

Furthermore, Markov's inequality (8.1) yields

$$\Pr\Big(|x - E[x]| \geq \varepsilon\Big) = \Pr\Big((x - E[x])^2 \geq \varepsilon^2\Big) \leq \frac{V[x]}{\varepsilon^2}, \tag{8.4}$$

which is called *Chebyshev's inequality* (Fig. 8.2). While Markov's inequality can only bound one-sided probabilities, Chebyshev's inequality allows us to bound two-sided probabilities. A lower bound can also be obtained similarly:

$$\Pr\Big(|x - E[x]| < \varepsilon\Big) \geq 1 - \frac{V[x]}{\varepsilon^2}. \tag{8.5}$$

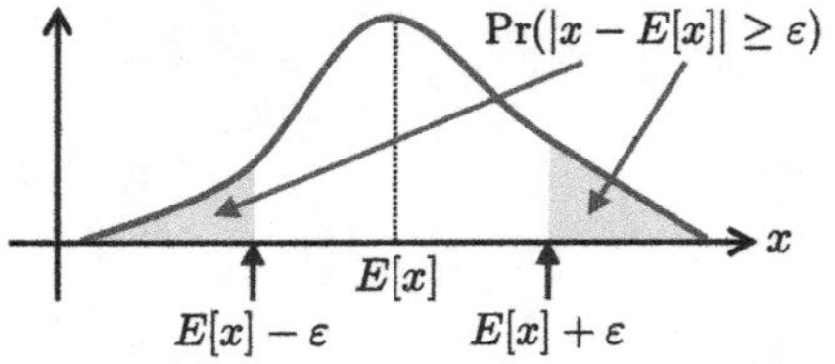

FIGURE 8.2

Chebyshev's inequality.

Chebyshev's inequality can be extended to an arbitrary interval $[a, b]$ as

$$\Pr\left(a < x < b\right) \geq 1 - \frac{V[x] + \left(E[x] - \frac{a+b}{2}\right)^2}{\left(\frac{b-a}{2}\right)^2},$$

which can be proved by applying Markov's inequality as

$$\begin{aligned}\Pr\left((x \leq a) \cup (b \leq x)\right) &= \Pr\left(\left|x - \frac{a+b}{2}\right| \geq \frac{b-a}{2}\right)\\ &= \Pr\left(\left(x - \frac{a+b}{2}\right)^2 \geq \left(\frac{b-a}{2}\right)^2\right)\\ &\leq \frac{E[(x - \frac{a+b}{2})^2]}{\left(\frac{b-a}{2}\right)^2} = \frac{V[x] + \left(E[x] - \frac{a+b}{2}\right)^2}{\left(\frac{b-a}{2}\right)^2}.\end{aligned}$$

Note that the above inequality is reduced to original Chebyshev's inequality (8.5) by setting

$$a = -\varepsilon + E[x] \quad \text{and} \quad b = \varepsilon + E[x].$$

8.3 INEQUALITIES FOR EXPECTATION

In this section, inequalities for the expectation are introduced.

8.3.1 JENSEN'S INEQUALITY

For all $\theta \in [0, 1]$ and all $a < b$, a real-valued function $h(x)$ that satisfies

$$h\left(\theta a + (1 - \theta)b\right) \leq \theta h(a) + (1 - \theta)h(b) \tag{8.6}$$

is called a *convex function* (see Fig. 8.3).

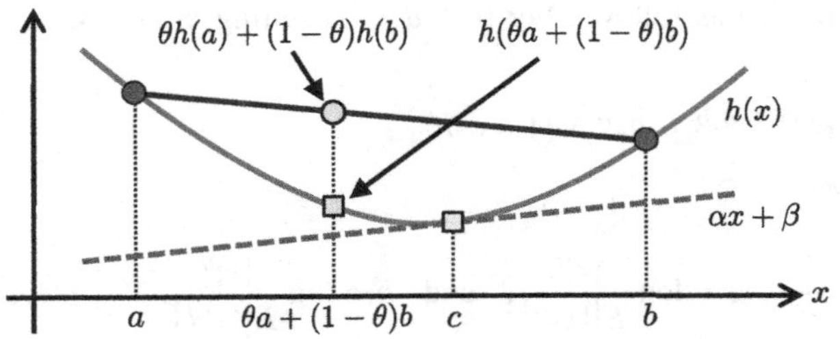

FIGURE 8.3

Convex function and tangent line.

If $h(x)$ is convex, for any c, there exists a tangent line

$$g(x) = \alpha x + \beta$$

such that it touches $h(x)$ at c and lower-bounds $h(x)$ (Fig. 8.3):

$$g(c) = h(c) \quad \text{and} \quad g(x) \leq h(x) \quad \text{for all } x.$$

Setting $c = E[x]$ yields

$$E\big[h(x)\big] \geq E\big[g(x)\big] = \alpha E[x] + \beta = g(E[x]) = h(E[x]),$$

which is called *Jensen's inequality*. In practice, computing the expectation $E\big[h(x)\big]$ is often hard due to nonlinear transformation $h(x)$. On the other hand, computing $h(E[x])$ may be straightforward because it is just a nonlinear transformation of the expectation. Thus, Jensen's inequality allows us to know a lower bound of $E\big[h(x)\big]$ even if it is hard to compute.

Jensen's inequality can be extended to multidimensional convex function $h(\boldsymbol{x})$:

$$E\big[h(\boldsymbol{x})\big] \geq h(E[\boldsymbol{x}]).$$

8.3.2 HÖLDER'S INEQUALITY AND SCHWARZ'S INEQUALITY

For scalars p and q such that

$$\frac{1}{p} + \frac{1}{q} = 1,$$

if random variables $|x|^p$ and $|y|^q$ possess the expectations, the following *Hölder's inequality* holds:

$$E\big[|xy|\big] \leq \left(E\big[|x|^p\big]\right)^{1/p} \left(E\big[|y|^q\big]\right)^{1/q}. \tag{8.7}$$

Hölder's inequality can be proved as follows. For $0 \le \theta \le 1$, setting $h(x) = e^x$ in Eq. (8.6) yields

$$e^{\theta a+(1-\theta)b} \le \theta e^a + (1-\theta)e^b.$$

Setting

$$\theta = \frac{1}{p}, \quad 1-\theta = \frac{1}{q}, \quad a = \log\frac{|x|^p}{E[|x|^p]}, \quad \text{and} \quad b = \log\frac{|y|^q}{E[|y|^q]}$$

yields

$$\frac{|xy|}{(E[|x|^p])^{1/p}(E[|y|^q])^{1/q}} \le \frac{1}{p}\frac{|x|^p}{E[|x|^p]} + \frac{1}{q}\frac{|y|^q}{E[|y|^q]}.$$

Taking the expectation of both sides yields Eq. (8.7) because the expectation of the right-hand side is 1.

Hölder's inequality for $p = q = 2$ is particularly referred to as *Schwarz's inequality*:

$$E[|xy|] \le \sqrt{E[|x|^2]}\sqrt{E[|y|^2]}.$$

8.3.3 MINKOWSKI'S INEQUALITY

For $p \ge 1$, *Minkowski's inequality* is given by

$$(E[|x+y|^p])^{1/p} \le (E[|x|^p])^{1/p} + (E[|y|^p])^{1/p}. \tag{8.8}$$

Minkowski's inequality can be proved as follows. A generic inequality

$$|x+y| \le |x| + |y|$$

yields

$$E[|x+y|^p] \le E[|x| \cdot |x+y|^{p-1}] + E[|y| \cdot |x+y|^{p-1}].$$

When $p = 1$, this immediately yields Eq.(8.8). When $p > 1$, applying Hölder's inequality to each term in the right-hand side yields

$$E[|x| \cdot |x+y|^{p-1}] \le (E[|x|^p])^{1/p}\left(E[|x+y|^{(p-1)q}]\right)^{1/q},$$
$$E[|y| \cdot |x+y|^{p-1}] \le (E[|y|^p])^{1/p}\left(E[|x+y|^{(p-1)q}]\right)^{1/q},$$

where $q = \frac{p}{p-1}$. Then

$$E[|x+y|^p] \le \left((E[|x|^p])^{1/p} + (E[|y|^p])^{1/p}\right)(E[|x+y|^p])^{1-1/p}$$

holds, and dividing the both side by $(E[|x+y|^p])^{1-1/p}$ yields Eq. (8.8).

8.3.4 KANTOROVICH'S INEQUALITY

For random variable x such that $0 < a \leq x \leq b$, *Kantorovich's inequality* is given by

$$E[x]E\left[\frac{1}{x}\right] \leq \frac{(a+b)^2}{4ab}. \tag{8.9}$$

Kantorovich's inequality can be proved as follows. A generic inequality

$$0 \leq (b-x)(x-a) = (a+b-x)x - ab$$

yields

$$\frac{1}{x} \leq \frac{a+b-x}{ab},$$

which yields

$$E[x]E\left[\frac{1}{x}\right] \leq \frac{E[x](a+b-E[x])}{ab}.$$

Completing the square as

$$\begin{aligned} E[x](a+b-E[x]) &= -\left(E[x] - \frac{a+b}{2}\right)^2 + \frac{(a+b)^2}{4} \\ &\leq \frac{(a+b)^2}{4} \end{aligned}$$

yields

$$\begin{aligned} E[x]E\left[\frac{1}{x}\right] &\leq \frac{-(E[x]-(a+b)/2)^2 + (a+b)^2/4}{ab} \\ &\leq \frac{(a+b)^2}{4ab}, \end{aligned}$$

which proves Eq. (8.9).

8.4 INEQUALITIES FOR THE SUM OF INDEPENDENT RANDOM VARIABLES

In this section, inequalities for the sum and average of independent random variables $x_1, \ldots, x_n$,

$$\widetilde{x} = \sum_{i=1}^{n} x_i \quad \text{and} \quad \overline{x} = \frac{1}{n}\sum_{i=1}^{n} x_i,$$

are introduced.

8.4.1 CHEBYSHEV'S INEQUALITY AND CHERNOFF'S INEQUALITY

For $\overline{x} - E[\overline{x}]$, Chebyshev's inequality (8.4) yields

$$\Pr\left(|\overline{x} - E[\overline{x}]| \geq \varepsilon\right) = \Pr\left((\overline{x} - E[\overline{x}])^2 \geq \varepsilon^2\right)$$
$$\leq \frac{V[\overline{x}]}{\varepsilon^2} = \frac{\sum_{i=1}^{n} V[x_i]}{\varepsilon^2}.$$

When $V[x_1] = \cdots = V[x_n] = \sigma^2$, this yields

$$\Pr\left(|\overline{x} - E[\overline{x}]| \geq \varepsilon\right) \leq \frac{\sigma^2}{n\varepsilon^2}.$$

This upper bound is proportional to $1/n$.

Similarly, for arbitrary positive t, Chernoff's inequality (8.3) yields

$$\Pr\left(\overline{x} - E[\overline{x}] \geq \varepsilon\right) \leq \exp(-t\varepsilon)\, E\left[\exp\left(t\sum_{i=1}^{n}(x_i - E[x_i])\right)\right]$$
$$= \exp(-t\varepsilon)\prod_{i=1}^{n} E\left[\exp\left(t(x_i - E[x_i])\right)\right]. \tag{8.10}$$

This upper bound is the product of the moment-generating functions of $x_i - E[x_i]$ for $i = 1, \ldots, n$, and therefore it is expected to decrease exponentially with respect to n.

8.4.2 HOEFFDING'S INEQUALITY AND BERNSTEIN'S INEQUALITY

For random variables x_i such that $a_i \leq x_i \leq b_i$ for $i = 1, \ldots, n$, applying Hoeffding's formula,

$$E\left[\exp\left(t(x_i - E[x_i])\right)\right] \leq \exp\left(\frac{t^2(b_i - a_i)^2}{8}\right),$$

to Chernoff's inequality (8.10) yields

$$\Pr\left(\overline{x} - E[\overline{x}] \geq \varepsilon\right) \leq \exp\left(\frac{t^2}{8}\sum_{i=1}^{n}(b_i - a_i)^2 - t\varepsilon\right).$$

Setting

$$t = \frac{4\varepsilon}{\sum_{i=1}^{n}(b_i - a_i)^2}$$

to minimize the above upper bound yields

$$\Pr\left(\overline{x} - E[\overline{x}] \geq \varepsilon\right) \leq \exp\left(-\frac{2\varepsilon^2}{\sum_{i=1}^{n}(b_i - a_i)^2}\right).$$

This is called *Hoeffding's inequality*. Its variant for sample average $\overline{x}$ is given as

$$\Pr\left(\overline{x} - E[\overline{x}] \geq \varepsilon\right) \leq \exp\left(-\frac{2n\varepsilon^2}{\frac{1}{n}\sum_{i=1}^{n}(b_i - a_i)^2}\right).$$

For random variables x_i such that $|x_i - E[x_i]| \leq a$ for $i = 1,\ldots,n$, *Bernstein's inequality* is given as

$$\Pr\left(\overline{x} - E[\overline{x}] \geq \varepsilon\right) \leq \exp\left(-\frac{\varepsilon^2}{2\sum_{i=1}^{n} V[x_i] + 2a\varepsilon/3}\right).$$

Its derivation will be explained in Section 8.4.3. Its variant for sample average $\overline{x}$ is given as

$$\Pr\left(\overline{x} - E[\overline{x}] \geq \varepsilon\right) \leq \exp\left(-\frac{n\varepsilon^2}{\frac{2}{n}\sum_{i=1}^{n} V[x_i] + 2a\varepsilon/3}\right).$$

When $V[x_1] = \cdots = V[x_n] = \varepsilon$, this yields

$$\Pr\left(\overline{x} - E[\overline{x}] \geq \varepsilon\right) \leq \exp\left(-\frac{n\varepsilon}{2 + 2a/3}\right).$$

Thus, for small positive ε, Bernstein's inequality $\exp(-n\varepsilon)$ gives a tighter upper bound than Hoeffding's inequality $\exp(-n\varepsilon^2)$. This is because Bernstein's inequality uses the variance of $V[x_i]$, while Hoeffding's inequality only uses the domain $[a_i, b_i]$ of each random variable x_i.

8.4.3 BENNETT'S INEQUALITY

For random variables x_i such that $|x_i - E[x_i]| \leq a$ for $i = 1,\ldots,n$, applying Bennett's formula,

$$E\left[\exp\left(t(x_i - E[x_i])\right)\right] \leq \exp\left(V[x_i]\frac{\exp(ta) - 1 - ta}{a^2}\right),$$

to Chernoff's inequality (8.10) yields

$$\Pr\left(\overline{x} - E[\overline{x}] \geq \varepsilon\right) \leq \exp\left(\sum_{i=1}^{n} V[x_i]\frac{\exp(ta) - 1 - ta}{a^2} - t\varepsilon\right).$$

Setting

$$t = \frac{1}{a}\log\left(\frac{a\varepsilon}{\sum_{i=1}^{n} V[x_i]} + 1\right)$$

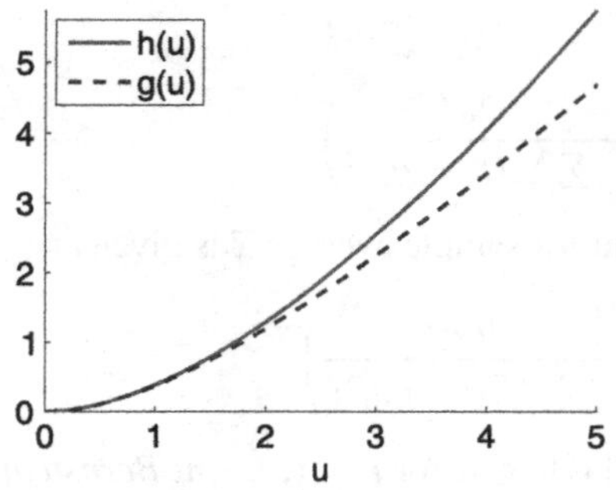

FIGURE 8.4

$h(u) = (1+u)\log(1+u) - u$ and $g(u) = \frac{u^2}{2+2u/3}$.

to minimize the above upper bound yields

$$\Pr\left(\overline{x} - E[\overline{x}] \geq \varepsilon\right) \leq \exp\left(-\frac{\sum_{i=1}^{n} V[x_i]}{a^2} h\left(\frac{a\varepsilon}{\sum_{i=1}^{n} V[x_i]}\right)\right),$$

where

$$h(u) = (1+u)\log(1+u) - u.$$

This is called *Bennett's inequality*.

For $u \geq 0$, the following inequality holds (Fig. 8.4):

$$h(u) \geq g(u) = \frac{u^2}{2 + 2u/3}.$$

Further upper-bounding Bennett's inequality by this actually gives Bernstein's inequality explained in Section 8.4.2. Thus, Bennett's inequality gives a tighter upper bound than Bernstein's inequality, although it is slightly more complicated than Bernstein's inequality.

CHAPTER

STATISTICAL ESTIMATION

9

CHAPTER CONTENTS

So far, various properties of random variables and probability distributions have been discussed. However, in practice, probability distributions are often unknown and only samples are available. In this chapter, an overview of *statistical estimation* for identifying an underlying probability distribution from samples is provided.

9.1 FUNDAMENTALS OF STATISTICAL ESTIMATION

A quantity estimated from samples is called an *estimator* and is denoted with a "hat." For example, when the expectation μ of a probability distribution is estimated by the sample average, its estimator is denoted as

$$\widehat{\mu} = \frac{1}{n}\sum_{i=1}^{n} \boldsymbol{x}_i.$$

An estimator is a function of samples $\{\boldsymbol{x}_i\}_{i=1}^n$ and thus is a random variable. On the other hand, if particular values are plugged in the estimator, the obtained value is called an *estimate*.

A set of probability mass/density functions described with a finite-dimensional parameter $\boldsymbol{\theta}$ is called a *parametric model* and is denoted by $g(\boldsymbol{x};\boldsymbol{\theta})$. In the notation $g(\boldsymbol{x};\boldsymbol{\theta})$, $\boldsymbol{x}$ before the semicolon is a random variable and $\boldsymbol{\theta}$ after the semicolon is a parameter. For example, a parametric model corresponding to the d-dimensional normal distribution,

$$g(\boldsymbol{x};\boldsymbol{\mu},\boldsymbol{\Sigma}) = \frac{1}{(2\pi)^{d/2}\sqrt{\det(\boldsymbol{\Sigma})}}\exp\left(-\frac{1}{2}(\boldsymbol{x}-\boldsymbol{\mu})^\top\boldsymbol{\Sigma}^{-1}(\boldsymbol{x}-\boldsymbol{\mu})\right),$$

has expectation vector $\boldsymbol{\mu}$ and variance-covariance matrix $\boldsymbol{\Sigma}$ as parameters.

An approach to statistical estimation by identifying the parameter in a parametric model is called a *parametric method*, while a *nonparametric method* does not use parametric models or uses a parametric model having infinitely many parameters.

Below, samples $\mathcal{D} = \{\boldsymbol{x}_i\}_{i=1}^n$ are assumed i.i.d. with $f(\boldsymbol{x})$ (see Section 7.3).

9.2 POINT ESTIMATION

Point estimation gives a best estimate of an unknown parameter from samples. Since methods of point estimation will be extensively explored in Part 3 and Part 4, only a brief overview is provided in this section.

9.2.1 PARAMETRIC DENSITY ESTIMATION

Maximum likelihood estimation determines the parameter value so that samples at hand, $\mathcal{D} = \{\boldsymbol{x}_i\}_{i=1}^n$, are generated most probably. The *likelihood* is the probability that the samples $\mathcal{D}$ are generated:

$$L(\boldsymbol{\theta}) = \prod_{i=1}^n g(\boldsymbol{x}_i;\boldsymbol{\theta}),$$

and maximum likelihood estimation maximizes the likelihood:

$$\widehat{\boldsymbol{\theta}}_{\mathrm{ML}} = \underset{\boldsymbol{\theta}}{\operatorname{argmax}}\, L(\boldsymbol{\theta}),$$

where $\operatorname{argmax}_{\boldsymbol{\theta}} L(\boldsymbol{\theta})$ is the maximizer of $L(\boldsymbol{\theta})$ with respect to $\boldsymbol{\theta}$. See Chapter 12, details of maximum likelihood estimation.

The parameter $\boldsymbol{\theta}$ is regarded as a deterministic variable in maximum likelihood estimation, while it is regarded as a random variable in *Bayesian inference*. Then the following probabilities can be considered:

$$\text{Prior probability: } p(\boldsymbol{\theta}),$$
$$\text{Likelihood: } p(\mathcal{D}|\boldsymbol{\theta}),$$
$$\text{Posterior probability: } p(\boldsymbol{\theta}|\mathcal{D}).$$

Typical Bayesian point-estimators are given as the posterior expectation or the posterior mode:

$$\text{Posterior expectation: } \int \boldsymbol{\theta} p(\boldsymbol{\theta}|\mathcal{D})\mathrm{d}\boldsymbol{\theta},$$
$$\text{posterior mode: } \underset{\boldsymbol{\theta}}{\operatorname{argmax}}\, p(\boldsymbol{\theta}|\mathcal{D}).$$

Estimating the posterior mode is often called *maximum a posteriori probability estimation*. The posterior probability can be computed by *Bayes' theorem* explained in Section 5.4 as

$$p(\boldsymbol{\theta}|\mathcal{D}) = \frac{p(\mathcal{D}|\boldsymbol{\theta})p(\boldsymbol{\theta})}{p(\mathcal{D})} = \frac{p(\mathcal{D}|\boldsymbol{\theta})p(\boldsymbol{\theta})}{\int p(\mathcal{D}|\boldsymbol{\theta}')p(\boldsymbol{\theta}')\mathrm{d}\boldsymbol{\theta}'}.$$

Thus, given the likelihood $p(\mathcal{D}|\boldsymbol{\theta})$ and the prior probability $p(\boldsymbol{\theta})$, the posterior probability $p(\boldsymbol{\theta}|\mathcal{D})$ can be computed. However, the posterior probability $p(\boldsymbol{\theta}|\mathcal{D})$ depends on subjective choice of the prior probability $p(\boldsymbol{\theta})$ and its computation can be cumbersome if the posterior probability $p(\boldsymbol{\theta}|\mathcal{D})$ has a complex profile. See Chapter 17, Section 17.3, and Section 19.3 for the details of Bayesian inference.

Maximum likelihood estimation is sometimes referred to as *frequentist inference* when it is contrasted to Bayesian inference.

9.2.2 NONPARAMETRIC DENSITY ESTIMATION

Kernel density estimation (KDE) is a nonparametric technique to approximate the probability density function $f(\boldsymbol{x})$ from samples $\mathcal{D} = \{\boldsymbol{x}_i\}_{i=1}^n$ as

$$\widehat{f}_{\mathrm{KDE}}(\boldsymbol{x}) = \frac{1}{n}\sum_{i=1}^{n} K(\boldsymbol{x},\boldsymbol{x}_i),$$

where $K(\boldsymbol{x},\boldsymbol{x}')$ is a *kernel function*. Typically, the Gaussian kernel function,

$$K(\boldsymbol{x},\boldsymbol{x}') = \frac{1}{(2\pi h^2)^{d/2}} \exp\left(-\frac{\|\boldsymbol{x}-\boldsymbol{x}'\|^2}{2h^2}\right),$$

is used, where $h > 0$ is the *bandwidth* of the Gaussian function and d denotes the dimensionality of $\boldsymbol{x}$, and $\|\boldsymbol{x}\| = \sqrt{\boldsymbol{x}^\top \boldsymbol{x}}$ denotes the *Euclidean norm*.

Nearest neighbor density estimation (NNDE) is another nonparametric method given by

$$\widehat{f}_{\mathrm{NNDE}}(\boldsymbol{x}) = \frac{k\Gamma(\frac{d}{2}+1)}{n\pi^{\frac{d}{2}}\|\boldsymbol{x}-\widetilde{\boldsymbol{x}}_k\|^d},$$

where $\widetilde{\boldsymbol{x}}_k$ denotes the kth nearest sample to $\boldsymbol{x}$ among $\boldsymbol{x}_1,\dots,\boldsymbol{x}_n$ and $\Gamma(\cdot)$ denotes the gamma function explained in Section 4.3.

See Chapter 16 for the derivation and properties of nonparametric density estimation.

9.2.3 REGRESSION AND CLASSIFICATION

Regression is a problem to estimate a function from d-dimensional input $\boldsymbol{x}$ to a real scalar output y based on input-output paired samples $\{(\boldsymbol{x}_i, y_i)\}_{i=1}^n$.

The method of *least squares* (LS) fits a regression model $r(\boldsymbol{x};\boldsymbol{\alpha})$ to data by minimizing the squared sum of residuals:

$$\widehat{\boldsymbol{\alpha}}_{\mathrm{LS}} = \operatorname*{argmin}_{\boldsymbol{\alpha}} \sum_{i=1}^{n} \left(y_i - r(\boldsymbol{x}_i;\boldsymbol{\alpha})\right)^2.$$

A nonparametric Gaussian kernel model is a popular choice as a regression model:

$$r(\boldsymbol{x};\boldsymbol{\alpha}) = \sum_{j=1}^{n} \alpha_j \exp\left(-\frac{\|\boldsymbol{x}-\boldsymbol{x}_j\|^2}{2h^2}\right),$$

where $h > 0$ is the *bandwidth* of the Gaussian kernel. To avoid overfitting to noisy samples, *regularization* (see Chapter 23) is effective:

$$\widehat{\boldsymbol{\alpha}}_{\mathrm{RLS}} = \operatorname*{argmin}_{\boldsymbol{\alpha}} \left[\sum_{i=1}^{n} \left(y_i - r(\boldsymbol{x}_i;\boldsymbol{\alpha})\right)^2 + \lambda\|\boldsymbol{\alpha}\|^2\right],$$

where $\lambda \geq 0$ is the regularization parameter to control the strength of regularization. The LS method is equivalent to maximum likelihood estimation if output y is modeled by the normal distribution with expectation $r(\boldsymbol{x};\boldsymbol{\alpha})$:

$$\frac{1}{\sigma\sqrt{2\pi}} \exp\left(-\frac{\left(y - r(\boldsymbol{x};\boldsymbol{\alpha})\right)^2}{2\sigma^2}\right).$$

Similarly, the regularized LS method is equivalent to Bayesian maximum *a posteriori* probability estimation if the normal prior probability,

$$\frac{1}{(2\pi\lambda^2)^{n/2}} \exp\left(-\frac{\|\boldsymbol{\alpha}\|^2}{2\lambda^2}\right),$$

is used for parameter $\boldsymbol{\alpha} = (\alpha_1,\ldots,\alpha_n)^\top$. See Chapter 22, Chapter 23, Chapter 24, and Chapter 25 for the details of regression.

When output value y takes c discrete *categorical* value, the function estimation problem is called *classification*. When $c = 2$, setting $y = \pm 1$ allows us to naively use (regularized) LS regression in classification. See Chapter 26, Chapter 27, Chapter 30, and Chapter 28 for the details of classification.

9.2.4 MODEL SELECTION

The performance of statistical estimation methods depends on the choice of tuning parameters such as the regularization parameters and the Gaussian bandwidth. Choosing such tuning parameter values based on samples is called *model selection*.

In the frequentist approach, *cross validation* is the most popular model selection method: First, samples $\mathcal{D} = \{\boldsymbol{x}_i\}_{i=1}^n$ are split into k disjoint subsets $\mathcal{D}_1,\ldots,\mathcal{D}_k$. Then statistical estimation is performed with $\mathcal{D}\backslash\mathcal{D}_j$ (i.e., all samples without

$\mathcal{D}_j$), and its estimation error (such as the log-likelihood in density estimation, the squared error in regression, and the misclassification rate in classification) for $\mathcal{D}_j$ is computed. This process is repeated for all $j = 1, \ldots, k$ and the model that minimizes the average estimation error is chosen as the most promising one. See Chapter 14 for the details of frequentist model selection.

In the Bayesian approach, the model $\mathcal{M}$ that maximizes the *marginal likelihood*,

$$p(\mathcal{D}|\mathcal{M}) = \int p(\mathcal{D}|\boldsymbol{\theta}, \mathcal{M}) p(\boldsymbol{\theta}|\mathcal{M}) \mathrm{d}\boldsymbol{\theta},$$

is chosen as the most promising one. This approach is called *type-II maximum likelihood estimation* or the *empirical Bayes* method. See Section 17.4 for the details of Bayesian model selection.

9.3 INTERVAL ESTIMATION

Since an estimator $\widehat{\theta}$ is a function of samples $\mathcal{D} = \{\boldsymbol{x}_i\}_{i=1}^n$, its value depends on the realizations of the samples. Thus, it would be practically more informative if not only a point-estimated value but also its reliability is provided. The interval that an estimator $\widehat{\theta}$ is included with probability at least $1 - \alpha$ is called the *confidence interval* with *confidence level* $1 - \alpha$. In this section, methods for estimating the confidence interval are explained.

9.3.1 INTERVAL ESTIMATION FOR EXPECTATION OF NORMAL SAMPLES

For one-dimensional i.i.d. samples $x_1, \ldots, x_n$ with normal distribution $N(\mu, \sigma^2)$, if the expectation μ is estimated by the sample average,

$$\widehat{\mu} = \frac{1}{n} \sum_{i=1}^{n} x_i,$$

the standardized estimator

$$z = \frac{\widehat{\mu} - \mu}{\sigma / \sqrt{n}}$$

follows the standard normal distribution $N(0, 1)$. Thus, the confidence interval of $\widehat{\mu}$ with confidence level $1 - \alpha$ can be obtained as

$$\left[\widehat{\mu} - \frac{\sigma}{\sqrt{n}} z_{\alpha/2}, \widehat{\mu} + \frac{\sigma}{\sqrt{n}} z_{\alpha/2} \right],$$

where $[-z_{\alpha/2}, +z_{\alpha/2}]$ corresponds to the middle $1 - \alpha$ probability mass of the standard normal density (see Fig. 9.1). However, to compute the confidence interval in practice, knowledge of the standard deviation σ is necessary.

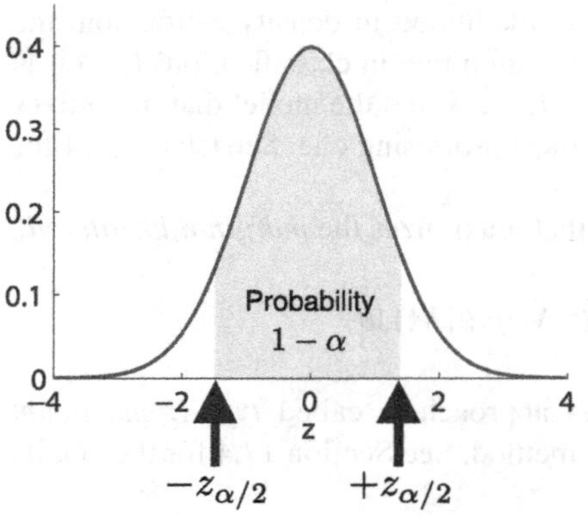

FIGURE 9.1

Confidence interval for normal samples.

When σ is unknown, it is estimated from samples as

$$\widehat{\sigma} = \sqrt{\frac{1}{n-1}\sum_{i=1}^{n}(x_i - \widehat{\mu})^2}.$$

In this case, an estimator standardized with $\widehat{\sigma}$,

$$t = \frac{\widehat{\mu} - \mu}{\widehat{\sigma}/\sqrt{n}},$$

follows the t-distribution with $n-1$ degrees of freedom (Section 4.6). For this reason, standardization with $\widehat{\sigma}$ is sometimes called *Studentization.*

The middle $1 - \alpha$ probability mass of the t-density (see Section 4.6) gives the confidence interval with confidence level $1 - \alpha$ as

$$\left[\widehat{\mu} - \frac{\widehat{\sigma}}{\sqrt{n}}t_{\alpha/2}, \widehat{\mu} + \frac{\widehat{\sigma}}{\sqrt{n}}t_{\alpha/2}\right],$$

where $[-t_{\alpha/2}, +t_{\alpha/2}]$ corresponds to the middle $1-\alpha$ probability mass of the t-density with $n-1$ degrees of freedom (as in Fig. 9.1). As shown in Fig. 4.9, the t-density has heavier tails than the normal density.

9.3.2 BOOTSTRAP CONFIDENCE INTERVAL

The above method for computing the confidence interval is applicable only to the average of normal samples. For statistics other than the expectation estimated from samples following a non-normal distribution, the probability distribution of the

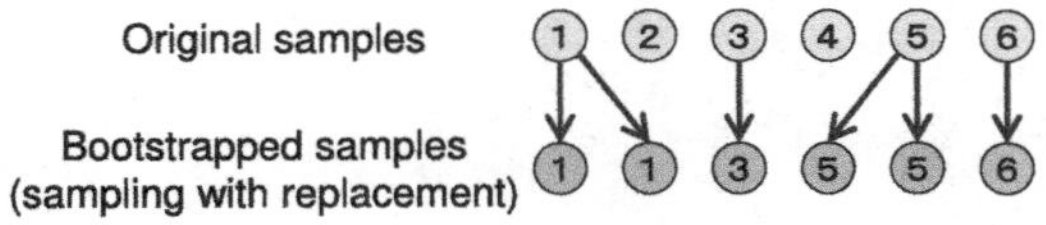

FIGURE 9.2

Bootstrap resampling by sampling with replacement.

estimator cannot be explicitly obtained in general. In such a situation, the use of *bootstrap* allows us to numerically compute the confidence interval.

In the bootstrap method, n pseudo samples $\mathcal{D}' = \{x'_i\}_{i=1}^n$ are gathered by *sampling with replacement* from the original set of samples $\mathcal{D} = \{x_i\}_{i=1}^n$. Because of sampling with replacement, some samples in the original set $\mathcal{D} = \{x_i\}_{i=1}^n$ may be selected multiple times and others may not be selected in $\mathcal{D}' = \{x'_i\}_{i=1}^n$ (Fig. 9.2). From the bootstrapped samples $\mathcal{D}' = \{x'_i\}_{i=1}^n$, an estimator $\widehat{\theta}'$ of the target statistic is computed. These resampling and estimation procedures are repeated many times and the histogram of the estimator $\widehat{\theta}'$ can be constructed. Extracting the middle $1-\alpha$ probability mass of the histogram (as in Fig. 9.1) gives the confidence interval $[-b_{\alpha/2}, +b_{\alpha/2}]$ with confidence level $1-\alpha$.

As illustrated above, the bootstrap method allows us to construct the confidence interval for any statistic and any probability distribution. Furthermore, not only the confidence interval but also any statistics such as the variance and higher-order moments of any estimator can be numerically evaluated by the bootstrap method. However, since the resampling and estimation procedures need to be repeated many times, the computation cost of the bootstrap method can be expensive.

9.3.3 BAYESIAN CREDIBLE INTERVAL

In Bayesian inference, the middle $1-\alpha$ probability mass of the posterior probability $p(\theta|\mathcal{D})$ corresponds to the confidence interval with confidence level $1-\alpha$. This is often referred to as the *Bayesian credible interval*. Thus, in Bayesian inference, the confidence interval can be naively obtained without additional computation. However, if the posterior probability $p(\theta|\mathcal{D})$ has a complex profile, computation of the confidence interval can be cumbersome. Moreover, dependency of the confidence interval on subjective choice of the prior probability $p(\theta)$ can be an issue in practice.

CHAPTER

HYPOTHESIS TESTING 10

CHAPTER CONTENTS

When tossing a coin 20 times, heads are obtained 17 times. Can we then conclude that the coin is biased? The framework of *hypothesis testing* allows us to answer this question statistically. In this chapter, the basic idea of hypothesis testing and standard tests is introduced.

10.1 FUNDAMENTALS OF HYPOTHESIS TESTING

The hypothesis that we want to test is called a *null hypothesis*, while the opposite is called the *alternative hypothesis*. In the above coin-toss example, the null hypothesis is that the coin is not biased (i.e., the probability of obtaining heads is $1/2$), while the alternative hypothesis is that the coin is biased (i.e., the probability of obtaining heads is not $1/2$). The null hypothesis and the alternative hypothesis are often denoted as H_0 and H_1, respectively.

In hypothesis testing, probability that the current samples are obtained under the null hypothesis is computed. If the probability, called the *p-value*, is less than the prespecified *significance level* α, then the null hypothesis is *rejected*; otherwise the null hypothesis is *accepted*. Conventionally, significance level α is set at either 5% or 1%.

As shown in Section 3.2, the probability of obtaining heads in coin tossing follows the binomial distribution. Thus, if the coin is not biased (i.e., the probability of obtaining heads is $1/2$), the probability that heads are obtained more than or equal to

17 times for 20 trials is given by

$$\left(\binom{20}{17}+\binom{20}{18}+\binom{20}{19}+\binom{20}{20}\right)\times\left(\frac{1}{2}\right)^{20}\approx 0.0013.$$

If significance level α is set at 0.01, 0.0013 is less than the significance level. Thus, the null hypothesis is rejected and the alternative hypothesis is accepted, and the coin is concluded to be biased. Note that the probabilities of observing heads more than or equal to 16, 15, and 14 times are 0.0059, 0.0207, and 0.0577, respectively. Thus, if heads is observed no more than 15 times, the null hypothesis is accepted under significance level $\alpha = 0.01$ and the coin is concluded not to be biased.

As illustrated above, when rejecting a null hypothesis by hypothesis testing, the null hypothesis is shown to seldom occur based on samples. On the other hand, when a null hypothesis is accepted, its validity is not actively proved—there is no strong enough evidence that the null hypothesis is wrong and thus the null hypothesis is accepted inevitably. Such a logic is called *proof by contradiction*.

A *two-sided test* is aimed at testing whether the observed value is equal to a target value. For example, if a computationally efficient algorithm of a machine learning method is developed, a two-sided test is used to confirm whether the same performance can still be obtained by the new algorithm. On the other hand, a *one-sided test* is aimed at testing whether the observed value is larger (or smaller) than a target value. If a new machine learning method is developed, a one-sided test is used to see whether superior performance can be obtained by the new method.

In hypothesis testing, a *test statistic* z that can be computed from samples is considered, and its probability distribution is computed under the null hypothesis. If the value $\widehat{z}$ of the test statistic computed from the current samples can occur only with low probability, then the null hypothesis is rejected; otherwise the null hypothesis is accepted. The region to which rejected $\widehat{z}$ belongs is called a *critical region*, and its threshold is called the *critical value*. The critical region in a two-sided test is the left and right tails of the probability mass/density of test statistic z, while that in a one-sided test is the left (or right) tail (see Fig. 10.1).

10.2 TEST FOR EXPECTATION OF NORMAL SAMPLES

For one-dimensional i.i.d. normal samples $x_1, \ldots, x_n$ with variance σ^2, a test for the null hypothesis that its expectation is μ is introduced.

Since the sample average,

$$\widehat{\mu} = \frac{1}{n}\sum_{i=1}^{n} x_i,$$

follows normal distribution $N(\mu, \sigma^2/n)$ under the null hypothesis that the expectation is μ, its standardization,

$$z = \frac{\widehat{\mu} - \mu}{\sqrt{\sigma^2/n}},$$

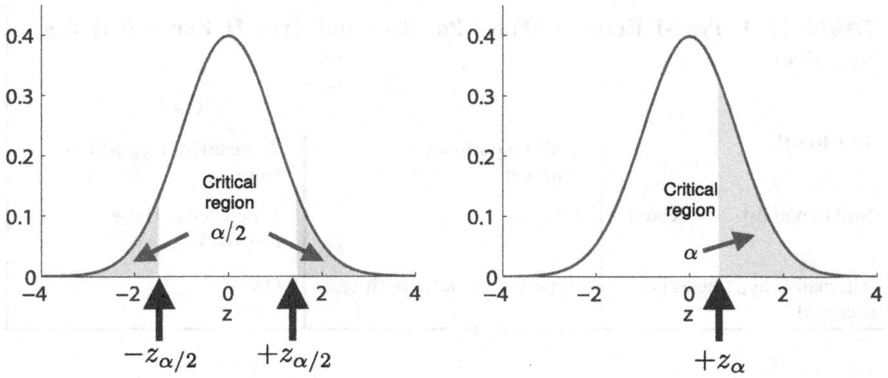

(a) Two-sided test (critical value $z_{\alpha/2}$) (b) One-sided test (critical value z_{α})

FIGURE 10.1

Critical region and critical value.

follows the standard normal distribution $N(0,1)$. The hypothesis test that uses the above z as a test statistic is called a *z-test*. The critical region and critical values are set in the same way as in Fig. 10.1.

When the variance σ^2 is unknown, it is replaced with an unbiased estimator,

$$\widehat{\sigma}^2 = \frac{1}{n-1}\sum_{i=1}^{n}(x_i - \widehat{\mu})^2.$$

Then the test statistic,

$$t = \frac{\widehat{\mu} - \mu}{\sqrt{\widehat{\sigma}^2/n}},$$

follows the t-distribution with $n-1$ degrees of freedom under the null hypothesis that the expectation is μ [7]. This is called a *t-test*.

10.3 NEYMAN-PEARSON LEMMA

The error that the correct null hypothesis is rejected is called the *type-I error* or *false positive*. In the coin-toss example, the type-I error corresponds to concluding that an unbiased coin is biased. On the other hand, the error that the incorrect null hypothesis is accepted is called the *type-II error* or *false negative*, which corresponds to concluding that a biased coin is unbiased (Table 10.1). The type-II error is denoted by β, and $1-\beta$ is called the *power* of the test. In the framework of hypothesis testing, the type-I error is set at α, and the type-II error is reduced (equivalently the power is increased) as much as possible.

Table 10.1 Type-I Error α (False Positive) and Type-II Error β (False Negative)

	Truth	
Test Result	**Null Hypothesis is Correct**	**Alternative Hypothesis is Correct**
Null hypothesis is accepted	OK	Type-II error (false negative)
Alternative hypothesis is accepted	Type-I error (false positive)	OK

Suppose that null hypothesis $\theta = \theta_0$ and alternative hypothesis $\theta = \theta_1$ are tested using samples $\mathcal{D} = \{x_1, \ldots, x_n\}$. Under the null hypothesis $\theta = \theta_0$, let η_α be a scalar such that

$$\Pr\left(\frac{L(\theta_0)}{L(\theta_1)} \leq \eta_\alpha\right) = \alpha,$$

where $L(\theta)$ is the likelihood. Then setting the critical value at η_α and the critical region as

$$\frac{L(\theta_0)}{L(\theta_1)} \leq \eta_\alpha$$

minimizes the type-II error (equivalently maximizes the power) subject to the constraint that the type-I error is fixed at α. This is called the *Neyman-Pearson lemma* and a test based on the ratio of likelihoods is called a *likelihood-ratio test*.

10.4 TEST FOR CONTINGENCY TABLES

In this section, a *goodness-of-fit test* and an *independence test* for contingency tables (see Table 10.2) are introduced. For discrete random variables $x \in \{1, \ldots, \ell\}$ and $y \in \{1, \ldots, m\}$, the *Pearson divergence* from $p_{x,y}$ to $q_{x,y}$ is considered below:

$$\sum_{x=1}^{\ell} \sum_{y=1}^{m} \frac{\left(p_{x,y} - q_{x,y}\right)^2}{q_{x,y}}. \tag{10.1}$$

In the goodness-of-fit test, the null hypothesis that the sample joint probability mass function $\widehat{f}(x,y) = c_{x,y}/n$ is equivalent to a target value $f(x,y)$ is tested. More specifically, Pearson divergence (10.1) for

$$p_{x,y} = \widehat{f}(x,y) \quad \text{and} \quad q_{x,y} = f(x,y)$$

Table 10.2 Contingency Table for $x \in \{1,\ldots,\ell\}$ and $y \in \{1,\ldots,m\}$. $c_{x,y}$ Denotes the Frequency of (x,y), $d_x = \sum_{y=1}^{m} c_{x,y}$, $e_y = \sum_{x=1}^{\ell} c_{x,y}$, and $n = \sum_{x=1}^{\ell}\sum_{y=1}^{m} c_{x,y}$

$x \setminus y$	1	$\cdots$	m	Total
1	$c_{1,1}$	$\cdots$	$c_{1,m}$	d_1
$\vdots$	$\vdots$	$\ddots$	$\vdots$	$\vdots$
ℓ	$c_{\ell,1}$	$\cdots$	$c_{\ell,m}$	d_ℓ
Total	e_1	$\cdots$	e_m	n

is used as a test statistic, and the critical region is computed based on the fact that the Pearson divergence follows the chi-squared distribution with $\ell m - 1$ degrees of freedom [7].

In the independence test, statistical independence between random variables x and y is tested by considering the null hypothesis that the sample joint probability mass function $\widehat{f}(x,y) = c_{x,y}/n$ is equivalent to the product of marginals $\widehat{g}(x)\widehat{h}(y)$, where

$$\widehat{g}(x) = \frac{d_x}{n} = \frac{1}{n}\sum_{y=1}^{m} c_{x,y},$$

$$\widehat{h}(y) = \frac{e_y}{n} = \frac{1}{n}\sum_{x=1}^{\ell} c_{x,y}.$$

More specifically, Pearson divergence (10.1) for

$$p_{x,y} = \widehat{f}(x,y) \quad \text{and} \quad q_{x,y} = \widehat{g}(x)\widehat{h}(y)$$

is used as a test statistic, and the critical region is computed based on the fact that the Pearson divergence follows the chi-squared distribution with $(\ell - 1)(m - 1)$ degrees of freedom when x and y follow multinomial distributions.

The test that uses the *Kullback-Leibler (KL) divergence* (see Section 14.2),

$$\sum_{x=1}^{\ell}\sum_{y=1}^{m} p_{x,y} \log \frac{p_{x,y}}{q_{x,y}},$$

instead of the Pearson divergence is called a *G-test*. This is a likelihood-ratio test and the KL divergence approximately follows the chi-squared distribution with $(\ell - 1)(m - 1)$ degrees of freedom.

A test whose test statistic (approximately) follows the chi-squared distribution under the null hypothesis is called a *chi-square test* [7].

10.5 TEST FOR DIFFERENCE IN EXPECTATIONS OF NORMAL SAMPLES

Let $\mathcal{D} = \{x_1, \ldots, x_n\}$ and $\mathcal{D}' = \{x'_1, \ldots, x'_{n'}\}$ be the i.i.d. samples with normal distributions $N(\mu, \sigma^2)$ and $N(\mu', \sigma^2)$, where the variance is common but the expectations can be different. In this section, a test for the difference in expectation, $\mu - \mu'$, is introduced.

10.5.1 TWO SAMPLES WITHOUT CORRESPONDENCE

Let $\widehat{\mu}$ and $\widehat{\mu}'$ be the sample averages for $\mathcal{D} = \{x_1, \ldots, x_n\}$ and $\mathcal{D}' = \{x'_1, \ldots, x'_{n'}\}$:

$$\widehat{\mu} = \frac{1}{n}\sum_{i=1}^{n} x_i \quad \text{and} \quad \widehat{\mu}' = \frac{1}{n'}\sum_{i=1}^{n'} x'_i.$$

Since $\mathcal{D} = \{x_1, \ldots, x_n\}$ and $\mathcal{D}' = \{x'_1, \ldots, x'_{n'}\}$ are statistically independent of each other, the variance of the difference in sample average, $\widehat{\mu} - \widehat{\mu}'$, is given by $\sigma^2(1/n + 1/n')$ under the null hypothesis $\mu = \mu'$. Thus, its standardization,

$$z_{\mathrm{u}} = \frac{\widehat{\mu} - \widehat{\mu}'}{\sqrt{\sigma^2(1/n + 1/n')}},$$

follows the standard normal distribution $N(0,1)$. The test that uses the above z_{u} as a test statistic is called an *unpaired z-test*. The critical region and critical values are set in the same way as in Fig. 10.1.

When the variance σ^2 is unknown, it is replaced with an unbiased estimator:

$$\widehat{\sigma}_{\mathrm{u}}^2 = \frac{\sum_{i=1}^{n}(x_i - \widehat{\mu})^2 + \sum_{i=1}^{n'}(x'_i - \widehat{\mu}')^2}{n + n' - 2}.$$

Then the test statistic,

$$t_{\mathrm{u}} = \frac{\widehat{\mu} - \widehat{\mu}'}{\sqrt{\widehat{\sigma}_{\mathrm{u}}^2(1/n + 1/n')}},$$

follows the t-distribution with $n + n' - 2$ degrees of freedom under the null hypothesis $\mu = \mu'$. This is called an *unpaired t-test*.

If the variances of $\mathcal{D} = \{x_1, \ldots, x_n\}$ and $\mathcal{D}' = \{x'_1, \ldots, x'_{n'}\}$ can be different, the variances, say σ^2 and σ'^2, are replaced with their unbiased estimators:

$$\widehat{\sigma}^2 = \frac{\sum_{i=1}^{n}(x_i - \widehat{\mu})^2}{n - 1},$$

$$\widehat{\sigma}'^2 = \frac{\sum_{i=1}^{n'}(x'_i - \widehat{\mu}')^2}{n' - 1}.$$

Then, under the null hypothesis $\mu = \mu'$, the test statistic,

$$t_{\mathrm{W}} = \frac{\widehat{\mu} - \widehat{\mu}'}{\sqrt{\widehat{\sigma}^2/n + \widehat{\sigma}'^2/n'}},$$

approximately follows the t-distribution with round(k) degrees of freedom, where

$$k = \frac{\left(\widehat{\sigma}^2/n + \widehat{\sigma}'^2/n'\right)^2}{\widehat{\sigma}^4/(n^2(n-1)) + \widehat{\sigma}'^4/(n'^2(n'-1))},$$

and round($\cdot$) rounds off the value to the nearest integer. This is called *Welch's t-test*.

Under the null hypothesis $\sigma^2 = \sigma'^2$, the test statistic,

$$F = \frac{\widehat{\sigma}^2}{\widehat{\sigma}'^2},$$

follows the F-distribution explained in Section 4.6. This is called the *F-test*, which allows us to test the equality of the variance.

10.5.2 TWO SAMPLES WITH CORRESPONDENCE

Suppose that two sets of samples, $\mathcal{D} = \{x_1, \ldots, x_n\}$ and $\mathcal{D}' = \{x'_1, \ldots, x'_{n'}\}$, have correspondence, i.e., for $n = n'$, the samples are paired as

$$\{(x_1, x'_1), \ldots, (x_n, x'_n)\}.$$

Then, whether the expectations of $\mathcal{D}$ and $\mathcal{D}'$ are equivalent can be tested by the unpaired z-test:

$$z_{\mathrm{u}} = \frac{\Delta\widehat{\mu}}{\sqrt{2\sigma^2/n}},$$

where $\Delta\widehat{\mu}$ is the average of the difference between the paired samples:

$$\Delta\widehat{\mu} = \frac{1}{n}\sum_{i=1}^{n}(x_i - x'_i) = \widehat{\mu} - \widehat{\mu}'.$$

In this situation, the power of the test (see Section 10.3) can be improved if positive correlation exists between the two samples. More specifically, under the null hypothesis $\mu - \mu' = 0$, the variance of $\Delta\widehat{\mu}$ is given by $2\sigma^2(1-\rho)/n$, where ρ is the correlation coefficient:

$$\rho = \frac{\mathrm{Cov}[x, x']}{\sqrt{V[x]}\sqrt{V[x']}}.$$

Then its standardization,

$$z_{\mathrm{p}} = \frac{\Delta\widehat{\mu}}{\sqrt{2\sigma^2(1-\rho)/n}},$$

follows the standard normal distribution $N(0,1)$. The test that uses the above z_{p} as a test statistic is called a *paired z-test*. If $\rho > 0$,

$$|z_{\mathrm{p}}| > |z_{\mathrm{u}}|$$

holds and thus the power of the test can be improved.

When the variance σ^2 is unknown, it is replaced with

$$\widehat{\sigma}_{\mathrm{p}}^2 = \frac{\sum_{i=1}^{n}(x_i - x_i' - \Delta\widehat{\mu})^2}{2(n-1)}. \tag{10.2}$$

Then the test statistic,

$$t_{\mathrm{p}} = \frac{\Delta\widehat{\mu}}{\sqrt{2\widehat{\sigma}_{\mathrm{p}}^2(1-\rho)/n}},$$

follows the *t*-distribution with $n-1$ degrees of freedom under the null hypothesis $\mu - \mu' = 0$. This is called a *paired t-test*.

The test statistic of the unpaired *t*-test for $n = n'$ can be expressed as

$$t_{\mathrm{u}} = \frac{\Delta\widehat{\mu}}{\sqrt{2\widehat{\sigma}_{\mathrm{u}}^2/n}},$$

where

$$\widehat{\sigma}_{\mathrm{u}}^2 = \frac{\sum_{i=1}^{n}\left((x_i - \widehat{\mu})^2 + (x_i' - \widehat{\mu}')^2\right)}{2(n-1)}.$$

$\widehat{\sigma}_{\mathrm{u}}^2$ and $\widehat{\sigma}_{\mathrm{p}}^2$ defined in Eq. (10.2) can be expressed by using $\widehat{\sigma}_{\mathrm{u}}^2$ as

$$\begin{aligned}
\widehat{\sigma}_{\mathrm{p}}^2 &= \frac{\sum_{i=1}^{n}(x_i - x_i' - \Delta\widehat{\mu})^2}{2(n-1)} \\
&= \frac{\sum_{i=1}^{n}((x_i - \mu) - (x_i' - \mu'))^2}{2(n-1)} \\
&= \frac{\sum_{i=1}^{n}(x_i - \mu)^2 + \sum_{i=1}^{n}(x_i' - \mu')^2 - 2\sum_{i=1}^{n}(x_i - \mu)(x_i' - \mu')}{2(n-1)} \\
&= \widehat{\sigma}_{\mathrm{u}}^2 - \widehat{\mathrm{Cov}}[x, x'],
\end{aligned}$$

where $\widehat{\mathrm{Cov}}[x, x']$ is the sample covariance given by

$$\widehat{\mathrm{Cov}}[x, x'] = \frac{1}{n-1}\sum_{i=1}^{n}(x_i - \mu)(x_i' - \mu').$$

If $\widehat{\mathrm{Cov}}[x, x'] > 0$,

$$|t_{\mathrm{p}}| > |t_{\mathrm{u}}|$$

holds and thus the power of the test can be improved.

Table 10.3 Wilcoxon Rank-Sum Test. In this Example, $r_1 = 3$, $r_2 = 5.5$, $r_3 = 1$, and the Rank-Sum is $r = 9.5$

$\mathcal{D}$	x_3		x_1		x_2		
$\mathcal{D}'$		x'_2		x'_4		x'_3	x'_1
Sample value	−2	0	1	3.5	7	7	7.1
Rank	1	2	3	4	5.5	5.5	7

10.6 NONPARAMETRIC TEST FOR RANKS

In the previous section, samples $\mathcal{D} = \{x_1,\ldots,x_n\}$ and $\mathcal{D}' = \{x'_1,\ldots,x'_{n'}\}$ were assumed to follow normal distributions. When samples do not follow the normal distributions, particularly in the presence of outliers, tests based on the normality may not be reliable. In this section, *nonparametric tests* that do not require parametric assumptions on the probability distributions are introduced.

10.6.1 TWO SAMPLES WITHOUT CORRESPONDENCE

Without loss of generality, assume $n \le n'$ below (if $n > n'$, just $\mathcal{D}$ and $\mathcal{D}'$ are swapped to satisfy $n \le n'$).

Let us merge all samples $x_1,\ldots,x_n$ and $x'_1,\ldots,x'_{n'}$ together and sort them in the ascending order. Let us denote the ranks of $x_1,\ldots,x_n$ in the set of $n + n'$ samples by $r_1,\ldots,r_n$. If there are ties, the mean rank is used. For example, if x_i the third smallest sample in $x_1,\ldots,x_n$ and $x'_1,\ldots,x'_{n'}$, r_i is set at 3; if the fifth and sixth smallest samples share the same value, their ranks are 5.5 (Table 10.3). The *Wilcoxon rank-sum test* uses the sum of the ranks of $x_1,\ldots,x_n$,

$$r = \sum_{i=1}^{n} r_i,$$

a test statistic.

Under the null hypothesis that $\mathcal{D}$ and $\mathcal{D}'$ follow the same probability distribution, the above test statistic r approximately follows the normal distribution with expectation and variance given by

$$\mu = \frac{n(n+n'+1)}{2},$$
$$\sigma^2 = \frac{nn'(n+n'+1)}{12}.$$

Since the standardized statistic $(r-\mu)/\sigma$ follows the standard normal distribution $N(0,1)$, setting the critical region and critical values as in Fig. 10.1 allows us to perform hypothesis testing.

The *Mann-Whitney U-test* is essentially the same as the Wilcoxon rank-sum test.

10.6.2 TWO SAMPLES WITH CORRESPONDENCE

For $n = n'$, suppose that two sets of samples $\mathcal{D} = \{x_1, \ldots, x_n\}$ and $\mathcal{D}' = \{x'_1, \ldots, x'_{n'}\}$ have correspondence as

$$\{(x_1, x'_1), \ldots, (x_n, x'_n)\}.$$

Let us remove pairs such that $x_i = x'_i$ and reduce the value of n accordingly (i.e., n denotes the number of sample pairs such that $x_i \neq x'_i$ below). Let us sort the sample pairs in the ascending order of $|x_i - x'_i|$, and let r_i be its rank. If there are ties, the mean rank is used in the same way as in the Wilcoxon rank-sum test. Then the *Wilcoxon signed-rank test* uses the sum of the ranks of samples such that $x_i - x'_i > 0$,

$$s = \sum_{i: x_i - x'_i > 0} r_i,$$

a test statistic.

Under the null hypothesis that $\mathcal{D}$ and $\mathcal{D}'$ follow the same probability distribution, the above test statistic s approximately follows the normal distribution with expectation and variance given by

$$\mu = \frac{n(n+1)}{4},$$
$$\sigma^2 = \frac{n(n+1)(2n+1)}{24}.$$

Since the standardized statistic $(s - \mu)/\sigma$ follows the standard normal distribution $N(0, 1)$, setting the critical region and critical values as in Fig. 10.1 allows us to perform hypothesis testing.

10.7 MONTE CARLO TEST

The test statistics introduced above all (approximately) follow the normal distribution, t-distribution, and chi-squared distribution under the null hypothesis. However, if a test statistic is more complicated, its distribution cannot be analytically derived even approximately. In such a situation, computing the value of a test statistic using samples generated by a *Monte Carlo method* and numerically obtaining the critical region are practically useful. The Monte Carlo method is a generic name of algorithms that use random numbers, and its name stems from the Monte Carlo Casino in Monaco. A test based on the Monte Carlo method is called a *Monte Carlo test*.

For testing whether the expectation of samples $\mathcal{D} = \{x_1, \ldots, x_n\}$ is μ (which was discussed in Section 10.2) by a Monte Carlo test, the bootstrap method introduced in Section 9.3.2 is used. More specifically, bootstrap resampling of $\mathcal{D} = \{x_1, \ldots, x_n\}$ and computing their average are repeated many times, and a histogram of the average is constructed. Then hypothesis testing can be approximately performed by verifying whether the target value μ is included in the critical region (see Fig. 10.1). The p-value can also be approximated from the histogram. As illustrated above, the bootstrap-based hypothesis test is highly general and can be applied to any statistic computed from samples following any probability distribution.

In the contingency table explained in Section 10.4, enumeration of all possible combinations allows us to obtain the probability distribution of any test statistic. For 2×2 contingency tables, computing the p-value by such an exhaustive way is called *Fisher's exact test*. In the Monte Carlo test, combinations are randomly generated and the test is performed approximately. Since this allows us to numerically approximate the p-value for contingency tables of arbitrary size and for arbitrary test statistic, it can be regarded as generalization of Fisher's exact test.

For testing the null hypothesis that $\mathcal{D} = \{x_1, \ldots, x_n\}$ and $\mathcal{D}' = \{x'_1, \ldots, x'_{n'}\}$ follow the same probability distributions, let us merge all samples $x_1, \ldots, x_n$ and $x'_1, \ldots, x'_{n'}$ together and partition them into two sets with sizes n and n'. Then enumeration of all possible partitions allows us to obtain the probability distribution of any test statistic. This is called a *permutation test*, and the Monte Carlo test can be regarded as its approximate implementation with a limited number of repetitions.

PART

GENERATIVE APPROACH TO STATISTICAL PATTERN RECOGNITION

3

The objective of pattern recognition is to classify a given pattern $\boldsymbol{x}$ to one of the pre-specified classes, y. For example, in hand-written digit recognition, pattern $\boldsymbol{x}$ is an image of hand-written digit and class y corresponds to the number the image represents. The number of classes is 10 (i.e., from "0" to "9"). Among various approaches, statistical pattern recognition tries to learn a classifier based on statistical properties of training samples. In Part 3, an approach to statistical pattern recognition based on estimation of the data-generating probability distribution.

After the problem of pattern recognition based on generative model estimation is formulated in Chapter 11, various statistical estimators are introduced. These methods are categorizes as either *parametric* or *non-parametric* and either *frequentist* or *Bayesian*.

First, a standard parametric frequentist method called *maximum likelihood estimation* is introduced in Chapter 12, its theoretical properties are investigated in Chapter 13, the issue of model selection is discussed in Chapter 14, and the algorithm for Gaussian mixture models called the *expectation–maximization algorithm* is introduced in Chapter 15. Then, non-parametric frequentist methods called *kernel density estimation* and *nearest neighbor density estimation* are introduced in Chapter 16.

The basic ideas of the parametric Bayesian approach is introduced in Chapter 17, its analytic approximation methods are discussed in Chapter 18, and its numerical approximation methods are introduced in Chapter 19. Then practical Bayesian inference algorithms for *Gaussian mixture models* and *topic models* are introduced in Chapter 20, which also includes a non-parametric Bayesian approach.

CHAPTER

PATTERN RECOGNITION VIA GENERATIVE MODEL ESTIMATION

11

CHAPTER CONTENTS

In this chapter, the framework of pattern recognition based on generative model estimation is first explained. Then, criteria for quantitatively evaluating the goodness of a classification algorithm are discussed.

11.1 FORMULATION OF PATTERN RECOGNITION

In this section, the problem of statistical pattern recognition is mathematically formulated.

Let $\boldsymbol{x}$ be a *pattern* (which is also called a *feature vector*, an *input variable*, an *independent variable*, an *explanatory variable*, an *exogenous variable*, a *predictor variable*, a *regressor*, and a *covariate*), which is a member of a subset $\mathcal{X}$ of the d-dimensional Euclidean space $\mathbb{R}^d$:

$$\boldsymbol{x} \in \mathcal{X} \subset \mathbb{R}^d,$$

$\mathcal{X}$ is called the *pattern space*. Let y be a *class* (which is also called a *category*, an *output variable*, a *target variable*, and a *dependent variable*) to which a pattern $\boldsymbol{x}$ belongs. Let c be the number of classes, i.e.,

$$y \in \mathcal{Y} = \{1, \ldots, c\}.$$

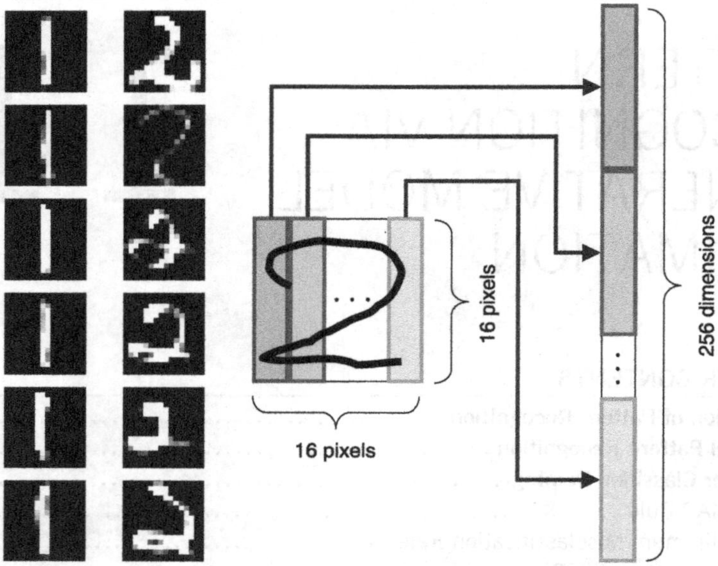

(a) Hand-written digit image

(b) Vectorization of two-dimensional image

FIGURE 11.1

Hand-written digit image and its vectorization.

In hand-written digit recognition, a scanned digit image is a pattern. If the scanned image consists of 16×16 pixels, pattern $\boldsymbol{x}$ is a 256-dimensional real vector which vertically stacks the pixel values as illustrated in Fig. 11.1. Rigorously speaking, pixel values are integers (e.g., 0–255), but they are regarded as real numbers here. When the pixel values are normalized to be in $[0, 1]$, the pattern space is given by $\mathcal{X} = [0, 1]^{256}$. Classes are the numbers "0," "1," ..., "9," and thus the number of classes is $c = 10$.

A classifier is a mapping from a pattern $\boldsymbol{x}$ to a class y. Such a mapping is called a *discrimination function* (see Fig. 11.2(a)) and is denoted by $f(\boldsymbol{x})$. A region to which patterns in class y belong is called a *decision region* (see Fig. 11.2(b)) and is denoted by $\mathcal{X}_y$. A boundary between decision regions is called a *decision boundary*. Thus, pattern recognition is equivalent to dividing the pattern space $\mathcal{X}$ into decision regions $\{\mathcal{X}_y\}_{y=1}^{c}$.

In practice, the discrimination function (or decision regions or decision boundaries) is unknown. Here, pattern $\boldsymbol{x}$ and class y are treated as *random variables* and learn the optimal discrimination function based on their statistical properties. Such an approach is called *statistical pattern recognition*.

Let us illustrate how hard directly constructing a discrimination function (or decision regions and decision boundaries) in hand-written digit recognition. Let the

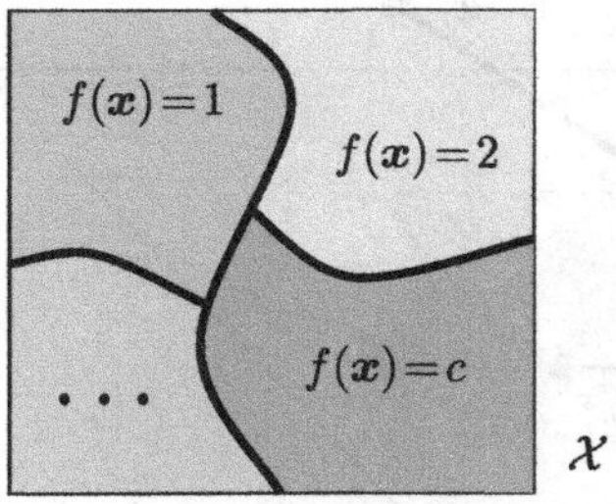

(a) Discrimination function

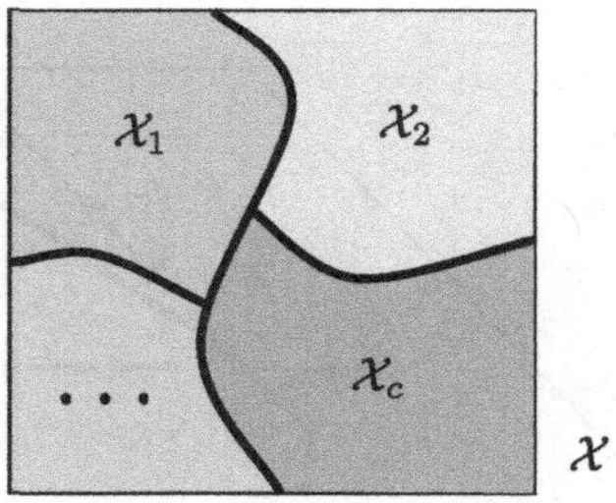

(b) Decision region and decision boundary

FIGURE 11.2

Constructing a classifier is equivalent to determine a discrimination function, decision regions, and decision boundaries.

number of pixels be 100 (=10 × 10) and each pixel takes an integer from 0 to 255. Then the number of possible images is

$$256^{100} = (2^8)^{100} = (2^{10})^{80} \approx (10^3)^{80} = 10^{240},$$

which is an astronomical number having 240 zeros after the first one. Therefore, even for a toy hand-written digit recognition example from tiny images with 10 × 10 pixels, just enumerating all possible images is not realistic. In practice, instead of just memorizing classes of all possible patterns, the class of unlearned patterns may be predicted from some learned patterns. The capability that unlearned patterns can be classified correctly is called the *generalization ability*. The objective of pattern recognition is to let a classifier being equipped with the generalization ability.

11.2 STATISTICAL PATTERN RECOGNITION

In this section, a statistical approach to pattern recognition is explained.

Suppose that pairs of patterns and their classes, called *training samples*, are available:

$$\{(\boldsymbol{x}_i, y_i) \mid \boldsymbol{x}_i \in \mathcal{X}, y_i \in \mathcal{Y}\}_{i=1}^{n},$$

where n denotes the number of training samples. Among the n training samples, the number of samples which belong to class y is denoted by n_y. Below, the training samples are assumed to be generated for $i = 1, \ldots, n$ as the following:

1. Class y_i is selected according to the *class-prior probability* $p(y)$.
2. For chosen class y_i, pattern $\boldsymbol{x}_i$ is generated according to the *class-conditional probability density* $p(\boldsymbol{x}|y = y_i)$.

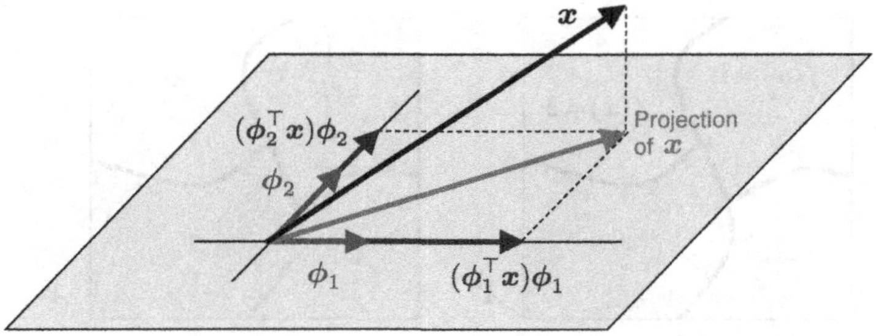

FIGURE 11.3

Dimensionality reduction onto a two-dimensional subspace by principal component analysis (see Section 35.2.1).

Then training samples $\{(\boldsymbol{x}_i, y_i)\}_{i=1}^n$ independently follow the joint probability density $p(\boldsymbol{x}, y)$, i.e., $\{(\boldsymbol{x}_i, y_i)\}_{i=1}^n$ are i.i.d. with $p(\boldsymbol{x}, y)$ (see Section 7.3). This i.i.d. assumption is one of the most fundamental presuppositions in statistical pattern recognition. Machine learning techniques when this i.i.d. assumption is violated are discussed in Chapter 33; see also [81, 101].

Given training samples, what is the best way to learn the discrimination function? Let us illustrate the distribution of patterns for the hand-written digit data shown in Fig. 11.1. Since the hand-written digit samples are 256-dimensional, their distribution cannot be directly visualized. Here, a dimensionality reduction method called *principal component analysis* (PCA) is used to reduce the dimensionality from 256 to 2. More specifically, the variance-covariance matrix of training samples $\{\boldsymbol{x}_i\}_{i=1}^n$ is eigendecomposed (see Fig. 6.2), and eigenvalues $\lambda_1 \geq \cdots \geq \lambda_d$ and corresponding eigenvectors $\boldsymbol{\phi}_1, \ldots, \boldsymbol{\phi}_d$ are obtained. Then each training sample $\boldsymbol{x}_i$ is transformed as $(\boldsymbol{\phi}_1^\top \boldsymbol{x}_i, \boldsymbol{\phi}_2^\top \boldsymbol{x}_i)^\top$, where the eigenvectors $\boldsymbol{\phi}_1, \ldots, \boldsymbol{\phi}_d$ are assumed to be normalized to have unit norm. Since $\boldsymbol{\phi}_j^\top \boldsymbol{x}_i$ corresponds to the length of projection of $\boldsymbol{x}_i$ along $\boldsymbol{\phi}_j$, the above transformation is the projection of $\boldsymbol{x}_i$ onto the subspace spanned by $\boldsymbol{\phi}_1$ and $\boldsymbol{\phi}_2$ (Fig. 11.3). As detailed in Section 35.2.1, PCA gives the best approximation to original data in a lower-dimensional subspace, and therefore it is often used for *data visualization*.

The PCA projection of the hand-written digit data shown in Fig. 11.1 is plotted in Fig. 11.4(a), showing that digit “2” is distributed more broadly than digit “1.” This well agrees with the intuition that the shape of “2” may have more individuality than the shape of “1.”

What is the best decision boundary for the samples plotted in Fig. 11.4(a)? Examples of decision boundaries are illustrated in Fig. 11.4(b). The decision boundary shown by the solid line is a complicated curve, but it can perfectly separate “1” and “2.” On the other hand, the decision boundaries shown by the dashed line and dashed-dotted line are much simpler, but some training samples are classified

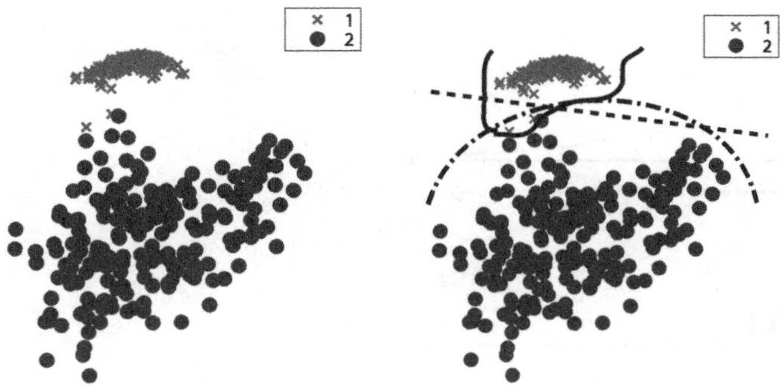

(a) Distribution of hand-written digits (b) Examples of decision boundary

FIGURE 11.4

Illustration of hand-written digit samples in the pattern space.

incorrectly. For the purpose of classifying training samples, the decision boundary shown by the solid line is better than those shown by the dashed line and dashed-dotted line. However, the true objective of pattern recognition is not only to classify training samples correctly but also to classify unlearned test samples given in the future, i.e., to acquire the generalization ability, as mentioned in Section 11.1.

11.3 CRITERIA FOR CLASSIFIER TRAINING

In order to equip a classifier with the generalization ability, it is important to define a criterion that quantitatively evaluate the goodness of a discrimination function (or decision regions or decision boundaries). In this section, three examples of such criteria are introduced and their relation is discussed.

11.3.1 MAP RULE

When deciding which class a given pattern belongs to, it would be natural to choose the one with the highest probability. This corresponds to choosing the class that maximizes the *class-posterior probability* $p(y|\boldsymbol{x})$, i.e., pattern $\boldsymbol{x}$ is classified into class $\widehat{y}$, where

$$\widehat{y} = \underset{y}{\operatorname{argmax}}\, p(y|\boldsymbol{x}).$$

Here, "argmax" indicates *the argument of the maximum*, i.e., the maximizer of an objective function. Such a decision rule is called the MAP rule. The MAP rule is equivalent to setting the decision regions as follows (Fig. 11.5):

$$\mathcal{X}_y = \{\boldsymbol{x} \mid p(y|\boldsymbol{x}) \geq p(y'|\boldsymbol{x}) \text{ for all } y' \neq y\}. \tag{11.1}$$

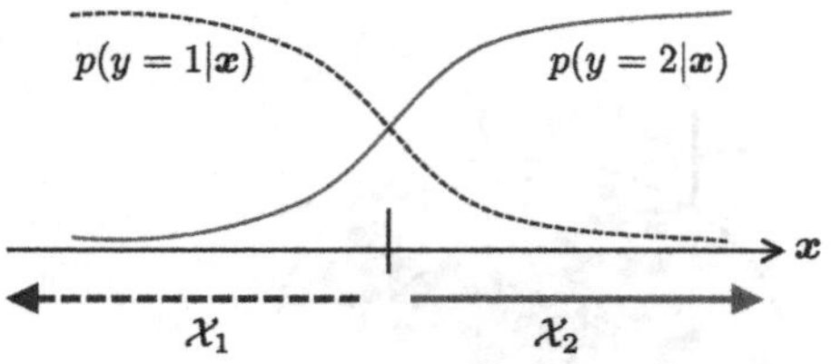

FIGURE 11.5

MAP rule.

11.3.2 MINIMUM MISCLASSIFICATION RATE RULE

Another natural idea is to choose the class with the lowest misclassification error, which is called the *minimum misclassification rate rule*.

Let $p_e(y \to y')$ be the probability that a pattern in class y is misclassified into class y'. Since $p_e(y \to y')$ is equivalent to the probability that pattern x in class y falls into decision region $\mathcal{X}_{y'}$ (see Fig. 11.6), it is given by

$$p_e(y \to y') = \int_{x \in \mathcal{X}_{y'}} p(x|y)\mathrm{d}x.$$

Then the probability that a pattern in class y is classified into an incorrect class, denoted by $p_e(y)$, is given as

$$\begin{aligned} p_e(y) &= \sum_{y' \neq y} p_e(y \to y') = \sum_{y' \neq y} \int_{x \in \mathcal{X}_{y'}} p(x|y)\mathrm{d}x \\ &= \sum_{y' \neq y} \int_{x \in \mathcal{X}_{y'}} p(x|y)\mathrm{d}x + \int_{x \in \mathcal{X}_y} p(x|y)\mathrm{d}x - \int_{x \in \mathcal{X}_y} p(x|y)\mathrm{d}x \\ &= 1 - \int_{x \in \mathcal{X}_y} p(x|y)\mathrm{d}x. \end{aligned}$$

The second term in the last equation,

$$\int_{x \in \mathcal{X}_y} p(x|y)\mathrm{d}x,$$

denotes the probability that a pattern in class y is classified into class y, i.e., the correct classification rate. Thus, the above equation shows the common fact that the misclassification rate is given by one minus the correct classification rate.

Finally, the overall misclassification rate, denoted by p_e, is given by the expectation of $p_e(y)$ over all classes:

$$p_e = \sum_{y=1}^{c} p_e(y)p(y).$$

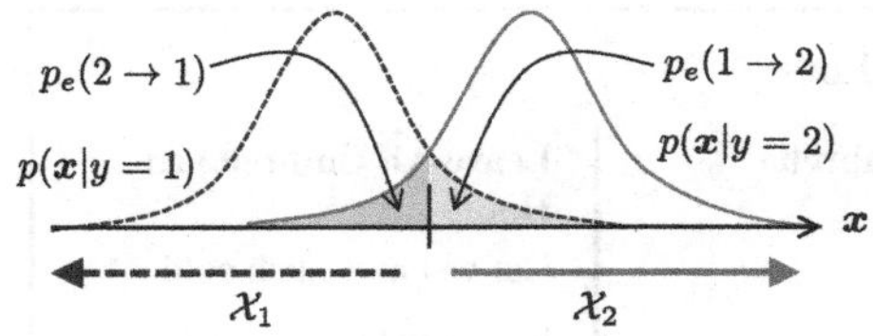

FIGURE 11.6

Minimum misclassification rate rule.

The minimum misclassification rate rule finds the classifier that minimizes the above p_e.

The overall misclassification rate p_e can be expressed as

$$
\begin{aligned}
p_e &= \sum_{y=1}^{c} \left(1 - \int_{x \in \mathcal{X}_y} p(x|y)\mathrm{d}x\right) p(y) \\
&= \sum_{y=1}^{c} p(y) - \sum_{y=1}^{c} \int_{x \in \mathcal{X}_y} p(x|y)p(y)\mathrm{d}x \\
&= 1 - \sum_{y=1}^{c} \int_{x \in \mathcal{X}_y} p(y,x)\mathrm{d}x = 1 - \sum_{y=1}^{c} \int_{x \in \mathcal{X}_y} p(y|x)p(x)\mathrm{d}x.
\end{aligned}
$$

Thus, minimizing p_e is equivalent to determining the decision regions $\{\mathcal{X}_y\}_{y=1}^{c}$ so that the second term, $\sum_{y=1}^{c} \int_{x \in \mathcal{X}_y} p(y|x)p(x)\mathrm{d}x$, is maximized. This can be achieved by setting $\mathcal{X}_y$ to be the set of all x such that

$$p(y|x) \geq p(y'|x) \quad \text{for all } y' \neq y.$$

This is actually equivalent to Eq. (11.1), and therefore minimizing the misclassification error is actually equivalent to maximizing the class-posterior probability.

11.3.3 BAYES DECISION RULE

According to the MAP rule (equivalently the minimum misclassification rate rule), when the probability of precipitation is 40%, the anticipated weather is no rain. If there will be no rain, then there is no need to carry an umbrella. However, perhaps many people will bring an umbrella with them if the probability of precipitation is 40%. This is because, the loss of no rain when an umbrella is carried (i.e., need to carry a slightly heavier bag) is much smaller than the loss of having rain when no umbrella is carried (i.e., getting wet with rain and catching a cold) in reality (see Table 11.1). In this way, choosing the class that has the smallest loss is called the *Bayes decision rule*.

Table 11.1 Example Of Asymmetric Loss

	Carry An Umbrella	Leave An Umbrella At Home
Rain	Avoid get wet	Get wet and catch cold
No rain	Bag is heavy	Bag is light

Let $\ell_{y,y'}$ be the *loss* that a pattern in class y is misclassified into class y'. Since the probability that pattern $\boldsymbol{x}$ belongs to class y is given by the class-posterior probability $p(y|\boldsymbol{x})$, the expected loss for classifying pattern $\boldsymbol{x}$ into class y' is given by

$$R(y'|\boldsymbol{x}) = \sum_{y=1}^{c} \ell_{y,y'} p(y|\boldsymbol{x}).$$

This is called the *conditional risk* for pattern $\boldsymbol{x}$.

In the Bayes decision rule, pattern $\boldsymbol{x}$ is classified into the class that incurs the minimum conditional risk. More specifically, pattern $\boldsymbol{x}$ is classified into class $\widehat{y}$, where

$$\widehat{y} = \underset{y}{\text{argmin}}\, R(y|\boldsymbol{x}).$$

This is equivalent to determining the decision regions $\{\mathcal{X}_y\}_{y=1}^{c}$ as

$$\mathcal{X}_y = \{\boldsymbol{x} \mid R(y|\boldsymbol{x}) \le R(y'|\boldsymbol{x}) \text{ for all } y' \neq y\}.$$

The expectation of the conditional risk for all $\boldsymbol{x}$ is called the *total risk*:

$$R = \int_{\mathcal{X}} R(\widehat{y}|\boldsymbol{x}) p(\boldsymbol{x}) \mathrm{d}\boldsymbol{x},$$

where $\widehat{y}$ is an output of a classifier. The value of the total risk for the Bayes decision rule is called the *Bayes risk*, and this is the lowest possible risk for the target classification problem. Note that the Bayes risk is not zero in general, meaning that the risk cannot be zero even with the optimally trained classifier.

Suppose the loss for correct classification is set at *zero* and the loss for incorrect classification is set at a positive constant ℓ:

$$\ell_{y,y'} = \begin{cases} 0 & (y = y'), \\ \ell & (y \neq y'). \end{cases} \tag{11.2}$$

Then the conditional risk is expressed as

$$R(y|\boldsymbol{x}) = \ell \sum_{y' \neq y} p(y'|\boldsymbol{x}) = \ell \left(\sum_{y'=1}^{c} p(y'|\boldsymbol{x}) - p(y|\boldsymbol{x}) \right) = \ell \,(1 - p(y|\boldsymbol{x})). \tag{11.3}$$

Since ℓ is just a proportional constant, minimization of Eq. (11.3) is equivalent to maximization of class-posterior probability $p(y|\boldsymbol{x})$. Thus, when loss $\ell_{y,y'}$ is given by Eq. (11.2), the Bayes decision rule is reduced to the MAP rule (and therefore the minimum misclassification rate rule, too).

11.3.4 DISCUSSION

Among the MAP rule, the minimum misclassification rate rule, and the Bayes decision rule, the Bayes decision rule seems to be natural and the most powerful. However, in practice, it is often difficult to precisely determine the loss $\ell_{y,y'}$, which makes the use of the Bayes decision rule not straightforward. For example, in the rain-umbrella example described in Table 11.1, it would be clear that the loss of no rain when an umbrella is carried is much smaller than the loss of having rain when no umbrella is carried. However, it is not immediately clear how small the loss of no rain when an umbrella is carried should be.

For this reason, in the following sections, we focus on the MAP rule (and therefore the minimum misclassification rate rule, too).

11.4 GENERATIVE AND DISCRIMINATIVE APPROACHES

Learning a classifier based on the MAP rule requires to find a maximizer of the class-posterior probability $p(y|\boldsymbol{x})$. Pattern recognition through estimation of the class-posterior probability $p(y|\boldsymbol{x})$ is called the *discriminative approach* and will be covered in Part 4.

Another approach is to use the Bayes' theorem explained in Section 5.4 to express the class-posterior probability $p(y|\boldsymbol{x})$ as

$$p(y|\boldsymbol{x}) = \frac{p(\boldsymbol{x}|y)p(y)}{p(\boldsymbol{x})}.$$

Since the denominator in the right-hand side, $p(\boldsymbol{x})$, is independent of class y, it can be ignored when the class-posterior probability $p(y|\boldsymbol{x})$ is maximized with respect to y:

$$p(y|\boldsymbol{x}) \propto p(\boldsymbol{x}|y)p(y),$$

where "$\propto$" means "proportional to." Statistical pattern recognition through estimation of the *class-conditional probability density* $p(\boldsymbol{x}|y)$ and the *class-prior probability* $p(y)$ is called the *generative approach* since

$$p(\boldsymbol{x}|y)p(y) = p(\boldsymbol{x}, y),$$

which is the data-generating probability distribution.

In the following chapters, the generative approach to statistical pattern recognition is explored. The class-prior probability $p(y)$ may simply be estimated by the ratio of training samples in class y, i.e.,

$$\widehat{p}(y) = \frac{n_y}{n}, \tag{11.4}$$

where n_y denotes the number of training samples in class y and n denotes the number of all training samples. On the other hand, the class-conditional probability $p(\boldsymbol{x}|y)$ is generally a high-dimensional probability density function, and therefore its estimation is not straightforward. In the following chapters, various approaches to estimating the class-conditional probability $p(\boldsymbol{x}|y)$ will be discussed.

For the sake of simplicity, the problem of estimating an unconditional probability density $p(\boldsymbol{x})$ from its i.i.d. training samples $\{\boldsymbol{x}_i\}_{i=1}^n$ is considered in the following chapters, because this allows us to estimate a conditional probability density $p(\boldsymbol{x}|y)$ by only using n_y samples in class y, $\{\boldsymbol{x}_i\}_{i:y_i=y}$, for density estimation.

Methods of probability density functions are categorized into *parametric* and *nonparametric* methods. The parametric methods seek the best approximation to the true probability density function from a parameterized family of probability density functions, called a parametric model. For example, the Gaussian model contains the mean vector and the variance-covariance matrix as parameters and they are estimated from training samples. Once a parametric model is considered, the problem of estimating a probability density function is reduced to the problem of learning a parameter in the model. On the other hand, methods that do not use parametric models are called nonparametric.

In the following chapters, various parametric and nonparametric methods will be introduced.

CHAPTER

MAXIMUM LIKELIHOOD ESTIMATION 12

CHAPTER CONTENTS

MLE is a generic method for parameter estimation proposed in the early twentieth century. Thanks to its excellent theoretical and practical properties, it is still one of the most popular techniques even now and it forms the basis of various advanced machine learning techniques. In this chapter, the definition of MLE, its application to Gaussian models, and its usage in pattern recognition are explained.

12.1 DEFINITION

In this section, the definition of MLE is provided.

A set of probability density functions specified by a finite number of parameters is called a *parametric model*. Let us denote a parametric model by $q(\boldsymbol{x};\boldsymbol{\theta})$, a parameter vector by $\boldsymbol{\theta}$, and the domain of parameters by Θ, respectively. Let b be the dimensionality of $\boldsymbol{\theta}$:

$$\boldsymbol{\theta} = (\theta^{(1)}, \ldots, \theta^{(b)})^\top.$$

In the notation $q(\boldsymbol{x};\boldsymbol{\theta})$, $\boldsymbol{x}$ before the semicolon is a random variable and $\boldsymbol{\theta}$ after the semicolon is a parameter.

A natural idea to specify the value of parameter $\boldsymbol{\theta}$ is to maximize the chance of obtaining the current training samples $\{\boldsymbol{x}\}_{i=1}^n$. To this end, let us consider the probability that the training samples $\{\boldsymbol{x}\}_{i=1}^n$ are produced under parameter $\boldsymbol{\theta}$. This probability viewed as a function of parameter $\boldsymbol{\theta}$ is called the *likelihood* and denoted by $L(\boldsymbol{\theta})$. Under the i.i.d. assumption (see Section 11.2), the likelihood is expressed as

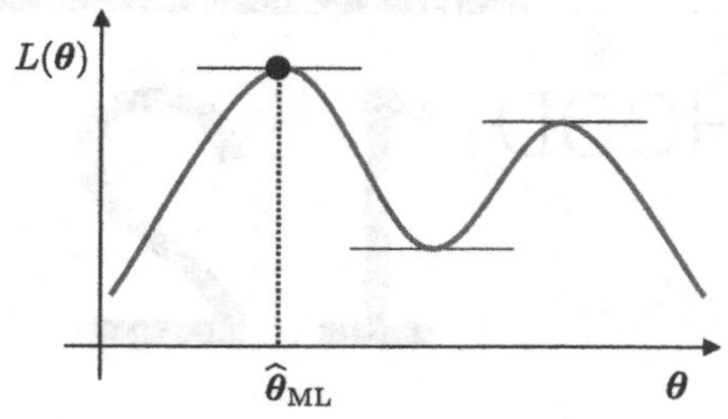

FIGURE 12.1

Likelihood equation, setting the derivative of the likelihood to zero, is a necessary condition for the maximum likelihood solution but is not a sufficient condition in general.

$$L(\boldsymbol{\theta}) = \prod_{i=1}^{n} q(\boldsymbol{x}_i; \boldsymbol{\theta}).$$

MLE finds the maximizer of the likelihood,

$$\widehat{\boldsymbol{\theta}}_{\mathrm{ML}} = \underset{\boldsymbol{\theta} \in \Theta}{\mathrm{argmax}}\ L(\boldsymbol{\theta}),$$

and then a density estimator is given by

$$\widehat{p}(\boldsymbol{x}) = q(\boldsymbol{x}; \widehat{\boldsymbol{\theta}}_{\mathrm{ML}}).$$

If parametric model $q(\boldsymbol{x}; \boldsymbol{\theta})$ is differentiable with respect to $\boldsymbol{\theta}$, $\widehat{\boldsymbol{\theta}}_{\mathrm{ML}}$ satisfies

$$\frac{\partial}{\partial \boldsymbol{\theta}} L(\boldsymbol{\theta}) \bigg|_{\boldsymbol{\theta} = \widehat{\boldsymbol{\theta}}_{\mathrm{ML}}} = \mathbf{0}_b, \tag{12.1}$$

where $\mathbf{0}_b$ denotes the b-dimensional zero vector, and $\frac{\partial}{\partial \boldsymbol{\theta}}$ denotes the *partial derivative* with respect to $\boldsymbol{\theta}$. The partial derivative with respect to vector $\boldsymbol{\theta}$ gives the b-dimensional vector whose ℓth element is given by $\frac{\partial}{\partial \theta^{(\ell)}}$:

$$\frac{\partial}{\partial \boldsymbol{\theta}} = \left(\frac{\partial}{\partial \theta^{(1)}}, \ldots, \frac{\partial}{\partial \theta^{(b)}} \right)^{\top}.$$

Eq. (12.1) is called the *likelihood equation* and is a *necessary condition* for the maximum likelihood solution. Note, however, that it is not generally a *sufficient condition*, i.e., the maximum likelihood solution always satisfies Eq. (12.1), but solving Eq. (12.1) does not necessarily give the solution (Fig. 12.1).

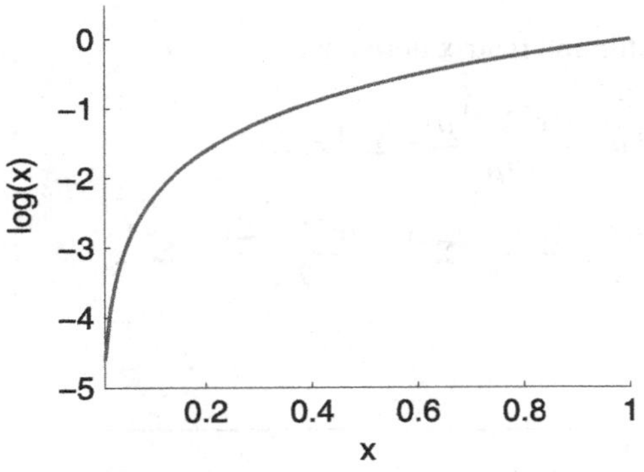

FIGURE 12.2

Log function is monotone increasing.

Since the log function is monotone increasing, the maximizer of the likelihood can also be obtained by maximizing the *log-likelihood* (Fig. 12.2):

$$\widehat{\theta}_{\mathrm{ML}} = \underset{\theta \in \Theta}{\mathrm{argmax}} \ \log L(\theta) = \underset{\theta \in \Theta}{\mathrm{argmax}} \left[\sum_{i=1}^{n} \log q(x_i; \theta) \right].$$

While the original likelihood contains the product of probability densities, the log-likelihood contains the sum of log probability densities, which is often easier to compute in practice. The likelihood equation for the log-likelihood is given by

$$\frac{\partial}{\partial \theta} \log L(\theta) \bigg|_{\theta = \widehat{\theta}_{\mathrm{ML}}} = \mathbf{0}_b.$$

Below, MLE for the Gaussian model is explained in detail. MLE for the Gaussian mixture model will be explained in Chapter 15.

12.2 GAUSSIAN MODEL

In Section 4.2 and Section 6.2, the Gaussian distribution was introduced. The *Gaussian model* corresponds to a parametric model for the Gaussian distribution and is given for d-dimensional pattern x as

$$q(x; \mu, \Sigma) = \frac{1}{(2\pi)^{\frac{d}{2}} \det(\Sigma)^{\frac{1}{2}}} \exp\left(-\frac{1}{2} (x - \mu)^{\top} \Sigma^{-1} (x - \mu) \right).$$

Here, d-dimensional vector μ and $d \times d$ matrix Σ are the parameters of the Gaussian model, and $\det(\cdot)$ denotes the determinant. μ and Σ correspond to the expectation

The following formulas hold for vector and matrix derivatives:

$$\frac{\partial \boldsymbol{\mu}^\top \boldsymbol{\Sigma}^{-1}\boldsymbol{\mu}}{\partial \boldsymbol{\mu}} = 2\boldsymbol{\Sigma}^{-1}\boldsymbol{\mu}, \quad \frac{\partial \boldsymbol{x}^\top \boldsymbol{\Sigma}^{-1}\boldsymbol{\mu}}{\partial \boldsymbol{\mu}} = \boldsymbol{\Sigma}^{-1}\boldsymbol{x},$$

$$\frac{\partial \boldsymbol{x}^\top \boldsymbol{\Sigma}^{-1}\boldsymbol{x}}{\partial \boldsymbol{\Sigma}} = -\boldsymbol{\Sigma}^{-1}\boldsymbol{x}\boldsymbol{x}^\top\boldsymbol{\Sigma}^{-1}, \quad \frac{\partial \log(\det(\boldsymbol{\Sigma}))}{\partial \boldsymbol{\Sigma}} = \boldsymbol{\Sigma}^{-1}, \quad \frac{\partial \mathrm{tr}(\widetilde{\boldsymbol{\Sigma}}^{-1}\boldsymbol{\Sigma})}{\partial \boldsymbol{\Sigma}} = \widetilde{\boldsymbol{\Sigma}}^{-1}.$$

FIGURE 12.3

Formulas for vector and matrix derivatives [80].

matrix and the variance-covariance matrix, respectively,

$$\boldsymbol{\mu} = E[\boldsymbol{x}] = \int \boldsymbol{x} q(\boldsymbol{x};\boldsymbol{\mu},\boldsymbol{\Sigma})\mathrm{d}\boldsymbol{x},$$

$$\boldsymbol{\Sigma} = V[\boldsymbol{x}] = \int (\boldsymbol{x}-\boldsymbol{\mu})(\boldsymbol{x}-\boldsymbol{\mu})^\top q(\boldsymbol{x};\boldsymbol{\mu},\boldsymbol{\Sigma})\mathrm{d}\boldsymbol{x}.$$

For i.i.d. training samples $\{\boldsymbol{x}_i\}_{i=1}^n$ following the Gaussian model $q(\boldsymbol{x};\boldsymbol{\mu},\boldsymbol{\Sigma})$, the log-likelihood is given by

$$\log L(\boldsymbol{\mu},\boldsymbol{\Sigma}) = -\frac{nd\log 2\pi}{2} - \frac{n\log(\det(\boldsymbol{\Sigma}))}{2} - \frac{1}{2}\sum_{i=1}^n (\boldsymbol{x}_i-\boldsymbol{\mu})^\top\boldsymbol{\Sigma}^{-1}(\boldsymbol{x}_i-\boldsymbol{\mu}).$$

The likelihood equation for the Gaussian model is given as

$$\begin{cases} \left.\dfrac{\partial}{\partial \boldsymbol{\mu}}\log L(\boldsymbol{\mu},\boldsymbol{\Sigma})\right|_{\boldsymbol{\mu}=\widehat{\boldsymbol{\mu}}_{\mathrm{ML}}} = \boldsymbol{0}_d, \\ \left.\dfrac{\partial}{\partial \boldsymbol{\Sigma}}\log L(\boldsymbol{\mu},\boldsymbol{\Sigma})\right|_{\boldsymbol{\Sigma}=\widehat{\boldsymbol{\Sigma}}_{\mathrm{ML}}} = \boldsymbol{O}_{d\times d}, \end{cases}$$

where $\boldsymbol{O}_{d\times d}$ denotes the $d\times d$ zero matrix.

For deriving the maximum likelihood solutions, formulas for vector and matrix derivatives shown in Fig. 12.3 are useful. Indeed, the partial derivatives of the log-likelihood with respect to vector $\boldsymbol{\mu}$ and matrix $\boldsymbol{\Sigma}$ are given by

$$\frac{\partial \log L}{\partial \boldsymbol{\mu}} = n\boldsymbol{\Sigma}^{-1}\boldsymbol{\mu} + \boldsymbol{\Sigma}^{-1}\sum_{i=1}^n \boldsymbol{x}_i,$$

$$\frac{\partial \log L}{\partial \boldsymbol{\Sigma}} = -\frac{n}{2}\boldsymbol{\Sigma}^{-1} + \frac{1}{2}\boldsymbol{\Sigma}^{-1}\left(\sum_{i=1}^n (\boldsymbol{x}_i-\boldsymbol{\mu})(\boldsymbol{x}_i-\boldsymbol{\mu})^\top\right)\boldsymbol{\Sigma}^{-1}.$$

Then the maximum likelihood estimators $\widehat{\boldsymbol{\mu}}_{\mathrm{ML}}$ and $\widehat{\boldsymbol{\Sigma}}_{\mathrm{ML}}$ are given as

$$\widehat{\boldsymbol{\mu}}_{\mathrm{ML}} = \frac{1}{n}\sum_{i=1}^{n} \boldsymbol{x}_i,$$

$$\widehat{\boldsymbol{\Sigma}}_{\mathrm{ML}} = \frac{1}{n}\sum_{i=1}^{n} (\boldsymbol{x}_i - \widehat{\boldsymbol{\mu}}_{\mathrm{ML}})(\boldsymbol{x}_i - \widehat{\boldsymbol{\mu}}_{\mathrm{ML}})^\top,$$

which correspond to the *sample mean* and the *sample variance-covariance matrix*. Here, we assumed that we have enough training samples so that $\widehat{\boldsymbol{\Sigma}}_{\mathrm{ML}}$ is invertible.

The above Gaussian model considered a generic variance-covariance matrix $\boldsymbol{\Sigma}$, but a slightly simplified one having no correlation can also be considered (see Fig. 6.1). This corresponds to restricting $\boldsymbol{\Sigma}$ to be a diagonal matrix:

$$\boldsymbol{\Sigma} = \mathrm{diag}\left((\sigma^{(1)})^2, \ldots, (\sigma^{(d)})^2\right).$$

Then the Gaussian model having no correlation can be expressed as

$$q(\boldsymbol{x}; \boldsymbol{\mu}, \sigma^{(1)}, \ldots, \sigma^{(d)}) = \prod_{j=1}^{d} \frac{1}{\sqrt{2\pi(\sigma^{(j)})^2}} \exp\left(-\frac{(x^{(j)} - \mu^{(j)})^2}{2(\sigma^{(j)})^2}\right),$$

where $x^{(j)}$ and $\mu^{(j)}$ denotes the jth elements of d-dimensional vectors $\boldsymbol{x}$ and $\boldsymbol{\mu}$, respectively. The maximum likelihood solution for $\sigma^{(j)}$ is given by

$$\widehat{\sigma}^{(j)}_{\mathrm{ML}} = \sqrt{\frac{1}{n}\sum_{i=1}^{n} (x_i^{(j)} - \mu_i^{(j)})^2}.$$

The Gaussian model can be further simplified if all variances $(\sigma^{(j)})^2$ are assumed to be equal. Denoting the common variance by σ^2, the Gaussian model is expressed as

$$q(\boldsymbol{x}; \boldsymbol{\mu}, \sigma) = \frac{1}{(2\pi\sigma^2)^{\frac{d}{2}}} \exp\left(-\frac{(\boldsymbol{x} - \boldsymbol{\mu})^\top(\boldsymbol{x} - \boldsymbol{\mu})}{2\sigma^2}\right).$$

Then the maximum likelihood solution for σ is given by

$$\widehat{\sigma}_{\mathrm{ML}} = \sqrt{\frac{1}{nd}\sum_{i=1}^{n} (\boldsymbol{x}_i - \boldsymbol{\mu})^\top(\boldsymbol{x}_i - \boldsymbol{\mu})} = \sqrt{\frac{1}{d}\sum_{j=1}^{d} (\widehat{\sigma}^{(j)}_{\mathrm{ML}})^2}.$$

A MATLAB code for MLE with one-dimensional Gaussian model is given in Fig. 12.4, and its behavior is illustrated in Fig. 12.5.

12.3 COMPUTING THE CLASS-POSTERIOR PROBABILITY

Let us come back to the pattern recognition problem and learn the class-conditional probability density $p(\boldsymbol{x}|y)$ by MLE with Gaussian models:

```
n=5; m=0; s=1; x=s*randn(n,1)+m; mh=mean(x); sh=std(x,1);
X=linspace(-4,4,100); Y=exp(-(X-m).^2./(2*s^2))/(2*pi*s);
Yh=exp(-(X-mh).^2./(2*sh^2))/(2*pi*sh);

figure(1); clf; hold on;
plot(X,Y,'r-',X,Yh,'b--',x,zeros(size(x)),'ko');
legend('True','Estimated');
```

FIGURE 12.4

MATLAB code for MLE with one-dimensional Gaussian model.

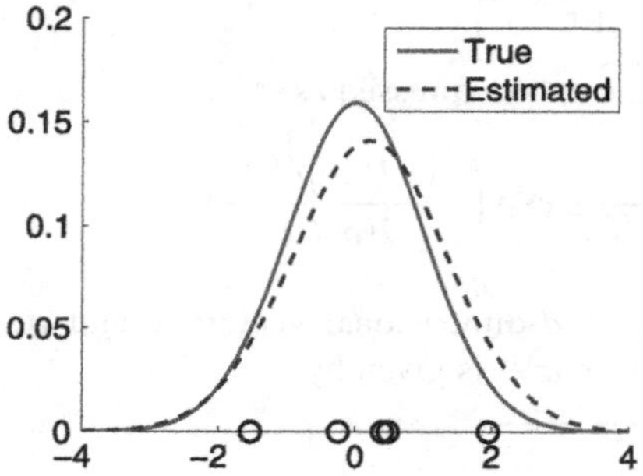

FIGURE 12.5

Example of MLE with one-dimensional Gaussian model.

$$\widehat{p}(\boldsymbol{x}|y) = \frac{1}{(2\pi)^{\frac{d}{2}}\det(\widehat{\boldsymbol{\Sigma}}_y)^{\frac{1}{2}}} \exp\left(-\frac{1}{2}(\boldsymbol{x}-\widehat{\boldsymbol{\mu}}_y)^\top \widehat{\boldsymbol{\Sigma}}_y^{-1}(\boldsymbol{x}-\widehat{\boldsymbol{\mu}}_y)\right).$$

Here, $\widehat{\boldsymbol{\mu}}_y$ and $\widehat{\boldsymbol{\Sigma}}_y$ are the maximum likelihood estimators of the expectation and variance-covariance matrices for class y:

$$\widehat{\boldsymbol{\mu}}_y = \frac{1}{n_y}\sum_{i:y_i=y}\boldsymbol{x}_i, \tag{12.2}$$

$$\widehat{\boldsymbol{\Sigma}}_y = \frac{1}{n_y}\sum_{i:y_i=y}(\boldsymbol{x}_i-\widehat{\boldsymbol{\mu}}_y)(\boldsymbol{x}_i-\widehat{\boldsymbol{\mu}}_y)^\top, \tag{12.3}$$

where n_y denotes the number of training samples in class y.

As shown in Fig. 12.2, the log function is monotone increasing and maximizing the log class-posterior probability is often more convenient than maximizing the plain class-posterior probability. From the Bayes' theorem explained in Section 5.4,

$$p(y|\boldsymbol{x}) = \frac{p(\boldsymbol{x}|y)p(y)}{p(\boldsymbol{x})},$$

the log class-posterior probability is expressed as

$$\log p(y|\boldsymbol{x}) = \log p(\boldsymbol{x}|y) + \log p(y) - \log p(\boldsymbol{x}).$$

As shown in Eq. (11.4), the class-prior probability is simply estimated by the ratio of training samples:

$$\widehat{p}(y) = \frac{n_y}{n}.$$

Then the log class-posterior probability $\log p(y|\boldsymbol{x})$ can be estimated as

$$\begin{aligned}\log \widehat{p}(y|\boldsymbol{x}) &= \log \widehat{p}(\boldsymbol{x}|y) + \log \widehat{p}(y) - \log p(\boldsymbol{x}) \\ &= -\frac{d}{2}\log(2\pi) - \frac{1}{2}\log(\det(\widehat{\boldsymbol{\Sigma}}_y)) - \frac{1}{2}(\boldsymbol{x} - \widehat{\boldsymbol{\mu}}_y)^\top \widehat{\boldsymbol{\Sigma}}_y^{-1}(\boldsymbol{x} - \widehat{\boldsymbol{\mu}}_y) \\ &\quad + \log\frac{n_y}{n} - \log p(\boldsymbol{x}) \\ &= -\frac{1}{2}(\boldsymbol{x} - \widehat{\boldsymbol{\mu}}_y)^\top \widehat{\boldsymbol{\Sigma}}_y^{-1}(\boldsymbol{x} - \widehat{\boldsymbol{\mu}}_y) - \frac{1}{2}\log(\det(\widehat{\boldsymbol{\Sigma}}_y)) + \log\frac{n_y}{n} + C,\end{aligned}$$

where C is a constant independent of y. As shown above, if $p(\boldsymbol{x}|y)$ is estimated by a Gaussian model, $\log \widehat{p}(y|\boldsymbol{x})$ becomes a quadratic function of $\boldsymbol{x}$.

The first term in the above equation,

$$(\boldsymbol{x} - \widehat{\boldsymbol{\mu}})^\top \widehat{\boldsymbol{\Sigma}}^{-1}(\boldsymbol{x} - \widehat{\boldsymbol{\mu}}),$$

is called the *Mahalanobis distance* between $\boldsymbol{x}$ and $\widehat{\boldsymbol{\mu}}$. The Mahalanobis distance regards the set of points on a *hyperellipsoid* specified by $\widehat{\boldsymbol{\Sigma}}$ as the same distance. To understand this, let us eigendecompose (see Fig. 6.2) the variance-covariance matrix $\widehat{\boldsymbol{\Sigma}}$ as

$$\widehat{\boldsymbol{\Sigma}}\boldsymbol{\phi} = \lambda\boldsymbol{\phi},$$

where $\{\lambda_j\}_{j=1}^d$ are the eigenvalues of $\widehat{\boldsymbol{\Sigma}}$ and $\{\boldsymbol{\phi}_j\}_{j=1}^d$ are the corresponding eigenvectors that are normalized to have a unit norm. Then $\widehat{\boldsymbol{\Sigma}}$ can be expressed as

$$\widehat{\boldsymbol{\Sigma}} = \sum_{j=1}^{d} \lambda_j \boldsymbol{\phi}_j \boldsymbol{\phi}_j^\top,$$

and $\widehat{\boldsymbol{\Sigma}}^{-1}$ is written explicitly as

$$\widehat{\boldsymbol{\Sigma}}^{-1} = \sum_{j=1}^{d} \frac{1}{\lambda_j} \boldsymbol{\phi}_j \boldsymbol{\phi}_j^\top.$$

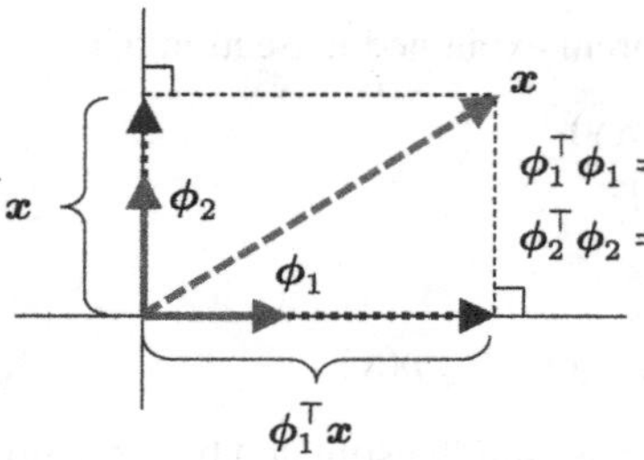

FIGURE 12.6

Orthogonal projection.

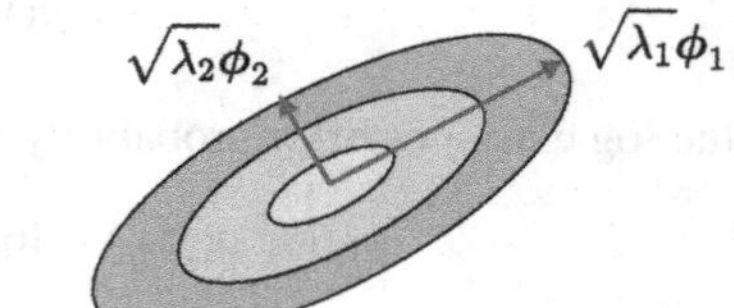

FIGURE 12.7

Mahalanobis distance having hyperellipsoidal contours.

Since $\{\boldsymbol{\phi}_j\}_{j=1}^d$ are mutually orthogonal, the Mahalanobis distance can be expressed as

$$(\boldsymbol{x}-\widehat{\boldsymbol{\mu}})^\top \widehat{\boldsymbol{\Sigma}}^{-1}(\boldsymbol{x}-\widehat{\boldsymbol{\mu}}) = \sum_{j=1}^d \frac{(\boldsymbol{\phi}_j^\top(\boldsymbol{x}-\widehat{\boldsymbol{\mu}}))^2}{\lambda_j},$$

$\boldsymbol{\phi}_j^\top(\boldsymbol{x}-\widehat{\boldsymbol{\mu}})$ corresponds to the length of the projection of $\boldsymbol{x}-\widehat{\boldsymbol{\mu}}$ onto $\boldsymbol{\phi}_j$ (see Fig. 12.6). Thus, the Mahalanobis distance is the squared sum of the projection lengths divided by λ_j, and the set of points having the same Mahalanobis distance (say, 1) satisfies

$$\sum_{j=1}^d \frac{1}{\lambda_j}(\boldsymbol{\phi}_j^\top(\boldsymbol{x}-\widehat{\boldsymbol{\mu}}))^2 = 1.$$

This represents the hyperellipsoid with the *principal axes* parallel to $\{\boldsymbol{\phi}_j\}_{j=1}^d$ and length $\{\sqrt{\lambda_j}\}_{j=1}^d$. When the dimensionality of $\boldsymbol{x}$ is $d = 2$, the hyperellipsoid is reduced to the ellipse (Fig. 12.7).

When the number of classes is $c = 2$, the decision boundary is the set of points having the same class-posterior probability (i.e., $1/2$):

$$p(y=1|\boldsymbol{x}) = p(y=2|\boldsymbol{x}). \tag{12.4}$$

Therefore, if $p(\boldsymbol{x}|y)$ is estimated by a Gaussian model, the decision boundary is a *quadratic hypersurface*.

12.4 FISHER'S LINEAR DISCRIMINANT ANALYSIS (FDA)

Suppose further that the variance-covariance matrix of $\boldsymbol{\Sigma}_y$ of each class is the same:

$$\boldsymbol{\Sigma}_1 = \cdots = \boldsymbol{\Sigma}_c = \boldsymbol{\Sigma},$$

where $\boldsymbol{\Sigma}$ is the common variance-covariance matrix that is independent of class y. Then the maximum likelihood solution $\widehat{\boldsymbol{\Sigma}}$ for the common variance-covariance matrix $\boldsymbol{\Sigma}$ is given by

$$\widehat{\boldsymbol{\Sigma}} = \frac{1}{n}\sum_{y=1}^{c}\sum_{i:y_i=y}(\boldsymbol{x}_i - \widehat{\boldsymbol{\mu}}_y)^\top(\boldsymbol{x}_i - \widehat{\boldsymbol{\mu}}_y) = \sum_{y=1}^{c}\frac{n_y}{n}\widehat{\boldsymbol{\Sigma}}_y. \tag{12.5}$$

This shows that $\widehat{\boldsymbol{\Sigma}}$ is the weighted average of each solution $\widehat{\boldsymbol{\Sigma}}_y$ (see Eq. (12.3)) according to the ratio of samples n_y/n.

With $\widehat{\boldsymbol{\Sigma}}$, the log class-posterior probability $\log p(y|\boldsymbol{x})$ can be estimated as

$$\begin{aligned}\log \widehat{p}(y|\boldsymbol{x}) &= -\frac{1}{2}\boldsymbol{x}^\top\widehat{\boldsymbol{\Sigma}}^{-1}\boldsymbol{x} + \boldsymbol{x}^\top\widehat{\boldsymbol{\Sigma}}^{-1}\widehat{\boldsymbol{\mu}}_y - \frac{1}{2}\widehat{\boldsymbol{\mu}}_y^\top\widehat{\boldsymbol{\Sigma}}^{-1}\widehat{\boldsymbol{\mu}}_y \\ &\quad - \frac{1}{2}\log(\det(\widehat{\boldsymbol{\Sigma}})) + \log\frac{n_y}{n} + C \\ &= \boldsymbol{x}^\top\widehat{\boldsymbol{\Sigma}}^{-1}\widehat{\boldsymbol{\mu}}_y - \frac{1}{2}\widehat{\boldsymbol{\mu}}_y^\top\widehat{\boldsymbol{\Sigma}}^{-1}\widehat{\boldsymbol{\mu}}_y + \log\frac{n_y}{n} + C',\end{aligned} \tag{12.6}$$

where C' is a constant independent of y. Thus, when the variance-covariance matrix is common to all classes, the log class-posterior probability is a linear function of $\boldsymbol{x}$.

Eq. (12.4) implies that the decision boundary when the number of classes is $c = 2$ is given by

$$\widehat{\boldsymbol{a}}^\top\boldsymbol{x} + \widehat{b} = 0,$$

where

$$\begin{aligned}\widehat{\boldsymbol{a}} &= \widehat{\boldsymbol{\Sigma}}^{-1}(\widehat{\boldsymbol{\mu}}_1 - \widehat{\boldsymbol{\mu}}_2), \\ \widehat{b} &= -\frac{1}{2}(\widehat{\boldsymbol{\mu}}_1^\top\widehat{\boldsymbol{\Sigma}}^{-1}\widehat{\boldsymbol{\mu}}_1 - \widehat{\boldsymbol{\mu}}_2^\top\widehat{\boldsymbol{\Sigma}}^{-1}\widehat{\boldsymbol{\mu}}_2) + \log\frac{n_1}{n_2}.\end{aligned}$$

This shows that the decision boundary is a *hyperplane*. This classification method is called *Fisher's linear discriminant analysis* (FDA) [41].

Let us compute the FDA solution for two-dimensional training samples plotted in Fig. 12.8, where "∘" and "×" denote training samples in class 1 and class 2 for $n = 300$ and $n_1 = n_2 = 150$. The Gaussian models with common variance-covariance matrix are used to approximate $p(\boldsymbol{x}|y = 1)$ and $p(\boldsymbol{x}|y = 2)$, and the following maximum likelihood estimators are obtained:

$$\widehat{\boldsymbol{\mu}}_1 = \begin{pmatrix}2\\0\end{pmatrix}, \quad \widehat{\boldsymbol{\mu}}_2 = \begin{pmatrix}-2\\0\end{pmatrix}, \quad \text{and} \quad \widehat{\boldsymbol{\Sigma}} = \begin{pmatrix}1 & 0\\0 & 9\end{pmatrix}.$$

Then the FDA solution is given by

$$\widehat{\boldsymbol{a}} = \begin{pmatrix}4\\0\end{pmatrix} \quad \text{and} \quad \widehat{b} = 0.$$

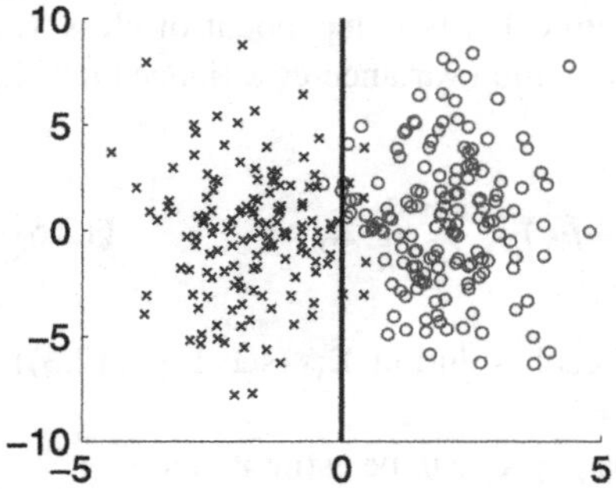

FIGURE 12.8

Linear discriminant analysis.

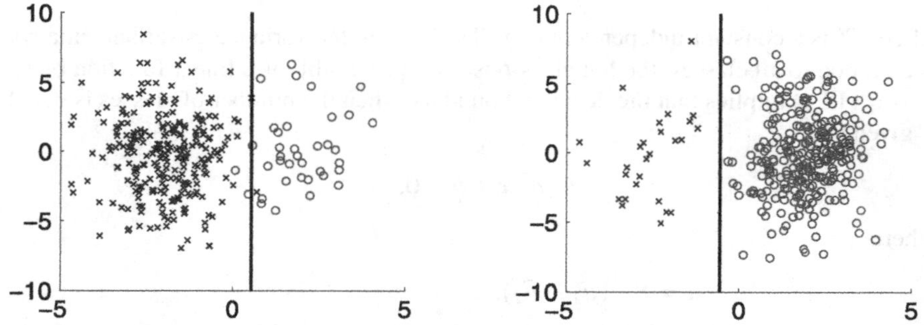

(a) $n_1/n_2 = 1/9$ (b) $n_1/n_2 = 9$

FIGURE 12.9

When the classwise sample ratio n_1/n_2 is changed.

The obtained decision boundary is plotted in Fig. 12.8, showing that the FDA solution separates the two Gaussian distributions in the middle.

If the classwise sample ratio n_1/n_2 is changed, the decision boundary yields

$$\widehat{\boldsymbol{a}} = \begin{pmatrix} 4 \\ 0 \end{pmatrix} \quad \text{and} \quad \widehat{b} = \log \frac{n_1}{n_2}.$$

As illustrated in Fig. 12.9, the decision boundary shifts depending on the classwise sample ratio n_1/n_2.

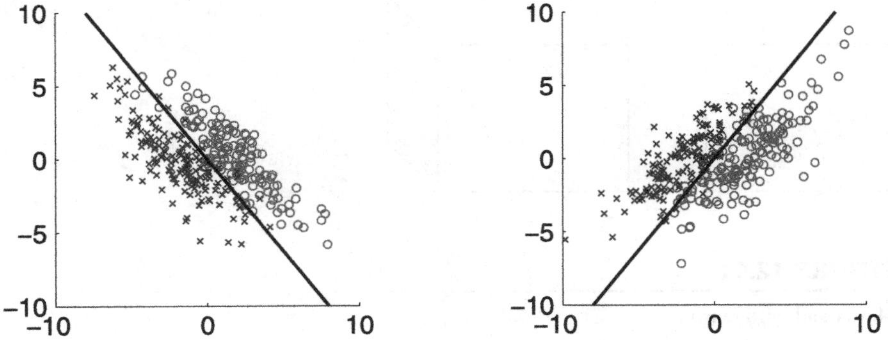

(a) $\beta = \pi/4$ (b) $\beta = -\pi/4$

FIGURE 12.10

When the classwise sample distributions are rotated.

Next, let us set $n_1/n_2 = 1$ again and change the common variance-covariance matrix $\widehat{\boldsymbol{\Sigma}}$ to

$$\widehat{\boldsymbol{\Sigma}} = \begin{pmatrix} 9 - 8\cos^2\beta & 8\sin\beta\cos\beta \\ 8\sin\beta\cos\beta & 9 - 8\sin^2\beta \end{pmatrix},$$

which corresponds to rotating the classwise sample distributions by angle β. Then the FDA solution is given by

$$\widehat{\boldsymbol{a}} = \begin{pmatrix} 1 + 8\cos^2\beta \\ -8\sin\beta\cos\beta \end{pmatrix} \quad \text{and} \quad \widehat{b} = 0,$$

where

$$\widehat{\boldsymbol{\Sigma}}^{-1} = \begin{pmatrix} 1 - \frac{8}{9}\sin^2\beta & -\frac{8}{9}\sin\beta\cos\beta \\ -\frac{8}{9}\sin\beta\cos\beta & 1 - \frac{8}{9}\cos^2\beta \end{pmatrix}$$

is used in the derivation. As illustrated in Fig. 12.10, the decision boundary rotates according to the rotation of the data set.

12.5 HAND-WRITTEN DIGIT RECOGNITION

In this section, FDA is applied to hand-written digit recognition using MATLAB. The digit data set available from

`http://www.ms.k.u-tokyo.ac.jp/sugi/data/digit.mat`

is used.

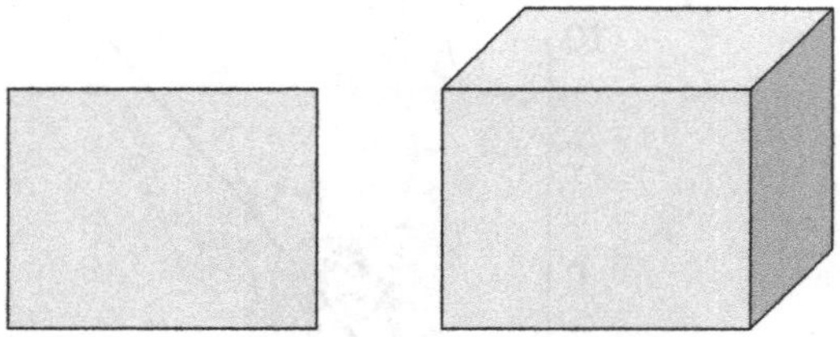

FIGURE 12.11

Matrix and third tensor.

12.5.1 PREPARATION

First, let us load the digit data in the memory space as follows.

```
> load digit.mat
```

The whos function shows that variable X contains digit data for training and variable T contains digit data for testing:

```
> whos

  Name      Size              Bytes    Class

  T         256x200x10        4096000  double
  X         256x500x10       10240000  double
```

Here, X and T are the third *tensors*, meaning that they are arrays taking three arguments (see Fig. 12.11). Each digit image is represented by a 256-dimensional vector, which is a vectorized expression of a 16×16 pixel image, as explained in Fig. 11.1(b). Each element take a real value in $[-1, 1]$, where -1 corresponds to black and 1 corresponds to white. X contains 5000 digit samples (500 samples for each digit from "0" to "9") and T contains 2000 digit samples (200 samples for each digit from "0" to "9").

For example, the 23rd training sample of digit "5" can be extracted into variable x as follows:

```
> x=X(:,23,5);
```

Note that digit "0" corresponds to 10 in the third argument. The extracted sample can be visualized as follows:

```
> imagesc(reshape(x,[16 16])')
```

12.5.2 IMPLEMENTING LINEAR DISCRIMINANT ANALYSIS

Let us implement FDA for classifying digits "1" and "2." First, let us clear the memory space and reload the digit data:

```
> clear all
> load digit.mat
```

Suppose that patterns of digits "1" and "2" follow the normal distribution with a common variance-covariance matrix. The sample means of each class can be obtained as follows (see Eq. (12.2)):

```
> mu1=mean(X(:,:,1),2);
> mu2=mean(X(:,:,2),2);
```

The common variance-covariance matrix can be obtained as follows (see Eq. (12.5)):

```
> S=(cov(X(:,:,1)')+cov(X(:,:,2)'))/2;
```

Then the class-posterior probability for a test pattern can be computed as follows (see Eq. (12.6)):

```
> t=T(:,1,2);
> invS=inv(S+0.000001*eye(256));
> p1=mu1'*invS*t-mu1'*invS*mu1/2;
> p2=mu2'*invS*t-mu2'*invS*mu2/2;
```

Note that irrelevant constants are ignored here, and `0.000001*eye(256)` is added to S to avoid numerical problems when it is inverted, where `eye(256)` denotes the 256×256 identity matrix. Investigating the difference of the class-posterior probabilities allows us to classify the test pattern:

```
> sign(p1-p2)
ans = -1
```

In this example, the test pattern is classified into class "2," meaning that the classification is successful. If `sign(p1-p2)` is 1, the test pattern is classified into class "1."

Classification results for all test patterns in class "2" can be obtained as follows:

```
> t=T(:,:,2);
> p1=mu1'*invS*t-mu1'*invS*mu1/2;
> p2=mu2'*invS*t-mu2'*invS*mu2/2;
> result=sign(p1-p2);
```

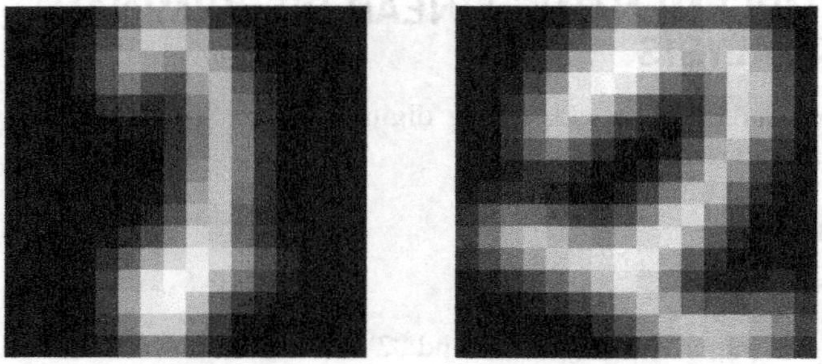

(a) 69th test pattern (b) 180th test pattern

FIGURE 12.12

Misclassified test patterns.

Note that p1 and p2 are the horizontal vectors. The classification results can be summarized as follows:

```
> sum(result==-1)
ans = 198
> sum(result~=-1)
ans = 2
```

This shows that, among 200 test patterns in class "2," 198 samples are correctly classified and 2 samples are misclassified. Thus, the correct classification rate is 2/200 = 99%. Misclassified samples can be found as follows:

```
> find(result~=-1)
ans =
   69  180
```

These test patterns are actually difficult to classify even for human (in particular, the 69th pattern), as illustrated in Fig. 12.12:

```
> imagesc(reshape(t(:,69),[16 16])')
> imagesc(reshape(t(:,180),[16 16])')
```

12.5.3 MULTICLASS CLASSIFICATION

Next, digit samples in classes "1," "2," and "3" are classified by FDA, under the assumption that the variance-covariance matrix is common to all classes. A MATLAB code is provided in Fig. 12.13. Here the classification results are summarized in the *confusion matrix*, which contains the number of times patterns in class y is categorized into class y' in the (y, y') element.

```
clear all
load digit.mat X T
[d,m,c]=size(T); c=3;

S=zeros(d,d);
for y=1:c
  mu(:,y)=mean(X(:,:,y),2);
  S=S+cov(X(:,:,y)')/c;
end
h=inv(S)*mu;
for y=1:c
  p(:,:,y)=h'*T(:,:,y)-repmat(sum(mu.*h)',[1,m])/2;
end
[pmax P]=max(p);
P=squeeze(P);
for y=1:c
  C(:,y)=sum(P==y);
end
C
```

FIGURE 12.13

MATLAB code for multiclass classification by FDA.

The following confusion matrix is obtained:

```
> C
C =
   199     1     0
     0   192     8
     0     2   198
```

This means that

- 199 test patterns in class "1" are correctly classified and the remaining 1 pattern is misclassified as "2."
- 192 test patterns in class "2" are correctly classified and the remaining 8 patterns are misclassified as "3."
- 198 test patterns in class "3" are correctly classified and the remaining 3 patterns are misclassified as "2."

If `c=3;` is commented out in the program in Fig. 12.13, all patterns in all 10 classes are classified. The obtained confusion matrix is shown in Fig. 12.14, showing

Predicted class

True class	1	2	3	4	5	6	7	8	9	0
1	199	0	0	0	1	0	0	0	0	0
2	0	169	8	8	1	2	4	8	0	0
3	0	0	182	1	5	0	2	8	1	1
4	2	2	0	182	0	1	0	3	10	0
5	0	0	21	4	162	1	0	4	4	4
6	1	2	0	1	5	185	0	3	0	3
7	2	0	1	5	1	0	181	0	9	1
8	0	1	16	6	6	0	1	164	3	3
9	1	0	0	8	0	0	7	2	182	0
0	0	0	3	0	0	4	0	1	0	192

FIGURE 12.14

Confusion matrix for 10-class classification by FDA. The correct classification rate is 1798/2000 = 89.9%.

that 1798 test samples out of 2000 samples are correctly classified. Thus, the correct classification rate is

$$1798/2000 = 89.9\%.$$

CHAPTER

PROPERTIES OF MAXIMUM LIKELIHOOD ESTIMATION 13

CHAPTER CONTENTS

In the previous chapter, the definition of MLE and its usage in pattern recognition were explained. In this chapter, the asymptotic behavior of MLE is theoretically investigated. Throughout this chapter, the true probability density function $p(\boldsymbol{x})$ is assumed to be included in the parametric model $q(\boldsymbol{x};\boldsymbol{\theta})$, i.e., there exists $\boldsymbol{\theta}^*$ such that $q(\boldsymbol{x};\boldsymbol{\theta}^*) = p(\boldsymbol{x})$.

13.1 CONSISTENCY

First, consistency of MLE is explored. Consistency is the property that the optimal solution $\boldsymbol{\theta}^*$ can be obtained asymptotically (i.e., in the limit that the number of training samples n tends to infinity), which would be a minimum requirement for reasonable estimators.

More precisely, let $\widehat{\boldsymbol{\theta}}_n$ be an estimator obtained from n i.i.d. training samples. Then, an estimator $\widehat{\boldsymbol{\theta}}_n$ is said to be *consistent* if the following property is satisfied for any $\boldsymbol{\theta}^* \in \Theta$ and any $\varepsilon > 0$:

$$\lim_{n\to\infty} \Pr(\|\widehat{\boldsymbol{\theta}}_n - \boldsymbol{\theta}^*\| \geq \varepsilon) = 0.$$

This means that $\widehat{\boldsymbol{\theta}}_n$ converges in probability to $\boldsymbol{\theta}^*$ (Section 7.3) and is denoted as

$$\widehat{\boldsymbol{\theta}}_n \xrightarrow{\text{p}} \boldsymbol{\theta}^*.$$

MLE was proved to be consistent under mild assumptions. For example, the consistency of the maximum likelihood estimator of the expectation for the one-dimensional Gaussian model, $\widehat{\mu}_{\text{ML}} = \frac{1}{n}\sum_{i=1}^n x_i$, was shown in Section 7.3.

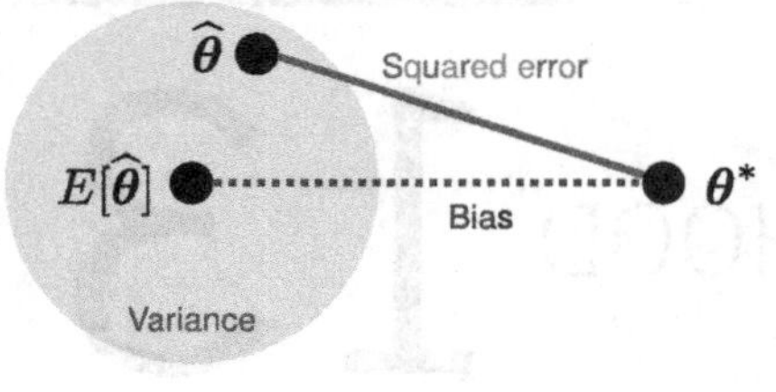

FIGURE 13.1

Bias-variance decomposition of expected squared error.

Next, let us investigate the relation between consistency and the *squared error* $\|\widehat{\theta} - \theta^*\|^2$. *Markov's inequality* shown in Section 8.2.1 gives the following upper bound:

$$\Pr(\|\widehat{\theta} - \theta^*\| \geq \varepsilon) = \Pr(\|\widehat{\theta} - \theta^*\|^2 \geq \varepsilon^2) \leq \frac{1}{\varepsilon^2}\mathbb{E}\left[\|\widehat{\theta} - \theta^*\|^2\right],$$

where $\mathbb{E}\left[\|\widehat{\theta} - \theta^*\|^2\right]$ is the *expected squared error* and $\mathbb{E}$ denotes the expectation over all training samples $\{x_i\}_{i=1}^n$ following i.i.d. with $p(x)$:

$$\mathbb{E}[\bullet] = \int \cdots \int \bullet\, p(x_1) \cdots p(x_n)\mathrm{d}x_1 \cdots \mathrm{d}x_n. \tag{13.1}$$

This upper bound shows that if the expected squared error of an estimator vanishes asymptotically, the estimator possesses consistency.

13.2 ASYMPTOTIC UNBIASEDNESS

The expected squared error can be decomposed as

$$\begin{aligned}
&\mathbb{E}\left[\|\widehat{\theta} - \theta^*\|^2\right] \\
&= \mathbb{E}\left[\|\widehat{\theta} - \mathbb{E}[\widehat{\theta}] + \mathbb{E}[\widehat{\theta}] - \theta^*\|^2\right] \\
&= \mathbb{E}\left[\|\widehat{\theta} - \mathbb{E}[\widehat{\theta}]\|^2\right] + \|\mathbb{E}[\widehat{\theta}] - \theta^*\|^2 + 2\mathbb{E}\left[(\widehat{\theta} - \mathbb{E}[\widehat{\theta}])^\top(\mathbb{E}[\widehat{\theta}] - \theta^*)\right] \\
&= \mathbb{E}\left[\|\widehat{\theta} - \mathbb{E}[\widehat{\theta}]\|^2\right] + \|\mathbb{E}[\widehat{\theta}] - \theta^*\|^2 + 2[(\mathbb{E}[\widehat{\theta}] - \mathbb{E}[\widehat{\theta}])^\top(\mathbb{E}[\widehat{\theta}] - \theta^*)] \\
&= \mathbb{E}\left[\|\widehat{\theta} - \mathbb{E}[\widehat{\theta}]\|^2\right] + \|\mathbb{E}[\widehat{\theta}] - \theta^*\|^2,
\end{aligned}$$

where the first term and second term are called the *variance term* and the *bias term*, respectively (Fig. 13.1).

An estimator $\widehat{\theta}$ is said to be *unbiased* if

$$\mathbb{E}[\widehat{\theta}] = \theta^*,$$

and an estimator $\widehat{\theta}_n$ is said to be *asymptotically unbiased* if

$$\mathbb{E}[\widehat{\theta}_n] \xrightarrow{\text{p}} \theta^*.$$

MLE was shown to be asymptotically unbiased under mild assumptions.

13.3 ASYMPTOTIC EFFICIENCY

Consistency and asymptotic unbiasedness are properties for infinitely many training samples. However, in practice, the number of training samples cannot be infinity, and therefore an estimator with consistency and asymptotic unbiasedness is not necessarily superior. Indeed, as explained in Section 13.2, not only the bias but also the variance has to be taken into account to reduce the expected squared error.

13.3.1 ONE-DIMENSIONAL CASE

Efficiency concerns the variance of an estimator. To explain the concept of efficiency more precisely, let us first consider a one-dimensional case where the parameter to be estimated, θ, is a scalar. Let $\mathbb{V}$ denote the variance operator over all training samples $\{(\boldsymbol{x}_i, y_i)\}_{i=1}^n$ following i.i.d. with $p(\boldsymbol{x})$:

$$\mathbb{V}(\bullet) = \mathbb{E}[(\bullet - \mathbb{E}[\bullet])^2], \tag{13.2}$$

where $\mathbb{E}$ is defined in Eq. (13.1). Then the variance of any unbiased estimator $\widehat{\theta}$ is lower-bounded as

$$\mathbb{V}(\widehat{\theta}) = \mathbb{E}[(\widehat{\theta} - \theta^*)^2] \geq \frac{1}{nF(\theta^*)},$$

which is called the *Cramér-Rao inequality* [31, 82]. Here, $F(\theta)$ is called *Fisher information*:

$$F(\theta) = \int \left(\frac{\partial}{\partial \theta} \log q(\boldsymbol{x}; \theta) \right)^2 q(\boldsymbol{x}; \theta) \mathrm{d}\boldsymbol{x}.$$

The partial derivative of $\log q(\boldsymbol{x}; \theta)$,

$$\frac{\partial}{\partial \theta} \log q(\boldsymbol{x}; \theta), \tag{13.3}$$

is often called the *Fisher score*. An unbiased estimator $\widehat{\theta}$ is said to be efficient if the Cramér-Rao lower bound is attained with strict equality:

$$\mathbb{V}(\widehat{\theta}) = \frac{1}{nF(\theta^*)}.$$

13.3.2 MULTIDIMENSIONAL CASES

To extend the above definition to multidimensional cases, let us define the *Fisher information matrix* $\boldsymbol{F}(\boldsymbol{\theta})$ as

$$F(\theta) = \int \left(\frac{\partial}{\partial \theta} \log q(x;\theta)\right)\left(\frac{\partial}{\partial \theta^\top} \log q(x;\theta)\right) q(x;\theta)\mathrm{d}x. \tag{13.4}$$

Here, $\frac{\partial}{\partial \theta}$ and $\frac{\partial}{\partial \theta^\top}$ denote the vertical and horizontal vectors of partial derivatives, respectively. That is, the (j, j')th element of $F(\theta)$ for $\theta = (\theta^{(1)}, \ldots, \theta^{(b)})^\top$ is given by

$$F_{j,j'}(\theta) = \int \left(\frac{\partial}{\partial \theta^{(j)}} \log q(x;\theta)\right)\left(\frac{\partial}{\partial \theta^{(j')}} \log q(x;\theta)\right) q(x;\theta)\mathrm{d}x.$$

Then a multidimensional version of the Cramér-Rao inequality is given for any unbiased estimator $\widehat{\theta}$ as

$$\mathbb{V}(\widehat{\theta}) = \mathbb{E}[(\widehat{\theta} - \theta^*)(\widehat{\theta} - \theta^*)^\top] \geq \frac{1}{n} F(\theta^*)^{-1}.$$

Here, the inequality $A \geq B$ for square matrices A and B means that $A - B$ is a *positive semidefinite matrix*: a matrix C is said to be positive semidefinite if it satisfies $\varphi^\top C \varphi \geq 0$ for any vector φ. An unbiased estimator $\widehat{\theta}$ is said to be efficient if the above multidimensional version of the Cramér-Rao lower bound is attained with strict equality:

$$\mathbb{V}(\widehat{\theta}) = \frac{1}{n} F(\theta^*)^{-1}.$$

The concept of efficiency is defined for unbiased estimators. Therefore, since MLE is asymptotically unbiased, but not generally unbiased with finite samples, MLE is not efficient. An estimator is said to be *asymptotic efficient* if the Cramér-Rao lower bound is attained asymptotically:

$$n\mathbb{V}(\widehat{\theta}) \xrightarrow{\text{p}} F(\theta^*)^{-1}.$$

MLE was shown to be asymptotically efficient under mild assumptions.

Suppose that $q(x;\theta)$ is twice differentiable. Then

$$\begin{aligned}
&\int \left(\frac{\partial^2}{\partial \theta \partial \theta^\top} \log q(x;\theta)\right) q(x;\theta)\mathrm{d}x \\
&= \int \frac{\frac{\partial^2}{\partial \theta \partial \theta^\top} q(x;\theta)}{q(x;\theta)} q(x;\theta)\mathrm{d}x - \int \frac{\frac{\partial}{\partial \theta} q(x;\theta) \frac{\partial}{\partial \theta^\top} q(x;\theta)}{q(x;\theta)^2} q(x;\theta)\mathrm{d}x \\
&= \int \frac{\partial^2}{\partial \theta \partial \theta^\top} q(x;\theta)\mathrm{d}x - \int \left(\frac{\partial}{\partial \theta} \log q(x;\theta)\right)\left(\frac{\partial}{\partial \theta^\top} \log q(x;\theta)\right) q(x;\theta)\mathrm{d}x \\
&= \frac{\partial^2}{\partial \theta \partial \theta^\top} \int q(x;\theta)\mathrm{d}x - F(\theta) = \frac{\partial^2}{\partial \theta \partial \theta^\top} 1 - F(\theta) = -F(\theta).
\end{aligned}$$

Therefore, the Fisher information matrix defined in Eq. (13.4) can be expressed as

$$F(\theta) = -\int \left(\frac{\partial^2}{\partial \theta \partial \theta^\top} \log q(x;\theta)\right) q(x;\theta)\mathrm{d}x. \tag{13.5}$$

The matrix

$$\frac{\partial^2}{\partial \boldsymbol{\theta} \partial \boldsymbol{\theta}^\top} \log q(\boldsymbol{x}; \boldsymbol{\theta})$$

is called the *Hessian matrix* of $\log q(\boldsymbol{x}; \boldsymbol{\theta})$, which plays an important role in optimization theory. Eq. (13.5) shows that the Fisher information matrix agrees with the expected negative Hessian matrix of $\log q(\boldsymbol{x}; \boldsymbol{\theta})$.

13.4 ASYMPTOTIC NORMALITY

If an estimator approximately follows the normal distribution when the number of training samples n is large, it is said to possess *asymptotic normality*. In this section, asymptotic normality of the maximum likelihood estimator is explained.

As explained in Section 7.4, the *central limit theorem* asserts that, for n one-dimensional i.i.d. samples $\{x_i\}_{i=1}^n$ having expectation 0 and variance 1, their mean $\overline{x}_n = \frac{1}{n}\sum_{i=1}^n x_i$ satisfies

$$\lim_{n\to\infty} \Pr(a \leq \sqrt{n}\overline{x}_n \leq b) = \frac{1}{\sqrt{2\pi}} \int_a^b \exp\left(-\frac{x^2}{2}\right) \mathrm{d}x.$$

This means that $\sqrt{n}\overline{x}_n$ asymptotically follows the standard normal distribution.

The central limit theorem can be extended to multidimensions as follows. Let $\{\boldsymbol{x}_i\}_{i=1}^n$ be the i.i.d. samples having expectation $\boldsymbol{0}_d$ and variance-covariance matrix $\boldsymbol{\Sigma}$, where d denotes the dimensionality of $\boldsymbol{x}_i$. Let $\overline{\boldsymbol{x}}_n$ be the sample average:

$$\overline{\boldsymbol{x}}_n = \frac{1}{n}\sum_{i=1}^n \boldsymbol{x}_i.$$

Then $\sqrt{n}\overline{\boldsymbol{x}}_n$ approximately follows the d-dimensional normal distribution with expectation $\boldsymbol{0}_d$ and variance-covariance matrix $\boldsymbol{\Sigma}$ (Section 6.2):

$$\sqrt{n}\overline{\boldsymbol{x}}_n \xrightarrow{\mathrm{d}} N(\boldsymbol{0}_d, \boldsymbol{\Sigma}),$$

where "d" denotes the convergence in distribution (see Section 7.4).

Based on the above central limit theorem, asymptotic normality of the maximum likelihood estimator is explained. Let $\boldsymbol{\xi}(\boldsymbol{\theta})$ be the sample average of the *Fisher score* (see Eq. (13.3)):

$$\boldsymbol{\xi}(\boldsymbol{\theta}) = \frac{1}{n}\sum_{i=1}^n \frac{\partial}{\partial \boldsymbol{\theta}} \log q(\boldsymbol{x}_i; \boldsymbol{\theta}).$$

Since the maximum likelihood estimator $\widehat{\boldsymbol{\theta}}_{\mathrm{ML}}$ maximizes the log-likelihood, it satisfies $\boldsymbol{\xi}(\widehat{\boldsymbol{\theta}}_{\mathrm{ML}}) = \boldsymbol{0}_b$. Then the *Taylor series expansion* (Fig. 2.8) of the left-hand side about the true parameter $\boldsymbol{\theta}^*$ yields

$$\boldsymbol{\xi}(\boldsymbol{\theta}^*) - \widehat{\boldsymbol{F}}(\boldsymbol{\theta}^*)(\widehat{\boldsymbol{\theta}}_{\mathrm{ML}} - \boldsymbol{\theta}^*) + \boldsymbol{r} = \boldsymbol{0}_b,$$

where $\boldsymbol{r}$ is the residual vector that contains higher-order terms and

$$\widehat{\boldsymbol{F}}(\boldsymbol{\theta}) = -\frac{1}{n}\sum_{i=1}^{n}\frac{\partial^2}{\partial\boldsymbol{\theta}\partial\boldsymbol{\theta}^\top}\log q(\boldsymbol{x}_i;\boldsymbol{\theta}).$$

Since $\widehat{\boldsymbol{F}}(\boldsymbol{\theta})$ is a sample approximation of Fisher information matrix (13.5), it converges in probability to the true Fisher information matrix in the limit $n \to \infty$:

$$\widehat{\boldsymbol{F}}(\boldsymbol{\theta}) \xrightarrow{\text{p}} \boldsymbol{F}(\boldsymbol{\theta}).$$

Also, consistency of MLE implies that the residual $\boldsymbol{r}$ can be ignored in the limit $n \to \infty$. Thus,

$$\sqrt{n}(\widehat{\boldsymbol{\theta}}_{\text{ML}} - \boldsymbol{\theta}^*) \xrightarrow{\text{p}} \boldsymbol{F}(\boldsymbol{\theta}^*)^{-1}\sqrt{n}\boldsymbol{\xi}(\boldsymbol{\theta}^*). \tag{13.6}$$

The true parameter $\boldsymbol{\theta}^*$ maximizes the expected log-likelihood $E[\log q(\boldsymbol{x};\boldsymbol{\theta})]$ with respect to $\boldsymbol{\theta}$, where E is the expectation operator over $p(\boldsymbol{x})$:

$$E[\bullet] = \int \bullet\, p(\boldsymbol{x})\mathrm{d}\boldsymbol{x}.$$

Then the first-order optimality (Fig. 12.1) yields

$$E\left[\frac{\partial}{\partial\boldsymbol{\theta}}\log q(\boldsymbol{x};\boldsymbol{\theta})\Big|_{\boldsymbol{\theta}=\boldsymbol{\theta}^*}\right] = \boldsymbol{0}_b.$$

Furthermore, Eq. (13.4) implies

$$V\left[\frac{\partial}{\partial\boldsymbol{\theta}}\log q(\boldsymbol{x};\boldsymbol{\theta})\Big|_{\boldsymbol{\theta}=\boldsymbol{\theta}^*}\right] = \boldsymbol{F}(\boldsymbol{\theta}^*),$$

where V denotes the variance operator over $p(\boldsymbol{x})$:

$$V[\bullet] = \int (\bullet - E[\bullet])(\bullet - E[\bullet])^\top p(\boldsymbol{x})\mathrm{d}\boldsymbol{x}.$$

Therefore, the central limit theorem asserts that $\sqrt{n}\boldsymbol{\xi}(\boldsymbol{\theta}^*)$ asymptotically follows the normal distribution with expectation $\boldsymbol{0}_b$ and variance-covariance matrix $\boldsymbol{F}(\boldsymbol{\theta}^*)$:

$$\sqrt{n}\boldsymbol{\xi}(\boldsymbol{\theta}^*) \xrightarrow{\text{d}} N(\boldsymbol{0}_b, \boldsymbol{F}(\boldsymbol{\theta}^*)).$$

Moreover, the variance-covariance matrix of $\boldsymbol{F}(\boldsymbol{\theta}^*)^{-1}\sqrt{n}\boldsymbol{\xi}(\boldsymbol{\theta}^*)$ is given by

$$\begin{aligned}\mathbb{V}[\boldsymbol{F}(\boldsymbol{\theta}^*)^{-1}\sqrt{n}\boldsymbol{\xi}(\boldsymbol{\theta}^*)] &= \boldsymbol{F}(\boldsymbol{\theta}^*)^{-1}V\left[\sqrt{n}\boldsymbol{\xi}(\boldsymbol{\theta}^*)\right]\boldsymbol{F}(\boldsymbol{\theta}^*)^{-1}\\ &= \boldsymbol{F}(\boldsymbol{\theta}^*)^{-1}\boldsymbol{F}(\boldsymbol{\theta}^*)\boldsymbol{F}(\boldsymbol{\theta}^*)^{-1} = \boldsymbol{F}(\boldsymbol{\theta}^*)^{-1},\end{aligned}$$

```
n=10; t=10000; s=1/12/n;
x=linspace(-0.4,0.4,100);
y=1/sqrt(2*pi*s)*exp(-x.^2/(2*s));
z=mean(rand(t,n)-0.5,2);

figure(1); clf; hold on
b=20; hist(z,b); c=max(z)-min(z);
h=plot(x,y*t/b*c,'r-');
```

FIGURE 13.2

MATLAB code for illustrating asymptotic normality of MLE.

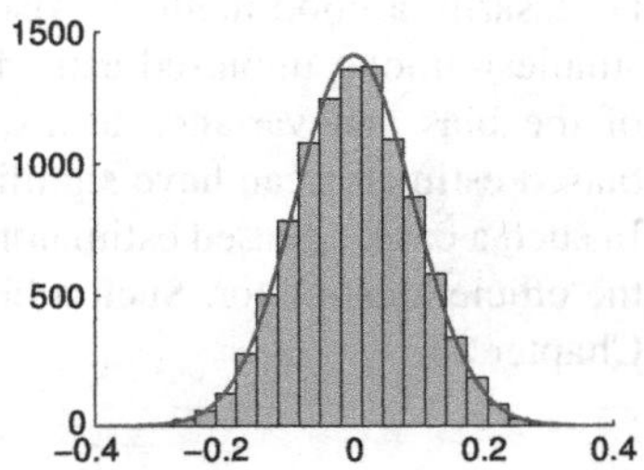

FIGURE 13.3

Example of asymptotic normality of MLE.

where $\mathbb{V}$ denotes the variance with respect to all training samples $\{\boldsymbol{x}_i\}_{i=1}^n$ drawn i.i.d. from $p(\boldsymbol{x})$ (see Eq. (13.2)). Therefore, Eq. (13.6) yields

$$\sqrt{n}(\widehat{\boldsymbol{\theta}}_{\mathrm{ML}} - \boldsymbol{\theta}^*) \xrightarrow{\mathrm{d}} N(\mathbf{0}_b, \boldsymbol{F}(\boldsymbol{\theta}^*)^{-1}).$$

This means that the maximum likelihood estimator $\widehat{\boldsymbol{\theta}}_{\mathrm{ML}}$ asymptotically follows the normal distribution with expectation $\boldsymbol{\theta}^*$ and variance-covariance matrix $\frac{1}{n}\boldsymbol{F}(\boldsymbol{\theta}^*)^{-1}$.

The above asymptotic normality of MLE implies that MLE is asymptotically unbiased. Furthermore, the variance-covariance matrix $\frac{1}{n}\boldsymbol{F}(\boldsymbol{\theta}^*)^{-1}$ vanishes asymptotically, meaning that the bias and variance terms explained in Section 13.2 also vanish asymptotically. Therefore, the expected squared error, which is the sum of the bias and variance terms, also vanishes asymptotically. Moreover, the above results show that the Cramér-Rao lower bound is attained asymptotically, meaning that MLE is asymptotically efficient.

A MATLAB code for illustrating the asymptotic normality of

$$\widehat{\mu}_{\mathrm{ML}} = \frac{1}{n}\sum_{i=1}^{n} x_i,$$

when $p(x)$ is the uniform distribution on $[-0.5, 0.5]$, is given in Fig. 13.2, and its behavior is illustrated in Fig. 13.3.

13.5 SUMMARY

MLE is consistent and asymptotically unbiased, and therefore its validity is theoretically guaranteed when the number of training samples is infinite. Furthermore, since MLE is asymptotically efficient, its high reliability is guaranteed even when the number of training samples is not infinite, but large.

However, when the number of training samples is not large, MLE is not necessarily a good method. Also, efficiency just guarantees that the variance is the smallest among unbiased estimators; the expected squared error, which is the sum of the bias and variance terms, is not guaranteed to be small. Indeed, a slightly biased estimator can have significantly smaller variance than the efficient estimator. In such a case, a biased estimator can have much smaller expected squared error than the efficient estimator. Such a biased estimator will be discussed in Chapter 17 and Chapter 23.

CHAPTER

MODEL SELECTION FOR MAXIMUM LIKELIHOOD ESTIMATION

14

CHAPTER CONTENTS

So far, a parametric model was assumed to be given and fixed. However, in practice, it is often necessary to choose a parametric model from some candidates. In this chapter, a data-driven method to choose an appropriate parametric model is explained.

14.1 MODEL SELECTION

As shown in Chapter 12, "Gaussian models" have several different variations.

(A) Variance-covariance matrix $\boldsymbol{\Sigma}$ is generic:

$$q(\boldsymbol{x};\boldsymbol{\mu},\boldsymbol{\Sigma}) = \frac{1}{(2\pi)^{\frac{d}{2}}\det(\boldsymbol{\Sigma})^{\frac{1}{2}}}\exp\left(-\frac{1}{2}(\boldsymbol{x}-\boldsymbol{\mu})^\top\boldsymbol{\Sigma}^{-1}(\boldsymbol{x}-\boldsymbol{\mu})\right).$$

(B) Variance-covariance matrix $\boldsymbol{\Sigma}$ is diagonal:

$$q(\boldsymbol{x};\boldsymbol{\mu},(\sigma^{(1)})^2,\ldots,(\sigma^{(d)})^2) = \prod_{j=1}^{d}\frac{1}{\sqrt{2\pi(\sigma^{(j)})^2}}\exp\left(-\frac{(x^{(j)}-\mu^{(j)})^2}{2(\sigma^{(j)})^2}\right).$$

(C) Variance-covariance matrix $\boldsymbol{\Sigma}$ is proportional to the identity matrix:

$$q(\boldsymbol{x};\boldsymbol{\mu},\sigma^2) = \frac{1}{(2\pi\sigma^2)^{\frac{d}{2}}}\exp\left(-\frac{(\boldsymbol{x}-\boldsymbol{\mu})^\top(\boldsymbol{x}-\boldsymbol{\mu})}{2\sigma^2}\right).$$

To obtain a good approximation to the true probability density function by a parametric method, a parametric model that (approximately) contains the true

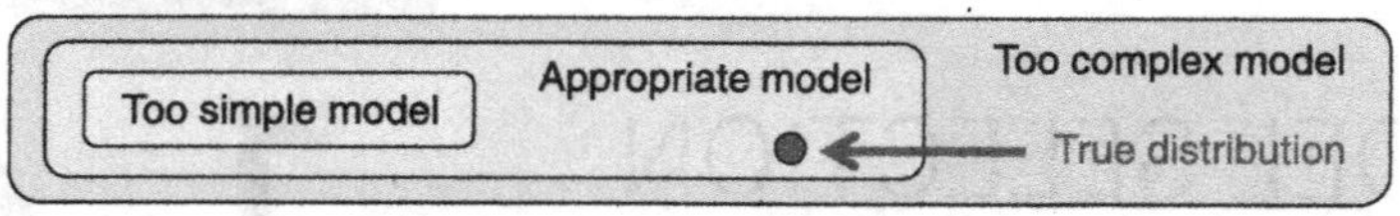

FIGURE 14.1

Model selection. Too simple model may not be expressive enough to represent the true probability distribution, while too complex model may cause unreliable parameter estimation.

probability density function may be required. From this viewpoint, a richer model that contains various probability density functions is preferable. On the other hand, as explained in Chapter 12, the good performance of MLE is guaranteed only when the number of training samples is large (relative to the number of parameters). This implies that a simple model that has a small number of parameters is desirable. Therefore, in practice, a model that fulfills these two conflicting requirements in a balanced way should be selected (Fig. 14.1).

However, since the appropriate balance depends on the unknown true probability distribution, let us consider the following data-driven procedure for model selection:

1. Prepare parametric models $\{q_m(\boldsymbol{x};\boldsymbol{\theta})\}_m$.
2. For each parametric model $q_m(\boldsymbol{x};\boldsymbol{\theta})$, compute maximum likelihood estimator $\widehat{\boldsymbol{\theta}}_m$ and obtain a density estimator $\widehat{p}_m(\boldsymbol{x})$ as
$$\widehat{p}_m(\boldsymbol{x}) = q_m(\boldsymbol{x};\widehat{\boldsymbol{\theta}}_m).$$
3. From the obtained density estimators $\{\widehat{p}_m(\boldsymbol{x})\}_m$, choose the one that is closest to the true probability density function $p(\boldsymbol{x})$.

The problem of finding the most promising model from a set of model candidates is called *model selection*. At a glance, finding the estimator that is closest to the true probability density function $p(\boldsymbol{x})$ in Step 3 is not possible because $p(\boldsymbol{x})$ is unknown. Below, two model selection approaches called the *Akaike information criterion* (AIC) and *cross validation* are introduced for coping with this problem.

14.2 KL DIVERGENCE

To perform model selection following the above procedure, a "closeness" measure between the true probability density function $p(\boldsymbol{x})$ and its estimator $\widehat{p}(\boldsymbol{x})$ is needed. Mathematically, for a set $\mathcal{X}$, function $g : \mathcal{X}\times\mathcal{X} \to \mathbb{R}$ that satisfies the following four conditions for any $x, y, z \in \mathcal{X}$ is called a *distance function*:

$$\text{Non-negativity: } g(x,y) \geq 0,$$
$$\text{Symmetry: } g(x,y) = g(y,x),$$
$$\text{Identity: } g(x,y) = 0 \Longleftrightarrow x = y,$$

Triangle inequality: $g(x,y) + g(y,z) \geq g(x,z)$.

In this section, a typical closeness measure called the *Kullback-Leibler divergence* (KL divergence) [64] is introduced.

The KL divergence from p to $\widehat{p}$ is defined as

$$\mathrm{KL}(p\|\widehat{p}) = E\left[\log \frac{p(\boldsymbol{x})}{\widehat{p}(\boldsymbol{x})}\right], \tag{14.1}$$

where E is the expectation operator over $p(\boldsymbol{x})$:

$$E[\bullet] = \int \bullet\, p(\boldsymbol{x})\mathrm{d}\boldsymbol{x}. \tag{14.2}$$

The KL divergence does not actually satisfy symmetric and triangle inequality, meaning that it is not mathematically a distance. Nevertheless, the KL divergence still satisfies non-negativity and identity. Therefore, even though it is not a proper distance, a smaller KL divergence would imply that $\widehat{p}$ is a better approximator to p.

Since the KL divergence $\mathrm{KL}(p\|\widehat{p})$ contains unknown true probability density function $p(\boldsymbol{x})$, it cannot be directly computed. Let us expand Eq. (14.1) as

$$\mathrm{KL}(p\|\widehat{p}) = E\left[\log p(\boldsymbol{x})\right] - E\left[\log \widehat{p}(\boldsymbol{x})\right],$$

where $E\left[\log p(\boldsymbol{x})\right]$ is the negative *entropy* of p that is a constant independent of $\widehat{p}$. In the context of model selection, the KL divergence is used for comparing different models, and therefore the constant term is irrelevant. For this reason, the first term is ignored and only the second term $E\left[\log \widehat{p}(\boldsymbol{x})\right]$, which is the expected log-likelihood, is considered below.

Even if the first term is ignored, the second term still contains the expectation over unknown p and thus it cannot be directly computed. The most naive approach to estimating the expectation would be the sample average, i.e., the expected log-likelihood is approximated by the average log-likelihood over i.i.d. samples $\{\boldsymbol{x}_i\}_{i=1}^n$ with $p(\boldsymbol{x})$:

$$E[\log \widehat{p}(\boldsymbol{x})] \approx \frac{1}{n}\sum_{i=1}^{n} \log \widehat{p}(\boldsymbol{x}_i).$$

Then, a larger log-likelihood approximately implies a smaller KL divergence, and therefore a model with a large log-likelihood would be close to the true probability density function p.

However, the naive use of the log-likelihood for model selection does not work well in practice. To illustrate this, let us consider model selection from the three Gaussian models introduced in the beginning of Section 14.1:

(A) Variance-covariance matrix $\boldsymbol{\Sigma}$ is generic.

(B) Variance-covariance matrix $\boldsymbol{\Sigma}$ is diagonal.

(C) Variance-covariance matrix $\boldsymbol{\Sigma}$ is proportional to the identity matrix.

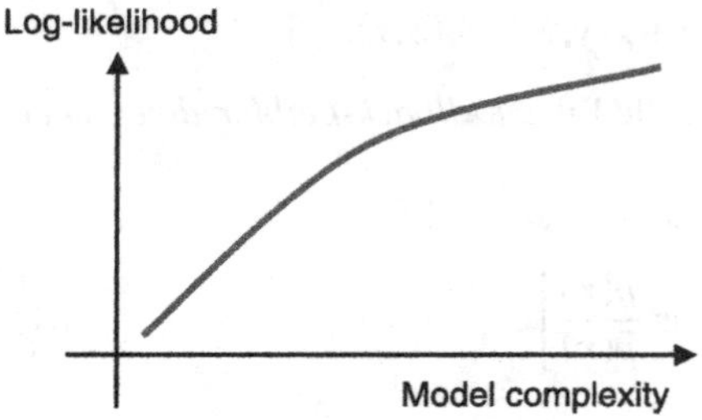

FIGURE 14.2

For nested models, log-likelihood is monotone nondecreasing as the model complexity increases.

These three models are nested, meaning that model (C) is a special case of model (B) and model (B) is a special case of model (A). This implies that the maximum likelihood solution in model (C) is also included in model (B), and therefore model (B) can achieve at least the same log-likelihood as model (C). Similarly, the maximum likelihood solutions in model (B) and model (C) are also included in model (A), and therefore model (A) can achieve at least the same log-likelihood as model (B) and model (C). For this reason, when model selection is performed for nested models based on the log-likelihood, the largest model (e.g., model (A) in the above Gaussian model examples) is always chosen (Fig. 14.2).

14.3 AIC

As explained above, the naive use of log-likelihood for model selection results in just always selecting the most complex model. This is caused by the fact that the average log-likelihood is not an accurate enough estimator of the expected log-likelihood. For appropriate model selection, therefore, a more accurate estimator of the expected log-likelihood is needed. The *Akaike information criterion*AIC gives a better estimator of the expected log-likelihood [1].

AIC is defined as

$$\mathrm{AIC} = -\sum_{i=1}^{n} \log q(\boldsymbol{x}_i; \widehat{\boldsymbol{\theta}}_{\mathrm{ML}}) + b, \tag{14.3}$$

where the first term is the negative log-likelihood and the second term is the number of parameters. Thus, AIC can be regarded as correcting the negative log-likelihood adding by the number of parameters.

Let us intuitively explain why AIC can yield better model selection. As illustrated in Fig. 14.2, the negative log-likelihood is monotone nondecreasing as the number of parameters increases. On the other hand, the number of parameters is monotone

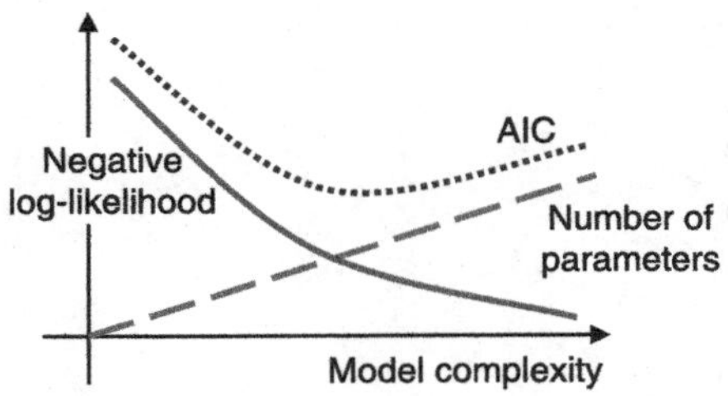

FIGURE 14.3

AIC is the sum of the negative log-likelihood and the number of parameters.

increasing as the number of parameters increases. Therefore, AIC is balancing these conflicting functions by summing them up. As a result, an appropriate model that achieves a reasonably large log-likelihood with a reasonably small number of parameters tends to have a small AIC value (Fig. 14.3).

In various areas of science and engineering, *Occam's razor*, also known as the *principle of parsimony*, is often used as a guiding principle. Occam's razor suggests choosing a simpler hypothesis among similar hypotheses, and AIC can be regarded as justifying the use of Occam's razor in statistical inference. More specifically, among the models that achieve similar log-likelihood values, AIC suggests choosing the one that has the smallest number of parameters.

Next, theoretical validity of AIC is investigated. Let $J(\boldsymbol{\theta})$ be the negative expected log-likelihood of $q(\boldsymbol{x};\boldsymbol{\theta})$,

$$J(\boldsymbol{\theta}) = -E[\log q(\boldsymbol{x};\boldsymbol{\theta})],$$

where E is the expectation operator over $p(\boldsymbol{x})$ (see Eq.(14.2)). Let $\boldsymbol{\theta}^*$ be the minimizer of $J(\boldsymbol{\theta})$:

$$\boldsymbol{\theta}^* = \underset{\boldsymbol{\theta}}{\mathrm{argmin}}\, J(\boldsymbol{\theta}).$$

According to *asymptotic theory*, the expectation of $J(\widehat{\boldsymbol{\theta}}_{\mathrm{ML}})$ can be expanded using $\boldsymbol{\theta}^*$ as follows [63, 107]:

$$\mathbb{E}[J] = -\mathbb{E}\left[\frac{1}{n}\sum_{i=1}^{n}\log q(\boldsymbol{x}_i;\widehat{\boldsymbol{\theta}}_{\mathrm{ML}})\right] + \frac{1}{n}\mathrm{tr}\left(\boldsymbol{Q}(\boldsymbol{\theta}^*)\boldsymbol{G}(\boldsymbol{\theta}^*)^{-1}\right) + o(n^{-1}), \qquad (14.4)$$

where $\mathbb{E}$ denotes the expectation over all training samples $\{\boldsymbol{x}_i\}_{i=1}^n$ following i.i.d. with $p(\boldsymbol{x})$:

$$\mathbb{E}[\bullet] = \int\cdots\int \bullet\, p(\boldsymbol{x}_1)\cdots p(\boldsymbol{x}_n)\mathrm{d}\boldsymbol{x}_1\cdots\mathrm{d}\boldsymbol{x}_n.$$

$f(n) = O(g(n))$ means that

$$\lim_{n\to\infty} \frac{f(n)}{g(n)} < \infty,$$

while $f(n) = o(g(n))$ means that

$$\lim_{n\to\infty} \frac{f(n)}{g(n)} = 0.$$

Intuitively, $O(n^{-1})$ denotes a term that has the same size as n^{-1}, while $o(n^{-1})$ denotes a term that is smaller than n^{-1}.

FIGURE 14.4

Big-o and small-o notations.

$o(n^{-1})$ denotes a term that is smaller than n^{-1} asymptotically (see Fig. 14.4). Matrices $\boldsymbol{Q}(\boldsymbol{\theta})$ and $\boldsymbol{G}(\boldsymbol{\theta})$ are defined as

$$\boldsymbol{Q}(\boldsymbol{\theta}) = E\left[\frac{\partial}{\partial\boldsymbol{\theta}}\log q(\boldsymbol{x};\boldsymbol{\theta})\frac{\partial}{\partial\boldsymbol{\theta}^\top}\log q(\boldsymbol{x};\boldsymbol{\theta})\bigg|_{\boldsymbol{\theta}}\right],$$

$$\boldsymbol{G}(\boldsymbol{\theta}) = -E\left[\frac{\partial^2}{\partial\boldsymbol{\theta}\partial\boldsymbol{\theta}^\top}\log q(\boldsymbol{x};\boldsymbol{\theta})\bigg|_{\boldsymbol{\theta}}\right].$$

$\boldsymbol{Q}(\boldsymbol{\theta})$ looks similar to the *Fisher information matrix* $\boldsymbol{F}(\boldsymbol{\theta})$ (see Eq. (13.4)):

$$\boldsymbol{F}(\boldsymbol{\theta}) = \int\left(\frac{\partial}{\partial\boldsymbol{\theta}}\log q(\boldsymbol{x};\boldsymbol{\theta})\right)\left(\frac{\partial}{\partial\boldsymbol{\theta}^\top}\log q(\boldsymbol{x};\boldsymbol{\theta})\right)q(\boldsymbol{x};\boldsymbol{\theta})\mathrm{d}\boldsymbol{x}.$$

However, $\boldsymbol{Q}(\boldsymbol{\theta})$ includes the expectation over true $p(\boldsymbol{x})$, while $\boldsymbol{F}(\boldsymbol{\theta})$ includes the expectation over model $q(\boldsymbol{x};\boldsymbol{\theta})$. $\boldsymbol{G}(\boldsymbol{\theta})$ is the negative Hessian matrix of $\log q(\boldsymbol{x};\boldsymbol{\theta})$ expected over true $p(\boldsymbol{x})$.

$\boldsymbol{G}(\boldsymbol{\theta})$ can be expanded as

$$\begin{aligned}
\boldsymbol{G}(\boldsymbol{\theta}) &= -E\left[\frac{\partial^2}{\partial\boldsymbol{\theta}\partial\boldsymbol{\theta}^\top}\log q(\boldsymbol{x};\boldsymbol{\theta})\right] \\
&= -E\left[\frac{\frac{\partial^2}{\partial\boldsymbol{\theta}\partial\boldsymbol{\theta}^\top}q(\boldsymbol{x};\boldsymbol{\theta})}{q(\boldsymbol{x};\boldsymbol{\theta})}\right] + E\left[\frac{\frac{\partial}{\partial\boldsymbol{\theta}}q(\boldsymbol{x};\boldsymbol{\theta})}{q(\boldsymbol{x};\boldsymbol{\theta})}\frac{\frac{\partial}{\partial\boldsymbol{\theta}^\top}q(\boldsymbol{x};\boldsymbol{\theta})}{q(\boldsymbol{x};\boldsymbol{\theta})}\right] \\
&= -E\left[\frac{\frac{\partial^2}{\partial\boldsymbol{\theta}\partial\boldsymbol{\theta}^\top}q(\boldsymbol{x};\boldsymbol{\theta})}{q(\boldsymbol{x};\boldsymbol{\theta})}\right] + E\left[\frac{\partial}{\partial\boldsymbol{\theta}}\log q(\boldsymbol{x};\boldsymbol{\theta})\frac{\partial}{\partial\boldsymbol{\theta}^\top}\log q(\boldsymbol{x};\boldsymbol{\theta})\right] \\
&= -\int\frac{\frac{\partial^2}{\partial\boldsymbol{\theta}\partial\boldsymbol{\theta}^\top}q(\boldsymbol{x};\boldsymbol{\theta})}{q(\boldsymbol{x};\boldsymbol{\theta})}p(\boldsymbol{x})\mathrm{d}\boldsymbol{x} + \boldsymbol{Q}(\boldsymbol{\theta}).
\end{aligned}$$

Now, suppose that the parametric model $q(\boldsymbol{x};\boldsymbol{\theta})$ contains the true probability density function $p(\boldsymbol{x})$, i.e., there exists $\boldsymbol{\theta}^*$ such that $q(\boldsymbol{x};\boldsymbol{\theta}^*) = p(\boldsymbol{x})$. Then $\boldsymbol{G}(\boldsymbol{\theta}^*)$ can be expressed as

$$\begin{aligned}\boldsymbol{G}(\boldsymbol{\theta}^*) &= -\int \frac{\partial^2}{\partial\boldsymbol{\theta}\partial\boldsymbol{\theta}^\top} q(\boldsymbol{x};\boldsymbol{\theta})\mathrm{d}\boldsymbol{x} + \boldsymbol{Q}(\boldsymbol{\theta}^*) = -\frac{\partial^2}{\partial\boldsymbol{\theta}\partial\boldsymbol{\theta}^\top}\int q(\boldsymbol{x};\boldsymbol{\theta})\mathrm{d}\boldsymbol{x} + \boldsymbol{Q}(\boldsymbol{\theta}^*)\\ &= -\frac{\partial^2}{\partial\boldsymbol{\theta}\partial\boldsymbol{\theta}^\top} 1 + \boldsymbol{Q}(\boldsymbol{\theta}^*) = \boldsymbol{Q}(\boldsymbol{\theta}^*).\end{aligned}$$

Thus, the second term in Eq. (14.4) can be expressed as

$$\mathrm{tr}\left(\boldsymbol{Q}(\boldsymbol{\theta}^*)\boldsymbol{G}(\boldsymbol{\theta}^*)^{-1}\right) = \mathrm{tr}\left(\boldsymbol{I}_b\right) = b,$$

which agrees with the second term in AIC, i.e., the number of parameters.

As explained above, AIC assumes that the parametric model $q(\boldsymbol{x};\boldsymbol{\theta})$ contains the true probability density function $p(\boldsymbol{x})$. The *Takeuchi information criterion* (TIC) is a generalization of AIC that removes this assumption [63, 107]. TIC is defined as

$$\mathrm{TIC} = -\sum_{i=1}^{n} \log q(\boldsymbol{x}_i;\widehat{\boldsymbol{\theta}}_{\mathrm{ML}}) + \mathrm{tr}\left(\widehat{\boldsymbol{Q}}(\widehat{\boldsymbol{\theta}}_{\mathrm{ML}})\widehat{\boldsymbol{G}}(\widehat{\boldsymbol{\theta}}_{\mathrm{ML}})^{-1}\right),$$

where $\widehat{\boldsymbol{Q}}(\widehat{\boldsymbol{\theta}}_{\mathrm{ML}})$ and $\widehat{\boldsymbol{G}}(\widehat{\boldsymbol{\theta}}_{\mathrm{ML}})$ are the sample approximations to $\boldsymbol{Q}(\widehat{\boldsymbol{\theta}}_{\mathrm{ML}})$ and $\boldsymbol{G}(\widehat{\boldsymbol{\theta}}_{\mathrm{ML}})$, respectively:

$$\begin{aligned}\widehat{\boldsymbol{Q}}(\widehat{\boldsymbol{\theta}}_{\mathrm{ML}}) &= \frac{1}{n}\sum_{i=1}^{n} \frac{\partial}{\partial\boldsymbol{\theta}}\log q(\boldsymbol{x}_i;\boldsymbol{\theta})\frac{\partial}{\partial\boldsymbol{\theta}^\top}\log q(\boldsymbol{x}_i;\boldsymbol{\theta})\bigg|_{\boldsymbol{\theta}=\widehat{\boldsymbol{\theta}}_{\mathrm{ML}}},\\ \widehat{\boldsymbol{G}}(\widehat{\boldsymbol{\theta}}_{\mathrm{ML}}) &= -\frac{1}{n}\sum_{i=1}^{n} \frac{\partial^2}{\partial\boldsymbol{\theta}\partial\boldsymbol{\theta}^\top}\log q(\boldsymbol{x}_i;\boldsymbol{\theta})\bigg|_{\boldsymbol{\theta}=\widehat{\boldsymbol{\theta}}_{\mathrm{ML}}}.\end{aligned}$$

The law of large numbers and consistency of $\widehat{\boldsymbol{\theta}}_{\mathrm{ML}}$ yield that $\widehat{\boldsymbol{Q}}(\widehat{\boldsymbol{\theta}}_{\mathrm{ML}})$ and $\widehat{\boldsymbol{G}}(\widehat{\boldsymbol{\theta}}_{\mathrm{ML}})$ converge in probability to $\boldsymbol{Q}(\boldsymbol{\theta}^*)$ and $\boldsymbol{G}(\boldsymbol{\theta}^*)$, respectively:

$$\widehat{\boldsymbol{Q}}(\widehat{\boldsymbol{\theta}}_{\mathrm{ML}}) \xrightarrow{\mathrm{p}} \boldsymbol{Q}(\boldsymbol{\theta}^*) \quad\text{and}\quad \widehat{\boldsymbol{G}}(\widehat{\boldsymbol{\theta}}_{\mathrm{ML}}) \xrightarrow{\mathrm{p}} \boldsymbol{G}(\boldsymbol{\theta}^*).$$

It is known that the expected TIC divided by n converges to J and its error has smaller order than n^{-1}:

$$\frac{1}{n}\mathbb{E}[\mathrm{TIC}] = \mathbb{E}[J(\widehat{\boldsymbol{\theta}}_{\mathrm{ML}})] + o(n^{-1}).$$

On the other hand, if the negative average log-likelihood is naively used as an estimator of $J(\widehat{\boldsymbol{\theta}}_{\mathrm{ML}})$,

$$\mathbb{E}\left[-\frac{1}{n}\sum_{i=1}^{n}\log q(\boldsymbol{x}_i;\widehat{\boldsymbol{\theta}}_{\mathrm{ML}})\right] = \mathbb{E}[J(\widehat{\boldsymbol{\theta}}_{\mathrm{ML}})] + O(n^{-1})$$

holds, where $O(n^{-1})$ denotes a term that has the same size as n^{-1} (see Fig. 14.4). Thus, TIC is a more accurate estimator of $J(\widehat{\boldsymbol{\theta}}_{\mathrm{ML}})$ than the negative average log-likelihood. Since TIC does not require the assumption that the parametric model $q(\boldsymbol{x};\boldsymbol{\theta})$ contains

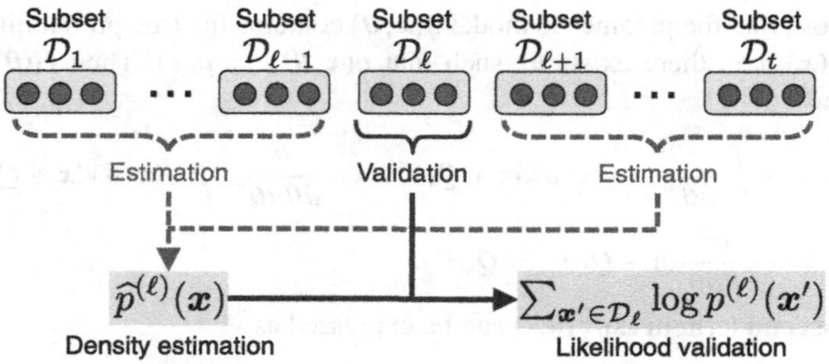

FIGURE 14.5

Cross validation.

the true probability density function $p(\boldsymbol{x})$, it is more general than AIC. However, since AIC is much simpler to implement, it seems to be more frequently used in practice than TIC.

14.4 CROSS VALIDATION

Although AIC is easy to implement, its validity is only guaranteed when the number of training samples is large. Here, a more flexible model selection method called *cross validation* is introduced.

The basic idea of cross validation is to split training samples $\mathcal{D} = \{\boldsymbol{x}_i\}_{i=1}^n$ into the estimation samples and the validation samples. The estimation samples are used for estimating the probability density function, and the validation samples are used for evaluating the validity of the estimated probability density function. If the log-likelihood is used as the evaluation measure, the model that gives the largest log-likelihood for the validation samples is chosen as the most promising one.

However, simply splitting the training samples into the estimation subset and the validation subset for model selection can cause strong dependency on data splitting. To mitigate this problem, cross validation splits the training samples into t disjoint subsets of (approximately) the same size (see Fig. 14.5). Then $(t-1)$ subsets are used for estimation and the remaining subset is used for validation. The choice of $(t-1)$ subsets has t possibilities and thus all t possibilities are investigated and the average log-likelihood is regarded as the final score for model selection.

The algorithm of cross validation is summarized in Fig. 14.6.

14.5 DISCUSSION

Let us interpret the statistical meaning of cross validation using the *KL divergence* from true probability density p to its estimator $\widehat{p}_j^{(\ell)}$:

1. Prepare candidates of models: $\{\mathcal{M}_j\}_j$.
2. Split training samples $\mathcal{D} = \{\boldsymbol{x}_i\}_{i=1}^n$ into t disjoint subsets of (approximately) the same size: $\{\mathcal{D}_\ell\}_{\ell=1}^t$.
3. For each model candidate $\mathcal{M}_j$
   (a) For each split $\ell = 1, \ldots, t$
      i. Obtain density estimator $\widehat{p}_j^{(\ell)}(\boldsymbol{x})$ using model $\mathcal{M}_j$ from all training samples without $\mathcal{D}_\ell$.
      ii. Compute the average log-likelihood $J_j^{(\ell)}$ of $\widehat{p}_j^{(\ell)}(\boldsymbol{x})$ for holdout samples $\mathcal{D}_\ell$:

$$J_j^{(\ell)} = \frac{1}{|\mathcal{D}_\ell|} \sum_{\boldsymbol{x}' \in \mathcal{D}_\ell} \log \widehat{p}_j^{(\ell)}(\boldsymbol{x}'),$$

      where $|\mathcal{D}_\ell|$ denotes the number of elements in set $\mathcal{D}_\ell$.
   (b) Compute the average log-likelihood J_j over all t splits:

$$J_j = \frac{1}{t} \sum_{\ell=1}^{t} J_j^{(\ell)}.$$

4. Choose the model $\mathcal{M}_{\widehat{j}}$ that maximizes the average log-likelihood:

$$\widehat{j} = \underset{j}{\mathrm{argmax}}\, J_j.$$

5. Obtain the final density estimator using chosen model $\mathcal{M}_{\widehat{j}}$, from all training samples $\{\boldsymbol{x}_i\}_{i=1}^n$.

FIGURE 14.6

Algorithm of likelihood cross validation.

$$\mathrm{KL}(p\|\widehat{p}) = E\left[\log \frac{p(\boldsymbol{x})}{\widehat{p}_j^{(\ell)}(\boldsymbol{x})}\right] = E\left[\log p(\boldsymbol{x})\right] - E\left[\log \widehat{p}_j^{(\ell)}(\boldsymbol{x})\right].$$

Since the first term is constant, let us ignore it and define the second term as J:

$$J = -E[\log \widehat{p}_j^{(\ell)}(\boldsymbol{x})].$$

The cross validation score $J_j^{(\ell)}$ (see Fig. 14.6) is known to be an almost unbiased estimator of J [118]. Thus, model selection by likelihood cross validation corresponds to finding the model that minimizes the KL divergence, which is the same objective as the AIC explained in Section 14.3.

Indeed, the likelihood cross validation and the AIC are known to perform similarly when the number of training samples is large [97]. However, for a small number of training samples, likelihood cross validation seems to be more reliable in practice. Furthermore, while the AIC is applicable only to model selection of maximum likelihood estimation under the KL divergence, cross validation is applicable to any estimation method and any error metric. For example, as shown in Fig. 16.17, cross validation can be applied to choosing tuning parameters in pattern recognition under the misclassification rate criterion. Thus, cross validation is a useful alternative to the AIC. A practical drawback of cross validation is that it is computationally expensive due to repeated estimation and validation. However, note that the cross validation procedure can be easily parallelized for multiple servers or cores.

CHAPTER

MAXIMUM LIKELIHOOD ESTIMATION FOR GAUSSIAN MIXTURE MODEL 15

CHAPTER CONTENTS

Fisher's linear discriminant analysis introduced in Chapter 12 is a simple and practical classification method. However, approximating class-conditional probability densities by the Gaussian models can be too restrictive in practice. In this chapter, a more expressive model called the *Gaussian mixture model* is introduced and its MLE is discussed.

15.1 GAUSSIAN MIXTURE MODEL

If patterns in a class are distributed in several clusters, approximating the class-conditional distribution by a single Gaussian model may not be appropriate. For example, Fig. 15.1(a) illustrates the situation where a *unimodal* distribution is approximated by a single Gaussian model, which results in accurate estimation. On the other hand, Fig. 15.1(b) illustrates the situation where a *multimodal* distribution is approximated by a single Gaussian model, which performs poorly even with a large number of training samples.

A *Gaussian mixture model*, defined by

$$q(\boldsymbol{x};\boldsymbol{\theta}) = \sum_{\ell=1}^{m} w_\ell N(\boldsymbol{x};\boldsymbol{\mu}_\ell,\boldsymbol{\Sigma}_\ell),$$

is suitable to approximate such multimodal distributions. Here, $N(\boldsymbol{x};\boldsymbol{\mu},\boldsymbol{\Sigma})$ denotes a Gaussian model with expectation $\boldsymbol{\mu}$ and variance-covariance matrix $\boldsymbol{\Sigma}$:

$$N(\boldsymbol{x};\boldsymbol{\mu},\boldsymbol{\Sigma}) = \frac{1}{(2\pi)^{d/2}\det(\boldsymbol{\Sigma})^{1/2}} \exp\left(-\frac{1}{2}(\boldsymbol{x}-\boldsymbol{\mu})^\top \boldsymbol{\Sigma}^{-1}(\boldsymbol{x}-\boldsymbol{\mu})\right).$$

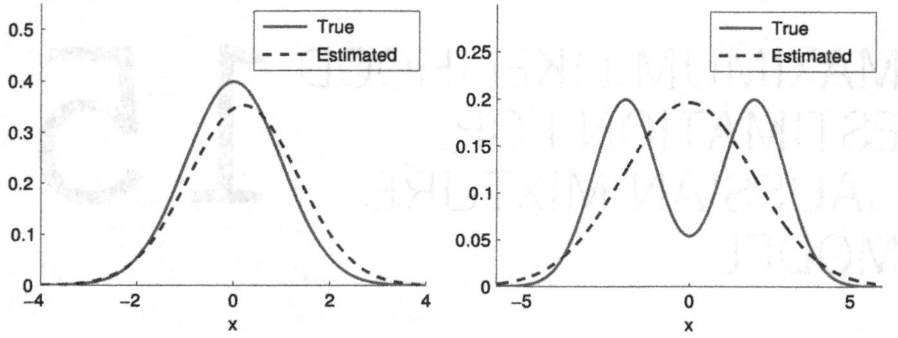

(a) Unimodal distribution (b) Multimodal distribution

FIGURE 15.1

MLE for Gaussian model.

Thus, a Gaussian mixture model is a *linear combination* of m Gaussian models weighted according to $\{w_\ell\}_{\ell=1}^m$. The parameter $\boldsymbol{\theta}$ of the Gaussian mixture model is given by

$$\boldsymbol{\theta} = (w_1, \ldots, w_m, \boldsymbol{\mu}_1, \ldots, \boldsymbol{\mu}_m, \boldsymbol{\Sigma}_1, \ldots, \boldsymbol{\Sigma}_m).$$

The Gaussian mixture model $q(\boldsymbol{x}; \boldsymbol{\theta})$ should satisfy the following condition to be a probability density function:

$$\forall \boldsymbol{x} \in \mathcal{X}, \ q(\boldsymbol{x}; \boldsymbol{\theta}) \geq 0 \ \text{ and } \ \int_{\mathcal{X}} q(\boldsymbol{x}; \boldsymbol{\theta}) \mathrm{d}\boldsymbol{x} = 1.$$

To this end, $\{w_\ell\}_{\ell=1}^m$ are imposed to be

$$w_1, \ldots, w_m \geq 0 \ \text{ and } \ \sum_{\ell=1}^m w_\ell = 1. \tag{15.1}$$

Fig. 15.2 illustrates an example of the Gaussian mixture model, which represents a multimodal distribution by linearly combining multiple Gaussian models.

15.2 MLE

The parameter $\boldsymbol{\theta}$ in the Gaussian mixture model is learned by MLE explained in Chapter 12. The likelihood is given by

$$L(\boldsymbol{\theta}) = \prod_{i=1}^n q(\boldsymbol{x}_i; \boldsymbol{\theta}), \tag{15.2}$$

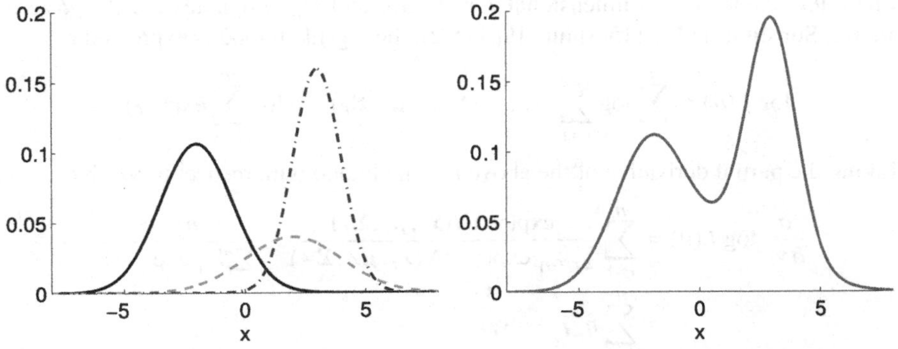

(a) Each Gaussian component (b) Gaussian mixture model

FIGURE 15.2

Example of Gaussian mixture model: $q(x) = 0.4N(x; -2, 1.5^2) + 0.2N(x; 2, 2^2) + 0.4N(x; 3, 1^2)$.

and MLE finds its maximizer with respect to $\boldsymbol{\theta}$. When the above likelihood is maximized for the Gaussian mixture model, the *constraints* given by Eq. (15.1) need to be satisfied:

$$\widehat{\boldsymbol{\theta}} = \operatorname*{argmax}_{\boldsymbol{\theta}} L(\boldsymbol{\theta})$$

$$\text{subject to} \quad w_1, \ldots, w_m \geq 0 \quad \text{and} \quad \sum_{\ell=1}^{m} w_\ell = 1.$$

Due to the constraints, the maximizer $\widehat{\boldsymbol{\theta}}$ cannot be simply obtained by setting the derivative of the likelihood to zero. Here, $w_1, \ldots, w_m$ are re-parameterized as

$$w_\ell = \frac{\exp(\gamma_\ell)}{\sum_{\ell'=1}^{m} \exp(\gamma_{\ell'})}, \tag{15.3}$$

and $\{\gamma_\ell\}_{\ell=1}^m$ are learned, which automatically fulfills Eq. (15.1).

The maximum likelihood solution $\widehat{\boldsymbol{\theta}}$ satisfies the following likelihood equation for $\log L(\boldsymbol{\theta})$:

$$\begin{cases} \left.\dfrac{\partial}{\partial \gamma_\ell} \log L(\boldsymbol{\theta})\right|_{\boldsymbol{\theta}=\widehat{\boldsymbol{\theta}}} = 0, \\ \left.\dfrac{\partial}{\partial \boldsymbol{\mu}_\ell} \log L(\boldsymbol{\theta})\right|_{\boldsymbol{\theta}=\widehat{\boldsymbol{\theta}}} = \mathbf{0}_d, \\ \left.\dfrac{\partial}{\partial \boldsymbol{\Sigma}_\ell} \log L(\boldsymbol{\theta})\right|_{\boldsymbol{\theta}=\widehat{\boldsymbol{\theta}}} = \boldsymbol{O}_{d\times d}, \end{cases} \tag{15.4}$$

where $\mathbf{0}_d$ denotes the d-dimensional zero vector and $\boldsymbol{O}_{d\times d}$ denotes the $d \times d$ zero matrix. Substituting Eq. (15.3) into Eq. (15.2), the log-likelihood is expressed as

$$\log L(\boldsymbol{\theta}) = \sum_{i=1}^{n} \log \sum_{\ell=1}^{m} \exp(\gamma_\ell) N(\boldsymbol{x}_i; \boldsymbol{\mu}_\ell, \boldsymbol{\Sigma}_\ell) - n \log \sum_{\ell=1}^{m} \exp(\gamma_\ell).$$

Taking the partial derivative of the above log-likelihood with respect to γ_ℓ gives

$$\begin{aligned}\frac{\partial}{\partial \gamma_\ell} \log L(\boldsymbol{\theta}) &= \sum_{i=1}^{n} \frac{\exp(\gamma_\ell) N(\boldsymbol{x}_i; \boldsymbol{\mu}_\ell, \boldsymbol{\Sigma}_\ell)}{\sum_{\ell'=1}^{m} \exp(\gamma_{\ell'}) N(\boldsymbol{x}_i; \boldsymbol{\mu}_{\ell'}, \boldsymbol{\Sigma}_{\ell'})} - \frac{n\gamma_\ell}{\sum_{\ell'=1}^{m} \exp(\gamma_{\ell'})} \\ &= \sum_{i=1}^{n} \eta_{i,\ell} - n w_\ell,\end{aligned}$$

where $\eta_{i,\ell}$ is defined as

$$\eta_{i,\ell} = \frac{w_\ell N(\boldsymbol{x}_i; \boldsymbol{\mu}_\ell, \boldsymbol{\Sigma}_\ell)}{\sum_{\ell'=1}^{m} w_{\ell'} N(\boldsymbol{x}_i; \boldsymbol{\mu}_{\ell'}, \boldsymbol{\Sigma}_{\ell'})}.$$

Similarly, taking the partial derivatives of the above log-likelihood with respect to $\boldsymbol{\mu}_\ell$ and $\boldsymbol{\Sigma}_\ell$ (see Fig. 12.3 for the derivative formulas) gives

$$\frac{\partial}{\partial \boldsymbol{\mu}_\ell} \log L(\boldsymbol{\theta}) = \sum_{i=1}^{n} \eta_{i,\ell} \boldsymbol{\Sigma}_\ell^{-1} (\boldsymbol{x}_i - \boldsymbol{\mu}_\ell),$$

$$\frac{\partial}{\partial \boldsymbol{\Sigma}_\ell} \log L(\boldsymbol{\theta}) = \frac{1}{2} \sum_{i=1}^{n} \eta_{i,\ell} \left(\boldsymbol{\Sigma}_\ell^{-1} (\boldsymbol{x}_i - \boldsymbol{\mu}_\ell)(\boldsymbol{x}_i - \boldsymbol{\mu}_\ell)^\top \boldsymbol{\Sigma}_\ell^{-1} - \boldsymbol{\Sigma}_\ell^{-1} \right).$$

Setting the above derivatives to zero shows that the maximum likelihood solution $\widehat{w}_\ell$, $\widehat{\boldsymbol{\mu}}_\ell$, and $\widehat{\boldsymbol{\Sigma}}_\ell$ should satisfy

$$\begin{cases} \widehat{w}_\ell = \dfrac{1}{n} \displaystyle\sum_{i=1}^{n} \widehat{\eta}_{i,\ell}, \\ \widehat{\boldsymbol{\mu}}_\ell = \dfrac{\sum_{i=1}^{n} \widehat{\eta}_{i,\ell} \boldsymbol{x}_i}{\sum_{i'=1}^{n} \widehat{\eta}_{i',\ell}}, \\ \widehat{\boldsymbol{\Sigma}}_\ell = \dfrac{\sum_{i=1}^{n} \widehat{\eta}_{i,\ell} (\boldsymbol{x}_i - \widehat{\boldsymbol{\mu}}_\ell)(\boldsymbol{x}_i - \widehat{\boldsymbol{\mu}}_\ell)^\top}{\sum_{i'=1}^{n} \widehat{\eta}_{i',\ell}}, \end{cases} \tag{15.5}$$

where $\widehat{\eta}_{i,\ell}$ is called the *responsibility* of the ℓth component for sample $\boldsymbol{x}_i$:

$$\widehat{\eta}_{i,\ell} = \frac{\widehat{w}_\ell N(\boldsymbol{x}_i; \widehat{\boldsymbol{\mu}}_\ell, \widehat{\boldsymbol{\Sigma}}_\ell)}{\sum_{\ell'=1}^{m} \widehat{w}_{\ell'} N(\boldsymbol{x}_i; \widehat{\boldsymbol{\mu}}_{\ell'}, \widehat{\boldsymbol{\Sigma}}_{\ell'})}. \tag{15.6}$$

In the above likelihood equation, variables are entangled in a complicated way and there is no known method to solve it analytically. Below, two *iterative algorithms* are introduced to numerically find a solution: the *gradient method* ad the *expectation-maximization (EM) algorithm*.

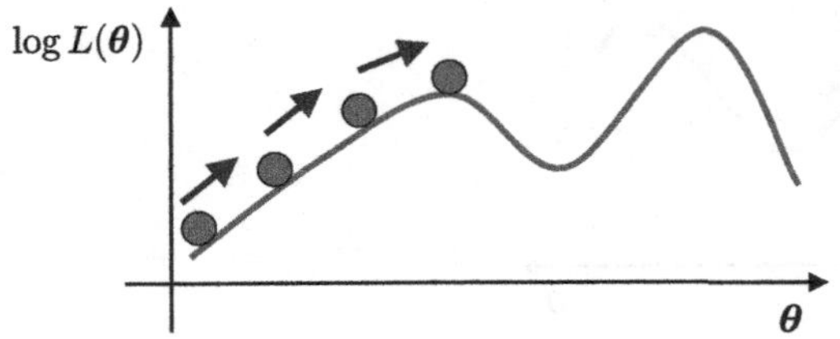

FIGURE 15.3

Schematic of gradient ascent.

1. Initialize the solution $\widehat{\boldsymbol{\theta}}$.
2. Compute the gradient of log-likelihood $\log L(\boldsymbol{\theta})$ at the current solution $\widehat{\boldsymbol{\theta}}$:

$$\frac{\partial}{\partial \boldsymbol{\theta}} \log L(\boldsymbol{\theta})\bigg|_{\boldsymbol{\theta}=\widehat{\boldsymbol{\theta}}}.$$

3. Update the parameter to go up the gradient:

$$\widehat{\boldsymbol{\theta}} \longleftarrow \widehat{\boldsymbol{\theta}} + \varepsilon \frac{\partial}{\partial \boldsymbol{\theta}} \log L(\boldsymbol{\theta})\bigg|_{\boldsymbol{\theta}=\widehat{\boldsymbol{\theta}}},$$

   where ε is a small positive scalar.
4. Iterate 2–3 until convergence.

FIGURE 15.4

Algorithm of gradient ascent.

15.3 GRADIENT ASCENT ALGORITHM

A *gradient method* is a generic and simple optimization approach that iteratively updates the parameter to go up (down in the case of minimization) the gradient of an *objective function* (Fig. 15.3). The algorithm of gradient ascent is summarized in Fig. 15.4. Under a mild assumption, a gradient ascent solution is guaranteed to be *local optimal*, which corresponds to a peak of a local mountain and the objective value cannot be increased by any local parameter update.

A stochastic variant of the gradient method is to randomly choose a sample and update the parameter to go up the gradient for the selected sample. Such a stochastic method, called the *stochastic gradient* algorithm, was also shown to produce to a local optimal solution [4].

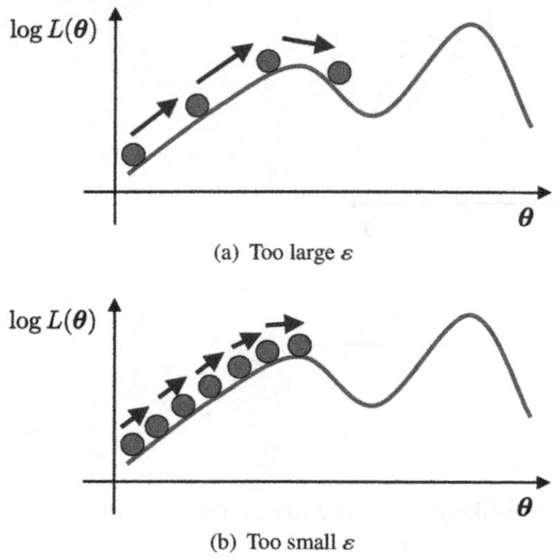

FIGURE 15.5

Step size ε in gradient ascent. The gradient flow can overshoot the peak if ε is large, while gradient ascent is slow if ε is too small.

Note that the (stochastic) gradient method does not only necessarily give the *global optimal solution* but also a local optimal solution, as illustrated in Fig. 15.3. Furthermore, its performance relies on the choice of the step size ε, which is not easy to determine in practice. If ε is large, gradient ascent is fast in the beginning, but the gradient flow can overshoot the peak (Fig. 15.5(a)). On the other hand, if ε is small, the peak may be found, but gradient ascent is slow in the beginning (Fig. 15.5(b)). To overcome this problem, starting from a large ε and then reducing ε gradually, called *simulated annealing*, would be useful. However, the choice of initial ε and the decreasing factor of ε is not straightforward in practice.

To mitigate the problem that only a local optimal solution can be found, it is practically useful to run the gradient algorithm multiple times from different initial solutions and choose the one that gives the best solution.

15.4 EM ALGORITHM

The difficulty of tuning the step size ε in the gradient method can be overcome by the EM algorithm [36]. The EM algorithm was originally developed for obtaining a maximum likelihood solution when input $\boldsymbol{x}$ is only partially observable. MLE for Gaussian mixture models can actually be regarded as learning from incomplete data, and the EM algorithm gives an efficient means to obtain a local optimal solution.

1. Initialize parameters $\{\widehat{w}_\ell, \widehat{\mu}_\ell, \widehat{\Sigma}_\ell\}_{\ell=1}^m$.
2. E-step: Compute responsibilities $\{\widehat{\eta}_{i,\ell}\}_{i=1,\ \ell=1}^{n \quad m}$ from current parameters $\{\widehat{w}_\ell, \widehat{\mu}_\ell, \widehat{\Sigma}_\ell\}_{\ell=1}^m$:

$$\widehat{\eta}_{i,\ell} \longleftarrow \frac{\widehat{w}_\ell N(\boldsymbol{x}_i; \widehat{\boldsymbol{\mu}}_\ell, \widehat{\boldsymbol{\Sigma}}_\ell)}{\sum_{\ell'=1}^m \widehat{w}_{\ell'} N(\boldsymbol{x}_i; \widehat{\boldsymbol{\mu}}_{\ell'}, \widehat{\boldsymbol{\Sigma}}_{\ell'})}.$$

3. M-step: Update parameters $\{\widehat{w}_\ell, \widehat{\mu}_\ell, \widehat{\Sigma}_\ell\}_{\ell=1}^m$ from current responsibilities $\{\widehat{\eta}_{i,\ell}\}_{i=1,\ \ell=1}^{n \quad m}$:

$$\widehat{w}_\ell \longleftarrow \frac{1}{n}\sum_{i=1}^n \widehat{\eta}_{i,\ell},$$

$$\widehat{\boldsymbol{\mu}}_\ell \longleftarrow \frac{\sum_{i=1}^n \widehat{\eta}_{i,\ell}\boldsymbol{x}_i}{\sum_{i'=1}^n \widehat{\eta}_{i',\ell}},$$

$$\widehat{\boldsymbol{\Sigma}}_\ell \longleftarrow \frac{\sum_{i=1}^n \widehat{\eta}_{i,\ell}(\boldsymbol{x}_i - \widehat{\boldsymbol{\mu}}_\ell)(\boldsymbol{x}_i - \widehat{\boldsymbol{\mu}}_\ell)^\top}{\sum_{i'=1}^n \widehat{\eta}_{i',\ell}},$$

4. Iterate 2–3 until convergence.

FIGURE 15.6

EM algorithm.

As summarized in Fig. 15.6, the EM algorithm consists of the E-step and the M-step, which correspond to updating the solution based on necessary condition (15.5) and computing its auxiliary variable (15.6) alternately.

The E-step and the M-step can be interpreted as follows:

E-step: A lower bound $b(\boldsymbol{\theta})$ of log-likelihood $\log L(\boldsymbol{\theta})$ that touches at current solution $\widehat{\boldsymbol{\theta}}$ is obtained:

$$\forall \boldsymbol{\theta}, \ \ \log L(\boldsymbol{\theta}) \geq b(\boldsymbol{\theta}), \ \ \text{and} \ \ \log L(\widehat{\boldsymbol{\theta}}) = b(\widehat{\boldsymbol{\theta}}).$$

Note that this lower-bounding step corresponds to computing the expectation over unobserved variables, which is why this step is called the E-step.

M-step: The maximizer $\widehat{\boldsymbol{\theta}}'$ of the lower bound $b(\boldsymbol{\theta})$ is obtained.

$$\widehat{\boldsymbol{\theta}}' = \underset{\boldsymbol{\theta}}{\text{argmax}}\ b(\boldsymbol{\theta}).$$

As illustrated in Fig. 15.7, iterating the E-step and the M-step increases the log-likelihood (precisely, the log-likelihood is monotone nondecreasing).

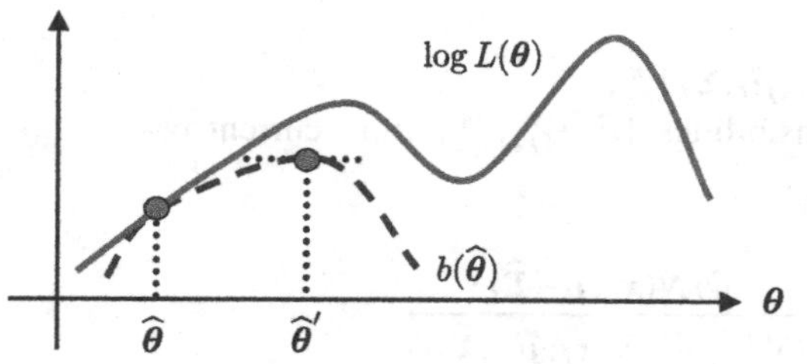

FIGURE 15.7

Maximizing the lower bound $b(\boldsymbol{\theta})$ of the log-likelihood $\log L(\boldsymbol{\theta})$.

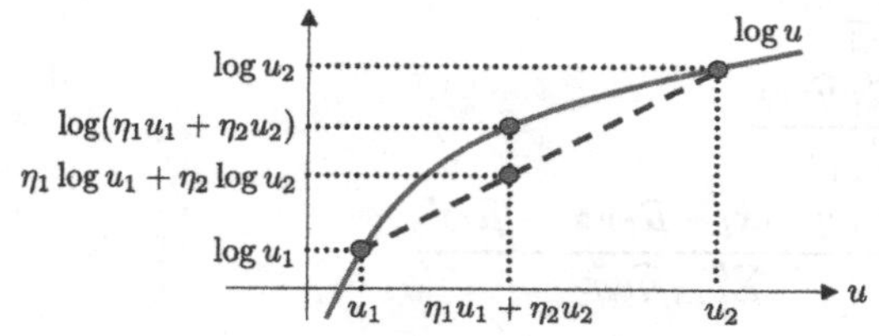

FIGURE 15.8

Jensen's inequality for $m = 2$. log is a concave function.

The lower bound in the E-step is derived based on *Jensen's inequality* explained in Section 8.3.1: for $\eta_1, \ldots, \eta_m \geq 0$ and $\sum_{\ell=1}^{m} \eta_\ell = 1$,

$$\log\left(\sum_{\ell=1}^{m} \eta_\ell u_\ell\right) \geq \sum_{\ell=1}^{m} \eta_\ell \log u_\ell. \tag{15.7}$$

For $m = 2$, Jensen's inequality is simplified as

$$\log(\eta_1 u_1 + \eta_2 u_2) \geq \eta_1 \log u_1 + \eta_2 \log u_2, \tag{15.8}$$

which can be intuitively understood by the *concavity* of the log function (see Fig. 15.8).

The log-likelihood $\log L(\boldsymbol{\theta})$ can be expressed by using the responsibility $\widehat{\eta}_{i,\ell}$ (see Eq. (15.6)) as

$$\begin{aligned}\log L(\boldsymbol{\theta}) &= \sum_{i=1}^{n} \log\left(\sum_{\ell=1}^{m} w_\ell N(\boldsymbol{x}_i; \boldsymbol{\mu}_\ell, \boldsymbol{\Sigma}_\ell)\right) \\ &= \sum_{i=1}^{n} \log\left(\sum_{\ell=1}^{m} \widehat{\eta}_{i,\ell} \frac{w_\ell N(\boldsymbol{x}_i; \boldsymbol{\mu}_\ell, \boldsymbol{\Sigma}_\ell)}{\widehat{\eta}_{i,\ell}}\right).\end{aligned} \tag{15.9}$$

By associating $w_\ell N(\boldsymbol{x}_i;\boldsymbol{\mu}_\ell,\boldsymbol{\Sigma}_\ell)/\widehat{\eta}_{i,\ell}$ in Eq. (15.9) with u_ℓ in Jensen's inequality (15.7), lower bound $b(\boldsymbol{\theta})$ of the log-likelihood $\log L(\boldsymbol{\theta})$ can be obtained as

$$\log L(\boldsymbol{\theta}) \geq \sum_{i=1}^{n}\sum_{\ell=1}^{m} \widehat{\eta}_{i,\ell} \log\left(\frac{w_\ell N(\boldsymbol{x}_i;\boldsymbol{\mu}_\ell,\boldsymbol{\Sigma}_\ell)}{\widehat{\eta}_{i,\ell}}\right) = b(\boldsymbol{\theta}).$$

This lower bound $b(\boldsymbol{\theta})$ touches $\log L(\boldsymbol{\theta})$ when $\boldsymbol{\theta} = \widehat{\boldsymbol{\theta}}$, because Eq. (15.6) implies

$$\begin{aligned} b(\widehat{\boldsymbol{\theta}}) &= \sum_{i=1}^{n}\left(\sum_{\ell=1}^{m} \widehat{\eta}_{i,\ell}\right) \log\left(\frac{\widehat{w}_\ell N(\boldsymbol{x}_i;\widehat{\boldsymbol{\mu}}_\ell,\widehat{\boldsymbol{\Sigma}}_\ell)}{\widehat{\eta}_{i,\ell}}\right) \\ &= \sum_{i=1}^{n} \log\left(\sum_{\ell'=1}^{m} \widehat{w}_{\ell'} N(\boldsymbol{x}_i;\widehat{\boldsymbol{\mu}}_{\ell'},\widehat{\boldsymbol{\Sigma}}_{\ell'})\right) \\ &= \log L(\widehat{\boldsymbol{\theta}}). \end{aligned}$$

The maximizer $\widehat{\boldsymbol{\theta}}'$ of the lower bound $b(\boldsymbol{\theta})$ in the M-step should satisfy

$$\begin{cases} \left.\dfrac{\partial}{\partial \gamma_\ell} b(\boldsymbol{\theta})\right|_{\boldsymbol{\theta}=\widehat{\boldsymbol{\theta}}'} = 0, \\ \left.\dfrac{\partial}{\partial \boldsymbol{\mu}_\ell} b(\boldsymbol{\theta})\right|_{\boldsymbol{\theta}=\widehat{\boldsymbol{\theta}}'} = \boldsymbol{0}_d, \\ \left.\dfrac{\partial}{\partial \boldsymbol{\Sigma}_\ell} b(\boldsymbol{\theta})\right|_{\boldsymbol{\theta}=\widehat{\boldsymbol{\theta}}'} = \boldsymbol{O}_{d\times d}, \end{cases}$$

from which the maximizer $\widehat{\boldsymbol{\theta}}'$ can be obtained as

$$\begin{cases} \widehat{w}'_\ell = \dfrac{1}{n}\displaystyle\sum_{i=1}^{n} \widehat{\eta}_{i,\ell}, \\ \widehat{\boldsymbol{\mu}}'_\ell = \dfrac{\sum_{i=1}^{n} \widehat{\eta}_{i,\ell}\boldsymbol{x}_i}{\sum_{i'=1}^{n} \widehat{\eta}_{i',\ell}}, \\ \widehat{\boldsymbol{\Sigma}}'_\ell = \dfrac{\sum_{i=1}^{n} \widehat{\eta}_{i,\ell}(\boldsymbol{x}_i - \widehat{\boldsymbol{\mu}}_\ell)(\boldsymbol{x}_i - \widehat{\boldsymbol{\mu}}_\ell)^\top}{\sum_{i'=1}^{n} \widehat{\eta}_{i',\ell}}. \end{cases}$$

The above explanation showed that the log-likelihood is monotone nondecreasing by iterating the E-step and the M-step. Furthermore, the EM algorithm was proved to produce a local optimal solution [120].

A MATLAB code for the EM algorithm is given in Fig. 15.9, and its behavior is illustrated in Fig. 15.10. Here, the mixture model of five Gaussian components is fitted to the mixture of two Gaussian distributions. As shown in Fig. 15.10, two of the five Gaussian components fit the true two Gaussian distributions well, and the remaining three Gaussian components are almost eliminated. Indeed, the learned

```
x=[2*randn(1,100)-5 randn(1,50); randn(1,100) randn(1,50)+3];
[d,n]=size(x);
m=5;
e=rand(n,m);
S=zeros(d,d,m);
for o=1:10000
  e=e./repmat(sum(e,2),[1 m]);
  g=sum(e);
  w=g/n;
  mu=(x*e)./repmat(g,[d 1]);
  for k=1:m
    t=x-repmat(mu(:,k),[1 n]);
    S(:,:,k)=(t.*repmat(e(:,k)',[d 1]))*t'/g(k);
    e(:,k)=w(k)*det(S(:,:,k))^(-1/2) ...
           *exp(-sum(t.*(S(:,:,k)\t))/2);
  end
  if o>1 && norm(w-w0)+norm(mu-mu0)+norm(S(:)-S0(:))<0.001
    break
  end
  w0=w;
  mu0=mu;
  S0=S;
end

figure(1); clf; hold on
plot(x(1,:),x(2,:),'ro');
v=linspace(0,2*pi,100);
for k=1:m
  [V,D]=eig(S(:,:,k));
  X=3*w(k)*V'*[cos(v)*D(1,1); sin(v)*D(2,2)];
  plot(mu(1,k)+X(1,:),mu(2,k)+X(2,:),'b-')
end
```

FIGURE 15.9

MATLAB code of EM algorithm for Gaussian mixture model.

mixing coefficients are given as

$$(\widehat{w}_1, \widehat{w}_2, \widehat{w}_3, \widehat{w}_4, \widehat{w}_5) = (0.09, 0.32, 0.05, 0.06, 0.49).$$

If the most responsible mixing component,

$$\widehat{y}_i = \underset{\ell}{\operatorname{argmax}}\, \widehat{\eta}_{i,\ell},$$

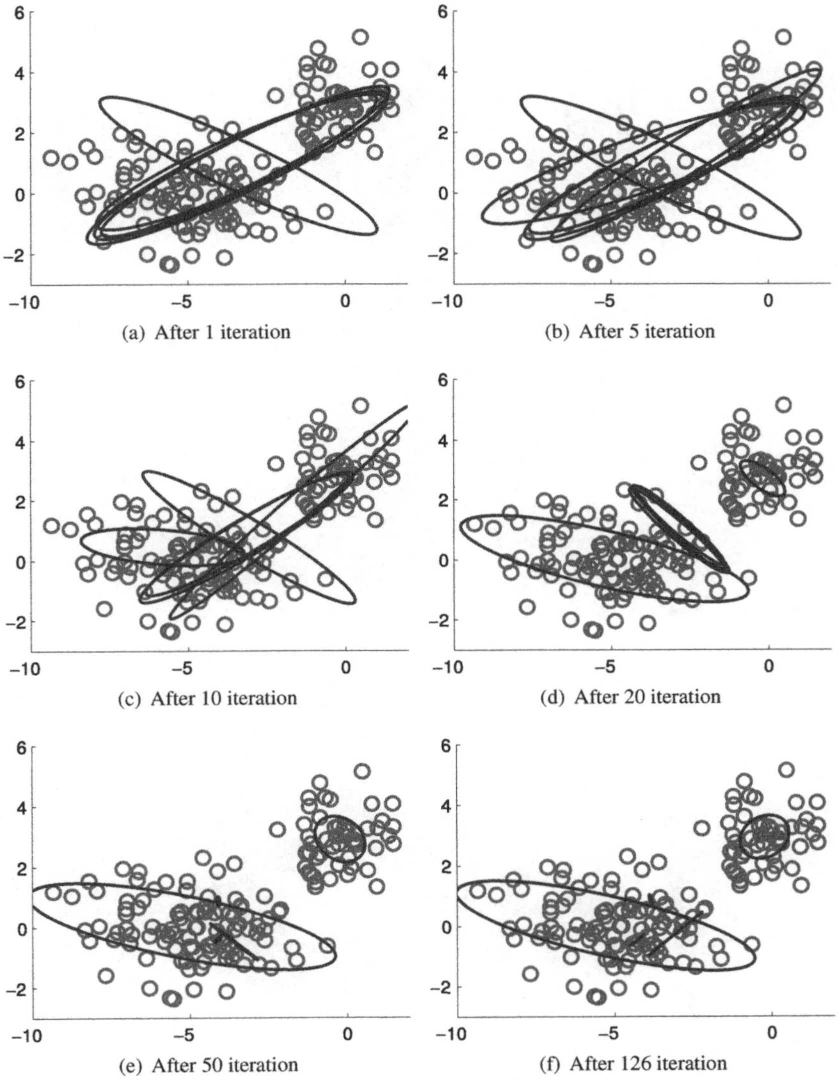

FIGURE 15.10

Example of EM algorithm for Gaussian mixture model. The size of ellipses is proportional to the mixing weights $\{w_\ell\}_{\ell=1}^m$.

is selected for each sample $\boldsymbol{x}_i$, density estimation with a mixture model can be regarded as *clustering*. Indeed, the EM algorithm for a Gaussian mixture model is reduced to the *k-means* clustering algorithm for $\boldsymbol{\Sigma}_\ell = \sigma_\ell^2 \boldsymbol{I}$. See Chapter 37 for details.

CHAPTER

NONPARAMETRIC ESTIMATION

16

CHAPTER CONTENTS

So far, estimation with parametric methods was discussed, which is useful when an appropriate parametric model is available (Fig. 16.1(a)). However, if the true probability density is highly complicated, it may be difficult to prepare an appropriate parametric model and then parametric methods do not work well (Fig. 16.1(b)). In this chapter, *nonparametric estimation* methods are introduced that do not use parametric models.

16.1 HISTOGRAM METHOD

The simplest nonparametric method would be the *histogram method*. In the histogram method, the pattern space $\mathcal{X}$ is partitioned into several *bins*. Then, in each bin, the probability density is approximated by a constant proportional to the number of training samples that fall into the bin. An example of the histogram method is illustrated in Fig. 16.2, and its MATLAB code is provided in Fig. 16.3. The graph shows that the profile of a complicated probability density function can be captured well by the histogram method.

However, the histogram method has several drawbacks:

- Probability densities become discontinuous across different bins.
- Appropriately determining the shape and size of bins is not straightforward (see Fig. 16.4).

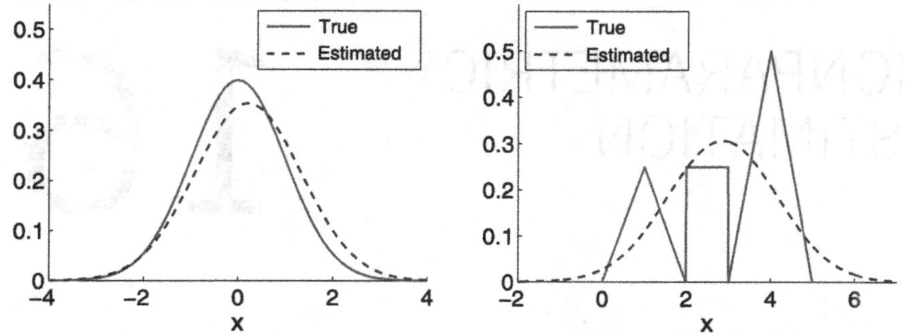

(a) When the true probability density is simple (b) When the true probability density is complicated

FIGURE 16.1

Examples of Gaussian MLE.

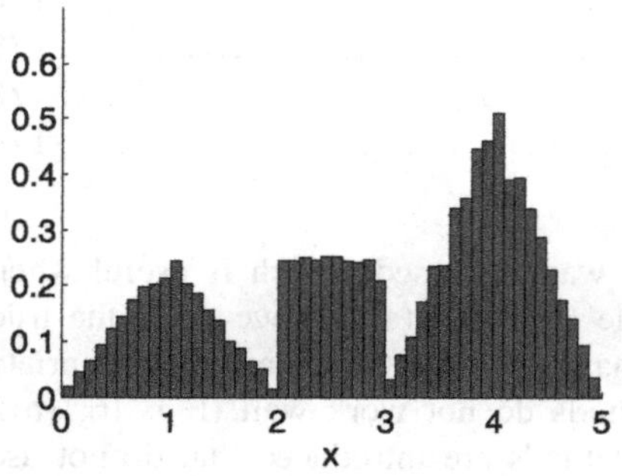

FIGURE 16.2

Example of histogram method.

- Naive splitting of the pattern space such as the equidistant grid exponentially grows the number of bins with respect to the input dimensionality d.

In this chapter, nonparametric methods that can address these issues are introduced.

16.2 PROBLEM FORMULATION

Let us consider the problem of estimating $p(\boldsymbol{x}')$, the value of probability density at point $\boldsymbol{x}'$. Let $\mathcal{R}$ be a region in the pattern space $\mathcal{X}$ that contains the point of interest

```
n=10000; x=myrand(n); s=0.1; b=[0:s:5];
figure(1); clf; hold on
a=histc(x,b); bar(b,a/s/n,'histc')
```

```
function x=myrand(n)

x=zeros(1,n); u=rand(1,n);
t=(0<=u & u<1/8); x(t)=sqrt(8*u(t));
t=(1/8<=u & u<1/4); x(t)=2-sqrt(2-8*u(t));
t=(1/4<=u & u<1/2); x(t)=1+4*u(t);
t=(1/2<=u & u<3/4); x(t)=3+sqrt(4*u(t)-2);
t=(3/4<=u & u<=1); x(t)=5-sqrt(4-4*u(t));
```

FIGURE 16.3

MATLAB code for inverse transform sampling (see Section 19.3.1) for probability density function shown in Fig. 16.1(b). The bottom function should be saved as "`myrand.m`."

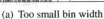

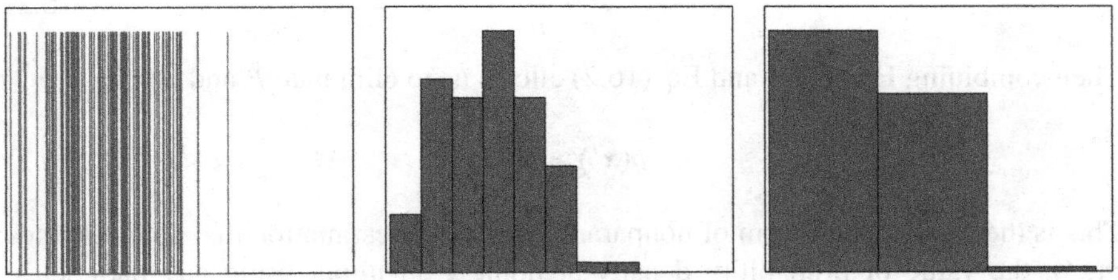

(a) Too small bin width (b) Appropriate bin width (c) Too large bin width

FIGURE 16.4

Choice of bin width in histogram method.

$\boldsymbol{x}'$, and let V be its *volume*:

$$V = \int_{\mathcal{R}} \mathrm{d}\boldsymbol{x}.$$

The probability P that a pattern $\boldsymbol{x}$ falls into region $\mathcal{R}$ is given by

$$P = \int_{\mathcal{R}} p(\boldsymbol{x}) \mathrm{d}\boldsymbol{x}.$$

These notations are summarized in Fig. 16.5.

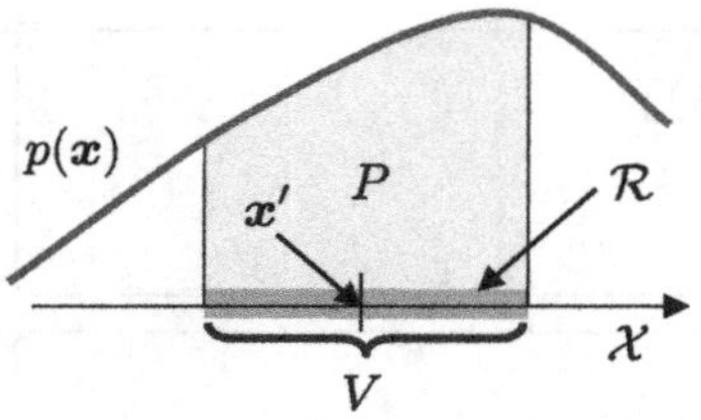

FIGURE 16.5

Notation of nonparametric methods.

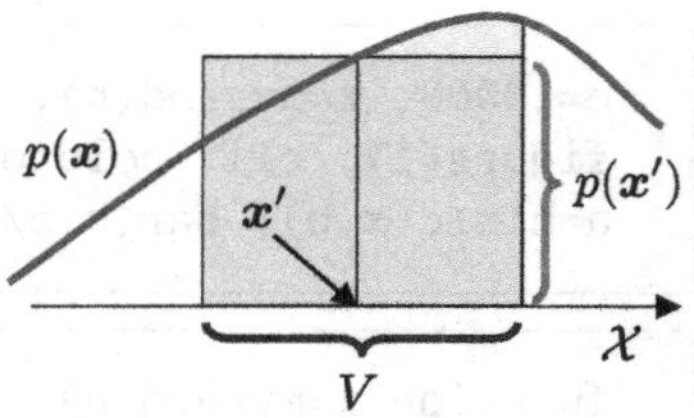

FIGURE 16.6

Probability P approximated by the size of rectangle.

The probability P may be approximated by two ways. One is to use the point of interest $\boldsymbol{x}'$ as follows (Fig. 16.6):

$$P \approx Vp(\boldsymbol{x}'). \tag{16.1}$$

The other is to use k, the number of training samples that fall into the region $\mathcal{R}$, as

$$P \approx \frac{k}{n}. \tag{16.2}$$

Then combining Eq. (16.1) and Eq. (16.2) allows us to eliminate P and obtain

$$p(\boldsymbol{x}') \approx \frac{k}{nV}. \tag{16.3}$$

This is the fundamental form of nonparametric density estimation that approximates $p(\boldsymbol{x}')$, the value of probability density at point $\boldsymbol{x}'$, without using any parametric model.

The accuracy of approximation (16.3) depends on the accuracy of Eq. (16.1) and Eq. (16.2), which can be controlled by the choice of region $\mathcal{R}$. Let us evaluate the accuracy of Eq. (16.1) and Eq. (16.2) in terms of the choice of $\mathcal{R}$. Eq. (16.1) is exact if the probability density is constant in the region $\mathcal{R}$. Thus, if the region $\mathcal{R}$ is smaller, Eq. (16.1) may be more accurate.

Next, let us evaluate the accuracy of Eq. (16.2). The probability that k points out of n training samples fall into the region $\mathcal{R}$ follows the *binomial distribution* (see Section 3.2). Its probability mass function is given by

$$\binom{n}{k} P^k (1-P)^{n-k}, \tag{16.4}$$

where $\binom{n}{k}$ is the binomial coefficient:

$$\binom{n}{k} = \frac{n!}{(n-k)!k!}.$$

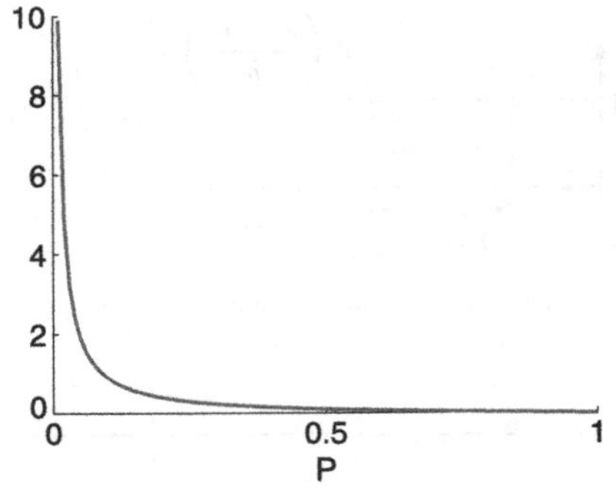

FIGURE 16.7

Normalized variance of binomial distribution.

The expectation and variance of binomial random variable k are given by

$$E[k] = nP \quad \text{and} \quad V[k] = nP(1 - P).$$

Then it can be easily confirmed that the expectation of k/n agrees with true P:

$$E\left[\frac{k}{n}\right] = P.$$

However, the fact that the expectation of k/n agrees with true P does not necessarily mean k/n is a good estimator of P. Indeed, if its variance is large, k/n may be a poor estimator of P. Here, let us normalize k by nP as

$$z = \frac{k}{nP},$$

so that the expectation of z is always 1 for any P:

$$E[z] = \frac{E[k]}{nP} = 1.$$

The variance of z is given by

$$V[z] = \frac{V[k]}{(nP)^2} = \frac{1 - P}{nP}.$$

If $V[z]$ is small, k/n will be a good approximator to P. Fig. 16.7 plots $V[z]$ as a function of P, showing that larger P gives smaller $V[z]$. P can be increased if region $\mathcal{R}$ is widen, and thus larger $\mathcal{R}$ makes Eq. (16.2) more accurate.

As explained above, Eq. (16.1) is more accurate if $\mathcal{R}$ is taken to be small, while Eq. (16.2) is more accurate if $\mathcal{R}$ is taken to be large. Thus, region $\mathcal{R}$ should be chosen

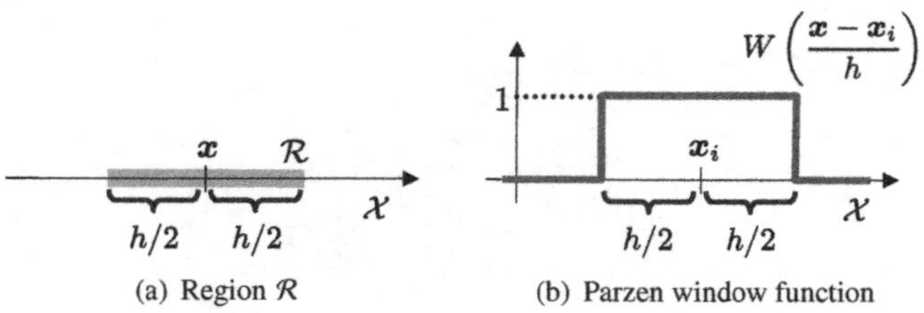

FIGURE 16.8

Parzen window method.

appropriately to improve the accuracy of approximation (16.3). In the following sections, two methods to determine region $\mathcal{R}$ based on training samples $\{x_i\}_{i=1}^n$ are introduced. In Section 16.3, the volume V of region $\mathcal{R}$ is fixed, and the number of training samples k that fall into $\mathcal{R}$ is determined from data. On the other hand, in Section 16.4, k is fixed, and the volume V of region $\mathcal{R}$ is determined from data.

16.3 KDE

In this section, the volume V of region $\mathcal{R}$ is fixed, and the number of training samples k that fall into $\mathcal{R}$ is determined from data.

16.3.1 PARZEN WINDOW METHOD

As region $\mathcal{R}$, let us consider the *hypercube* with edge length h centered at $\boldsymbol{x}$ in region $\mathcal{R}$ (Fig. 16.8(a)). Its volume V is given by

$$V = h^d, \tag{16.5}$$

where d is the dimensionality of the pattern space. The number of training samples falling into region $\mathcal{R}$ is expressed as

$$k = \sum_{i=1}^{n} W\left(\frac{\boldsymbol{x} - \boldsymbol{x}_i}{h}\right), \tag{16.6}$$

where $W(\boldsymbol{x})$ is called the *Parzen window function* defined for

$$\boldsymbol{x} = (x^{(1)}, \ldots, x^{(d)})^\top$$

as follows (Fig. 16.8(b)):

$$W(\boldsymbol{x}) = \begin{cases} 1 & \max_{i=1,\ldots,d} |x^{(i)}| \le \frac{1}{2}, \\ 0 & \text{otherwise.} \end{cases}$$

h is called the *bandwidth* of the Parzen window function.

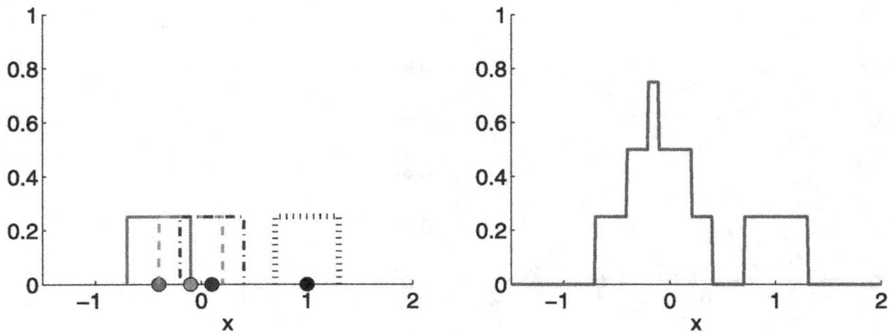

(a) Each Parzen window function

(b) Parzen window estimator

FIGURE 16.9

Example of Parzen window method.

Substituting Eq. (16.5) and Eq. (16.6) into Eq. (16.3) gives the following density estimator:

$$\widehat{p}_{\text{Parzen}}(\boldsymbol{x}) = \frac{1}{nh^d} \sum_{i=1}^{n} W\left(\frac{\boldsymbol{x} - \boldsymbol{x}_i}{h}\right).$$

This estimator called the *Parzen window method* and its numerical behavior are illustrated in Fig. 16.9. The result resembles that of the histogram method, but the bin widths are determined adaptively based on the training samples. However, discontinuity of estimated densities across different bins still remains in the Parzen window method.

16.3.2 SMOOTHING WITH KERNELS

The problem of discontinuity can be effectively overcome by KDE, which uses a smooth *kernel function* $K(\boldsymbol{x})$ instead of the Parzen window function:

$$\widehat{p}_{\text{KDE}}(\boldsymbol{x}) = \frac{1}{nh^d} \sum_{i=1}^{n} K\left(\frac{\boldsymbol{x} - \boldsymbol{x}_i}{h}\right).$$

Note that the kernel function should satisfy

$$\forall \boldsymbol{x} \in \mathcal{X},\ \ K(\boldsymbol{x}) \geq 0, \quad \text{and} \quad \int_{\mathcal{X}} K(\boldsymbol{x}) \mathrm{d}\boldsymbol{x} = 1.$$

The *Gaussian kernel* is a popular choice as a kernel function:

$$K(\boldsymbol{x}) = \frac{1}{(2\pi)^{\frac{d}{2}}} \exp\left(-\frac{\boldsymbol{x}^\top \boldsymbol{x}}{2}\right),$$

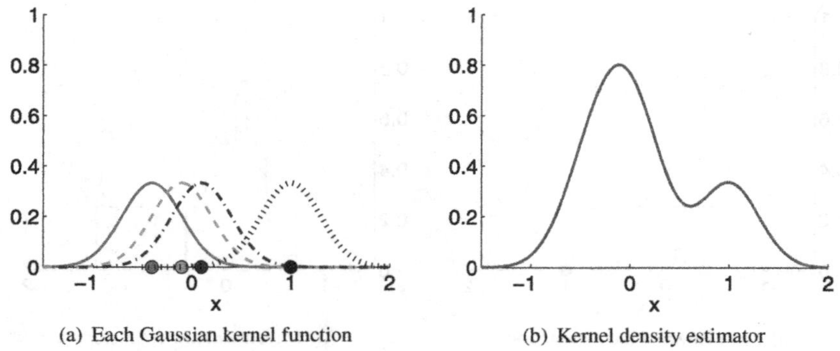
1
0.8
0.6
0.4
0.2
0
−1
0
1
2
x
(a) Each Gaussian kernel function
1
0.8
0.6
0.4
0.2
0
−1
0
1
2
x
(b) Kernel density estimator

FIGURE 16.10

Example of Gaussian KDE. Training samples are the same as those in Fig. 16.9.

where the bandwidth h corresponds to the standard deviation of the Gaussian density function. An example of Gaussian KDE is illustrated in Fig. 16.10, showing that a nice smooth density estimator is obtained.

A generalized KDE,

$$\widehat{p}_{\mathrm{KDE}}(\boldsymbol{x}) = \frac{1}{n\,\det(\boldsymbol{H})}\sum_{i=1}^{n} K\left(\boldsymbol{H}^{-1}(\boldsymbol{x}-\boldsymbol{x}_i)\right), \tag{16.7}$$

may also be considered, where $\boldsymbol{H}$ is the $d \times d$ positive definite matrix called the *bandwidth matrix*. If $K(\boldsymbol{x})$ is the Gaussian function, $\boldsymbol{H}\boldsymbol{H}^\top$ corresponds to the variance-covariance matrix of the Gaussian density function.

16.3.3 BANDWIDTH SELECTION

The estimator $\widehat{p}_{\mathrm{KDE}}(\boldsymbol{x})$ obtained by KDE depends on the bandwidth h (Fig. 16.11). Here, data-driven methods to choose h are introduced.

For generalized KDE (16.7), let us consider a diagonal bandwidth matrix $\boldsymbol{H}$:

$$\boldsymbol{h} = \mathrm{diag}\left(h^{(1)}, \ldots, h^{(d)}\right),$$

where d denotes the dimensionality of input $\boldsymbol{x}$. When the true probability distribution is Gaussian, the optimal bandwidth is given asymptotically as follows [90, 93]:

$$\widehat{h}^{(j)} = \left(\frac{4}{(d+2)n}\right)^{\frac{1}{d+4}} \sigma^{(j)},$$

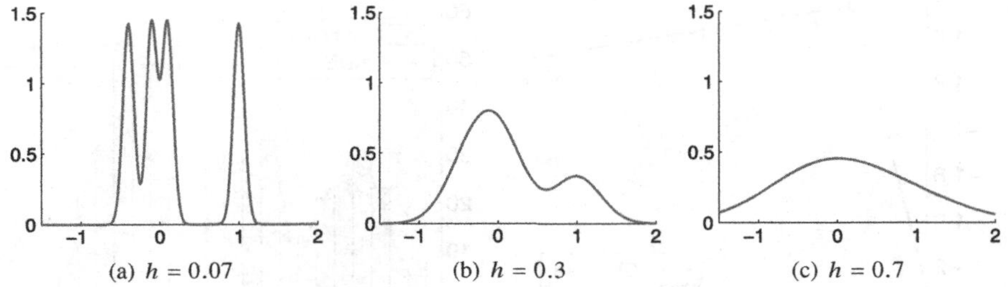

FIGURE 16.11

Choice of kernel bandwidth h in KDE.

```
n=500; x=myrand(n); x2=x.^2; hs=[0.01 0.1 0.5]; t=5;
d2=repmat(x2,[n 1])+repmat(x2',[1 n])-2*x'*x;
v=mod(randperm(n),t)+1;
for i=1:length(hs)
  hh=2*hs(i)^2; P=exp(-d2/hh)/sqrt(pi*hh);
  for j=1:t
    s(j,i)=mean(log(mean(P(v~=j,v==j))));
  end
end
[dum,a]=max(mean(s)); h=hs(a); hh=2*h^2;
ph=mean(exp(-d2/hh)/(sqrt(pi*hh)));
figure(1); clf; plot(x,ph,'r*'); h
```

FIGURE 16.12

MATLAB code for Gaussian KDE with bandwidth selected by likelihood cross validation. A random number generator "`myrand.m`" shown in Fig. 16.3 is used.

where $\sigma^{(j)}$ denotes the standard deviation of the jth element of $\boldsymbol{x}$. Since $\sigma^{(j)}$ may be unknown in practice, it is estimated from samples as

$$\widehat{\sigma}^{(j)} = \sqrt{\frac{1}{n-1}\sum_{i=1}^{n}\left(x_i^{(j)} - \frac{1}{n}\sum_{i=1}^{n} x_i^{(j)}\right)^2}.$$

This is called *Silverman's bandwidth selector*.

Although Silverman's bandwidth selector is easy to implement, its validity is only guaranteed when the true probability distribution is Gaussian. For more flexible model selection, *cross validation* explained in Section 14.4 is highly useful. A

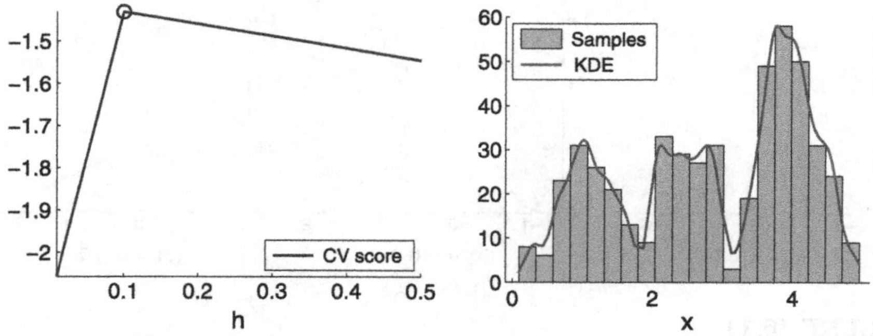

(a) CV score

(b) Estimated density

FIGURE 16.13

Example of Gaussian KDE with bandwidth selected by likelihood cross validation.

MATLAB code of likelihood cross validation for Gaussian KDE is provided in Fig. 16.12, and its behavior is illustrated in Fig. 16.13.

16.4 NNDE

In KDE explained above, the volume V of region $\mathcal{R}$ is fixed, and the number of training samples k that fall into region $\mathcal{R}$ is determined from data. In this section, an alternative method is introduced, where k is fixed and the volume V of region $\mathcal{R}$ is determined from data.

16.4.1 NEAREST NEIGHBOR DISTANCE

As region $\mathcal{R}$, let us consider the *hypersphere* with radius r centered at $\boldsymbol{x}$. Then the volume V of region $\mathcal{R}$ is given by

$$V = \frac{\pi^{\frac{d}{2}} r^d}{\Gamma(\frac{d}{2}+1)},$$

where $\Gamma(\cdot)$ is the *gamma function* (see Section 4.3).

Setting the radius r at the minimum number such that k training samples are included in the hypersphere, Eq. (16.3) immediately gives the following density estimator:

$$\widehat{p}_{\mathrm{KNN}}(\boldsymbol{x}) = \frac{k\Gamma(\frac{d}{2}+1)}{n\pi^{\frac{d}{2}} r^d}. \tag{16.8}$$

This is called NNDE.

k controls the smoothness of the density estimator, and it would be natural to increase k as the number of training samples n grows. Indeed, to guarantee

```
n=500; x=myrand(n); x2=x.^2;
ks=[10 50 100]; t=5; g=gamma(3/2);
d2=repmat(x2,[n 1])+repmat(x2',[1 n])-2*x'*x;
v=mod(randperm(n),t)+1;
for j=1:t
  S=sort(d2(v~=j,v==j));
  for i=1:length(ks)
    k=ks(i); r=sqrt(S(k+1,:));
    s(j,i)=mean(log(k*g./(sum(v~=j)*sqrt(pi)*r)));
  end
end
[dum,a]=max(mean(s)); k=ks(a);
m=1000; X=linspace(0,5,m);
D2=repmat(X.^2,[n 1])+repmat(x2',[1 m])-2*x'*X;
S=sort(D2); r=sqrt(S(k+1,:))'; Ph=k*g./(n*sqrt(pi)*r);
figure(1); clf; plot(X,Ph,'r*'); k
```

FIGURE 16.14

MATLAB code for NNDE with the number of nearest neighbors selected by likelihood cross validation. A random number generator "myrand.m" shown in Fig. 16.3 is used.

consistency of $\widehat{p}_{\mathrm{KNN}}(\boldsymbol{x})$ (i.e., $\widehat{p}_{\mathrm{KNN}}(\boldsymbol{x})$ converges to $p(\boldsymbol{x})$ as n tends to infinity), k should satisfy the following conditions [70]:

$$\lim_{n\to\infty} k = \infty \quad \text{and} \quad \lim_{n\to\infty} \frac{k}{n} = 0.$$

For example, $k = \sqrt{n}$ satisfy the above condition.

However, *likelihood cross validation* explained in Section 14.4 is more useful in practice. A MATLAB code of likelihood cross validation for NNDE is provided in Fig. 16.14, and its behavior is illustrated in Fig. 16.15.

16.4.2 NEAREST NEIGHBOR CLASSIFIER

Finally, NNDE is applied to estimating the class-conditional probability density $p(\boldsymbol{x}|y)$, and pattern recognition is performed based on the *MAP rule* (Section 11.3.1).

Pattern Recognition with Nearest Neighbor Distance

From Bayes' theorem (see Section 5.4), the class-posterior probability $p(y|\boldsymbol{x})$ is expressed as

$$p(y|\boldsymbol{x}) = \frac{p(\boldsymbol{x}|y)p(y)}{\sum_{y'=1}^{c} p(\boldsymbol{x}|y')p(y')} \propto p(\boldsymbol{x}|y)p(y).$$

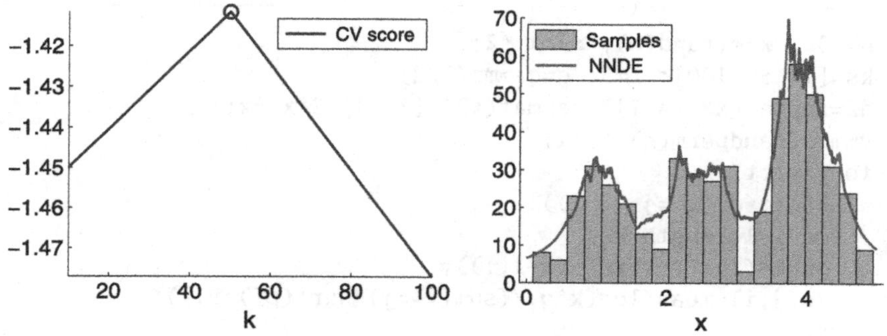

(a) CV score (b) Estimated density

FIGURE 16.15

Example of NNDE with the number of nearest neighbors selected by likelihood cross validation.

If NNDE with $k = 1$ is used, the class-conditional probability density $p(\boldsymbol{x}|y)$ is estimated as

$$p(\boldsymbol{x}|y) \approx \frac{\Gamma(\frac{d}{2}+1)}{n_y \pi^{\frac{d}{2}} r_y^d},$$

where r_y denotes the distance between input pattern $\boldsymbol{x}$ and the nearest training sample among training samples in class y, and n_y denotes the number of training samples in class y. If the class-prior probability $p(y)$ is approximated as

$$p(y) \approx \frac{n_y}{n},$$

the class-posterior probability $p(y|\boldsymbol{x})$ is approximated as

$$p(y|\boldsymbol{x}) \approx \frac{\Gamma(\frac{d}{2}+1)}{n_y \pi^{\frac{d}{2}} r_y^d} \frac{n_y}{n} \propto \frac{1}{r_y^d}.$$

This means that, to perform pattern recognition based on NNDE, only r_y^d is necessary. Since the class y that maximizes $1/r_y^d$ is the class that minimizes r_y, the input pattern $\boldsymbol{x}$ is simply classified into the same class as the nearest training sample in NNDE-based pattern recognition. This pattern recognition method is called the *nearest neighbor classifier*.

Fig. 16.16 illustrates an example of the nearest neighbor classifier, showing that the decision boundaries can be obtained from the *Voronoi diagram*.

k-Nearest Neighbor Classifier

The nearest neighbor classifier is intuitive and easy to implement, but it is not robust against *outliers*. For example, one of the training samples has an incorrect class label; the decision regions contain an isolated region (see Fig. 16.16).

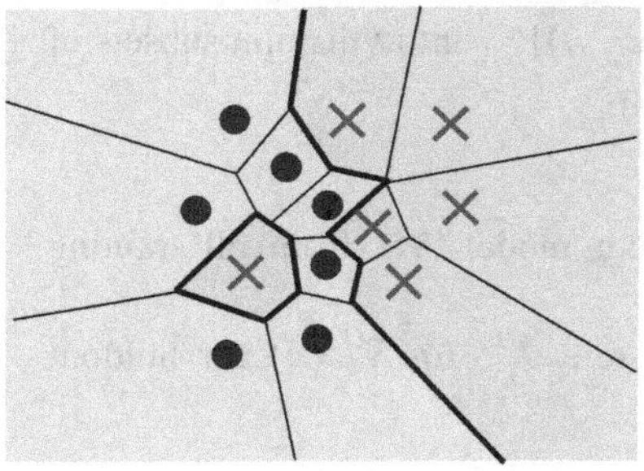

FIGURE 16.16

Example of nearest neighbor classifier.

To mitigate this problem, NNDE for $k > 1$ is useful:

$$\widehat{p}(\boldsymbol{x}|y) = \frac{k}{n_y V_y},$$

where n_y denotes the number of training samples in class y and V_y denotes the volume of the minimum hypersphere centered at $\boldsymbol{x}$ that contains k training samples.

However, the above method requires computation of hyperspheres for each class. A simpler implementation would be to find a hypersphere for training samples in all classes and to estimate the marginal probability $p(\boldsymbol{x})$ as

$$\widehat{p}(\boldsymbol{x}) = \frac{k}{nV}.$$

Since $p(\boldsymbol{x})$ can be decomposed as

$$p(\boldsymbol{x}) = \sum_{y=1}^{c} p(\boldsymbol{x}|y)p(y),$$

estimating the class-prior probability $p(y)$ by

$$\widehat{p}(y) = \frac{n_y}{n}$$

yields

$$\frac{k}{nV} \approx \sum_{y=1}^{c} p(\boldsymbol{x}|y)\frac{n_y}{n}.$$

Finally, the class-posterior probability $(\boldsymbol{x}|y)$ is estimated as

$$\widehat{p}(\boldsymbol{x}|y) = \frac{k_y}{n_y V}, \tag{16.9}$$

1. Split labeled training samples $\mathcal{Z} = \{(\boldsymbol{x}_i, y_i)\}_{i=1}^n$ into t disjoint subsets of (approximately) the same size: $\{\mathcal{Z}_\ell\}_{\ell=1}^t$.
2. For each model candidate $\mathcal{M}_j$
   (a) For each split $\ell = 1, \ldots, t$
      i. Obtain a classifier $\widehat{y}_j^{(\ell)}(\boldsymbol{x})$ using model $\mathcal{M}_j$ from all training samples without $\mathcal{Z}_\ell$.
      ii. Compute the misclassification rate $J_j^{(\ell)}$ for $\widehat{y}_j^{(\ell)}(\boldsymbol{x})$ for holdout samples $\mathcal{Z}_\ell$:
      $$J_j^{(\ell)} = \frac{1}{|\mathcal{Z}_\ell|} \sum_{(\boldsymbol{x}', y') \in \mathcal{Z}_\ell} I(\widehat{y}_j^{(\ell)}(\boldsymbol{x}') \neq y'),$$
      where $|\mathcal{Z}_\ell|$ denotes the number of elements in the set $\mathcal{Z}_\ell$ and $I(e) = 1$ if condition e is true and $I(e) = 0$ otherwise.
   (b) Compute the average misclassification rate J_j over all t splits:
   $$J_j = \frac{1}{t} \sum_{\ell=1}^{t} J_j^{(\ell)}.$$
3. Choose the model $\mathcal{M}_{\widehat{j}}$ that minimizes the average misclassification rate:
$$\widehat{j} = \underset{j}{\operatorname{argmin}}\, J_j.$$
4. Obtain a classifier using chosen model $\mathcal{M}_{\widehat{j}}$, from all training samples $\{(\boldsymbol{x}_i, y_i)\}_{i=1}^n$.

FIGURE 16.17

Algorithm of cross validation for misclassification rate.

where k_y denotes the number of training samples in the hypersphere that belong to class y. This allows us to construct a classifier in a simple way, which is called the *k-nearest neighbor classifier*.

Cross Validation for Misclassification Rate

The tuning parameter k in the k-nearest neighbor classifier can be chosen by *likelihood cross validation* for NNDE. However, in the context of pattern recognition, it would be more direct to choose k by cross validation in terms of the *misclassification rate*. The algorithm of cross validation for the misclassification rate is summarized in Fig. 16.17.

```
load digit.mat X T; [d,m,c]=size(X); X=reshape(X,[d m*c]);
Y=reshape(repmat([1:c],[m 1]),[1 m*c]);
ks=[1:10]; t=5; v=mod(randperm(m*c),t)+1;
for i=1:t
  Yh=knn(X(:,v~=i),Y(v~=i),X(:,v==i),ks);
  s(i,:)=mean(Yh~=repmat(Y(v==i),[length(ks) 1]),2);
end
[dum,a]=min(mean(s)); k=ks(a); [d,r,c]=size(T);
T=reshape(T,[d r*c]); U=reshape(knn(X,Y,T,k),[r c]);
for i=1:c, C(:,i)=sum(U==i); end, C, sum(diag(C))/sum(sum(C))
```

```
function U=knn(X,Y,T,ks)
m=size(T,2); D2=repmat(sum(T.^2,1),[size(X,2) 1]);
D2=D2+repmat(sum(X.^2,1)',[1 m])-2*X'*T; [dum,z]=sort(D2,1);
for i=1:length(ks)
  k=ks(i);
  for j=1:m
    Z=sort(Y(z(1:k,j))); g=find([1 Z(1:end-1)~=Z(2:end)]);
    [dum,a]= max([g(2:end) k+1]-g); U(i,j)=Z(g(a));
end, end
```

FIGURE 16.18

MATLAB code for k-nearest neighbor classifier with k chosen by cross validation. The bottom function should be saved as "knn.m."

Predicted class (columns) / True class (rows)

	1	2	3	4	5	6	7	8	9	0
1	200	0	0	0	0	0	0	0	0	0
2	0	193	1	0	0	0	1	4	1	0
3	0	0	195	0	3	0	0	1	1	0
4	0	0	0	191	1	2	0	0	6	0
5	0	3	4	0	187	0	1	1	2	2
6	0	2	0	0	2	195	0	0	0	1
7	0	0	1	2	0	0	192	2	3	0
8	0	1	4	1	3	0	0	186	2	3
9	0	0	0	3	0	0	1	1	195	0
0	0	1	1	0	0	0	0	0	0	198

FIGURE 16.19

Confusion matrix for 10-class classification by k-nearest neighbor classifier. $k = 1$ was chosen by cross validation for misclassification rate. The correct classification rate is 1932/2000 = 96.6%.

A MATLAB code for 10-class hand-written digit recognition (see Section 12.5) by the k-nearest neighbor classifier with k chosen by cross validation in terms of the misclassification rate is provided in Fig. 16.18. The obtained *confusion matrix* is shown in Fig. 16.19, where $k = 1$ was chosen by cross validation. The confusion matrix shows that 1932 test samples out of 2000 samples are correctly classified, meaning that the correct classification rate is

$$1932/2000 = 96.6\%.$$

On the other hand, as shown in Section 12.5, the correct classification rate for the same data set by FDA is

$$1798/2000 = 89.9\%.$$

Therefore, for this hand-written digit recognition experiment, the k-nearest neighbor classifier works much better than FDA.

CHAPTER

BAYESIAN INFERENCE 17

CHAPTER CONTENTS

In the framework of MLE, parameter $\boldsymbol{\theta}$ in parametric model $q(\boldsymbol{x};\boldsymbol{\theta})$ was treated as a deterministic variable. In this chapter, *Bayesian inference* [15] is introduced which handles parameter $\boldsymbol{\theta}$ as a random variable.

17.1 BAYESIAN PREDICTIVE DISTRIBUTION

In this section, the basic idea of Bayesian inference is explained.

17.1.1 DEFINITION

If $\boldsymbol{\theta}$ is regarded as a random variable, the following probabilities can be determined:

$$p(\boldsymbol{\theta}),\ p(\boldsymbol{\theta}|\mathcal{D}),\ p(\mathcal{D}|\boldsymbol{\theta}),\quad \text{and}\quad p(\mathcal{D},\boldsymbol{\theta}),$$

where

$$\mathcal{D} = \{\boldsymbol{x}_i\}_{i=1}^{n}.$$

$p(\boldsymbol{\theta}|\mathcal{D})$ is called the *posterior probability* of parameter $\boldsymbol{\theta}$ given training samples $\mathcal{D}$, while $p(\boldsymbol{\theta})$ is called the *prior probability* of $\boldsymbol{\theta}$ before observing training samples $\mathcal{D}$. $p(\mathcal{D}|\boldsymbol{\theta})$ denotes the *likelihood*, which is the same quantity as the one used in MLE (see Section 12.1), but it is regarded as a conditional probability in the Bayesian framework:

$$p(\mathcal{D}|\boldsymbol{\theta}) = \prod_{i=1}^{n} q(\boldsymbol{x}_i|\boldsymbol{\theta}).$$

Note that the parametric model $q(\boldsymbol{x}|\boldsymbol{\theta})$ is also represented as a conditional probability in the Bayesian framework. $p(\mathcal{D},\boldsymbol{\theta})$ denotes the *joint probability* of training samples $\mathcal{D}$ and parameter $\boldsymbol{\theta}$.

The joint probability $p(\mathcal{D},\boldsymbol{\theta})$ can be expressed as

$$p(\mathcal{D},\boldsymbol{\theta}) = p(\mathcal{D}|\boldsymbol{\theta})p(\boldsymbol{\theta}),$$

and its marginalization over $\boldsymbol{\theta}$ gives

$$\int p(\mathcal{D},\boldsymbol{\theta})\mathrm{d}\boldsymbol{\theta} = p(\mathcal{D}).$$

Thus, the marginal probability $p(\mathcal{D})$ can be expressed as

$$p(\mathcal{D}) = \int \left(\prod_{i=1}^{n} q(\boldsymbol{x}_i|\boldsymbol{\theta})\right) p(\boldsymbol{\theta})\mathrm{d}\boldsymbol{\theta}.$$

The solution of Bayesian inference $\widehat{p}_{\mathrm{Bayes}}(\boldsymbol{x})$, called the *Bayesian predictive distribution*, is given as the expectation of model $q(\boldsymbol{x}|\boldsymbol{\theta})$ over the posterior probability $p(\boldsymbol{\theta}|\mathcal{D})$:

$$\widehat{p}_{\mathrm{Bayes}}(\boldsymbol{x}) = \int q(\boldsymbol{x}|\boldsymbol{\theta})p(\boldsymbol{\theta}|\mathcal{D})\mathrm{d}\boldsymbol{\theta}. \tag{17.1}$$

Since the posterior probability $p(\boldsymbol{\theta}|\mathcal{D})$ can be expressed by using *Bayes' theorem* (see Section 5.4) as

$$\begin{aligned} p(\boldsymbol{\theta}|\mathcal{D}) &= \frac{p(\mathcal{D}|\boldsymbol{\theta})p(\boldsymbol{\theta})}{p(\mathcal{D})} \\ &= \frac{\prod_{i=1}^{n} q(\boldsymbol{x}_i|\boldsymbol{\theta})p(\boldsymbol{\theta})}{\int \prod_{i=1}^{n} q(\boldsymbol{x}_i|\boldsymbol{\theta}')p(\boldsymbol{\theta}')\mathrm{d}\boldsymbol{\theta}'}, \end{aligned} \tag{17.2}$$

the Bayesian predictive distribution $\widehat{p}_{\mathrm{Bayes}}(\boldsymbol{x})$ can be expressed as

$$\widehat{p}_{\mathrm{Bayes}}(\boldsymbol{x}) = \int q(\boldsymbol{x}|\boldsymbol{\theta}) \frac{\prod_{i=1}^{n} q(\boldsymbol{x}_i|\boldsymbol{\theta})p(\boldsymbol{\theta})}{\int \prod_{i=1}^{n} q(\boldsymbol{x}_i|\boldsymbol{\theta}')p(\boldsymbol{\theta}')\mathrm{d}\boldsymbol{\theta}'}\mathrm{d}\boldsymbol{\theta}. \tag{17.3}$$

This shows that the Bayesian predictive distribution $\widehat{p}_{\mathrm{Bayes}}(\boldsymbol{x})$ can actually be computed *without any learning*, if parametric model $q(\boldsymbol{x}|\boldsymbol{\theta})$ and prior probability $p(\boldsymbol{\theta})$ are specified.

17.1.2 COMPARISON WITH MLE

In MLE, unknown true probability density $p(\boldsymbol{x})$ is approximated by a parametric model $q(\boldsymbol{x};\boldsymbol{\theta})$ with a single parameter estimate $\widehat{\boldsymbol{\theta}}_{\mathrm{ML}}$. On the other hand, in Bayesian inference, infinitely many parameters are simultaneously considered and their

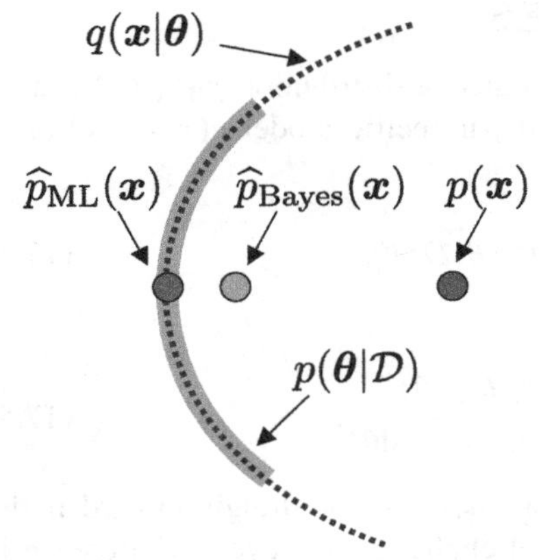

FIGURE 17.1

Bayes vs. MLE. The maximum likelihood solution $\widehat{p}_{\text{ML}}$ is always confined in the parametric model $q(\boldsymbol{x};\boldsymbol{\theta})$, while the Bayesian predictive distribution $\widehat{p}_{\text{Bayes}}(\boldsymbol{x})$ generally pops out from the model.

average over model $q(\boldsymbol{x}|\boldsymbol{\theta})$ weighted according to the posterior probability $p(\boldsymbol{\theta}|\mathcal{D})$ is used as a density estimator. Let us intuitively explain the difference between Bayesian inference and MLE using Fig. 17.1.

A parametric model $q(\boldsymbol{x}|\boldsymbol{\theta})$ is a set of probability density functions and it is denoted by a dotted line in Fig. 17.1. In practice, parametric model $q(\boldsymbol{x}|\boldsymbol{\theta})$ is more or less *misspecified*, meaning that the true probability density $p(\boldsymbol{x})$ is not exactly included in the parametric model $q(\boldsymbol{x}|\boldsymbol{\theta})$. MLE finds the probability density function in the parametric model that maximizes the likelihood, $\widehat{p}_{\text{ML}}(\boldsymbol{x})$, which is equivalent to finding the projection of $p(\boldsymbol{x})$ onto the model under the empirical KL divergence (see Section 14.2). On the other hand, in Bayesian inference, by taking the expectation of parametric model $q(\boldsymbol{x}|\boldsymbol{\theta})$ over the posterior probability $p(\boldsymbol{\theta}|\mathcal{D})$, the solution $\widehat{p}_{\text{Bayes}}(\boldsymbol{x})$ is not generally confined in the model. In the illustration in Fig. 17.1, the solution $\widehat{p}_{\text{Bayes}}(\boldsymbol{x})$ pops out from the model to the right-hand side, and consequently it is closer to the true probability density $p(\boldsymbol{x})$ than $\widehat{p}_{\text{ML}}(\boldsymbol{x})$.

The fundamental difference between Bayesian inference and MLE lies in whether parameter $\boldsymbol{\theta}$ is handled as a deterministic or random variable. However, in reality, more significant philosophical difference is involved. More specifically, prior probability $p(\boldsymbol{\theta})$ can contain subjective knowledge in Bayesian inference, which can arbitrarily change the solution. On the other hand, MLE is objective and its solution is purely computed from data. When non-Bayesian inference is contrasted to Bayesian inference, it is sometimes referred to as *frequentist inference*.

17.1.3 COMPUTATIONAL ISSUES

As explained in Section 17.1, the Bayesian predictive distribution $\widehat{p}_{\text{Bayes}}(\boldsymbol{x})$ can be computed without any learning in principle, if parametric model $q(\boldsymbol{x}|\boldsymbol{\theta})$ and prior probability $p(\boldsymbol{\theta})$ are specified:

$$\widehat{p}_{\text{Bayes}}(\boldsymbol{x}) = \int q(\boldsymbol{x}|\boldsymbol{\theta})p(\boldsymbol{\theta}|\mathcal{D})\mathrm{d}\boldsymbol{\theta}, \tag{17.4}$$

where

$$p(\boldsymbol{\theta}|\mathcal{D}) = \frac{\prod_{i=1}^{n} q(\boldsymbol{x}_i|\boldsymbol{\theta})p(\boldsymbol{\theta})}{\int \prod_{i=1}^{n} q(\boldsymbol{x}_i|\boldsymbol{\theta}')p(\boldsymbol{\theta}')\mathrm{d}\boldsymbol{\theta}'}. \tag{17.5}$$

However, computation of the two integrations above is not straightforward if the dimension of $\boldsymbol{\theta}$ is high. Thus, a main technical challenge in Bayesian inference is how to efficiently compute high-dimensional integrations.

To easily handle the integration in Eq. (17.4), it is preferable to obtain the posterior probability $p(\boldsymbol{\theta}|\mathcal{D})$ *analytically*. One possibility is to choose the prior probability $p(\boldsymbol{\theta})$ so that the parametric form of the posterior probability $p(\boldsymbol{\theta}|\mathcal{D})$ can be explicitly obtained. Such a prior choice will be explained in Section 17.2. For nonconjugate choice of prior probabilities, analytic approximation techniques of the integration in Eq. (17.5) will be discussed in Chapter 18.

For handling the integration in Eq. (17.4), i.e., the expectation of parametric model $q(\boldsymbol{x}|\boldsymbol{\theta})$ over posterior probability $p(\boldsymbol{\theta}|\mathcal{D})$, the simplest approximation scheme would be to use a single point $\widehat{\boldsymbol{\theta}}$ taken from the posterior probability. Such a single-point approximation will be introduced in Section 17.3. Techniques for numerically approximating the posterior expectation will be discussed in Chapter 19.

17.2 CONJUGATE PRIOR

As discussed above, it is convenient to analytically handle the integration in Eq. (17.5). If the prior probability $p(\boldsymbol{\theta})$ is chosen so that the posterior probability $p(\boldsymbol{\theta}|\mathcal{D})$ takes the same parametric form as the prior probability $p(\boldsymbol{\theta})$, the posterior probability $p(\boldsymbol{\theta}|\mathcal{D})$ can be analytically obtained just by specifying its parameters. Such a prior choice is called a *conjugate prior* for the likelihood $p(\mathcal{D}|\boldsymbol{\theta})$.

Let us illustrate an example of the conjugate prior for the Gaussian model with expectation 0 and variance σ^2, where the inverse of the variance $\tau = \sigma^{-2}$, called the *precision*, is regarded as a parameter (i.e., the parameter is $\theta = \tau$). Then the parametric model is expressed as

$$q(x|\tau) = \sqrt{\frac{\tau}{2\pi}} \exp\left(-\frac{\tau x^2}{2}\right). \tag{17.6}$$

For this model, let us employ the *gamma distribution* (see Section 4.3) as the prior probability for precision τ:

$$p(\tau; \alpha, \beta) \propto \tau^{\alpha-1} e^{-\beta\tau}. \tag{17.7}$$

For parametric model (17.6) combined with prior probability (17.7), the posterior probability is again the gamma distribution:

$$\begin{aligned} p(\tau|\mathcal{D}) &\propto \prod_{i=1}^{n} q(x_i|\tau)p(\tau;\alpha,\beta) \\ &\propto \tau^{n/2}\exp\left(-\frac{\tau}{2}\sum_{i=1}^{n} x_i^2\right)\tau^{\alpha-1}e^{-\beta\tau} \\ &= \tau^{\widetilde{\alpha}-1}e^{-\widetilde{\beta}\tau}, \end{aligned}$$

where the posterior parameters $\widetilde{\alpha}$ and $\widetilde{\beta}$ are given as

$$\widetilde{\alpha} = \alpha + \frac{n}{2} \quad \text{and} \quad \widetilde{\beta} = \beta + \frac{\sum_{i=1}^{n} x_i^2}{2}.$$

Thus, the posterior probability can be obtained analytically just by computing the posterior parameters $\widetilde{\alpha}$ and $\widetilde{\beta}$.

As shown above, conjugate priors are extremely useful from the viewpoint of computation. However, conjugate priors depend on the parametric form of likelihood $p(\mathcal{D}|\boldsymbol{\theta})$, and they may not be available depending on $p(\mathcal{D}|\boldsymbol{\theta})$. Furthermore, the meaning of choosing conjugate priors is not clear from the viewpoint of statistical inference. For general nonconjugate priors, analytic approximation techniques of the posterior probability will be introduced in Chapter 18.

17.3 MAP ESTIMATION

Given the posterior probability $p(\boldsymbol{\theta}|\mathcal{D})$ analytically, the next step is to compute the posterior expectation:

$$\widehat{p}_{\mathrm{Bayes}}(\boldsymbol{x}) = \int q(\boldsymbol{x}|\boldsymbol{\theta})p(\boldsymbol{\theta}|\mathcal{D})\mathrm{d}\boldsymbol{\theta}.$$

In this section, MAP estimation is introduced, which approximates the above integration by a single point $\widehat{\boldsymbol{\theta}}_{\mathrm{MAP}}$:

$$\widehat{p}_{\mathrm{MAP}}(\boldsymbol{x}) = q(\boldsymbol{x}|\widehat{\boldsymbol{\theta}}_{\mathrm{MAP}}),$$

where $\widehat{\boldsymbol{\theta}}_{\mathrm{MAP}}$ is the maximizer of the posterior probability $p(\boldsymbol{\theta}|\mathcal{D})$ (see Fig. 17.2):

$$\widehat{\boldsymbol{\theta}}_{\mathrm{MAP}} = \underset{\boldsymbol{\theta}}{\mathrm{argmax}}\, p(\boldsymbol{\theta}|\mathcal{D}).$$

Since MAP estimation approximates the target density by a single point $\widehat{\boldsymbol{\theta}}_{\mathrm{MAP}}$, its property is actually close to MLE. Indeed, the MAP solution $\widehat{\boldsymbol{\theta}}_{\mathrm{MAP}}$ can be expressed as

$$\widehat{\boldsymbol{\theta}}_{\mathrm{MAP}} = \underset{\boldsymbol{\theta}}{\mathrm{argmax}}\left(\sum_{i=1}^{n} \log q(\boldsymbol{x}_i|\boldsymbol{\theta}) + \log p(\boldsymbol{\theta})\right), \tag{17.8}$$

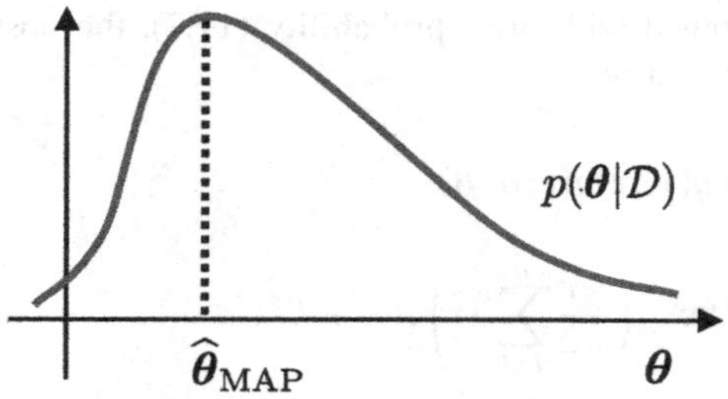

FIGURE 17.2

MAP estimation.

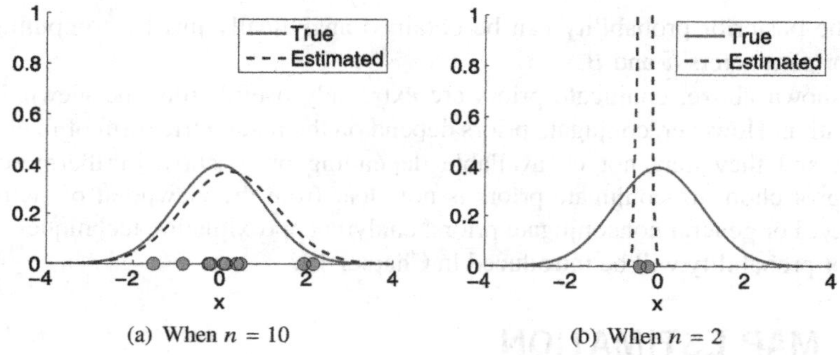

FIGURE 17.3

Example of MLE for Gaussian model. When the number of training samples, n, is small, MLE tends to overfit the samples.

and thus it actually minimizes the sum of the log-likelihood and an additional term $\log p(\theta)$.

As explained in Chapter 13, MLE tends to *overfit* the training samples if the sample size is small (Fig. 17.3). The additional term $\log p(\theta)$ in Eq. (17.8) can work as a penalty to mitigate overfitting. For this reason, MAP estimation is also referred to as *penalized MLE*. MAP estimation tries to increase not only the likelihood but also the prior probability, and therefore the solution tends to be biased toward the parameter having a larger prior probability. Penalizing the objective function in this way is also called *regularization* (see Chapter 23 for details).

Let us specifically compute the MAP solution for the Gaussian model with expectation μ and variance-covariance matrix I_d (i.e., the parameter is $\theta = \mu$):

$$q(x|\mu) = \frac{1}{(2\pi)^{\frac{d}{2}}} \exp\left(-\frac{(x-\mu)^\top(x-\mu)}{2}\right).$$

Let us consider the following Gaussian prior:

$$p(\boldsymbol{\mu};\beta) = \frac{1}{(2\pi\beta^2)^{\frac{d}{2}}}\exp\left(-\frac{\boldsymbol{\mu}^\top\boldsymbol{\mu}}{2\beta^2}\right), \tag{17.9}$$

which prefers $\boldsymbol{\mu}$ closer to the origin. For this setup, the penalized log-likelihood is given by

$$\begin{aligned}\mathrm{PL}(\boldsymbol{\mu}) &= \sum_{i=1}^{n}\log q(\boldsymbol{x}_i|\boldsymbol{\mu}) + \log p(\boldsymbol{\mu}) \\ &= -\frac{nd}{2}\log(2\pi) - \frac{1}{2}\sum_{i=1}^{n}\|\boldsymbol{x}_i-\boldsymbol{\mu}\|^2 - \frac{d}{2}\log(2\pi\beta^2) - \frac{1}{2\beta^2}\|\boldsymbol{\mu}\|^2.\end{aligned}$$

Taking the derivative of $\mathrm{PL}(\boldsymbol{\mu})$ and setting it to zero yield

$$\frac{\partial}{\partial\boldsymbol{\mu}}\mathrm{PL}(\boldsymbol{\mu}) = \sum_{i=1}^{n}(\boldsymbol{x}_i-\boldsymbol{\mu}) - \frac{1}{\beta^2}\boldsymbol{\mu} = \boldsymbol{0}_d,$$

from which the MAP solution $\widehat{\boldsymbol{\mu}}_{\mathrm{MAP}}$ is obtained as

$$\widehat{\boldsymbol{\mu}}_{\mathrm{MAP}} = \frac{1}{n+\beta^{-2}}\sum_{i=1}^{n}\boldsymbol{x}_i.$$

On the other hand, the maximum likelihood solution $\widehat{\boldsymbol{\mu}}_{\mathrm{MLE}}$ for this model is given by

$$\widehat{\boldsymbol{\mu}}_{\mathrm{MLE}} = \frac{1}{n}\sum_{i=1}^{n}\boldsymbol{x}_i.$$

$\beta > 0$ implies

$$\left|\widehat{\mu}^{(j)}_{\mathrm{MAP}}\right| = \frac{1}{n+\beta^{-2}}\left|\sum_{i=1}^{n}x_i^{(j)}\right| < \frac{1}{n}\left|\sum_{i=1}^{n}x_i^{(j)}\right| = \left|\widehat{\mu}^{(j)}_{\mathrm{MLE}}\right|,$$

where $x_i^{(j)}$, $\widehat{\mu}^{(j)}_{\mathrm{MAP}}$, and $\widehat{\mu}^{(j)}_{\mathrm{MLE}}$ denote the jth elements of vectors $\boldsymbol{x}_i$, $\widehat{\boldsymbol{\mu}}_{\mathrm{MAP}}$, and $\widehat{\boldsymbol{\mu}}_{\mathrm{MLE}}$, respectively. Thus, the MAP solution $\widehat{\boldsymbol{\mu}}_{\mathrm{MAP}}$ is always closer to the origin than the maximum likelihood solution $\widehat{\boldsymbol{\mu}}_{\mathrm{MLE}}$, which the Gaussian prior (17.9) favors.

A MATLAB code for penalized MLE with one-dimensional Gaussian model is given in Fig. 17.4, and its behavior is illustrated in Fig. 17.5. This demonstrates that, if the prior probability is chosen properly (i.e., $\beta \approx 1$), MAP estimation can give a better solution than MLE.

In MAP estimation, the maximizer of the posterior probability (i.e., the *mode*) was used. An alternative idea is to use the mean of the posterior probability:

$$\widehat{p}(\boldsymbol{x}) = q(\boldsymbol{x}|\overline{\boldsymbol{\theta}}),$$

```
n=12; mu=0.5; x=randn(n,1)+mu;
bs=[0.01:0.01:3]; bl=length(bs);
MLE=mean(x);
for i=1:bl
  MAP(i)=sum(x)/(n+bs(i).^(-2));
end

figure(1); clf; hold on;
plot(bs,mu*ones(1,bl),'k:');
plot(bs,MAP,'r-');
plot(bs,MLE*ones(1,bl),'b-.');
xlabel('\beta'); ylabel('\mu');
legend('True','MAP','MLE',4);
```

FIGURE 17.4

MATLAB code for penalized MLE with one-dimensional Gaussian model.

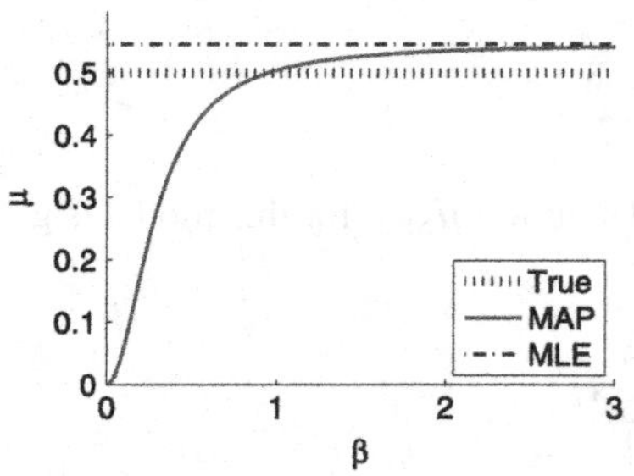

FIGURE 17.5

Example of MAP estimation with one-dimensional Gaussian model.

where

$$\overline{\theta} = \int \theta p(\theta|\mathcal{D})\mathrm{d}\theta.$$

If the posterior probability $p(\theta|\mathcal{D})$ is a popular probability distribution, its expectation may be known analytically and then the posterior expectation $\overline{\theta}$ can also be obtained analytically.

However, single-point approximation of the Bayesian predictive distribution $\widehat{p}_{\mathrm{Bayes}}(x)$ loses the distinctive feature of Bayesian inference that the solution $\widehat{p}_{\mathrm{Bayes}}(x)$

can pop out from the parametric model $q(\boldsymbol{x}|\boldsymbol{\theta})$ (see Fig. 17.1). To enjoy Bayesianity, it is essential to compute the integration at least approximately. In Chapter 19, techniques for numerically approximating the posterior expectation will be discussed.

17.4 BAYESIAN MODEL SELECTION

In Bayesian inference, the prior probability of parameters is utilized. If such prior knowledge is not available, the prior probability has to be determined by a user. Since the solution of Bayesian inference depends on the choice of the prior probability (see Fig. 17.5), it must be determined in an objective and appropriate way. In this section, choice of prior probabilities and models in the Bayesian framework is addressed.

Suppose that the prior probability is parameterized by $\boldsymbol{\beta}$:

$$p(\boldsymbol{\theta}; \boldsymbol{\beta}),$$

where $\boldsymbol{\beta}$ is called the *hyperparameter* and this should be distinguished from ordinary parameter $\boldsymbol{\theta}$ in the parametric model $q(\boldsymbol{x}|\boldsymbol{\theta})$. As explained in Section 14.3, model selection of MLE is possible by *cross validation*, and it can also be applied to Bayesian inference. Below, an alternative approach that is specific to Bayesian inference is introduced.

MLE is aimed at setting parameter $\boldsymbol{\theta}$ so that training samples $\{\boldsymbol{x}\}_{i=1}^{n}$ at hand are most typically generated. The fundamental idea of Bayesian model selection is to apply MLE to hyperparameter $\boldsymbol{\beta}$. More specifically, the probability that training samples $\mathcal{D} = \{\boldsymbol{x}\}_{i=1}^{n}$ at hand are generated is expressed as

$$p(\mathcal{D}; \boldsymbol{\beta}) = \int \prod_{i=1}^{n} q(\boldsymbol{x}_i|\boldsymbol{\theta}) p(\boldsymbol{\theta}; \boldsymbol{\beta}) \mathrm{d}\boldsymbol{\theta} = \mathrm{ML}(\boldsymbol{\beta}). \tag{17.10}$$

Eq. (17.10) viewed as a function of $\boldsymbol{\beta}$ is called the *marginal likelihood*, which is also referred to as the *evidence* and its negative log is called the *free energy*.

The method of determining hyperparameter $\boldsymbol{\beta}$ so that the marginal likelihood is maximized is called *empirical Bayes*, *type-II MLE*, or *evidence maximization*:

$$\boldsymbol{\beta}_{EB} = \underset{\boldsymbol{\beta}}{\mathrm{argmax}}\, \mathrm{ML}(\boldsymbol{\beta}).$$

In addition to the hyperparameter, parametric model $q(\boldsymbol{x}|\boldsymbol{\theta})$ may also be selected by empirical Bayes. That is, among a set of model candidates, the one that maximizes the marginal likelihood may be selected as the most promising one.

Let us specifically compute the marginal likelihood for one-dimensional Gaussian parametric model with expectation μ and variance 1,

$$q(x|\mu) = \frac{1}{\sqrt{2\pi}} \exp\left(-\frac{(x-\mu)^2}{2}\right), \tag{17.11}$$

and a Gaussian prior probability with expectation 0 and variance β^2:

$$p(\mu;\beta) = \frac{1}{\sqrt{2\pi\beta^2}} \exp\left(-\frac{\mu^2}{2\beta^2}\right). \tag{17.12}$$

The marginal likelihood can be expressed as

$$\begin{aligned}\mathrm{ML}(\beta) &= \int \prod_{i=1}^{n} q(x_i|\mu)p(\mu;\beta)\mathrm{d}\mu \\ &= (2\pi)^{-\frac{n}{2}}(2\pi\beta^2)^{-\frac{1}{2}} \int \exp\left(-\frac{1}{2}\sum_{i=1}^{n}(x_i-\mu)^2 - \frac{\mu^2}{2\beta^2}\right)\mathrm{d}\mu.\end{aligned}$$

For

$$\widehat{\mu}_{\mathrm{MAP}} = \frac{1}{n+\beta^{-2}}\sum_{i=1}^{n} x_i,$$

completing the square (see Eq. (4.2)) yields

$$\begin{aligned}\mathrm{ML}(\beta) &= (2\pi)^{-\frac{n}{2}}(2\pi\beta^2)^{-\frac{1}{2}} \exp\left(-\frac{1}{2}\sum_{i=1}^{n}(x_i-\widehat{\mu}_{\mathrm{MAP}})^2 - \frac{\widehat{\mu}_{\mathrm{MAP}}^2}{2\beta^2}\right) \\ &\quad\times \int \exp\left(-\frac{(\mu-\widehat{\mu}_{\mathrm{MAP}})^2}{2(n+\beta^{-2})^{-1}}\right)\mathrm{d}\mu \\ &= (2\pi)^{-\frac{n}{2}}(n\beta^2+1)^{-\frac{1}{2}} \exp\left(-\frac{1}{2}\sum_{i=1}^{n}(x_i-\widehat{\mu}_{\mathrm{MAP}})^2 - \frac{1}{2\beta^2}\widehat{\mu}_{\mathrm{MAP}}^2\right),\end{aligned}$$

where *Gaussian integral* shown in Fig. 4.1,

$$\int \exp\left(-\frac{(\mu-\widehat{\mu}_{\mathrm{MAP}})^2}{2(n+\beta^{-2})^{-1}}\right)\mathrm{d}\mu = \left(\frac{2\pi}{n+\beta^{-2}}\right)^{\frac{1}{2}},$$

is used. Consequently, the log marginal likelihood is expressed as

$$\begin{aligned}\log \mathrm{ML}(\beta) &= -\frac{n}{2}\log 2\pi - \frac{1}{2}\log(n\beta^2+1) \\ &\quad - \frac{1}{2}\sum_{i=1}^{n}(x_i-\widehat{\mu}_{\mathrm{MAP}})^2 - \frac{\widehat{\mu}_{\mathrm{MAP}}^2}{2\beta^2}.\end{aligned}$$

A MATLAB code of empirical Bayes for parametric model (17.11) and prior probability (17.12) is given in Fig. 17.6, and its behavior is illustrated in Fig. 17.7. In this example, $\beta_{EB} = 2.19$ is chosen as the empirical Bayes solution, and true $\mu = 0.5$ was estimated by the MAP method as $\widehat{\mu}_{\mathrm{MAP}} = 0.537$, which is slightly better than MLE $\widehat{\mu}_{\mathrm{MLE}} = 0.546$.

```
n=12; mu=0.5; x=randn(n,1)+mu;
bs=[0.01:0.01:3]; bl=length(bs);
MLE=mean(x);
for i=1:bl
  bb=bs(i)^(-2); MAP(i)=sum(x)/(n+bb);
  logML(i)=-n/2*log(2*pi)-sum((x-MAP(i)).^2)/2 ...
           -MAP(i)^2/(2*bb)-log(n*bb+1);
end
[dummy,c]=max(logML);

figure(1); clf; hold on;
plot(bs,logML,'r-');
plot(bs(c),logML(c),'ro');
xlabel('\beta'); legend('log ML',4);

figure(2); clf; hold on;
plot(bs,mu*ones(1,bl),'k:','LineWidth',5);
plot(bs,MAP,'r-','LineWidth',2);
plot(bs,MLE*ones(1,bl),'b-.','LineWidth',2);
plot(bs(c),MAP(c),'ro','LineWidth',4,'MarkerSize',10);
xlabel('\beta'); ylabel('\mu');
legend('True','MAP','MLE',4);
```

FIGURE 17.6

MATLAB code for empirical Bayes.

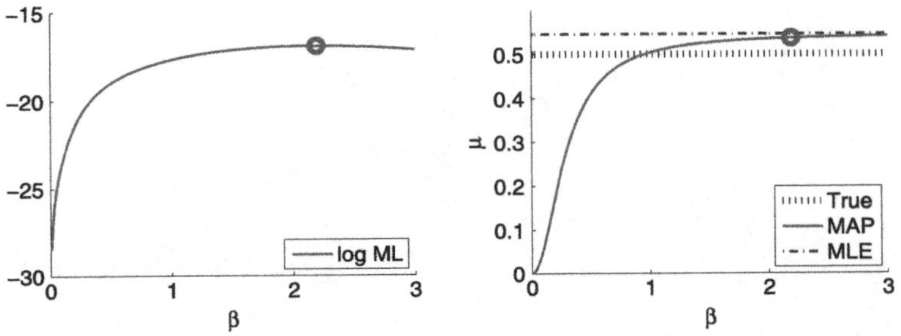

(a) Log marginal likelihood (b) MLE vs. MAP

FIGURE 17.7

Example of empirical Bayes.

CHAPTER

ANALYTIC APPROXIMATION OF MARGINAL LIKELIHOOD 18

CHAPTER CONTENTS

As discussed in Section 17.2, the use of conjugate priors allows us to avoid the explicit computation of the integration in the marginal likelihood:

$$\mathrm{ML}(\boldsymbol{\beta}) = \int \prod_{i=1}^{n} q(\boldsymbol{x}_i|\boldsymbol{\theta})p(\boldsymbol{\theta};\boldsymbol{\beta})\mathrm{d}\boldsymbol{\theta}.$$

In this chapter, general choices of prior probabilities are considered and analytic approximation methods of the marginal likelihood are introduced.

18.1 LAPLACE APPROXIMATION

In this section, *Laplace approximation* is introduced, which allows us to analytically approximate the integration of any twice-differentiable non-negative function $f(\boldsymbol{x})$:

$$\int f(\boldsymbol{\theta})\mathrm{d}\boldsymbol{\theta}.$$

18.1.1 APPROXIMATION WITH GAUSSIAN DENSITY

Let $\widehat{\boldsymbol{\theta}}$ be the maximizer of $f(\boldsymbol{\theta})$ with respect to $\boldsymbol{\theta}$:

$$\widehat{\boldsymbol{\theta}} = \underset{\boldsymbol{\theta}}{\mathrm{argmax}}\, f(\boldsymbol{\theta}).$$

Let us apply the *Taylor series expansion* (Fig. 2.8) to $\log f(\boldsymbol{\theta})$ about the maximizer $\widehat{\boldsymbol{\theta}}$:

$$\log f(\boldsymbol{\theta}) = \log f(\widehat{\boldsymbol{\theta}}) + (\boldsymbol{\theta} - \widehat{\boldsymbol{\theta}})^\top \frac{\partial}{\partial \boldsymbol{\theta}} \log f(\boldsymbol{\theta})\bigg|_{\boldsymbol{\theta}=\widehat{\boldsymbol{\theta}}} + \frac{1}{2}(\boldsymbol{\theta} - \widehat{\boldsymbol{\theta}})^\top \boldsymbol{H}(\boldsymbol{\theta} - \widehat{\boldsymbol{\theta}}) + \cdots, \tag{18.1}$$

where $\boldsymbol{H}$ is the *Hessian matrix*:

$$\boldsymbol{H} = \frac{\partial^2}{\partial \boldsymbol{\theta} \partial \boldsymbol{\theta}^\top} \log f(\boldsymbol{\theta})\bigg|_{\boldsymbol{\theta}=\widehat{\boldsymbol{\theta}}}.$$

Since $\widehat{\boldsymbol{\theta}}$ is the maximizer of $\log f(\boldsymbol{\theta})$,

$$\frac{\partial}{\partial \boldsymbol{\theta}} \log f(\boldsymbol{\theta})\bigg|_{\boldsymbol{\theta}=\widehat{\boldsymbol{\theta}}} = \boldsymbol{0}_b$$

holds and thus the first-order term in Eq. (18.1) is zero.

Let $\log \widehat{f}(\boldsymbol{\theta})$ be Eq. (18.1) up to the second-order terms (Fig. 18.1):

$$\log \widehat{f}(\boldsymbol{\theta}) = \log f(\widehat{\boldsymbol{\theta}}) + \frac{1}{2}(\boldsymbol{\theta} - \widehat{\boldsymbol{\theta}})^\top \boldsymbol{H}(\boldsymbol{\theta} - \widehat{\boldsymbol{\theta}}).$$

Exponentiating both sides yields

$$\widehat{f}(\boldsymbol{\theta}) = f(\widehat{\boldsymbol{\theta}}) \exp\left(\frac{1}{2}(\boldsymbol{\theta} - \widehat{\boldsymbol{\theta}})^\top \boldsymbol{H}(\boldsymbol{\theta} - \widehat{\boldsymbol{\theta}})\right).$$

Recalling that the integration of the normal density is 1,

$$\frac{1}{(2\pi)^{\frac{b}{2}} \det(-\boldsymbol{H})^{\frac{1}{2}}} \int \exp\left(-\frac{1}{2}(\boldsymbol{\theta} - \widehat{\boldsymbol{\theta}})^\top(-\boldsymbol{H})(\boldsymbol{\theta} - \widehat{\boldsymbol{\theta}})\right) \mathrm{d}\boldsymbol{\theta} = 1,$$

integration of $\widehat{f}(\boldsymbol{\theta})$ yields

$$\int \widehat{f}(\boldsymbol{\theta}) \mathrm{d}\boldsymbol{\theta} = f(\widehat{\boldsymbol{\theta}}) \sqrt{\frac{(2\pi)^b}{\det(-\boldsymbol{H})}} \approx \int f(\boldsymbol{\theta}) \mathrm{d}\boldsymbol{\theta},$$

where b denotes the dimensionality of $\boldsymbol{\theta}$. This is Laplace approximation to $\int f(\boldsymbol{\theta}) \mathrm{d}\boldsymbol{\theta}$.

Since approximating $\log f(\boldsymbol{\theta})$ by a quadratic function corresponds to approximating $f(\boldsymbol{\theta})$ by an unnormalized Gaussian function, Laplace approximation is quite accurate if $f(\boldsymbol{\theta})$ is close to Gaussian. For this reason, Laplace approximation is also referred to as *Gaussian approximation*.

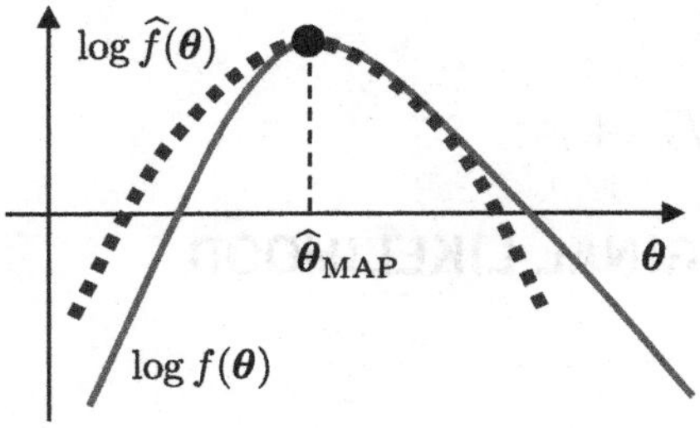

FIGURE 18.1

Laplace approximation.

18.1.2 ILLUSTRATION

Let us illustrate how to compute Laplace approximation to $\int f(\theta)\mathrm{d}\theta$, where

$$f(\theta) = N(\theta; 0, 1^2) + N(\theta; 0, 2^2),$$
$$N(\theta; \mu, \sigma^2) = \frac{1}{\sqrt{2\pi\sigma^2}} \exp\left(-\frac{(\theta-\mu)^2}{2\sigma^2}\right).$$

Note that $N(\theta; \mu, \sigma^2)$ is a normal density and thus the true value is 2.

The maximizer of $f(\theta)$ is given by

$$\widehat{\theta} = \underset{\theta}{\mathrm{argmax}}\, f(\theta) = 0.$$

Then

$$f(\widehat{\theta}) = \frac{1}{\sqrt{2\pi}} \exp\left(-\frac{\widehat{\theta}^2}{2}\right) + \frac{1}{\sqrt{8\pi}} \exp\left(-\frac{\widehat{\theta}^2}{8}\right) = \frac{3}{2\sqrt{2\pi}},$$

$$f'(\widehat{\theta}) = \frac{\partial}{\partial\theta} \log f(\theta)\bigg|_{\theta=\widehat{\theta}} = -\widehat{\theta}N(\widehat{\theta}; 0, 1^2) - \frac{\widehat{\theta}}{4}N(\widehat{\theta}; 0, 2^2) = 0,$$

$$f''(\widehat{\theta}) = \frac{\partial^2}{\partial\theta^2} \log f(\theta)\bigg|_{\theta=\widehat{\theta}} = (\widehat{\theta}^2 - 1)N(\widehat{\theta}; 0, 1^2) + \left(\frac{\widehat{\theta}^2}{16} - \frac{1}{4}\right)N(\widehat{\theta}; 0, 2^2)$$
$$= -\frac{9}{8\sqrt{2\pi}},$$

which yield

$$H = \frac{\partial^2}{\partial\theta^2} \log f(\theta)\bigg|_{\theta=\widehat{\theta}} = \frac{f''(0)f(0) - f'(0)^2}{f(0)^2} = -\frac{3}{4}.$$

Thus, the Laplace approximation is given by

$$f(0)\sqrt{\frac{2\pi}{-H}} = \sqrt{3} \approx 1.732.$$

18.1.3 APPLICATION TO MARGINAL LIKELIHOOD APPROXIMATION

For the marginal likelihood,

$$\mathrm{ML}(\boldsymbol{\beta}) = \int \prod_{i=1}^{n} q(\boldsymbol{x}_i|\boldsymbol{\theta})p(\boldsymbol{\theta};\boldsymbol{\beta})\mathrm{d}\boldsymbol{\theta}.$$

Laplace approximation with respect to $\boldsymbol{\theta}$ yields

$$\mathrm{ML}(\boldsymbol{\beta}) \approx \prod_{i=1}^{n} q(\boldsymbol{x}_i|\widehat{\boldsymbol{\theta}}_{\mathrm{MAP}})p(\widehat{\boldsymbol{\theta}}_{\mathrm{MAP}};\boldsymbol{\beta})\sqrt{\frac{(2\pi)^b}{\det(-\boldsymbol{H})}},$$

where b denotes the dimensionality of $\boldsymbol{\theta}$ (i.e., the number of parameters), and

$$\widehat{\boldsymbol{\theta}}_{\mathrm{MAP}} = \underset{\boldsymbol{\theta}}{\mathrm{argmax}}\left[\sum_{i=1}^{n}\log q(\boldsymbol{x}_i|\boldsymbol{\theta}) + \log p(\boldsymbol{\theta};\boldsymbol{\beta})\right],$$

$$\boldsymbol{H} = \frac{\partial^2}{\partial\boldsymbol{\theta}\partial\boldsymbol{\theta}^\top}\left(\sum_{i=1}^{n}\log q(\boldsymbol{x}_i|\boldsymbol{\theta}) + \log p(\boldsymbol{\theta};\boldsymbol{\beta})\right)\Bigg|_{\boldsymbol{\theta}=\widehat{\boldsymbol{\theta}}_{\mathrm{MAP}}}. \tag{18.2}$$

If the number of training samples, n, is large, the *central limit theorem* (see Section 7.4) asserts that the posterior probability $p(\boldsymbol{\theta}|\mathcal{D})$ converges in distribution to the Gaussian distribution. Therefore, the Laplace-approximated marginal likelihood would be accurate when a large number of training samples are available.

18.1.4 BAYESIAN INFORMATION CRITERION (BIC)

The logarithm of the Laplace-approximated marginal likelihood is given by

$$\log \mathrm{ML}(\boldsymbol{\beta}) \approx \sum_{i=1}^{n}\log q(\boldsymbol{x}_i|\widehat{\boldsymbol{\theta}}_{\mathrm{MAP}}) + \log p(\widehat{\boldsymbol{\theta}}_{\mathrm{MAP}};\boldsymbol{\beta}) + \frac{b}{2}\log(2\pi) - \frac{1}{2}\log\left(\det(-\boldsymbol{H})\right). \tag{18.3}$$

Let us further approximate this under the assumption that the number of training samples, n, is large.

The first term $\sum_{i=1}^{n}\log q(\boldsymbol{x}_i|\widehat{\boldsymbol{\theta}}_{\mathrm{MAP}})$ in Eq. (18.3) has asymptotic order n (see Fig. 14.4):

$$\sum_{i=1}^{n}\log q(\boldsymbol{x}_i|\widehat{\boldsymbol{\theta}}_{\mathrm{MAP}}) = O(n).$$

On the other hand, the second term $\log p(\widehat{\boldsymbol{\theta}}_{\text{MAP}}; \boldsymbol{\beta})$ and the third term $\frac{b}{2}\log(2\pi)$ are independent of n:

$$\log p(\widehat{\boldsymbol{\theta}}_{\text{MAP}}; \boldsymbol{\beta}) = O(1) \quad \text{and} \quad \frac{b}{2}\log(2\pi) = O(1).$$

According to the law of large numbers (see Section 7.3), the Hessian matrix $\boldsymbol{H}$ defined by Eq. (18.2) divided by n converges in probability to $\widetilde{\boldsymbol{H}}$:

$$\frac{1}{n}\boldsymbol{H} = \frac{\partial^2}{\partial\boldsymbol{\theta}\partial\boldsymbol{\theta}^\top}\left(\frac{1}{n}\sum_{i=1}^{n}\log q(\boldsymbol{x}_i|\boldsymbol{\theta}) + \frac{1}{n}\log p(\boldsymbol{\theta}; \boldsymbol{\beta})\right)\Bigg|_{\boldsymbol{\theta}=\widehat{\boldsymbol{\theta}}_{\text{MAP}}} \xrightarrow{\text{p}} \widetilde{\boldsymbol{H}},$$

where

$$\widetilde{\boldsymbol{H}} = \frac{\partial^2}{\partial\boldsymbol{\theta}\partial\boldsymbol{\theta}^\top}\left(E\left[\log q(\boldsymbol{x}|\boldsymbol{\theta})\right]\right)\Bigg|_{\boldsymbol{\theta}=\widehat{\boldsymbol{\theta}}_{\text{MAP}}}.$$

Since $\widetilde{\boldsymbol{H}}$ is a $b \times b$ matrix,

$$\det(n\widetilde{\boldsymbol{H}}) = n^b \det(\widetilde{\boldsymbol{H}})$$

holds and therefore

$$\frac{1}{2}\log\left(\det(-\boldsymbol{H})\right) \xrightarrow{\text{p}} \frac{b}{2}\log n + \frac{1}{2}\log\left(\det(-\widetilde{\boldsymbol{H}})\right)$$

is obtained. The first term $\frac{b}{2}\log n$ is proportional to $\log n$, while the second term $\frac{1}{2}\log\left(\det(-\widetilde{\boldsymbol{H}})\right)$ is independent of n:

$$\frac{b}{2}\log n = O(\log n) \quad \text{and} \quad \frac{1}{2}\log\left(\det(-\widetilde{\boldsymbol{H}})\right) = O(1).$$

Here, suppose that n is large enough so that terms with $O(1)$ can be ignored. Then only $\sum_{i=1}^{n}\log q(\boldsymbol{x}_i|\widehat{\boldsymbol{\theta}}_{\text{MAP}})$ and $\frac{b}{2}\log n$ remain. Furthermore, the difference between the MAP solution $\widehat{\boldsymbol{\theta}}_{\text{MAP}}$ and the maximum likelihood solution $\widehat{\boldsymbol{\theta}}_{\text{MLE}}$ is known to be $O(n^{-1})$:

$$\sum_{i=1}^{n}\log q(\boldsymbol{x}_i|\widehat{\boldsymbol{\theta}}_{\text{MAP}}) = \sum_{i=1}^{n}\log q(\boldsymbol{x}_i|\widehat{\boldsymbol{\theta}}_{\text{MLE}}) + O(n^{-1}).$$

Consequently, Laplace approximation of the log marginal likelihood given by Eq. (18.3) can be further approximated as

$$\log \text{ML}(\boldsymbol{\beta}) \approx \sum_{i=1}^{n}\log q(\boldsymbol{x}_i|\widehat{\boldsymbol{\theta}}_{\text{MLE}}) - \frac{b}{2}\log n.$$

The negative of the right-hand side is called the BIC:

$$\mathrm{BIC} = -\sum_{i=1}^{n} \log q(\boldsymbol{x}_i|\widehat{\boldsymbol{\theta}}_{\mathrm{MLE}}) + \frac{b}{2}\log n.$$

BIC is quite simple and thus is popularly used in model selection. However, BIC is no longer dependent on prior probabilities and thus cannot be used for setting the prior probability. It is also known that BIC is equivalent to the *minimum description length* (MDL) criterion [84], which was derived in a completely different framework.

BIC is similar to AIC explained in Section 14.3, but the second term is different:

$$\mathrm{AIC} = -\sum_{i=1}^{n} \log q(\boldsymbol{x}_i;\widehat{\boldsymbol{\theta}}_{\mathrm{MLE}}) + b.$$

When $n > e^2 \approx 7.39$, BIC has a stronger penalty than AIC and thus a simpler model would be chosen. However, since AIC and BIC are derived in completely different frameworks (KL divergence approximation and the marginal likelihood approximation), it cannot be simply concluded which one is more superior than the other.

18.2 VARIATIONAL APPROXIMATION

When the integrand $\prod_{i=1}^{n} q(\boldsymbol{x}_i|\boldsymbol{\theta})p(\boldsymbol{\theta};\boldsymbol{\beta})$ is not close to Gaussian, Laplace-approximated marginal likelihood may not be accurate. In this section, *variational approximation* is introduced, which finds the best approximation to the marginal likelihood in a limited function class that is easier to compute [71].

18.2.1 VARIATIONAL BAYESIAN EM (VBEM) ALGORITHM

The marginal likelihood can be expressed as

$$\mathrm{ML}(\boldsymbol{\beta}) = p(\mathcal{D};\boldsymbol{\beta}) = \iint p(\mathcal{D},\boldsymbol{\eta},\boldsymbol{\theta};\boldsymbol{\beta})\mathrm{d}\boldsymbol{\eta}\mathrm{d}\boldsymbol{\theta},$$

where $\boldsymbol{\eta}$ is called a *latent variable*. Let us consider probability density functions for $\boldsymbol{\eta}$ and $\boldsymbol{\theta}$, denoted by $q(\boldsymbol{\eta})$ and $r(\boldsymbol{\theta})$, called the *trial distributions*. Then Jensen's inequality (see Section 8.3.1) gives the following lower bound:

$$\begin{aligned}\log \mathrm{ML}(\boldsymbol{\beta}) &= \log \iint p(\mathcal{D},\boldsymbol{\eta},\boldsymbol{\theta};\boldsymbol{\beta})\mathrm{d}\boldsymbol{\eta}\mathrm{d}\boldsymbol{\theta}\\ &= \log \iint q(\boldsymbol{\eta})r(\boldsymbol{\theta})\frac{p(\mathcal{D},\boldsymbol{\eta},\boldsymbol{\theta};\boldsymbol{\beta})}{q(\boldsymbol{\eta})r(\boldsymbol{\theta})}\mathrm{d}\boldsymbol{\eta}\mathrm{d}\boldsymbol{\theta} \geq -F(q,r),\end{aligned}$$

where

$$F(q,r) = \iint q(\boldsymbol{\eta})r(\boldsymbol{\theta})\log\frac{q(\boldsymbol{\eta})r(\boldsymbol{\theta})}{p(\mathcal{D},\boldsymbol{\eta},\boldsymbol{\theta};\boldsymbol{\beta})}\mathrm{d}\boldsymbol{\eta}\mathrm{d}\boldsymbol{\theta}$$

is a functional of probability density functions q and r called the *variational free energy* (while $-\log \mathrm{ML}(\boldsymbol{\beta})$ is called the *free energy*). If q and r are chosen to minimize the variational free energy, a good approximator to the log marginal likelihood may be obtained.

Setting the partial derivatives of the variational free energy at zero,

$$\frac{\partial}{\partial q}F(q,r) = 0 \quad \text{and} \quad \frac{\partial}{\partial r}F(q,r) = 0,$$

yields that the solutions should satisfy

$$q(\boldsymbol{\eta}) \propto \exp\left(\int r(\boldsymbol{\theta}) \log p(\mathcal{D},\boldsymbol{\eta}|\boldsymbol{\theta};\boldsymbol{\beta})\mathrm{d}\boldsymbol{\theta}\right), \tag{18.4}$$

$$r(\boldsymbol{\theta}) \propto p(\boldsymbol{\theta})\exp\left(\int q(\boldsymbol{\eta}) \log p(\mathcal{D},\boldsymbol{\eta}|\boldsymbol{\theta};\boldsymbol{\beta})\mathrm{d}\boldsymbol{\eta}\right). \tag{18.5}$$

Since no method is known to analytically solve Eq. (18.4) and Eq. (18.5), these equations are used as updating formulas like the EM algorithm (see Section 15.4). Such an EM-like algorithm is called the VBEM algorithm, and Eq. (18.4) and Eq. (18.5) are, respectively, called the VB-E step and VB-M step.

The variational free energy can be expressed by using the KL divergence (see Section 14.2) as

$$F(q,r) = \mathrm{KL}\big(q(\boldsymbol{\eta})r(\boldsymbol{\theta})\big\|p(\boldsymbol{\eta},\boldsymbol{\theta}|\mathcal{D};\boldsymbol{\beta})\big) - \log \mathrm{ML}(\boldsymbol{\beta}).$$

This shows that reducing the variational free energy $F(q,r)$ corresponds to reducing $\mathrm{KL}\big(q(\boldsymbol{\eta})r(\boldsymbol{\theta})\big\|p(\boldsymbol{\eta},\boldsymbol{\theta}|\mathcal{D};\boldsymbol{\beta})\big)$, and therefore the VBEM algorithm can be regarded as approximating $p(\boldsymbol{\eta},\boldsymbol{\theta}|\mathcal{D};\boldsymbol{\beta})$ by $q(\boldsymbol{\eta})r(\boldsymbol{\theta})$.

For this reason, $r(\boldsymbol{\theta})$ obtained by the VBEM algorithm may be a good approximation to the posterior probability $p(\boldsymbol{\theta}|\mathcal{D};\boldsymbol{\beta})$.

18.2.2 RELATION TO ORDINARY EM ALGORITHM

As explained in Section 15.4, the ordinary EM algorithm maximized a lower bound of the likelihood. Here, it is shown that the ordinary EM algorithm can also be interpreted as a variational approximation. *Jensen's inequality* (Section 8.3.1) yields the following lower bound of the log-likelihood:

$$\begin{aligned}\log p(\mathcal{D}|\boldsymbol{\theta}) &= \log \int p(\mathcal{D},\boldsymbol{\eta}|\boldsymbol{\theta})\mathrm{d}\boldsymbol{\eta} \\ &= \log \int q(\boldsymbol{\eta}|\mathcal{D},\boldsymbol{\theta}')\frac{p(\mathcal{D},\boldsymbol{\eta}|\boldsymbol{\theta})}{q(\boldsymbol{\eta}|\mathcal{D},\boldsymbol{\theta}')}\mathrm{d}\boldsymbol{\eta} \geq b(q,\boldsymbol{\theta}),\end{aligned}$$

where

$$b(q,\boldsymbol{\theta}) = \int q(\boldsymbol{\eta}|\mathcal{D},\boldsymbol{\theta}')\log\frac{p(\mathcal{D},\boldsymbol{\eta}|\boldsymbol{\theta})}{q(\boldsymbol{\eta}|\mathcal{D},\boldsymbol{\theta}')}\mathrm{d}\boldsymbol{\eta}.$$

$\frac{\partial}{\partial q} b(q, \boldsymbol{\theta}) = 0$ yields $q(\boldsymbol{\eta}|\mathcal{D}, \boldsymbol{\theta}') = p(\boldsymbol{\eta}|\mathcal{D}, \boldsymbol{\theta})$, which is the E-step of the ordinary EM algorithm. Then finding $\boldsymbol{\theta}'$ that satisfies $\frac{\partial}{\partial \boldsymbol{\theta}} b(q, \boldsymbol{\theta})|_{\boldsymbol{\theta}=\boldsymbol{\theta}'} = \mathbf{0}$ is the M-step of the ordinary EM algorithm.

The relation between the ordinary EM algorithm and the VBEM algorithm can be further elucidated by the use of *Dirac's delta function* $\delta(\cdot)$, which satisfies for any function $g : \mathbb{R} \to \mathbb{R}$ and any real number κ,

$$\int_{-\infty}^{\infty} g(\tau)\delta(\kappa - \tau)\mathrm{d}\tau = g(\kappa).$$

Thus, *convolution* with Dirac's delta function allows us to extract the value of an arbitrary function $g(\cdot)$ at an arbitrary point κ. Dirac's delta function $\delta(\cdot)$ can be expressed as the limit of the normal density:

$$\delta(\tau) = \lim_{\sigma \to 0} \frac{1}{\sqrt{2\pi\sigma^2}} \exp\left(-\frac{\tau^2}{2\sigma^2}\right).$$

For multidimensional $\boldsymbol{\tau} = (\tau_1, \ldots, \tau_d)^\top$, Dirac's delta function is defined in an elementwise manner as

$$\delta(\boldsymbol{\tau}) = \delta(\tau_1) \times \cdots \times \delta(\tau_d).$$

Dirac's delta function, setting

$$r(\boldsymbol{\theta}') = \delta(\boldsymbol{\theta}' - \boldsymbol{\theta})$$

in the VB-E step yields

$$q(\boldsymbol{\eta}) \propto p(\mathcal{D}, \boldsymbol{\eta}|\boldsymbol{\theta}) \propto p(\boldsymbol{\eta}|\mathcal{D}, \boldsymbol{\theta}),$$

which agrees with the ordinary E-step.

CHAPTER

NUMERICAL APPROXIMATION OF PREDICTIVE DISTRIBUTION

19

CHAPTER CONTENTS

MAP estimation explained in Section 17.3 is a useful alternative to MLE due to its simplicity and ability to mitigate overfitting. However, since only a single parameter value,

$$\widehat{\boldsymbol{\theta}}_{\mathrm{MAP}} = \underset{\boldsymbol{\theta}}{\operatorname{argmax}}\, p(\boldsymbol{\theta}|\mathcal{D}),$$

is used, the distinctive feature of Bayesian inference that infinitely many parameters are considered is lost. Therefore, the obtained density estimator is always included in the parametric model (Fig. 17.1). In this chapter, algorithms for numerically approximating the Bayesian predictive distribution,

$$\widehat{p}_{\mathrm{Bayes}}(\boldsymbol{x}) = \int q(\boldsymbol{x}|\boldsymbol{\theta})p(\boldsymbol{\theta}|\mathcal{D})\mathrm{d}\boldsymbol{\theta}, \tag{19.1}$$

are introduced.

19.1 MONTE CARLO INTEGRATION

The *Monte Carlo method* is a generic name of algorithms that use random numbers, and its name stems from the Monte Carlo Casino in Monaco. In this section, the method of *Monte Carlo integration* is introduced to numerically approximate the integration in Eq. (19.1).

More specifically, Monte Carlo integration approximates the expectation of a function $g(\boldsymbol{\theta})$,

$$\int g(\boldsymbol{\theta})p(\boldsymbol{\theta})\mathrm{d}\boldsymbol{\theta}, \tag{19.2}$$

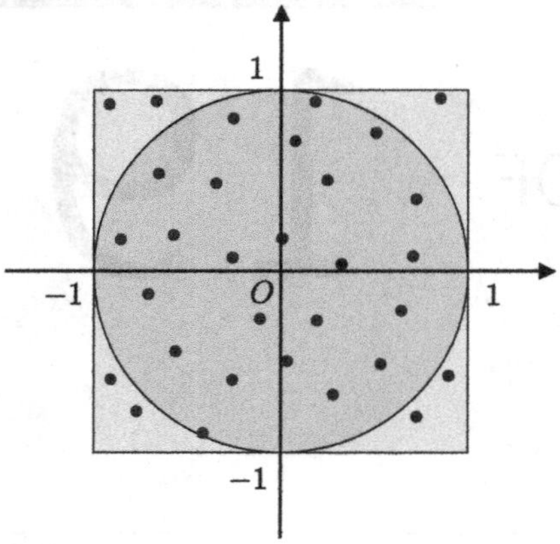

FIGURE 19.1

Numerical computation of π by Monte Carlo integration.

by the average over i.i.d. samples $\{\boldsymbol{\theta}_i\}_{i=1}^n$ following $p(\boldsymbol{\theta})$ as

$$\frac{1}{n}\sum_{i=1}^{n} g(\boldsymbol{\theta}_i). \tag{19.3}$$

By the law of large numbers (see Section 7.3), consistency of Monte Carlo integration is guaranteed. Namely, in the limit $n \to \infty$, Eq. (19.3) converges in probability to Eq. (19.2).

As illustration, let us approximate $\pi \approx 3.14$ by Monte Carlo integration. Let us consider a 2×2 square and its inscribed circle of radius 1 (Fig. 19.1). Let $g(x, y)$ be the function defined by

$$g(x,y) = \begin{cases} 1 & x^2 + y^2 \leq 1, \\ 0 & \text{otherwise}, \end{cases}$$

and let $p(x, y)$ be the uniform distribution on $[-1, 1]^2$:

$$p(x,y) = \begin{cases} 1/4 & -1 \leq x, y \leq 1, \\ 0 & \text{otherwise}. \end{cases}$$

Then the area of the inscribed circle, which is equal to π, is expressed as

$$\pi = 4 \iint g(x,y)p(x,y)\mathrm{d}x\mathrm{d}y.$$

```
n=10000000; x=rand(n,2)*2-1; pih=4*mean(sum(x.^2,2)<=1)
```

FIGURE 19.2

MATLAB code for numerically computing π by Monte Carlo integration.

Let us draw n i.i.d. samples $\{(x_i, y_i)\}_{i=1}^n$ following the uniform distribution $p(x, y)$ and compute the above expectation by Monte Carlo integration:

$$\pi \approx \frac{4}{n}\sum_{i=1}^{n} g(x_i, y_i) = \frac{4n'}{n},$$

where n' denotes the number of samples included in the circle. In the limit $n \to \infty$, $4n'/n$ converges in probability to π.

A MATLAB code for numerically computing π by Monte Carlo integration is given in Fig. 19.2, which tends to give 3.14.

19.2 IMPORTANCE SAMPLING

To perform Monte Carlo integration for approximating the Bayesian predictive distribution given by Eq. (19.1), random samples need to be generated following the posterior probability $p(\boldsymbol{\theta}|\mathcal{D})$. Techniques to generate random samples from an arbitrary probability distribution will be discussed in Section 19.3. In this section, another approach called *importance sampling* is introduced, which approximates the expectation over any (complicated) target probability density $p(\boldsymbol{\theta})$ based on another (simple) probability density $p'(\boldsymbol{\theta})$ such as the normal distribution. $p'(\boldsymbol{\theta})$ is called a *proxy distribution*.

In importance sampling, samples $\{\boldsymbol{\theta}'_{i'}\}_{i'=1}^{n'}$ drawn i.i.d. from $p'(\boldsymbol{\theta})$ are first generated. Then the expectation of a function $g(\boldsymbol{\theta})$ over any $p(\boldsymbol{\theta})$ is approximated by the average over $\{\boldsymbol{\theta}'_{i'}\}_{i'=1}^{n'}$ weighted according to the *importance* $p(\boldsymbol{\theta})/p'(\boldsymbol{\theta})$:

$$\int g(\boldsymbol{\theta})p(\boldsymbol{\theta})\mathrm{d}\boldsymbol{\theta} = \int \left(g(\boldsymbol{\theta})\frac{p(\boldsymbol{\theta})}{p'(\boldsymbol{\theta})}\right)p'(\boldsymbol{\theta})\mathrm{d}\boldsymbol{\theta} \approx \frac{1}{n'}\sum_{i'=1}^{n'} g(\boldsymbol{\theta}_{i'})\frac{p(\boldsymbol{\theta}_{i'})}{p'(\boldsymbol{\theta}_{i'})}.$$

As the name stands for, $p(\boldsymbol{\theta})/p'(\boldsymbol{\theta})$ indicates how important a sample drawn from $p'(\boldsymbol{\theta})$ is in $p(\boldsymbol{\theta})$. By the law of large numbers (see Section 7.3), consistency of importance sampling is guaranteed. Namely, in the limit $n \to \infty$, the importance-weighted average $\frac{1}{n'}\sum_{i'=1}^{n'} g(\boldsymbol{\theta}_{i'})\frac{p(\boldsymbol{\theta}_{i'})}{p'(\boldsymbol{\theta}_{i'})}$ converges in probability to the true expectation $\int g(\boldsymbol{\theta})p(\boldsymbol{\theta})\mathrm{d}\boldsymbol{\theta}$.

Let us compute the expectation of function $g(\theta) = \theta^2$ over the standard *Laplace distribution* (see Section 4.5) by importance sampling using the standard normal distribution as a proxy:

```
n=10000000; s=3; x=randn(n,1)*s; x2=x.^2; ss=2*s^2;
t=mean(x2.*(exp(-abs(x))/2)./(exp(-x2./ss)/sqrt(ss*pi)))
```

FIGURE 19.3

MATLAB code for importance sampling.

$$\int \theta^2 p(\theta)\mathrm{d}\theta \approx \frac{1}{n'}\sum_{i'=1}^{n'} \theta_{i'}^2 \frac{p(\theta_{i'})}{p'(\theta_{i'})}, \tag{19.4}$$

where $p(\theta)$ and $p'(\theta)$ are the standard Laplace and Gaussian densities:

$$p(\theta) = \frac{1}{2}\exp(-|\theta|) \quad \text{and} \quad p'(\theta) = \frac{1}{\sqrt{2\pi\sigma^2}}\exp\left(-\frac{\theta^2}{2\sigma^2}\right).$$

Since the left-hand side of Eq. (19.4) is actually the variance of the standard Laplace distribution, La(0, 1), its true value is 2. A MATLAB code for computing the right-hand side of Eq. (19.4) is given in Fig. 19.3, which tends to give 2.00.

Although importance sampling is guaranteed to be consistent, its variance can be large depending on the choice of a proxy distribution. This implies that a large number of samples may be needed to obtain a reliable value by importance sampling.

19.3 SAMPLING ALGORITHMS

In this section, methods for generating random samples from an arbitrary probability distribution are introduced, based on a random sample generator for the uniform distribution or the normal distribution (e.g. the rand or the randn functions in MATLAB). By directly generating i.i.d. samples $\{\boldsymbol{\theta}_i\}_{i=1}^n$ from $p(\boldsymbol{\theta})$, the expectation of a function $g(\boldsymbol{\theta})$ can be approximated as

$$\int g(\boldsymbol{\theta})p(\boldsymbol{\theta})\mathrm{d}\boldsymbol{\theta} \approx \frac{1}{n}\sum_{i=1}^{n} g(\boldsymbol{\theta}_i).$$

19.3.1 INVERSE TRANSFORM SAMPLING

Inverse transform sampling generates a one-dimensional random sample θ that follows a probability distribution with density $p(\theta)$ based on a uniform random variable u on $[0,1]$ [62] and the *cumulative distribution function* of $p(\theta)$. The cumulative distribution function of $p(\theta)$, denoted by $P(\theta)$, is defined as follows (Fig. 19.4):

$$P(\theta) = \int_{-\infty}^{\theta} p(u)\mathrm{d}u.$$

Let $\theta = P^{-1}(u)$ be the *inverse function* of $u = P(\theta)$. Then, for the uniform random variable u on $[0,1]$, $\theta = P^{-1}(u)$ has probability density $p(\theta)$ (Fig. 19.5). Thus, for n

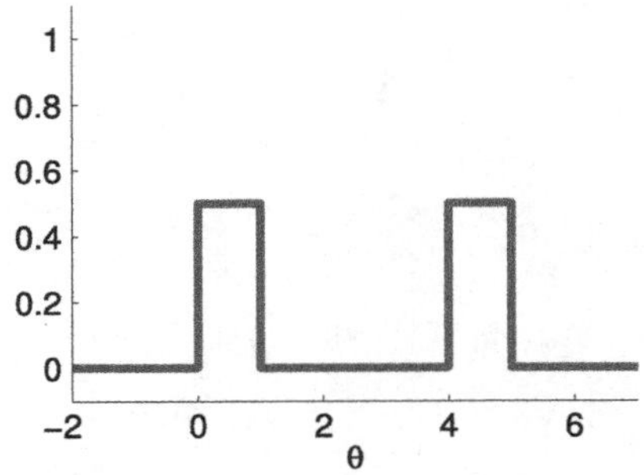

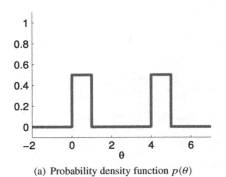

(a) Probability density function $p(\theta)$

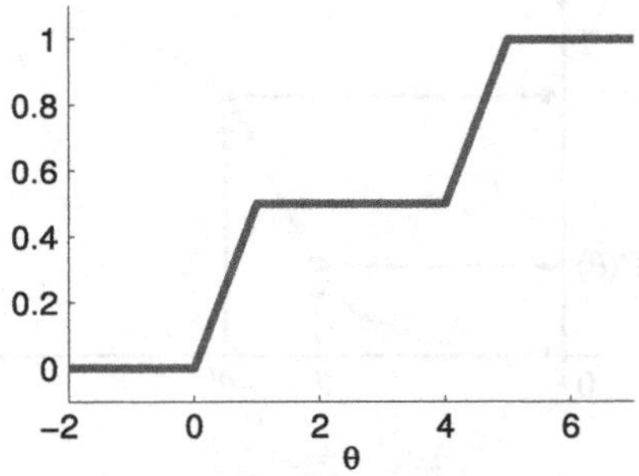

(b) Cumulative distribution function $P(\theta)$

FIGURE 19.4

Examples of probability density function $p(\theta)$ and its cumulative distribution function $P(\theta)$. Cumulative distribution function is monotone nondecreasing and satisfies $\lim_{\theta\to-\infty} P(\theta) = 0$ and $\lim_{\theta\to\infty} P(\theta) = 1$.

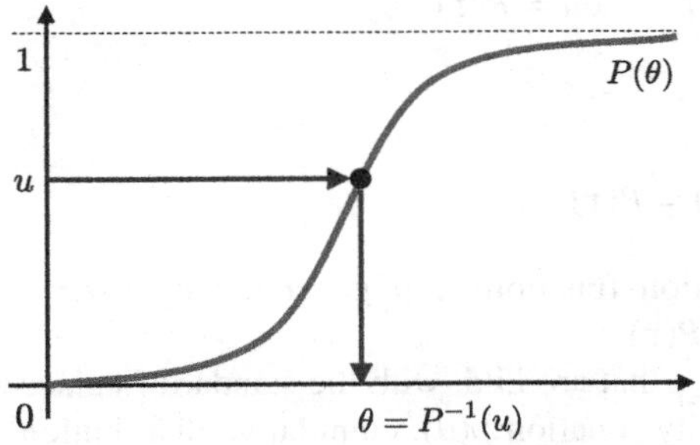

FIGURE 19.5

Inverse transform sampling.

uniform random variables $\{u_i\}_{i=1}^n$ on $[0, 1]$,

$$\{\theta_i \mid \theta_i = P^{-1}(u_i)\}_{i=1}^n$$

are i.i.d. with $p(\theta)$.

The validity of the above algorithm can be proved as follows. Since $\theta = P^{-1}(u)$, for any τ,

$$\Pr(\theta \leq \tau) = \Pr(P^{-1}(u) \leq \tau).$$

As illustrated in Fig. 19.6, $\theta \leq \theta'$ implies $P(\theta) \leq P(\theta')$, and therefore

$$\Pr(P^{-1}(u) \leq \tau) = \Pr(u \leq P(\tau)).$$

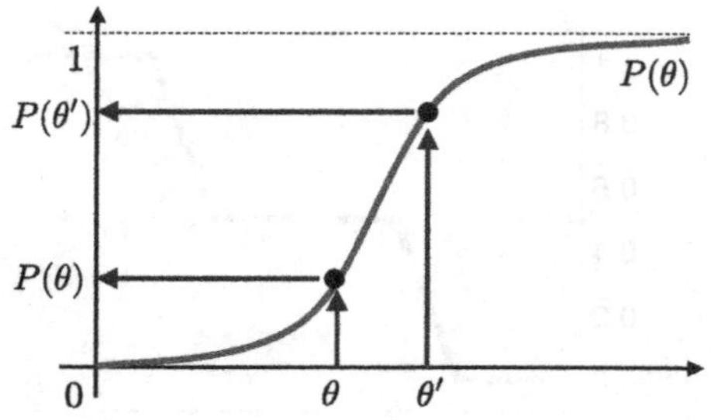

FIGURE 19.6

$\theta \leq \theta'$ implies $P(\theta) \leq P(\theta')$.

Furthermore, since u follows the uniform distribution on $[0, 1]$,

$$\Pr(u \leq P(\tau)) = \int_0^{P(\tau)} du = P(\tau)$$

holds and therefore

$$\Pr(\theta \leq \tau) = P(\tau).$$

This means that the cumulative distribution function of θ generated by inverse transform sampling agrees with the target $P(\tau)$.

Let us generate random samples $\{\theta_i\}_{i=1}^n$ that are i.i.d. with the standard Laplace distribution La(0, 1). The probability density function $p(\theta)$, cumulative distribution function $P(\theta)$, and its inverse function $P^{-1}(u)$ are given as follows (see Fig. 19.7):

$$p(\theta) = \frac{1}{2}e^{-|\theta|}, \quad P(\theta) = \frac{1}{2}\left(1 + \text{sign}(\theta)\left(1 - e^{-|\theta|}\right)\right),$$
$$P^{-1}(u) = -\text{sign}\left(u - \frac{1}{2}\right)\log\left(1 - 2\left|u - \frac{1}{2}\right|\right),$$

where sign(θ) denotes the *sign function*:

$$\text{sign}(\theta) = \begin{cases} 1 & (\theta > 0), \\ 0 & (\theta = 0), \\ -1 & (\theta < 0). \end{cases}$$

A MATLAB code for inverse transform sampling is given in Fig. 19.8, and its behavior is illustrated in Fig. 19.9.

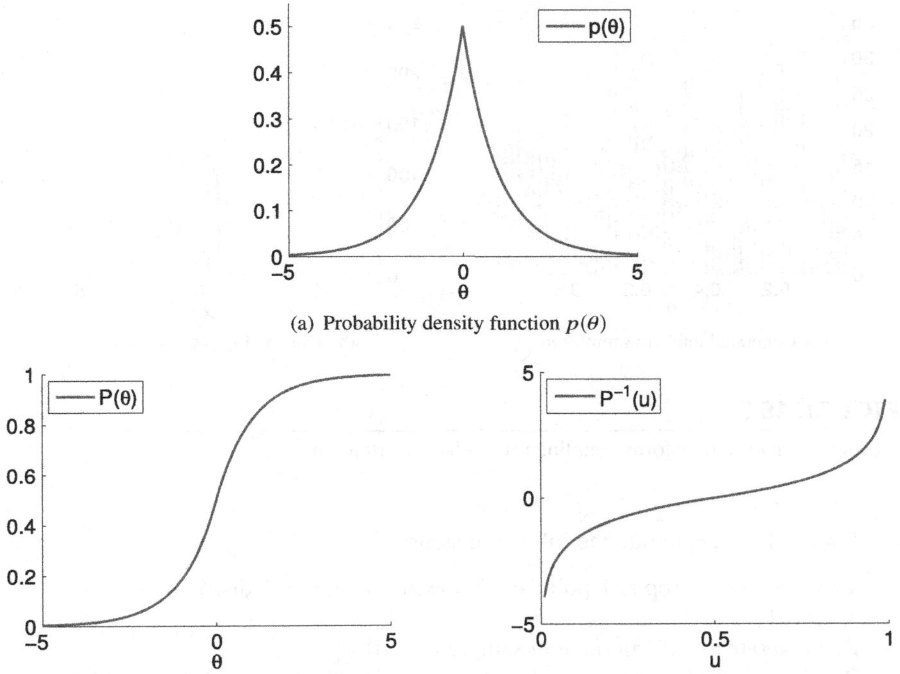

(a) Probability density function $p(\theta)$

(b) Cumulative distribution function $P(\theta)$

(c) Inverse cumulative distribution function $P^{-1}(u)$

FIGURE 19.7

Laplace distribution.

```
n=10000; u=rand(1,n); y=-sign(u-1/2).*log(1-2*abs(u-1/2));
figure(1); clf; hist(u,50);
figure(2); clf; hist(y,linspace(-8,8,30));
```

FIGURE 19.8

MATLAB code for inverse transform sampling.

As shown above, inverse transform sampling is a simple algorithm to generate samples following an arbitrary distribution. However, it can be applied only to one-dimensional distributions. Furthermore, the inverse cumulative distribution function $\theta = P^{-1}(u)$ needs to be explicitly computed, which can be difficult depending on the probability distributions.

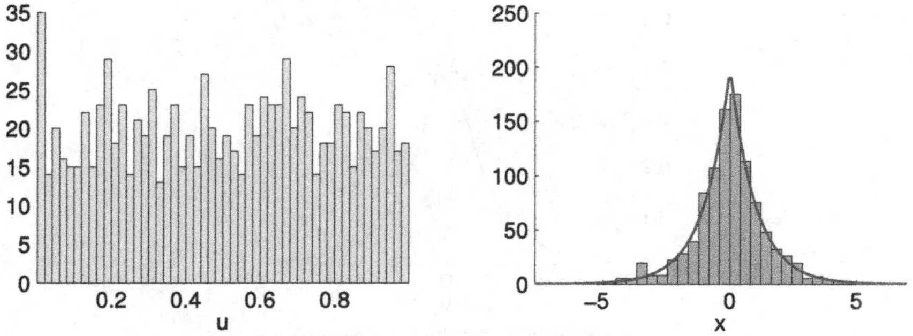

(a) Generated uniform samples u (b) Obtained Laplace samples

FIGURE 19.9

Example of inverse transform sampling for Laplace distribution.

For $i = 1, \ldots, n$, iterate the following steps:

1. Generate a proposal point $\boldsymbol{\theta}'$ following a proposal distribution $p'(\boldsymbol{\theta})$.
2. Generate a uniform random sample v on $[0, \kappa]$.
3. If $v > p(\boldsymbol{\theta}')/p'(\boldsymbol{\theta}')$, reject the proposal point $\boldsymbol{\theta}'$ and go to 1.
4. Accept the proposal point $\boldsymbol{\theta}'$ and set $\boldsymbol{\theta}_i \longleftarrow \boldsymbol{\theta}'$.
5. Increase the index: $i \longleftarrow i + 1$.

FIGURE 19.10

Algorithm of rejection sampling.

19.3.2 REJECTION SAMPLING

Rejection sampling is a computer-intensive approach to generating random samples following $p(\boldsymbol{\theta})$ based on samples drawn i.i.d. from another density $p'(\boldsymbol{\theta})$ [117]. $p'(\boldsymbol{\theta})$ is called a *proposal distribution*, and a simple distribution such as the uniform distribution or the normal distribution is usually chosen.

The assumption required in rejection sampling is that an upper bound of the probability density ratio $p(\boldsymbol{\theta})/p'(\boldsymbol{\theta})$ exists and is known:

$$\max_{\boldsymbol{\theta}} \left[\frac{p(\boldsymbol{\theta})}{p'(\boldsymbol{\theta})} \right] \leq \kappa < \infty.$$

In rejection sampling, a sample $\boldsymbol{\theta}'$ is generated following a proposal distribution $p'(\boldsymbol{\theta})$, which is called a *proposal point*. Then from the uniform distribution on $[0, \kappa]$,

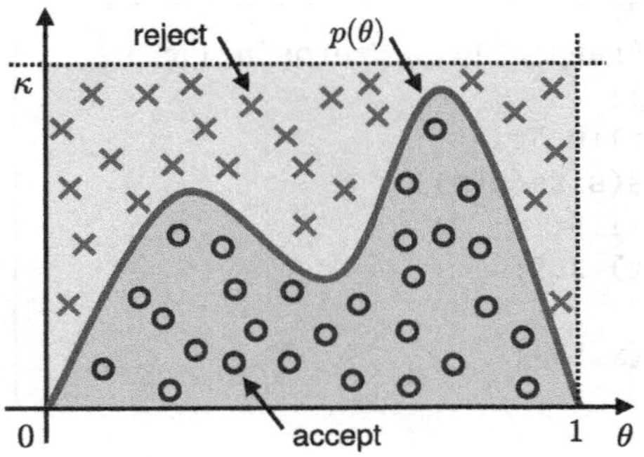

FIGURE 19.11

Illustration of rejection sampling when the proposal distribution is uniform.

a sample v is generated, which is used for evaluating the proposal point $\boldsymbol{\theta}'$. If

$$v \leq \frac{p(\boldsymbol{\theta}')}{p'(\boldsymbol{\theta}')},$$

the proposal point $\boldsymbol{\theta}'$ is *accepted*, i.e. set $\boldsymbol{\theta}_i = \boldsymbol{\theta}'$ and $i \leftarrow i + 1$. Otherwise, the proposal point $\boldsymbol{\theta}'$ is *rejected*. This procedure is repeated until n points are accepted. Then the accepted samples $\{\boldsymbol{\theta}_i\}_{i=1}^n$ have probability density $p(\boldsymbol{\theta})$. The algorithm of rejection sampling is summarized in Fig. 19.10.

The idea of rejection sampling can be more easily understood if $p(\boldsymbol{\theta})$ is defined on a finite set $\mathcal{X}$ (say $[0, 1]$) and the uniform distribution on $\mathcal{X}$ is used as the proposal distribution $p'(\boldsymbol{\theta})$. As illustrated in Fig. 19.11, samples generated below/above the curve of $p(\boldsymbol{\theta})$ are accepted/rejected in this setup, which results in samples having probability density $p(\boldsymbol{\theta})$.

A MATLAB code for rejection sampling for the probability density function,

$$p(\theta) = \begin{cases} \frac{1}{4}\theta & (0 \leq \theta < 1), \\ \frac{1}{2} - \frac{1}{4}\theta & (1 \leq \theta < 2), \\ \frac{1}{4} & (2 \leq \theta < 3), \\ -\frac{3}{2} + \frac{1}{2}\theta & (3 \leq \theta < 4), \\ \frac{5}{2} - \frac{1}{2}\theta & (4 \leq \theta \leq 5), \end{cases}$$

is given in Fig. 19.12, and its behavior is illustrated in Fig. 19.13.

Rejection sampling only requires the upper bound κ, while inverse transform sampling needs analytic computation of the inverse cumulative distribution function.

```
n=10000; u=5*rand(n,1); v=0.6*rand(n,1); y=zeros(n,1);
t=(0<=u & u<1); y(t)=0.25*u(t);
t=(1<=u & u<2); y(t)=-0.25*u(t)+0.5;
t=(2<=u & u<3); y(t)=0.25*ones(size(u(t)));
t=(3<=u & u<4); y(t)=0.5*u(t)-1.5;
t=(4<=u & u<=5); y(t)=-0.5*u(t)+2.5;
x=u(v<=y);
figure(1); clf; hold on; hist(x,50);
```

FIGURE 19.12

MATLAB code for rejection sampling.

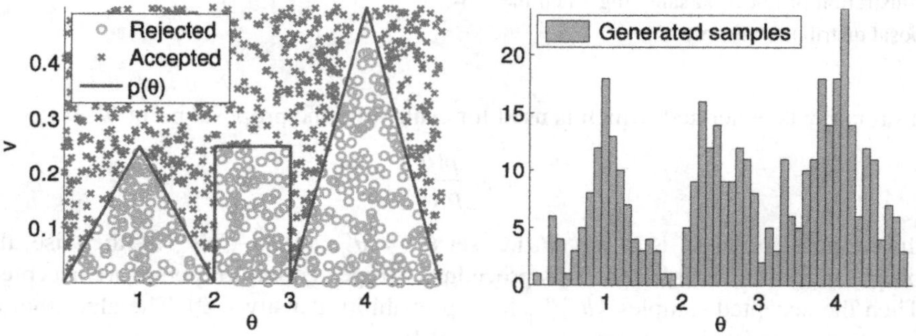

(a) Generated uniform samples (θ, v) (b) Obtained samples following $p(\theta)$

FIGURE 19.13

Example of rejection sampling.

Thus, it is much easier to implement in practice. Furthermore, rejection sampling is immediately applicable to multidimensional probability distributions.

However, depending on the profile of $p(\boldsymbol{\theta})$, the probability of accepting proposal points can be small and then rejection sampling is computationally expensive. For example, when $p(\boldsymbol{\theta})$ has a sharp profile, κ tends to be large and then the acceptance rate is small (see Fig. 19.14). Furthermore, depending on the choice of proposal distribution $p'(\boldsymbol{\theta})$, κ can be infinity and then rejection sampling is not applicable.

19.3.3 MARKOV CHAIN MONTE CARLO (MCMC) METHOD

Rejection sampling cannot be used when the upper bound κ is unknown, while MCMC methods do not have such a restriction. A *stochastic process* is a random

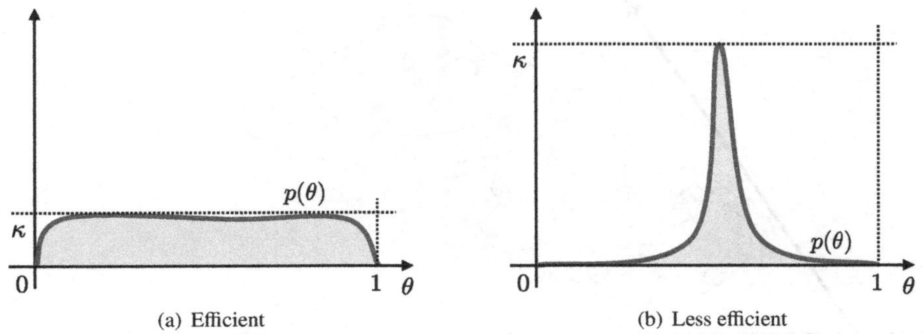

FIGURE 19.14

Computational efficiency of rejection sampling. (a) When the upper bound of the probability density, κ, is small, proposal points are almost always accepted and thus rejection sampling is computationally efficient. (b) When κ is large, most of the proposal points will be rejected and thus rejection sampling is computationally expensive.

variable that changes over time, and a Markov chain $\boldsymbol{\theta}_1, \ldots, \boldsymbol{\theta}_n$ is a stochastic process where sample $\boldsymbol{\theta}_i$ at time i depends only on the previous sample $\boldsymbol{\theta}_{i-1}$.

Metropolis-Hastings sampling

Metropolis-Hastings sampling is a MCMC method [52, 72] that generates samples following a Markov chain. Thus, the proposal distribution to draw the next sample $\boldsymbol{\theta}_{i+1}$ depends on the current sample $\boldsymbol{\theta}_i$:

$$p'(\boldsymbol{\theta}|\boldsymbol{\theta}_i).$$

A proposal point $\boldsymbol{\theta}'$ generated from the above proposal distribution $p'(\boldsymbol{\theta}|\boldsymbol{\theta}_i)$ is evaluated by a uniform random variable v on $[0,1]$. More specifically, if

$$v \leq \frac{p(\boldsymbol{\theta}')p'(\boldsymbol{\theta}_i|\boldsymbol{\theta}')}{p(\boldsymbol{\theta}_i)p'(\boldsymbol{\theta}'|\boldsymbol{\theta}_i)},$$

then the proposal point $\boldsymbol{\theta}'$ is accepted and set

$$\boldsymbol{\theta}_{i+1} = \boldsymbol{\theta}'.$$

Otherwise, the proposal point $\boldsymbol{\theta}'$ is rejected and set

$$\boldsymbol{\theta}_{i+1} = \boldsymbol{\theta}_i.$$

Note the difference from rejection sampling that the previous value is used when the proposal point is rejected. The initial value $\boldsymbol{\theta}_0$ needs to be chosen by a user.

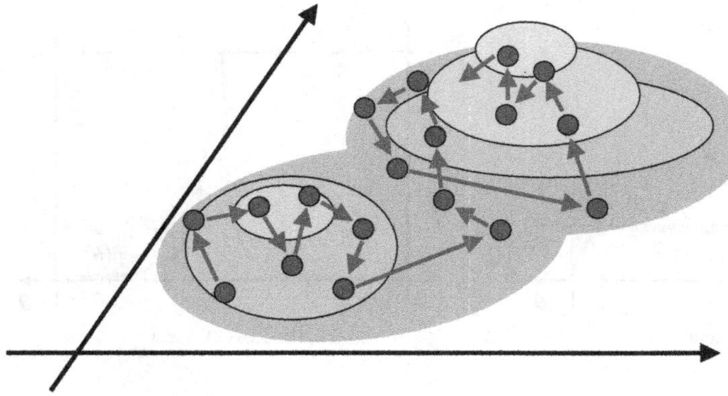

FIGURE 19.15

Random walk.

As a proposal distribution $p'(\boldsymbol{\theta}|\boldsymbol{\theta}_i)$, the Gaussian distribution with expectation $\boldsymbol{\theta}_i$ and variance-covariance matrix $\sigma^2 \boldsymbol{I}_b$ may be used:

$$p'(\boldsymbol{\theta}|\boldsymbol{\theta}_i) = \frac{1}{(2\pi\sigma^2)^{b/2}} \exp\left(-\frac{(\boldsymbol{\theta} - \boldsymbol{\theta}_i)^\top (\boldsymbol{\theta} - \boldsymbol{\theta}_i)}{2\sigma^2}\right),$$

where b denotes the dimensionality of $\boldsymbol{\theta}$. Since the Gaussian distribution is symmetric, i.e.

$$p'(\boldsymbol{\theta}|\boldsymbol{\theta}_i) = p'(\boldsymbol{\theta}_i|\boldsymbol{\theta}),$$

the threshold for rejection is simplified as

$$\frac{p(\boldsymbol{\theta}')}{p(\boldsymbol{\theta}_i)}.$$

When a local probability distribution such as the Gaussian distribution is used as the proposal distribution, the proposal point $\boldsymbol{\theta}'$ is generated around the previous parameter $\boldsymbol{\theta}_i$, and therefore points $\boldsymbol{\theta}_1, \boldsymbol{\theta}_2, \ldots$ move gradually. Such a random sequence is called a *random walk* (Fig. 19.15).

A MATLAB code for Metropolis-Hastings sampling is given in Fig. 19.16, and its behavior is illustrated in Fig. 19.17.

Unlike rejection sampling, Metropolis-Hastings sampling can be used without knowing the upper bound κ. Furthermore, even when the target probability density $p(\boldsymbol{\theta})$ is not explicitly available, Metropolis-Hastings sampling can still be employed, as long as $p(\boldsymbol{\theta})$ is known up to the *normalization term*. More specifically, suppose that $p(\boldsymbol{\theta})$ is unknown, but its unnormalized counterpart $\widetilde{p}(\boldsymbol{\theta})$ is available:

$$p(\boldsymbol{\theta}) = \frac{\widetilde{p}(\boldsymbol{\theta})}{Z},$$

```
n=100000; t=zeros(2,n);
for i=2:n
  u=randn(2,1)+t(:,i-1);
  if rand<=pdf(u)/pdf(t(:,i-1))
    t(:,i)=u;
  else
    t(:,i)=t(:,i-1);
  end
end

figure(1); clf; hold on
plot(t(1,1:500),t(2,1:500),'b-');

figure(2); clf; hold on
c=4; m=30; Q=zeros(m,m);
a=linspace(-c,c,m); b=-c+2*c/m*[1:m];
for k=1:n
  x=find(ceil(t(1,k)-b)==0,1);
  y=find(ceil(t(2,k)-b)==0,1);
  Q(x,y)=Q(x,y)+1;
end
surf(a,a,Q'/(2*c/m)^2/n);

figure(3); clf; hold on
P=zeros(m,m);
for X=1:m, for Y=1:m
  P(X,Y)=pdf([a(X);a(Y)]);
end, end
surf(a,a,P');
```

```
function p=pdf(x)

S1=[0.5 0; 0 4]; m1=x-[0; -2]; w1=0.7;
S2=[4 0; 0 1]; m2=x-[2; 0]; w2=0.3;
p=w1*sqrt(det(S1))/(2*pi)*exp(-m1'*S1*m1/2) ...
  +w2*sqrt(det(S2))/(2*pi)*exp(-m2'*S2*m2/2);
```

FIGURE 19.16

MATLAB code for Metropolis-Hastings sampling. The bottom function should be saved as "pdf.m."

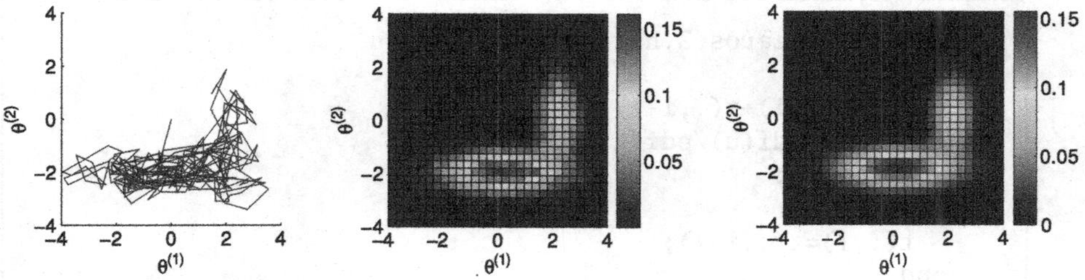

(a) Trajectory of generated samples (only first 500 samples are plotted) (b) Histogram of generated samples (c) True probability density

FIGURE 19.17

Example of Metropolis-Hastings sampling.

where

$$Z = \int \widetilde{p}(\boldsymbol{\theta})\mathrm{d}\boldsymbol{\theta}$$

is the normalization constant which is unknown. Even in this situation, the rejection threshold can be computed as

$$\frac{p(\boldsymbol{\theta}')}{p(\boldsymbol{\theta}_i)} = \frac{\widetilde{p}(\boldsymbol{\theta}')/Z}{\widetilde{p}(\boldsymbol{\theta}_i)/Z} = \frac{\widetilde{p}(\boldsymbol{\theta}')}{\widetilde{p}(\boldsymbol{\theta}_i)},$$

and therefore Metropolis-Hastings sampling can still be executed without any modification.

However, in the same way as rejection sampling, Metropolis-Hastings sampling is less efficient when the acceptance rate is low.

Gibbs sampling

Gibbs sampling is another MCMC method that generates samples in an elementwise manner in multidimensional probability distributions [47].

More specifically, at time step i, for $j = 1,\ldots,d$, the proposal point of the jth dimension, $\theta_i^{(j)}$, is generated from the proposal distribution that depends on the previous sample $\boldsymbol{\theta}_{i-1} = (\theta_{i-1}^{(1)},\ldots,\theta_{i-1}^{(d)})^\top$:

$$p(\theta^{(j)}|\theta_i^{(1)},\ldots,\theta_i^{(j-1)},\theta_{i-1}^{(j+1)},\ldots,\theta_{i-1}^{(d)}).$$

Note that the above conditional density is different from

$$p(\theta^{(j)}|\theta_{i-1}^{(1)},\ldots,\theta_{i-1}^{(j-1)},\theta_{i-1}^{(j+1)},\ldots,\theta_{i-1}^{(d)}),$$

```
n=100000; r=0.5; s=sqrt(1-r^2); t=zeros(2,n);
for i=2:n
  t(1,i)=s*randn+r*t(2,i-1); t(2,i)=s*randn+r*t(1,i);
end

figure(1); clf; hold on
u1=repmat(t(1,:),[2,1]); u1=u1(:);
u2=repmat(t(2,:),[2,1]); u2=u2(:);
plot(u1([2:200 200]),u2(1:200),'b-');
figure(2); clf; hold on
c=4; m=30; Q=zeros(m,m);
a=linspace(-c,c,m); b=-c+2*c/m*[1:m];
for k=1:n
  x=find(ceil(t(1,k)-b)==0,1);
  y=find(ceil(t(2,k)-b)==0,1);
  Q(x,y)=Q(x,y)+1;
end
surf(a,a,Q'/(2*c/m)^2/n);
figure(3); clf; hold on
P=zeros(m,m);
for X=1:m, for Y=1:m
  P(X,Y)=exp(-(a(X)^2-2*r*a(X)*a(Y)+a(Y)^2)/2/s)/(2*pi*s);
end, end
surf(a,a,P');
```

FIGURE 19.18

MATLAB code for Gibbs sampling.

i.e. not all values of the previous sample $\boldsymbol{\theta}_{i-1}$ are used, but the new values $\theta_i^{(1)}, \ldots, \theta_i^{(j-1)}$ are used up to the $(j-1)$th dimension.

Let us illustrate a more specific algorithm of Gibbs sampling for the two-dimensional Gaussian distribution:

$$p(\boldsymbol{\theta}) = N(\boldsymbol{\theta}; \mathbf{0}_2, \boldsymbol{\Sigma}_\rho),$$

where, for $-1 < \rho < 1$,

$$\boldsymbol{\Sigma}_\rho = \begin{pmatrix} 1 & \rho \\ \rho & 1 \end{pmatrix}.$$

For $\boldsymbol{\theta} = (\theta^{(1)}, \theta^{(2)})^\top$, the conditional density is given by

$$p(\theta^{(1)}|\theta^{(2)}) = N(\theta^{(1)}; \rho\theta^{(2)}, 1-\rho^2).$$

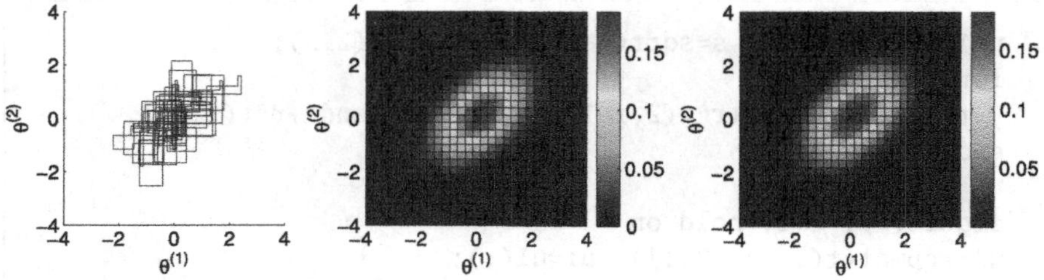

(a) Trajectory of generated samples (only first 100 samples are plotted) (b) Histogram of generated samples (c) True probability density

FIGURE 19.19

Example of Gibbs sampling.

A MATLAB code of Gibbs sampling for the above Gaussian distribution is given in Fig. 19.18, and its behavior is illustrated in Fig. 19.19.

A variant of Gibbs sampling, which generates samples not only in a dimensionwise manner but also for some dimensions together, is called *blocked Gibbs sampling* [95]. Another variant that marginalizes some of the conditioning variables $\theta_i^{(1)}, \ldots, \theta_i^{(j-1)}, \theta_{i-1}^{(j+1)}, \ldots, \theta_{i-1}^{(d)}$ is called *collapsed Gibbs sampling* [68].

Since Gibbs sampling does not involve proposal distributions and rejection, it is computationally efficient. However, it requires sampling from the conditional distribution, which may not always be straightforward in practice.

Discussions

Samples $\{\boldsymbol{\theta}_i\}_{i=1}^n$ generated by rejection sampling are guaranteed to be mutually independent and have density $p(\boldsymbol{\theta})$ for finite n. On the other hand, samples $\{\boldsymbol{\theta}_i\}_{i=1}^n$ generated by MCMC methods are generally dependent on each other due to the incremental nature of Markov chains. Moreover, samples generated in an early stage may depend on the choice of initial value $\boldsymbol{\theta}_0$. For these reasons, samples generated by MCMC methods have density $p(\boldsymbol{\theta})$ only in the limit $n \to \infty$.

To mitigate these problems, it is common to discard samples generated in an early stage to decrease the dependency on the initial value (which is often referred to as *burn-in*) and adopt samples of only every m time steps to weaken mutual dependency between samples.

CHAPTER

BAYESIAN MIXTURE MODELS

20

CHAPTER CONTENTS

In this chapter, within the Bayesian inference framework explained in Chapter 17, practical Bayesian inference algorithms for mixture models are presented. First, algorithms based on variational approximation and Gibbs sampling for Gaussian mixture models are introduced. Then a variational Bayesian algorithm for topic models is explained.

20.1 GAUSSIAN MIXTURE MODELS

In this section, Bayesian inference algorithms for *Gaussian mixture models* (see Section 15.1) are introduced.

20.1.1 BAYESIAN FORMULATION

Let us consider the mixture of m Gaussian models:

$$q(\boldsymbol{x}|\mathcal{W},\mathcal{M},\mathcal{S}) = \sum_{\ell=1}^{m} w_\ell N(\boldsymbol{x}|\boldsymbol{\mu}_\ell,\boldsymbol{S}_\ell^{-1}),$$

where $\mathcal{W} = (w_1,\ldots,w_m)$, $\mathcal{M} = (\boldsymbol{\mu}_1,\ldots,\boldsymbol{\mu}_m)$, $\mathcal{S} = (\boldsymbol{S}_1,\ldots,\boldsymbol{S}_m)$, and $N(\boldsymbol{x}|\boldsymbol{\mu},\boldsymbol{S}^{-1})$ denotes the Gaussian density with expectation $\boldsymbol{\mu}$ and variance-covariance matrix $\boldsymbol{S}^{-1}$ (or precision matrix $\boldsymbol{S}$):

$$N(\boldsymbol{x}|\boldsymbol{\mu},\boldsymbol{S}^{-1}) = \frac{\sqrt{\det(\boldsymbol{S})}}{(2\pi)^{d/2}} \exp\left(-\frac{1}{2}(\boldsymbol{x}-\boldsymbol{\mu})^\top \boldsymbol{S}(\boldsymbol{x}-\boldsymbol{\mu})\right). \quad (20.1)$$

For training samples $\mathcal{D} = \{\boldsymbol{x}_1, \ldots, \boldsymbol{x}_n\}$ drawn independently from the true density $p(\boldsymbol{x})$, the likelihood $p(\mathcal{D}|\mathcal{W}, \mathcal{M}, \mathcal{S})$ is given by

$$p(\mathcal{D}|\mathcal{W}, \mathcal{M}, \mathcal{S}) = \prod_{i=1}^{n} q(\boldsymbol{x}_i|\mathcal{W}, \mathcal{M}, \mathcal{S}).$$

For mixing weights $\mathcal{W}$, the *symmetric Dirichlet distribution* (see Section 6.3) is considered as the prior probability:

$$p(\mathcal{W}; \alpha_0) = \mathrm{Dir}(\boldsymbol{W}; \alpha_0) \propto \prod_{\ell=1}^{m} w_\ell^{\alpha_0 - 1},$$

where $\mathrm{Dir}(\boldsymbol{W}; \alpha)$ denotes the symmetric Dirichlet density with concentration parameter α. Note that the Dirichlet distribution is *conjugate* (see Section 17.2) for the discrete distribution given by Eq. (20.3).

For Gaussian expectations $\mathcal{M}$ and Gaussian precision matrices $\mathcal{S}$, the product of the normal distribution and the Wishart distribution (see Section 6.4), called the *normal-Wishart distribution*, is considered as the prior probability:

$$\begin{aligned} &p(\mathcal{M}, \mathcal{S}; \beta_0, \boldsymbol{W}_0, \nu_0) \\ &= \prod_{\ell=1}^{m} N(\boldsymbol{\mu}_\ell|\boldsymbol{0}, (\beta_0 \boldsymbol{S}_\ell)^{-1}) W(\boldsymbol{S}_\ell; \boldsymbol{W}_0, \nu_0) \\ &\propto \prod_{\ell=1}^{m} \det(\boldsymbol{S}_\ell)^{\frac{\nu_0 - d}{2} - 1} \exp\left(-\frac{\beta_0}{2}\boldsymbol{\mu}_\ell^\top \boldsymbol{S}_\ell \boldsymbol{\mu}_\ell - \frac{1}{2}\mathrm{tr}\left(\boldsymbol{W}_0^{-1}\boldsymbol{S}_\ell\right)\right), \end{aligned}$$

where $W(\boldsymbol{S}; \boldsymbol{W}, \nu)$ denotes the Wishart density with ν degrees of freedom:

$$W(\boldsymbol{S}; \boldsymbol{W}, \nu) = \frac{\det(\boldsymbol{S})^{\frac{\nu - d - 1}{2}} \exp\left(-\frac{1}{2}\mathrm{tr}\left(\boldsymbol{W}^{-1}\boldsymbol{S}\right)\right)}{\det(2\boldsymbol{W})^{\frac{\nu}{2}} \Gamma_d(\frac{\nu}{2})}.$$

Here, d denotes the dimensionality of input $\boldsymbol{x}$ and $\Gamma_d(\cdot)$ denotes the d-dimensional *gamma function* defined by Eq. (6.2):

$$\Gamma_d\left(\frac{\nu}{2}\right) = \int_{\mathbb{S}_d^+} \det(\boldsymbol{S})^{\frac{\nu - d - 1}{2}} \exp\left(-\mathrm{tr}(\boldsymbol{S})\right) \mathrm{d}\boldsymbol{S},$$

where $\mathbb{S}_d^+$ denotes the set of all $d \times d$ positive symmetric matrices. Note that the above *normal-Wishart distribution* is conjugate for the multivariate normal distribution with unknown expectation and unknown precision matrix.

The above formulation is summarized in Fig. 20.1. By the Bayes' theorem, the posterior probability $p(\mathcal{W}, \mathcal{M}, \mathcal{S}|\mathcal{D})$ is given as

$$\begin{aligned} &p(\mathcal{W}, \mathcal{M}, \mathcal{S}|\mathcal{D}; \alpha_0, \beta_0, \boldsymbol{W}_0, \nu_0) \\ &= \frac{p(\mathcal{D}|\mathcal{W}, \mathcal{M}, \mathcal{S}) p(\mathcal{W}; \alpha_0) p(\mathcal{M}, \mathcal{S}; \beta_0, \boldsymbol{W}_0, \nu_0)}{p(\mathcal{D}; \alpha_0, \beta_0, \boldsymbol{W}_0, \nu_0)}, \end{aligned} \quad (20.2)$$

which is not computationally tractable. Below, practical approximate inference methods are introduced.

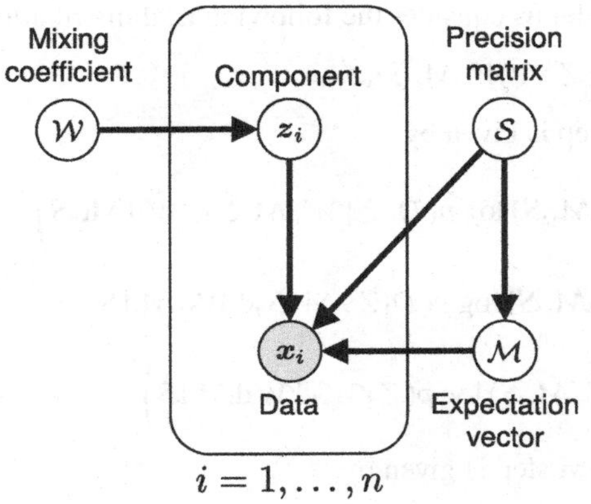

FIGURE 20.1

Variational Bayesian formulation of Gaussian mixture model.

To derive approximate inference methods, let us consider *latent variables*,

$$\mathcal{Z} = \{z_1, \ldots, z_n\},$$

for training samples $\mathcal{D} = \{\boldsymbol{x}_1, \ldots, \boldsymbol{x}_n\}$, where each $z_i = (z_{i,1}, \ldots, z_{i,m})^\top$ is an m-dimensional vector that indicates mixture component selection. Specifically, if the kth component is selected, the kth element of z_i is 1 and all other elements are 0. The probability of observing $\mathcal{Z}$ given mixing weights $\mathcal{W} = (w_1, \ldots, w_m)$ can be expressed as

$$p(\mathcal{Z}|\mathcal{W}) = \prod_{i=1}^{n} \prod_{\ell=1}^{m} w_\ell^{z_{i,\ell}}. \tag{20.3}$$

20.1.2 VARIATIONAL INFERENCE

Here, the *variational technique* introduced in Section 18.2 is employed to approximately compute the posterior probability $p(\mathcal{W}, \mathcal{M}, \mathcal{S}|\mathcal{D})$ given by Eq. (20.2).

The marginal likelihood can be expressed as

$$\begin{aligned}
&\mathrm{ML}(\alpha_0, \beta_0, \boldsymbol{W}_0, \nu_0)\\
&= p(\mathcal{D}; \alpha_0, \beta_0, \boldsymbol{W}_0, \nu_0)\\
&= \iiiint p(\mathcal{D}, \mathcal{Z}, \mathcal{W}, \mathcal{M}, \mathcal{S}; \alpha_0, \beta_0, \boldsymbol{W}_0, \nu_0) \mathrm{d}\mathcal{Z} \mathrm{d}\mathcal{W} \mathrm{d}\mathcal{M} \mathrm{d}\mathcal{S}\\
&= \iiiint p(\mathcal{D}|\mathcal{Z}, \mathcal{M}, \mathcal{S}) p(\mathcal{Z}|\mathcal{W}) p(\mathcal{W}; \alpha_0)\\
&\qquad \times p(\mathcal{M}, \mathcal{S}; \beta_0, \boldsymbol{W}_0, \nu_0) \mathrm{d}\mathcal{Z} \mathrm{d}\mathcal{W} \mathrm{d}\mathcal{M} \mathrm{d}\mathcal{S}.
\end{aligned}$$

For the above marginal likelihood, let us consider the following trial distributions:

$$q(\mathcal{Z})r(\mathcal{W},\mathcal{M},\mathcal{S}).$$

Then, from Eq. (18.4), the VB-E step is given by

$$\begin{aligned}
q(\mathcal{Z}) &\propto \exp\left(\iiint r(\mathcal{W},\mathcal{M},\mathcal{S}) \log p(\mathcal{D},\mathcal{Z}|\mathcal{W},\mathcal{M},\mathcal{S}) \mathrm{d}\mathcal{W}\mathrm{d}\mathcal{M}\mathrm{d}\mathcal{S}\right) \\
&= \exp\left(\iiint r(\mathcal{W},\mathcal{M},\mathcal{S}) \log p(\mathcal{D}|\mathcal{Z},\mathcal{M},\mathcal{S}) \mathrm{d}\mathcal{W}\mathrm{d}\mathcal{M}\mathrm{d}\mathcal{S}\right. \\
&\quad \left. + \iiint r(\mathcal{W},\mathcal{M},\mathcal{S}) \log p(\mathcal{Z}|\mathcal{W}) \mathrm{d}\mathcal{W}\mathrm{d}\mathcal{M}\mathrm{d}\mathcal{S}\right).
\end{aligned} \tag{20.4}$$

Similarly, from Eq. (18.5), the VB-M step is given by

$$\begin{aligned}
r(\mathcal{W},\mathcal{M},\mathcal{S}) &\propto p(\mathcal{W},\mathcal{M},\mathcal{S};\alpha_0,\beta_0,\boldsymbol{W}_0,\nu_0) \\
&\quad \times \exp\left(\int q(\mathcal{Z}) \log p(\mathcal{D},\mathcal{Z}|\mathcal{W},\mathcal{M},\mathcal{S}) \mathrm{d}\mathcal{Z}\right) \\
&= p(\mathcal{W};\alpha_0) \exp\left(\int q(\mathcal{Z}) \log p(\mathcal{Z}|\mathcal{W}) \mathrm{d}\mathcal{Z}\right) \\
&\quad \times p(\mathcal{M},\mathcal{S};\beta_0,\boldsymbol{W}_0,\nu_0) \exp\left(\int q(\mathcal{Z}) \log p(\mathcal{D}|\mathcal{Z},\mathcal{M},\mathcal{S}) \mathrm{d}\mathcal{Z}\right).
\end{aligned} \tag{20.5}$$

Combining Eq. (20.4) and Eq. (20.5) yields the VBEM algorithm described in Fig. 20.2 (see Section 10.2.1 of reference [15] for details).

As mentioned in Section 18.2.1, $r(\mathcal{W},\mathcal{M},\mathcal{S})$ obtained by the VBEM algorithm may be a good approximation to the posterior probability $p(\mathcal{W},\mathcal{M},\mathcal{S}|\mathcal{D})$:

$$r(\mathcal{W},\mathcal{M},\mathcal{S}) = \mathrm{Dir}(\mathcal{W}|\widehat{\alpha}) \prod_{\ell=1}^{m} N(\boldsymbol{\mu}_\ell|\widehat{\boldsymbol{h}}_\ell,(\widehat{\beta}_\ell \boldsymbol{S}_\ell)^{-1}) W(\boldsymbol{S}_\ell|\widehat{\boldsymbol{W}}_\ell,\widehat{\nu}_\ell).$$

The expectations of the marginals of $r(\mathcal{W},\mathcal{M},\mathcal{S})$ can be obtained as

$$\widehat{w}_\ell = \frac{\widehat{\alpha}_\ell}{\sum_{\ell'=1}^{m} \widehat{\alpha}_{\ell'}}, \quad \widehat{\boldsymbol{\mu}}_\ell = \widehat{\boldsymbol{h}}_\ell, \quad \text{and} \quad \widehat{\boldsymbol{S}}_\ell = \widehat{\nu}_\ell \widehat{\boldsymbol{W}}_\ell,$$

which may be used as the most plausible parameter values. Then the following density estimator is obtained:

$$\widehat{p}(\boldsymbol{x}) = \sum_{\ell=1}^{m} \widehat{w}_\ell N(\boldsymbol{x}|\widehat{\boldsymbol{\mu}}_\ell,\widehat{\boldsymbol{S}}_\ell^{-1}).$$

A MATLAB code for computing this VBEM-based density estimator is given in Fig. 20.3, and its behavior is illustrated in Fig. 20.4. Here, the mixture model of five Gaussian components is fitted to the mixture of two Gaussian distributions. As shown in Fig. 20.4, two out of five Gaussian components fit the true two Gaussian distributions well, and the remaining three Gaussian components are almost eliminated—the learned mixing coefficients are given as

1. Initialize parameters $\{\widehat{\alpha}_\ell, \widehat{\boldsymbol{h}}_\ell, \widehat{\beta}_\ell, \widehat{\boldsymbol{W}}_\ell, \widehat{\nu}_\ell\}_{\ell=1}^m$.
2. VB-E step: Compute the distribution of $\mathcal{Z} = \{z_1, \ldots, z_n\}$ from current solution $\{\widehat{\alpha}_\ell, \widehat{\boldsymbol{h}}_\ell, \widehat{\beta}_\ell, \widehat{\boldsymbol{W}}_\ell, \widehat{\nu}_\ell\}_{\ell=1}^m$:

$$q(\mathcal{Z}) = \prod_{i=1}^{n}\prod_{\ell=1}^{m} \widehat{\eta}_{i,\ell}^{z_{i,\ell}}, \quad \text{where } \widehat{\eta}_{i,\ell} = \frac{\widehat{\rho}_{i,\ell}}{\sum_{\ell'=1}^{m} \widehat{\rho}_{i,\ell'}}.$$

$\{\widehat{\eta}_{i,\ell}\}_{i=1,\ell=1}^{n\ \ \ m}$ are the responsibilities computed as

$$\widehat{\rho}_{i,\ell} = \exp\Bigg(\psi(\widehat{\alpha}_\ell) - \psi\left(\sum_{\ell'=1}^{m} \widehat{\alpha}_{\ell'}\right) + \frac{1}{2}\sum_{j=1}^{d} \psi\left(\frac{\widehat{\nu}_\ell + 1 - j}{2}\right) + \frac{1}{2}\log\det(\widehat{\boldsymbol{W}}_\ell) - \frac{d}{2\widehat{\beta}_\ell} - \frac{\widehat{\nu}_\ell}{2}(\boldsymbol{x}_i - \widehat{\boldsymbol{h}}_\ell)^\top \widehat{\boldsymbol{W}}_\ell(\boldsymbol{x}_i - \widehat{\boldsymbol{h}}_\ell)\Bigg),$$

where $\psi(\alpha)$ denotes the *digamma function* defined as the log-derivative of the gamma function $\Gamma(\alpha)$:

$$\psi(\alpha) = \frac{\mathrm{d}}{\mathrm{d}\alpha}\log\Gamma(\alpha) = \frac{\Gamma'(\alpha)}{\Gamma(\alpha)}.$$

3. VB-M step: Compute the joint distribution of $\mathcal{W} = (w_1, \ldots, w_m)$, $\mathcal{M} = (\boldsymbol{\mu}_1, \ldots, \boldsymbol{\mu}_m)$, and $\mathcal{S} = (\boldsymbol{S}_1, \ldots, \boldsymbol{S}_m)$ from current responsibilities $\{\widehat{\eta}_{i,\ell}\}_{i=1,\ell=1}^{n\ \ \ m}$:

$$r(\mathcal{W}, \mathcal{M}, \mathcal{S}) = \mathrm{Dir}(\mathcal{W}|\widehat{\alpha}) \prod_{\ell=1}^{m} N(\boldsymbol{\mu}_\ell|\widehat{\boldsymbol{h}}_\ell, (\widehat{\beta}_\ell \boldsymbol{S}_\ell)^{-1}) W(\boldsymbol{S}_\ell|\widehat{\boldsymbol{W}}_\ell, \widehat{\nu}_\ell),$$

where

$$\widehat{\gamma}_\ell = \sum_{i=1}^{n} \widehat{\eta}_{i,\ell}, \quad \widehat{\boldsymbol{c}}_\ell = \frac{1}{\widehat{\gamma}_\ell}\sum_{i=1}^{n} \widehat{\eta}_{i,\ell}\boldsymbol{x}_i, \quad \widehat{\boldsymbol{h}}_\ell = \frac{\widehat{\gamma}_\ell}{\widehat{\beta}_\ell}\widehat{\boldsymbol{c}}_\ell,$$

$$\widehat{\alpha}_\ell = \alpha_0 + \widehat{\gamma}_\ell, \quad \widehat{\beta}_\ell = \beta_0 + \widehat{\gamma}_\ell, \quad \widehat{\nu}_\ell = \nu_0 + \widehat{\gamma}_\ell, \quad \text{and}$$

$$\widehat{\boldsymbol{W}}_\ell = \left(\boldsymbol{W}_0^{-1} + \sum_{i=1}^{n} \widehat{\eta}_{i,\ell}(\boldsymbol{x}_i - \widehat{\boldsymbol{c}}_\ell)(\boldsymbol{x}_i - \widehat{\boldsymbol{c}}_\ell)^\top + \frac{\beta_0\widehat{\gamma}_\ell}{\beta_0 + \widehat{\gamma}_\ell}\widehat{\boldsymbol{c}}_\ell\widehat{\boldsymbol{c}}_\ell^\top\right)^{-1}.$$

4. Iterate 2–3 until convergence.

FIGURE 20.2

VBEM algorithm for Gaussian mixture model. $(\alpha_0, \beta_0, \boldsymbol{W}_0, \nu_0)$ are hyperparameters.

```
x=[2*randn(1,100)-5 randn(1,50); randn(1,100) randn(1,50)+3];
[d,n]=size(x); m=5; e=rand(n,m); W=zeros(d,d,m); b0=1;
for o=1:10000
  e=e./repmat(sum(e,2),[1 m]);
  g=sum(e); a=1+g; b=b0+g; nu=3+g; w=a/sum(a);
  xe=x*e; c=xe./repmat(g,[d 1]); h=xe./repmat(b,[d 1]);
  for k=1:m
    t1=x-repmat(c(:,k),[1 n]); t2=x-repmat(h(:,k),[1 n]);
    W(:,:,k)=inv(eye(d)+(t1.*repmat(e(:,k)',[d 1]))*t1' ...
                +c(:,k)*c(:,k)'*b0*g(k)/(b0+g(k)));
    t3=sum(psi((nu(k)+1-[1:d])/2))+log(det(W(:,:,k)));
    e(:,k)=exp(t3/2+psi(a(k))-psi(sum(a))-d/2/b(k) ...
                -sum(t2.*(W(:,:,k)*t2))*nu(k)/2);
  end
  if o>1 && norm(w-w0)+norm(h-h0)+norm(W(:)-W0(:))<0.001
    break
  end
  w0=w; h0=h; W0=W;
end

figure(1); clf; hold on
plot(x(1,:),x(2,:),'ro'); v=linspace(0,2*pi,100);
for k=1:m
  [V,D]=eig(nu(k)*W(:,:,k));
  X=3*w(k)*V'*[cos(v)/D(1,1); sin(v)/D(2,2)];
  plot(h(1,k)+X(1,:),h(2,k)+X(2,:),'b-')
end
```

FIGURE 20.3

MATLAB code of VBEM algorithm for Gaussian mixture model.

$$(\widehat{w}_1, \widehat{w}_2, \widehat{w}_3, \widehat{w}_4, \widehat{w}_5) = (0.01, 0.33, 0.01, 0.01, 0.65).$$

As explained in Section 18.2.1, the negative of the variational free energy,

$$F(q,r) = \iiiint q(\mathcal{Z})r(\mathcal{W},\mathcal{M},\mathcal{S}) \times \log \frac{q(\mathcal{Z})r(\mathcal{W},\mathcal{M},\mathcal{S})}{p(\mathcal{D},\mathcal{Z},\mathcal{W},\mathcal{M},\mathcal{S};\alpha_0,\beta_0,\mathbf{W}_0,\nu_0)} \mathrm{d}\mathcal{Z}\mathrm{d}\mathcal{W}\mathrm{d}\mathcal{M}\mathrm{d}\mathcal{S},$$

gives a lower bound of the log marginal likelihood, $\mathrm{ML}(\alpha_0, \beta_0, \mathbf{W}_0, \nu_0)$. Based on the dependency in Fig. 20.1, the variational free energy can be expressed as

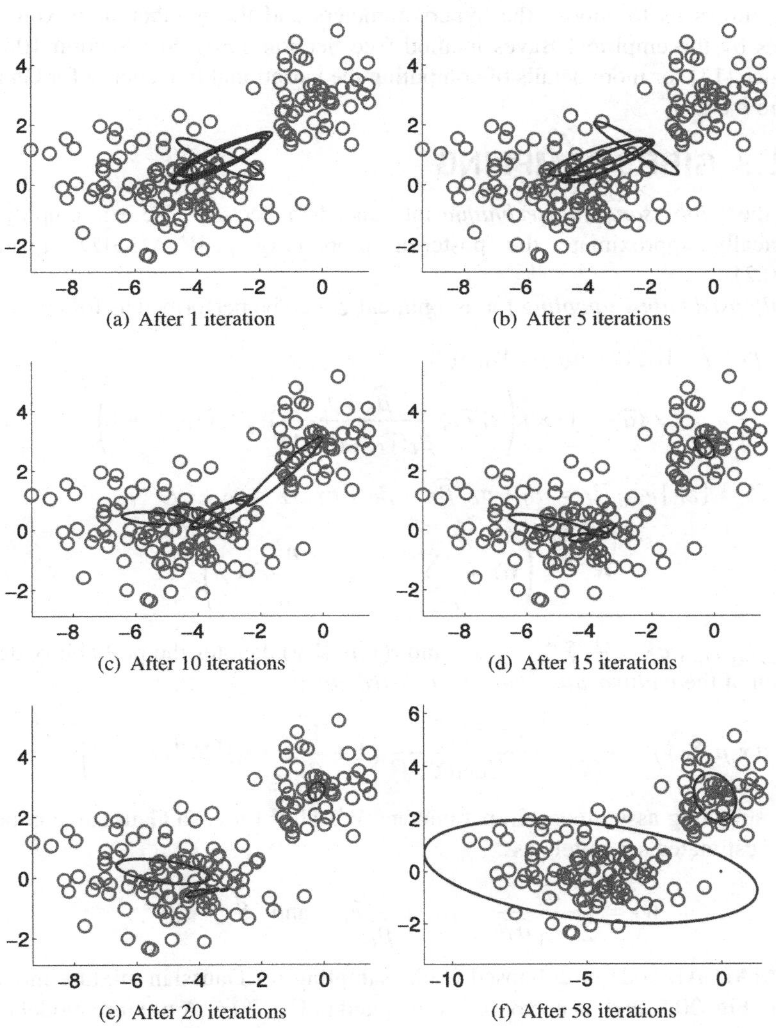

FIGURE 20.4

Example of VBEM algorithm for Gaussian mixture model. The size of ellipses is proportional to the mixing weights $\{w_\ell\}_{\ell=1}^m$. A mixture model of five Gaussian components is used here, but three components have mixing coefficient close to zero and thus they are almost eliminated.

$$\begin{aligned} F(q,r) = \iiiint & q(\mathcal{Z})r(\mathcal{W},\mathcal{M},\mathcal{S})\Big(\log q(\mathcal{Z}) + \log r(\mathcal{W},\mathcal{M},\mathcal{S}) \\ & - \log p(\mathcal{D}|\mathcal{Z},\mathcal{M},\mathcal{S}) - \log p(\mathcal{Z}|\mathcal{W}) - \log p(\mathcal{W};\alpha_0) \\ & - \log p(\mathcal{M},\mathcal{S};\beta_0,\mathbf{W}_0,\nu_0)\Big)\mathrm{d}\mathcal{Z}\mathrm{d}\mathcal{W}\mathrm{d}\mathcal{M}\mathrm{d}\mathcal{S}, \end{aligned}$$

which allows us to choose the hyperparameters and the number of mixing components by the empirical Bayes method (see Section 17.4). See Section 10.2.2 of reference [15] for more details of computing the variational free energy for Gaussian mixture models.

20.1.3 GIBBS SAMPLING

Next, the *Gibbs sampling technique* introduced in Section 19.3.3 is employed to numerically approximate the posterior probability $p(\mathcal{W},\mathcal{M},\mathcal{S}|\mathcal{D})$ given by Eq. (20.2).

Collapsed Gibbs sampling for assignment z_i can be performed as follows [74]:

$$p(z_{i,\ell}=1|\widetilde{\mathcal{Z}},\mathcal{D};\alpha_0,\beta_0,\boldsymbol{W}_0,\nu_0)$$
$$\propto (\widehat{\alpha}_\ell - 1)\times t\left(\boldsymbol{x}_i\middle|\widehat{\boldsymbol{c}}_\ell,\frac{\widehat{\beta}_\ell+1}{\widehat{\beta}_\ell(\widehat{\nu}_\ell-d+1)}\widehat{\boldsymbol{W}}_\ell^{-1},\widehat{\nu}_\ell-d+1\right), \tag{20.6}$$

where $\widetilde{\mathcal{Z}}=\{z_{i'}\}_{i'\neq i}$, $\widehat{\alpha}_\ell=\alpha_0+n_\ell$, $\widehat{\beta}_\ell=\beta_0+n_\ell$, $\widehat{\nu}_\ell=\nu_0+n_\ell$,

$$\widehat{\boldsymbol{W}}_\ell^{-1}=\left(\boldsymbol{W}_0+\sum_{i:z_{i,\ell}=1}\boldsymbol{x}_i\boldsymbol{x}_i^\top-\frac{n_\ell^2}{\widehat{\beta}_\ell}\widehat{\boldsymbol{c}}_\ell\widehat{\boldsymbol{c}}_\ell^\top\right)^{-1},$$

$n_\ell=\sum_{i=1}^n z_{i,\ell}$, $\widehat{\boldsymbol{c}}_\ell=\frac{1}{n_\ell}\sum_{i:z_{i,\ell}=1}^n\boldsymbol{x}_i$, and $t(\boldsymbol{x}|\boldsymbol{\mu},\boldsymbol{\Sigma},\nu)$ denotes the probability density function of the *multivariate Student's t-distribution*:

$$t(\boldsymbol{x}|\boldsymbol{\mu},\boldsymbol{\Sigma},\nu)=\frac{\Gamma\left(\frac{\nu+d}{2}\right)}{\Gamma(\frac{\nu}{2})(\nu\pi)^{d/2}\det(\boldsymbol{\Sigma})^{1/2}}\left(1+\frac{1}{\nu}(\boldsymbol{x}-\boldsymbol{\mu})^\top\boldsymbol{\Sigma}^{-1}(\boldsymbol{x}-\boldsymbol{\mu})\right)^{-\frac{\nu+d}{2}}.$$

After estimating assignment $\mathcal{Z}$, parameters $\mathcal{W},\mathcal{M},\mathcal{S}$ for each Gaussian component may be estimated separately as

$$\widehat{w}_\ell=\frac{\widehat{\alpha}_\ell}{\sum_{\ell'=1}^m\widehat{\alpha}_{\ell'}},\quad \widehat{\boldsymbol{\mu}}_\ell=\frac{n_\ell}{\widehat{\beta}_\ell}\widehat{\boldsymbol{c}}_\ell,\quad\text{and}\quad \widehat{\boldsymbol{S}}_\ell=\widehat{\nu}_\ell\widehat{\boldsymbol{W}}_\ell.$$

A MATLAB code of collapsed Gibbs sampling for Gaussian mixture models is given in Fig. 20.5, and its behavior is illustrated in Fig. 20.6. A mixture model of five Gaussian components is used here, but only two components remain and no samples belong to the remaining three components after Gibbs sampling.

The *Dirichlet process*, denoted by $\mathrm{DP}(\alpha_0,p_0)$, is a probability distribution of probability distributions, and a Dirichlet distribution is generated from a Dirichlet process. p_0 is called the *base distribution* which corresponds to the mean of the Dirichlet distribution, while α_0 is called the *concentration parameter* which corresponds to the *precision* (i.e., the inverse variance) of the Dirichlet distribution. The *Dirichlet process mixture* is a mixture model using the Dirichlet process as a prior probability and is equivalent to considering an infinite-dimensional Dirichlet distribution. Indeed, taking the limit $m\to\infty$ for $\alpha_0=\alpha_0'/m$ in Eq. (20.6) gives a *collapsed Gibbs sampling* procedure for the Dirichlet process mixture model.

```
x=[2*randn(1,100)-5 randn(1,50); randn(1,100) randn(1,50)+3];
[d,n]=size(x); m=5; z=mod(randperm(n),m)+1;
a0=1; b0=1; n0=1; W0=eye(d);
for o=1:100
  for i=1:n
    g=(1:n~=i); X=x(:,g); Z=z(g);
    for k=1:m
      p(k)=0; e=(Z==k); t=sum(e);
      if t~=0
        u=n0+t-d+1; b=b0+t; c=sum(X(:,e),2); xi=x(:,i)-c/t;
        W=inv((b+1)/b/u*(W0+X(:,e)*X(:,e)'-c*c'/b));
        p(k)=(a0+t-1)*gamma((u+d)/2)/gamma(u/2)*u^(-d/2) ...
             *sqrt(det(W))*(1+xi'*W*xi/u)^(-(u+d)/2);
    end, end
    z(i)=find(cumsum(p/sum(p))>rand,1);
end, end

figure(1); clf; hold on
plot(x(1,:),x(2,:),'ro'); v=linspace(0,2*pi,100);
for k=1:m
  e=(z==k); t=sum(e); nu(k)=n0+t; u=nu(k)-d+1; b=b0+t;
  c=sum(x(:,e),2); w(k)=a0+t; h(:,k)=c/b; W(:,:,k)=zeros(d);
  if t~=0
    W(:,:,k)=inv((W0+x(:,e)*x(:,e)'-c*c'/b));
  end
end
w=w./sum(w);
for k=1:m
  [V,D]=eig(nu(k)*W(:,:,k));
  X=3*w(k)*V'*[cos(v)/D(1,1); sin(v)/D(2,2)];
  plot(h(1,k)+X(1,:),h(2,k)+X(2,:),'b-')
end
```

FIGURE 20.5

MATLAB code of collapsed Gibbs sampling for Gaussian mixture model.

20.2 LATENT DIRICHLET ALLOCATION (LDA)

One of the most successful applications of the Bayesian generative approach is *topic modeling* called LDA [16]. In this section, an implementation of LDA based on *Gibbs sampling* is explained [51].

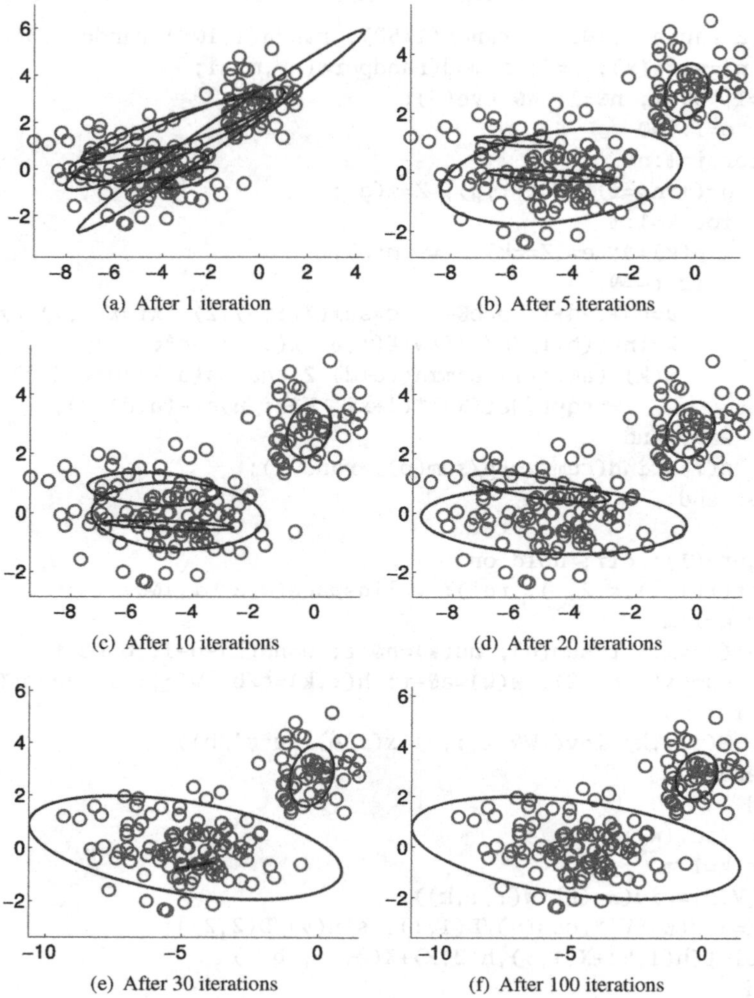

FIGURE 20.6

Example of collapsed Gibbs sampling for Gaussian mixture model. A mixture model of five Gaussian components is used here, but only two components remain and no samples belong to the remaining three components.

20.2.1 TOPIC MODELS

A topic model is a generative model for a set of D documents,

$$\mathcal{W} = \{\boldsymbol{w}_1, \ldots, \boldsymbol{w}_D\},$$

where the dth document $\boldsymbol{w}_d = (w_{d,1},\ldots,w_{d,n_d})$ is a sequence of n_d words in a corpus with W unique words. More specifically, each word $w_{n,d} \in \{1,\ldots,W\}$ is supposed to be given by the mixture of T topics as

$$p(w_{n,d}|\boldsymbol{\theta}^{(d)},\boldsymbol{\phi}^{(1)},\ldots,\boldsymbol{\phi}^{(T)}) = \sum_{t_{n,d}=1}^{T} p(t_{n,d}|\boldsymbol{\theta}^{(d)})p(w_{n,d}|t_{n,d},\boldsymbol{\phi}^{(t_{n,d})}),$$

where $t_{n,d} \in \{1,\ldots,T\}$ is a latent variable that indicates the topic of $w_{n,d}$. For topic $t = 1,\ldots,T$, the probability,

$$p(t|\boldsymbol{\theta}^{(d)}) = \theta_t^{(d)} \geq 0,$$

denotes the discrete probability of topic t with parameter $\boldsymbol{\theta}^{(d)} = (\theta_1^{(d)},\ldots,\theta_T^{(d)})^\top$ such that $\sum_{t=1}^{T}\theta_t^{(d)} = 1$. For $w = 1,\ldots,W$, the probability,

$$p(w|t,\boldsymbol{\phi}^{(t)}) = \phi_w^{(t)} \geq 0,$$

denotes the discrete probability of observing word w under topic t with parameter $\boldsymbol{\phi}^{(t)} = (\phi_1^{(t)},\ldots,\phi_W^{(t)})^\top$ such that $\sum_{w=1}^{W}\phi_w^{(t)} = 1$. This generative model means that, for each word w in document d, topic t is chosen following the discrete probability with parameter $\boldsymbol{\theta}^{(d)}$ and then word w is chosen following the discrete probability with parameter $\boldsymbol{\phi}^{(t)}$.

The goal of topic modeling is to specify the multinomial parameters $\boldsymbol{\theta}^{(d)}$ and $\boldsymbol{\phi}^{(t)}$ from data $\mathcal{W}$.

20.2.2 BAYESIAN FORMULATION

Here, instead of directly learning the parameters $\boldsymbol{\theta}^{(d)}$ and $\boldsymbol{\phi}^{(t)}$, let us consider the posterior probability of the set $\mathcal{T}$ of all topics for all documents $\mathcal{W}$:

$$\mathcal{T} = \{\boldsymbol{t}_1,\ldots,\boldsymbol{t}_D\},$$

where $\boldsymbol{t}_d = (t_{d,1},\ldots,t_{d,n_d})$.

For the multinomial parameters $\boldsymbol{\theta}^{(d)}$ and $\boldsymbol{\phi}^{(t)}$, the *symmetric Dirichlet distributions* (see Section 6.3) are considered as the prior probabilities:

$$p(\boldsymbol{\theta}^{(d)};\alpha) = \mathrm{Dir}(\boldsymbol{\theta}^{(d)};\alpha) \quad \text{and} \quad p(\boldsymbol{\phi}^{(t)};\beta) = \mathrm{Dir}(\boldsymbol{\phi}^{(t)};\beta),$$

where $\mathrm{Dir}(\cdot\,;\alpha)$ denotes the symmetric Dirichlet density with concentration parameter α. Note that the Dirichlet distributions are *conjugate* (see Section 17.2) for discrete distributions.

The posterior probability of topics $\mathcal{T}$ is given by

$$p(\mathcal{T}|\mathcal{W};\alpha,\beta) = \frac{p(\mathcal{W},\mathcal{T};\alpha,\beta)}{\sum_{\mathcal{T}} p(\mathcal{W},\mathcal{T};\alpha,\beta)}, \tag{20.7}$$

where the marginal likelihood $p(\mathcal{W}, \mathcal{T}; \alpha, \beta)$ can be expressed *analytically* as

$$p(\mathcal{W}, \mathcal{T}; \alpha, \beta)$$
$$= \prod_{d=1}^{D} \prod_{n=1}^{n_d} \int \theta_{t_{n,d}}^{(d)} \mathrm{Dir}(\boldsymbol{\theta}^{(d)}; \alpha) \mathrm{d}\boldsymbol{\theta}^{(d)} \int \phi_{w_{n,d}}^{(t_{n,d})} \mathrm{Dir}(\boldsymbol{\phi}^{(t_{n,d})}; \beta) \mathrm{d}\boldsymbol{\phi}^{(t_{n,d})}$$
$$= \left(\frac{\Gamma(T\alpha)}{\Gamma(\alpha)^T}\right)^D \prod_{d=1}^{D} \frac{\prod_{t=1}^{T} \Gamma(n_t^{(d)} + \alpha)}{\Gamma(\sum_{t=1}^{T} n_t^{(d)} + T\alpha)} \left(\frac{\Gamma(W\beta)}{\Gamma(\beta)^W}\right)^T \prod_{t=1}^{T} \frac{\prod_{w=1}^{W} \Gamma(n_t^{(w)} + \beta)}{\Gamma(\sum_{w=1}^{W} n_t^{(w)} + W\beta)},$$

where $n_t^{(d)}$ denotes the number of times a word from document d is assigned to topic t, $n_t^{(w)}$ denotes the number of times word w is assigned to topic t, and $\Gamma(\cdot)$ denotes the *gamma function* (see Section 4.3).

20.2.3 GIBBS SAMPLING

Let us use *Gibbs sampling* (see Section 19.3.3) to approximate the posterior probability $p(\mathcal{T}|\mathcal{W}; \alpha, \beta)$.

Let $\widetilde{\mathcal{W}}$ and $\widetilde{\mathcal{T}}$ be defined in the same way as $\mathcal{W}$ and $\mathcal{T}$, but $w_{n,d}$ and $t_{n,d}$ are excluded. Then the conditional probability needed for Gibbs sampling is given by

$$\begin{aligned} p(t_{n,d}|\widetilde{\mathcal{T}}, \mathcal{W}) &= p(t_{n,d}|\widetilde{\mathcal{T}}, w_{n,d}, \widetilde{\mathcal{W}}) \\ &\propto p(t_{n,d}|\widetilde{\mathcal{T}}) p(w_{n,d}|t_{n,d}, \widetilde{\mathcal{T}}, \widetilde{\mathcal{W}}). \end{aligned} \tag{20.8}$$

The first term in Eq. (20.8) can be expressed as

$$p(t_{n,d}|\widetilde{\mathcal{T}}) = \int p(t_{n,d}|\boldsymbol{\theta}^{(d)}) p(\boldsymbol{\theta}^{(d)}|\widetilde{\mathcal{T}}) \mathrm{d}\boldsymbol{\theta}^{(d)}, \tag{20.9}$$

where Bayes' theorem yields

$$p(\boldsymbol{\theta}^{(d)}|\widetilde{\mathcal{T}}) \propto p(\widetilde{\mathcal{T}}|\boldsymbol{\theta}^{(d)}) p(\boldsymbol{\theta}^{(d)}; \alpha).$$

Here, the prior probability $p(\boldsymbol{\theta}^{(d)}; \alpha)$ is chosen to be symmetric Dirichlet distribution with concentration parameter α, which is conjugate for the discrete distribution $p(\widetilde{\mathcal{T}}|\boldsymbol{\theta}^{(d)})$. Then the posterior probability $p(\boldsymbol{\theta}^{(d)}|\widetilde{\mathcal{T}})$ is also the symmetric Dirichlet distribution with concentration parameter $\widetilde{n}_{t_{n,d}}^{(d)} + \alpha$, where $\widetilde{n}_{t_{n,d}}^{(d)}$ is defined in the same way as $n_{t_{n,d}}^{(d)}$, but $w_{n,d}$ and $t_{n,d}$ are excluded. Then Eq. (20.9) can be computed analytically as

$$p(t_{n,d}|\widetilde{\mathcal{T}}) = \int \theta_{t_{n,d}}^{(d)} \mathrm{Dir}(\boldsymbol{\theta}^{(d)}; \widetilde{n}_{t_{n,d}}^{(d)} + \alpha) \mathrm{d}\boldsymbol{\theta}^{(d)} = \frac{\widetilde{n}_{t_{n,d}}^{(d)} + \alpha}{\sum_{t=1}^{T} \widetilde{n}_t^{(d)} + T\alpha}.$$

Similarly, the second term in Eq. (20.8) can be expressed as

$$p(w_{n,d}|t_{n,d}, \widetilde{\mathcal{T}}, \widetilde{\mathcal{W}}) = \int p(w_{n,d}|t_{n,d}, \boldsymbol{\phi}^{(t_{n,d})}) p(\boldsymbol{\phi}^{(t_{n,d})}|\widetilde{\mathcal{T}}, \widetilde{\mathcal{W}}) \mathrm{d}\boldsymbol{\phi}^{(t_{n,d})}, \tag{20.10}$$

where Bayes' theorem yields

$$p(\boldsymbol{\phi}^{(t_{n,d})}|\widetilde{\mathcal{T}},\widetilde{\mathcal{W}}) \propto p(\widetilde{\mathcal{W}}|\boldsymbol{\phi}^{(t_{n,d})},\widetilde{\mathcal{T}})p(\boldsymbol{\phi}^{(t_{n,d})};\beta).$$

Here, the prior probability $p(\boldsymbol{\phi}^{(t_{n,d})};\beta)$ is chosen to be the symmetric Dirichlet distribution with concentration parameter β, which is conjugate for the discrete distribution $p(\widetilde{\mathcal{W}}|\boldsymbol{\phi}^{(t_{n,d})},\widetilde{\mathcal{T}})$. Then the posterior probability $p(\boldsymbol{\phi}^{(t_{n,d})}|\widetilde{\mathcal{T}},\widetilde{\mathcal{W}})$ is also the symmetric Dirichlet distribution with concentration parameter $\widetilde{n}_{t_{n,d}}^{(w_{n,d})}+\beta$, where $\widetilde{n}_{t_{n,d}}^{(w_{n,d})}$ is defined in the same way as $n_{t_{n,d}}^{(w_{n,d})}$, but $w_{n,d}$ and $t_{n,d}$ are excluded. Then Eq. (20.10) can be computed analytically as

$$\begin{aligned} p(w_{n,d}|t_{n,d},\widetilde{\mathcal{T}},\widetilde{\mathcal{W}}) &= \int \phi_{w_{n,d}}^{(t_{n,d})}\mathrm{Dir}(\boldsymbol{\phi}^{(t_{n,d})};\widetilde{n}_{t_{n,d}}^{(w_{n,d})}+\beta)\mathrm{d}\boldsymbol{\phi}^{(t_{n,d})} \\ &= \frac{\widetilde{n}_{t_{n,d}}^{(w_{n,d})}+\beta}{\sum_{w=1}^{W}\widetilde{n}_{t_{n,d}}^{(w)}+W\beta}. \end{aligned}$$

Summarizing the above equations, Eq. (20.8) can be expressed as

$$p(t_{n,d}|\widetilde{\mathcal{T}},\mathcal{W}) \propto \frac{\widetilde{n}_{t_{n,d}}^{(d)}+\alpha}{\sum_{t=1}^{T}\widetilde{n}_{t}^{(d)}+T\alpha} \times \frac{\widetilde{n}_{t_{n,d}}^{(w_{n,d})}+\beta}{\sum_{w=1}^{W}\widetilde{n}_{t_{n,d}}^{(w)}+W\beta},$$

which shows that Gibbs sampling can be efficiently performed just by counting the number of words.

Finally, with estimated topics $\widehat{\mathcal{T}}$, the solutions $\widehat{\theta}_t^{(d)}$ and $\widehat{\phi}_t^{(w)}$ for any $w \in \{1,\ldots,W\}$, $t \in \{1,\ldots,T\}$, and $d \in \{1,\ldots,D\}$ can be computed as

$$\widehat{\theta}_t^{(d)} = \frac{\widehat{n}_{t_{n,d}}^{(d)}+\alpha}{\sum_{t=1}^{T}\widehat{n}_{t}^{(d)}+T\alpha} \quad \text{and} \quad \widehat{\phi}_t^{(w)} = \frac{\widehat{n}_{t_{n,d}}^{(w_{n,d})}+\beta}{\sum_{w=1}^{W}\widehat{n}_{t_{n,d}}^{(w)}+W\beta}.$$

PART 4

DISCRIMINATIVE APPROACH TO STATISTICAL MACHINE LEARNING

As discussed in Chapter 11, the problem of statistical pattern recognition is formulated as the problem of estimating the class-posterior probability $p(y|\boldsymbol{x})$. In the generative approach explored in Part 3, the problem of estimating the class-posterior probability $p(y|\boldsymbol{x})$ is replaced with the problem of estimating the joint probability $p(\boldsymbol{x}, y)$ based on the following equality:

$$\underset{y}{\text{argmax}}\, p(y|\boldsymbol{x}) = \underset{y}{\text{argmax}}\, p(\boldsymbol{x}, y).$$

Generative model estimation is the most general approach in statistical machine learning, because knowing the data generating model is equivalent to knowing everything about the data. Therefore, any kind of data analysis is possible through generative model estimation.

When a good parametric model is available, *maximum likelihood estimation* or *Bayesian methods* will be highly useful for generative model estimation (see Part 3). However, without strong prior knowledge on parametric models, estimating the generative model is statistically a hard problem. Non-parametric methods introduced in Chapter 16 could be used if prior knowledge on generative models is not available, but non-parametric methods tend to perform poorly when the dimensionality of data is not small.

In Part 4, an alternative approach to generative model estimation called the *discriminative approach* is explored. In the discriminative approach, the class-posterior probability $p(y|\boldsymbol{x})$ is directly modeled. More specifically, $p(y|\boldsymbol{x})$ is regarded as the sum of a function $f(\boldsymbol{x})$ and some noise, and the problem of *function approximation* from input–output paired samples $\{(\boldsymbol{x}_i, y_i)\}_{i=1}^n$ is considered. Such a problem formulation is called *supervised learning*, where output y is regarded as supervision from the (noisy) oracle. The supervised learning problem is called *regression* if the output y is continuous, and is called *classification* if the output y is categorical.

After introducing standard function models used for regression and classification in Chapter 21, various regression and classification techniques will be introduced. The most fundamental regression technique called the *least squares* method is introduced in Chapter 22, its constraint (or regularized) variants for avoiding overfitting are introduced in Chapter 23. In Chapter 24 and Chapter 25, more advanced regression techniques considering *sparsity* and *robustness* will be discussed, respectively.

In Chapter 26, it is shown that the least squares regression method can also be used for classification, and various issues specific to classification will be discussed. In Chapter 27, a powerful classification algorithm based on the *maximum margin principle* called the *support vector machine* and its robust variant is introduced. In Chapter 28, a probabilistic pattern recognition method that directly learns the class-posterior probability $p(y|\boldsymbol{x})$ called *logistic regression* is introduced. Finally, in Chapter 29, classification of sequence data is discussed.

CHAPTER

LEARNING MODELS 21

CHAPTER CONTENTS

Through Part 4, let us consider the *supervised learning* problem of approximating a function $y = f(\boldsymbol{x})$ from its input-output paired training samples $\{(\boldsymbol{x}_i, y_i)\}_{i=1}^{n}$ (see Section 1.2.1). The input $\boldsymbol{x}$ is assumed to be a real-valued d-dimensional vector, and the output y is a real scalar in the case of *regression* and a categorical value $\{1, \ldots, c\}$ in the case of *classification*, where c denotes the number of classes. In this chapter, various *models* for supervised learning are introduced.

21.1 LINEAR-IN-PARAMETER MODEL

For simplicity, let us begin with a one-dimensional learning target function f. The simplest model for approximating f would be the *linear-in-input model* $\theta \times x$. Here, θ denotes a scalar parameter and the target function is approximated by learning the parameter θ. Although the linear-in-input model is mathematically easy to handle, it can only approximate a linear function (i.e., a straight line) and thus its expression power is limited (Fig. 21.1).

The *linear-in-parameter model* is an extension of the linear-in-input model that allows approximation of nonlinear functions:

$$f_{\boldsymbol{\theta}}(x) = \sum_{j=1}^{b} \theta_j \phi_j(x),$$

where $\phi_j(x)$ and θ_j are a *basis function* and its parameter, respectively, and b denotes the number of basis functions. The linear-in-parameter model may be compactly expressed as

$$f_{\boldsymbol{\theta}}(x) = \boldsymbol{\theta}^\top \boldsymbol{\phi}(x),$$

where

$$\boldsymbol{\phi}(x) = (\phi_1(x), \ldots, \phi_b(x))^\top,$$

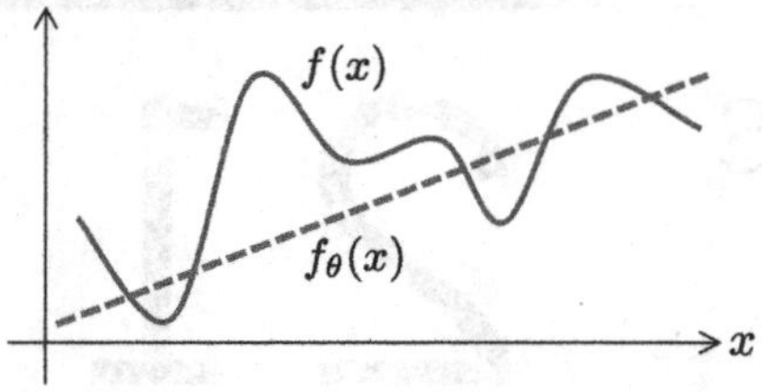

FIGURE 21.1

Linear-in-input model cannot approximate nonlinear functions.

$$\boldsymbol{\theta} = (\theta_1, \ldots, \theta_b)^\top,$$

and $^\top$ denotes the transpose. The linear-in-parameter is still linear in terms of parameters $\boldsymbol{\theta}$, but it can express nonlinear functions, e.g., by using *polynomial functions*

$$\boldsymbol{\phi}(x) = (1, x, x^2, \ldots, x^{b-1})^\top,$$

or *sinusoidal functions* for $b = 2m + 1$

$$\boldsymbol{\phi}(x) = (1, \sin x, \cos x, \sin 2x, \cos 2x, \ldots, \sin mx, \cos mx)^\top,$$

as basis functions.

The linear-in-parameter can be naturally extended to d-dimensional input vector $\boldsymbol{x} = (x^{(1)}, \ldots, x^{(d)})^\top$ as

$$f_{\boldsymbol{\theta}}(\boldsymbol{x}) = \sum_{j=1}^{b} \theta_j \phi_j(\boldsymbol{x}) = \boldsymbol{\theta}^\top \boldsymbol{\phi}(\boldsymbol{x}). \tag{21.1}$$

Below, methods for constructing multidimensional basis functions from one-dimensional basis functions are discussed.

The *multiplicative model* expresses multidimensional basis functions by the product of one-dimensional basis functions as

$$f_{\boldsymbol{\theta}}(\boldsymbol{x}) = \sum_{j_1=1}^{b'} \cdots \sum_{j_d=1}^{b'} \theta_{j_1,\ldots,j_d} \phi_{j_1}(x^{(1)}) \cdots \phi_{j_d}(x^{(d)}),$$

where b' denotes the number of parameters in each dimension. Since all possible combinations of one-dimensional basis functions are considered in the multiplicative model, it can express complicated functions, as illustrated in Fig. 21.2(a). However,

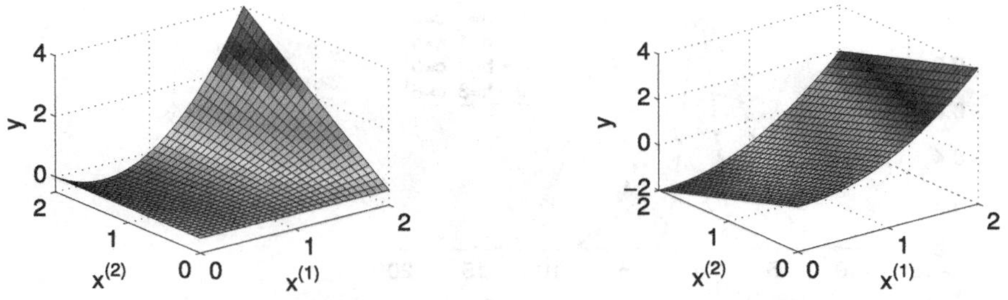

(a) Multiplicative model

(b) Additive model

FIGURE 21.2

Multidimensional basis functions. The multiplicative model is expressive, but the number of parameters grows exponentially in input dimensionality. On the other hand, in the additive model, the number of parameters grows only linearly in input dimensionality, but its expression power is limited.

the total number of parameters is $(b')^d$, which grows exponentially as input dimensionality d increases. For example, when $b' = 10$ and $d = 100$, the total number of parameters is given by

$$10^{100} = 1\underbrace{000\cdots000}_{100},$$

which is an astronomical number having 100 zeros after the first one and cannot be handled in computers. Such exponential growth in input dimensionality is often referred to as the *curse of dimensionality*.

The *additive model* expresses multidimensional basis functions by the sum of one-dimensional basis functions as

$$f_{\boldsymbol{\theta}}(\boldsymbol{x}) = \sum_{k=1}^{d} \sum_{j=1}^{b'} \theta_{k,j} \phi_j(x^{(k)}).$$

In the additive model, the total number of parameters is $b'd$, which grows only linearly as input dimensionality d increases. For example, when $b' = 10$ and $d = 100$, the total number of parameters is $10 \times 100 = 1000$ and thus it can be easily handled by standard computers. However, since only the sum of one-dimensional basis functions is considered in the additive model, it cannot express complicated functions, as illustrated in Fig. 21.2(b).

21.2 KERNEL MODEL

In the linear-in-parameter model introduced above, basis functions are fixed to, e.g., polynomial functions or sinusoidal functions without regard to training samples

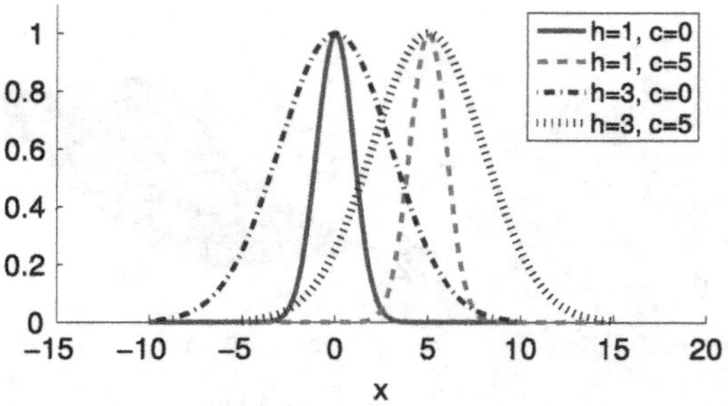

FIGURE 21.3

Gaussian kernel with bandwidth h and center c.

$\{(\boldsymbol{x}_i, y_i)\}_{i=1}^n$. In this section, the *kernel model* is introduced, which uses training input samples $\{\boldsymbol{x}_i\}_{i=1}^n$ for basis function design.

Let us consider a bivariate function called the *kernel function* $K(\cdot,\cdot)$. The kernel model is defined as the linear combination of $\{K(\boldsymbol{x},\boldsymbol{x}_j)\}_{j=1}^n$:

$$f_{\boldsymbol{\theta}}(\boldsymbol{x}) = \sum_{j=1}^{n} \theta_j K(\boldsymbol{x},\boldsymbol{x}_j). \tag{21.2}$$

As a kernel function, the *Gaussian kernel* would be the most popular choice:

$$K(\boldsymbol{x},\boldsymbol{c}) = \exp\left(-\frac{\|\boldsymbol{x}-\boldsymbol{c}\|^2}{2h^2}\right),$$

where $\|\cdot\|$ denotes the ℓ_2-norm:

$$\|\boldsymbol{x}\| = \sqrt{\boldsymbol{x}^\top \boldsymbol{x}}.$$

h and $\boldsymbol{c}$ are, respectively, called the *Gaussian bandwidth* and the *Gaussian center* (see Fig. 21.3).

In the Gaussian kernel model, Gaussian functions are located at training input samples $\{\boldsymbol{x}_i\}_{i=1}^n$ and their height $\{\theta_i\}_{i=1}^n$ is learned (Fig. 21.4). Since the Gaussian function is *local* around its center, the Gaussian kernel model can approximate the learning target function only in the vicinity of training input samples (Fig. 21.5). This is quite different from the multiplicative model which tries to approximate the learning target function over the entire input space.

In the kernel model, the number of parameters is given by the number of training samples, n, which is independent of the dimensionality d of the input variable $\boldsymbol{x}$. Thus, even when d is large, the kernel model can be easily handled in standard computers, as long as n is not too large. Even if n is very large, only a subset $\{\boldsymbol{c}_j\}_{j=1}^b$

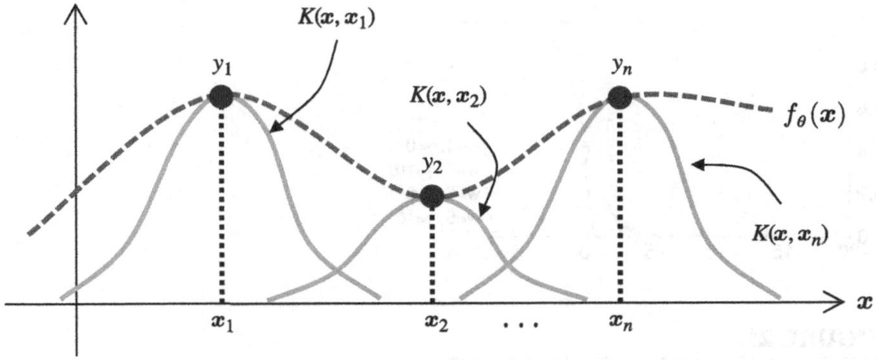

FIGURE 21.4

One-dimensional Gaussian kernel model. Gaussian functions are located at training input samples $\{x_i\}_{i=1}^n$ and their height $\{\theta_i\}_{i=1}^n$ is learned.

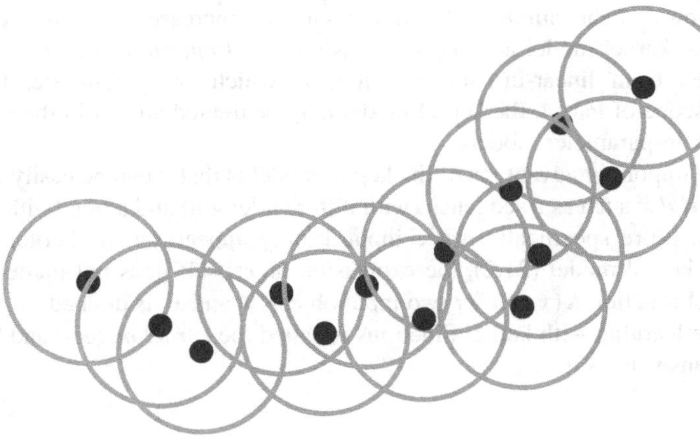

FIGURE 21.5

Two-dimensional Gaussian kernel model. The curse of dimensionality is mitigated by only approximating the learning target function in the vicinity of training input samples.

of training input samples $\{x_i\}_{i=1}^n$ may be used as kernel centers for reducing the computation costs:

$$f_\theta(x) = \sum_{j=1}^{b} \theta_j K(x, c_j).$$

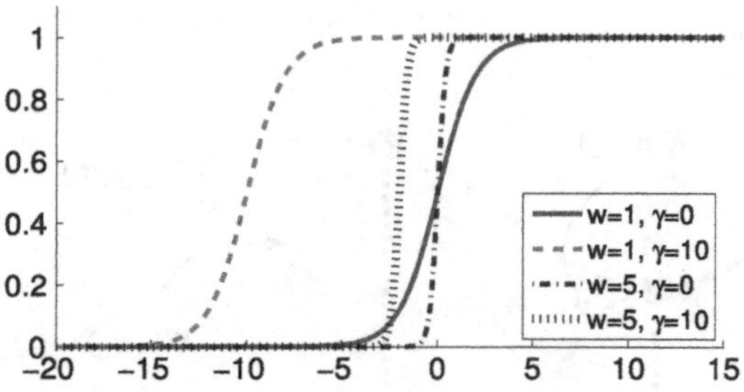

FIGURE 21.6

Sigmoidal function.

Since kernel model (21.2) is linear in terms of the parameter $\boldsymbol{\theta} = (\theta_1,\ldots,\theta_n)^\top$, it can also be regarded as a linear-in-parameter model. However, basis functions $\{K(\boldsymbol{x},\boldsymbol{x}_j)\}_{j=1}^n$ depend on training input samples $\{\boldsymbol{x}_i\}_{i=1}^n$ and the number of basis function grows as the number of training samples increases. For this reason, in statistics, the kernel model is categorized as a *nonparametric model* and is clearly differentiated from linear-in-parameter models which are parametric. However, within the scope of Part 4, the kernel model may be treated almost in the same way as the linear-in-parameter models.

Another important advantage of the kernel model is that it can be easily extended to *nonvectorial* $\boldsymbol{x}$ such as a sequence (with different length) and a tree (with different depth) [46]. More specifically, since input $\boldsymbol{x}$ only appears in the kernel function $K(\boldsymbol{x},\boldsymbol{x}')$ in kernel model (21.2), the expression of $\boldsymbol{x}$ itself does not matter as long as the kernel function $K(\boldsymbol{x},\boldsymbol{x}')$ for two input objects $\boldsymbol{x}$ and $\boldsymbol{x}'$ is defined.

Machine learning with kernel functions is called the *kernel method* and has been studied extensively [89].

21.3 HIERARCHICAL MODEL

A model that is nonlinear in terms of parameters is referred to as a *nonlinear model.* Among nonlinear models, a three-layer *hierarchical model* is a popular choice:

$$f_{\boldsymbol{\theta}}(\boldsymbol{x}) = \sum_{j=1}^{b} \alpha_j \phi(\boldsymbol{x};\boldsymbol{\beta}_j),$$

where $\phi(\boldsymbol{x};\boldsymbol{\beta})$ is a basis function parameterized by $\boldsymbol{\beta}$. A hierarchical model is linear in terms of parameter $\boldsymbol{\alpha} = (\alpha_1,\ldots,\alpha_b)^\top$, as in Eq. (21.1). However, in terms of the basis parameters $\{\boldsymbol{\beta}_j\}_{j=1}^b$, the hierarchical model is nonlinear.

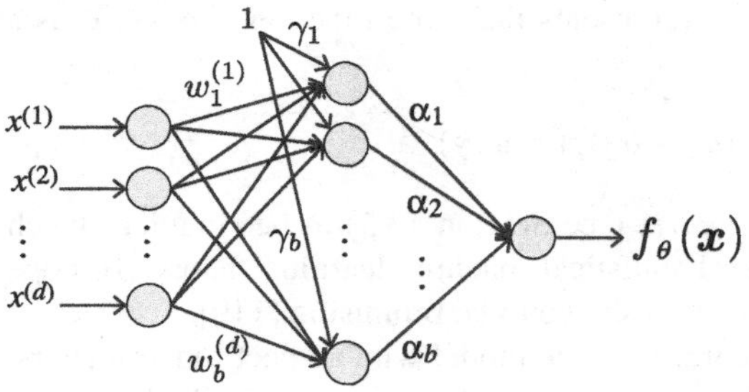

FIGURE 21.7

Hierarchical model as a three-layered network.

As a basis function, the *sigmoidal function* (see Fig. 21.6),

$$\phi(\boldsymbol{x};\boldsymbol{\beta}) = \frac{1}{1+\exp(-\boldsymbol{x}^\top \boldsymbol{w} - \gamma)}, \quad \boldsymbol{\beta} = (\boldsymbol{w}^\top, \gamma)^\top,$$

and the *Gaussian function* (see Fig. 21.3),

$$\phi(\boldsymbol{x};\boldsymbol{\beta}) = \exp\left(-\frac{\|\boldsymbol{x}-\boldsymbol{c}\|^2}{2h^2}\right), \quad \boldsymbol{\beta} = (\boldsymbol{c}^\top, h)^\top,$$

are standard choices. The sigmoidal function is motivated by the activity of neurons in human's brain. For this reason, a hierarchical model is also called a *neural network* and a basis function is called an *activation function*. To reduce the computation cost, the *rectified linear function*,

$$\phi(\boldsymbol{x};\boldsymbol{\beta}) = \max(\boldsymbol{x}^\top \boldsymbol{w} + \gamma, 0), \quad \boldsymbol{\beta} = (\boldsymbol{w}^\top, \gamma)^\top,$$

is also popularly used, although it is not differentiable at zero.

On the other hand, the Gaussian function is essentially the same as the Gaussian kernel introduced in Section 21.2, but the Gaussian bandwidth and center are also learned in the hierarchical model. For this reason, a neural network model with the Gaussian activation function is expected to be more expressive than the Gaussian kernel model.

A hierarchical model can be expressed by a three-layered network, as illustrated in Fig. 21.7. A notable characteristic of the neural network model is that a mapping between a parameter and a function is not necessarily one-to-one. In particular, different parameter values can yield the same function. For example, a neural network model with $b = 2$,

$$f_{\boldsymbol{\theta}}(\boldsymbol{x}) = \alpha_1 \phi(\boldsymbol{x}; \boldsymbol{w}_1, \gamma_1) + \alpha_2 \phi(\boldsymbol{x}; \boldsymbol{w}_2, \gamma_2)$$

for $w_1 = w_2 = w$ and $\gamma_1 = \gamma_2 = \gamma$, represents the same function if $\alpha_1 + \alpha_2$ is a constant:

$$f_{\boldsymbol{\theta}}(\boldsymbol{x}) = (\alpha_1 + \alpha_2)\phi(\boldsymbol{x}; \boldsymbol{w}, \gamma).$$

This causes the *Fisher information matrix* (see Section 13.3) to be singular, which makes it impossible to apply standard statistical machine learning theory. To cope with this problem, *Bayesian inference* was shown to be promising [119].

Due to the nonlinearity of the neural network model with respect to parameters, parameter learning is usually carried out by a gradient descent method called the *error back-propagation* algorithm [85]. However, only a local optimal solution may be found by the gradient method (see Section 15.3). To cope with this problem, running the gradient-based learning procedure multiple times from different initial solutions is effective. Furthermore, *pretraining* of each layer by an unsupervised learning method is demonstrated to be useful in practice [55].

CHAPTER

LEAST SQUARES REGRESSION

22

CHAPTER CONTENTS

Let us consider a regression problem of learning a real-valued function $y = f(\boldsymbol{x})$ defined on d-dimensional input space from input-output paired training samples $\{(\boldsymbol{x}_i, y_i)\}_{i=1}^n$. In practice, the training output values $\{y_i\}_{i=1}^n$ may be noisy observations of the true values $\{f(\boldsymbol{x}_i)\}_{i=1}^n$. In this chapter, the most fundamental regression technique called LS *regression* is introduced.

22.1 METHOD OF LS

Let $\boldsymbol{\theta}$ be the parameter of model $f_{\boldsymbol{\theta}}(\boldsymbol{x})$. The method of LS, learns the parameter $\boldsymbol{\theta}$ so that the squared difference between training output $\{y_i\}_{i=1}^n$ and model output $f_{\boldsymbol{\theta}}(\boldsymbol{x}_i)$ is minimized:

$$\widehat{\boldsymbol{\theta}}_{\mathrm{LS}} = \underset{\boldsymbol{\theta}}{\operatorname{argmin}}\, J_{\mathrm{LS}}(\boldsymbol{\theta}),$$

where

$$J_{\mathrm{LS}}(\boldsymbol{\theta}) = \frac{1}{2}\sum_{i=1}^{n}\left(y_i - f_{\boldsymbol{\theta}}(\boldsymbol{x}_i)\right)^2. \tag{22.1}$$

Since the squared error $(y_i - f_{\boldsymbol{\theta}}(\boldsymbol{x}_i))^2$ is the ℓ_2-norm of the *residual*

$$y_i - f_{\boldsymbol{\theta}}(\boldsymbol{x}_i),$$

the LS method is also referred to as *ℓ_2-loss minimization.*

22.2 SOLUTION FOR LINEAR-IN-PARAMETER MODEL

For a linear-in-parameter model (see Section 21.1),

$$f_{\boldsymbol{\theta}}(\boldsymbol{x}) = \sum_{j=1}^{b} \theta_i \phi_i(\boldsymbol{x}) = \boldsymbol{\theta}^\top \boldsymbol{\phi}(\boldsymbol{x}),$$

the training squared error J_{LS} is expressed as

$$J_{\text{LS}}(\boldsymbol{\theta}) = \frac{1}{2}\|\boldsymbol{y} - \boldsymbol{\Phi}\boldsymbol{\theta}\|^2,$$

where

$$\boldsymbol{y} = (y_1, \ldots, y_n)^\top$$

is the n-dimensional vector consisting of training output values and $\boldsymbol{\Phi}$ is the $n \times b$ matrix called *design matrix*:

$$\boldsymbol{\Phi} = \begin{pmatrix} \phi_1(\boldsymbol{x}_1) & \cdots & \phi_b(\boldsymbol{x}_1) \\ \vdots & \ddots & \vdots \\ \phi_1(\boldsymbol{x}_n) & \cdots & \phi_b(\boldsymbol{x}_n) \end{pmatrix}.$$

The partial derivative of the training squared error J_{LS} with respect to parameter $\boldsymbol{\theta}$ is given by

$$\nabla_{\boldsymbol{\theta}} J_{\text{LS}} = \left(\frac{\partial J_{\text{LS}}}{\partial \theta_1}, \ldots, \frac{\partial J_{\text{LS}}}{\partial \theta_b} \right)^\top = \boldsymbol{\Phi}^\top \boldsymbol{y} - \boldsymbol{\Phi}^\top \boldsymbol{\Phi}\boldsymbol{\theta}.$$

Setting this to zero shows that the solution should satisfy

$$\boldsymbol{\Phi}^\top \boldsymbol{\Phi}\boldsymbol{\theta} = \boldsymbol{\Phi}^\top \boldsymbol{y}.$$

Then the LS solution $\widehat{\boldsymbol{\theta}}_{\text{LS}}$ is given by

$$\widehat{\boldsymbol{\theta}}_{\text{LS}} = \boldsymbol{\Phi}^\dagger \boldsymbol{y},$$

where $\boldsymbol{\Phi}^\dagger$ is the *generalized inverse* (see Fig. 22.1) of the design matrix $\boldsymbol{\Phi}$. When $\boldsymbol{\Phi}^\top \boldsymbol{\Phi}$ is invertible, the generalized inverse $\boldsymbol{\Phi}^\dagger$ is expressed as

$$\boldsymbol{\Phi}^\dagger = (\boldsymbol{\Phi}^\top \boldsymbol{\Phi})^{-1} \boldsymbol{\Phi}^\top.$$

A MATLAB code of LS regression for the linear-in-parameter model with sinusoidal basis functions,

$$\boldsymbol{\phi}(x) = \left(1, \sin\frac{x}{2}, \cos\frac{x}{2}, \sin\frac{2x}{2}, \cos\frac{2x}{2}, \ldots, \sin\frac{15x}{2}, \cos\frac{15x}{2}\right)^\top,$$

is provided in Fig. 22.3. In the program, instead of explicitly obtaining the generalized inverse $\boldsymbol{\Phi}^\dagger$ by the `pinv` function, the equation $\boldsymbol{\Phi}\boldsymbol{\theta} = \boldsymbol{y}$ is directly solved by

A matrix $\boldsymbol{X}$ is called the *generalized inverse* of real matrix $\boldsymbol{A}$, if it satisfies the following four conditions:

$$\boldsymbol{AXA} = \boldsymbol{A},$$
$$\boldsymbol{XAX} = \boldsymbol{X},$$
$$(\boldsymbol{XA})^\top = \boldsymbol{XA},$$
$$(\boldsymbol{AX})^\top = \boldsymbol{AX}.$$

The generalized inverse of $\boldsymbol{A}$ is often denoted by $\boldsymbol{A}^\dagger$ and is also referred to as the *Moore-Penrose pseudoinverse*. The ordinary inverse is defined only for full-rank square matrices, while the generalized inverse is defined for singular or even nonsquare matrices. In MATLAB, the generalized inverse can be obtained by the `pinv` function. A generalized inverse of a scalar a is given by

$$a^\dagger = \begin{cases} 1/a & (a \neq 0), \\ 0 & (a = 0), \end{cases}$$

and a generalized inverse of a $d \times m$ matrix $\boldsymbol{A}$ is given by

$$\boldsymbol{A}^\dagger = \sum_{k=1}^{\min(d,m)} \kappa_k^\dagger \boldsymbol{\psi}_k \boldsymbol{\phi}_k^\top,$$

where $\boldsymbol{\psi}_k$, $\boldsymbol{\phi}_k$, and κ_k are a *left singular vector*, a *right singular vector*, and a *singular value* of $\boldsymbol{A}$, respectively (see Fig. 22.2). When $\boldsymbol{A}$ is square and full rank, its generalized inverse $\boldsymbol{A}^\dagger$ is reduced to the ordinary inverse $\boldsymbol{A}^{-1}$.

FIGURE 22.1

Generalized inverse.

`t=p\y`, which is computationally more efficient. The behavior of LS regression is illustrated in Fig. 22.4, showing that a complicated nonlinear function can be nicely approximated.

A LS method where the loss function for the ith training sample is weighted according to $w_i \geq 0$ is referred to as *weighted LS*:

$$\min_{\boldsymbol{\theta}} \frac{1}{2} \sum_{i=1}^{n} w_i \big(y_i - f_{\boldsymbol{\theta}}(\boldsymbol{x}_i)\big)^2.$$

Singular value decomposition is an extension *eigenvalue decomposition* (see Fig. 6.2) to nonsquare matrices. For $d \times m$ matrix $\boldsymbol{A}$, a d-dimensional nonzero vector $\boldsymbol{\psi}$, an m-dimensional nonzero vector $\boldsymbol{\varphi}$, and a non-negative scalar κ such that

$$\boldsymbol{A}\boldsymbol{\varphi} = \kappa\boldsymbol{\psi}$$

are called a *left singular vector*, a *right singular vector*, and a *singular value* of $\boldsymbol{A}$, respectively. Generally, there exist c singular values $\kappa_1, \ldots, \kappa_c$, where

$$c = \min(d, m).$$

Singular vectors $\boldsymbol{\varphi}_1, \ldots, \boldsymbol{\varphi}_c$ and $\boldsymbol{\psi}_1, \ldots, \boldsymbol{\psi}_c$ corresponding to singular values $\kappa_1, \ldots, \kappa_c$ are mutually orthogonal and are usually normalized, i.e., they are *orthonormal* as

$$\boldsymbol{\varphi}_k^\top \boldsymbol{\varphi}_{k'} = \begin{cases} 1 & (k = k') \\ 0 & (k \neq k') \end{cases} \quad \text{and} \quad \boldsymbol{\psi}_k^\top \boldsymbol{\psi}_{k'} = \begin{cases} 1 & (k = k'), \\ 0 & (k \neq k'). \end{cases}$$

A matrix $\boldsymbol{A}$ can be expressed by using its singular vectors and singular values as

$$\boldsymbol{A} = \sum_{k=1}^{c} \kappa_k \boldsymbol{\psi}_k \boldsymbol{\varphi}_k^\top.$$

In MATLAB, singular value decomposition can be performed by the svd function.

FIGURE 22.2

Singular value decomposition.

The solution of weighted LS is given as

$$(\boldsymbol{\Phi}^\top \boldsymbol{W} \boldsymbol{\Phi})^\dagger \boldsymbol{\Phi}^\top \boldsymbol{W} \boldsymbol{y},$$

where $\boldsymbol{W}$ is the diagonal matrix consisting of weights:

$$\boldsymbol{W} = \operatorname{diag}(w_1, \ldots, w_n).$$

The kernel model introduced in Section 21.2 is also linear in terms of parameters:

$$f_{\boldsymbol{\theta}}(\boldsymbol{x}) = \sum_{j=1}^{n} \theta_j K(\boldsymbol{x}, \boldsymbol{x}_j). \tag{22.2}$$

```
n=50; N=1000; x=linspace(-3,3,n)'; X=linspace(-3,3,N)';
pix=pi*x; y=sin(pix)./(pix)+0.1*x+0.05*randn(n,1);

p(:,1)=ones(n,1); P(:,1)=ones(N,1);
for j=1:15
  p(:,2*j)=sin(j/2*x); p(:,2*j+1)=cos(j/2*x);
  P(:,2*j)=sin(j/2*X); P(:,2*j+1)=cos(j/2*X);
end
t=p\y; F=P*t;

figure(1); clf; hold on; axis([-2.8 2.8 -0.5 1.2]);
plot(X,F,'g-'); plot(x,y,'bo');
```

FIGURE 22.3

MATLAB code for LS regression.

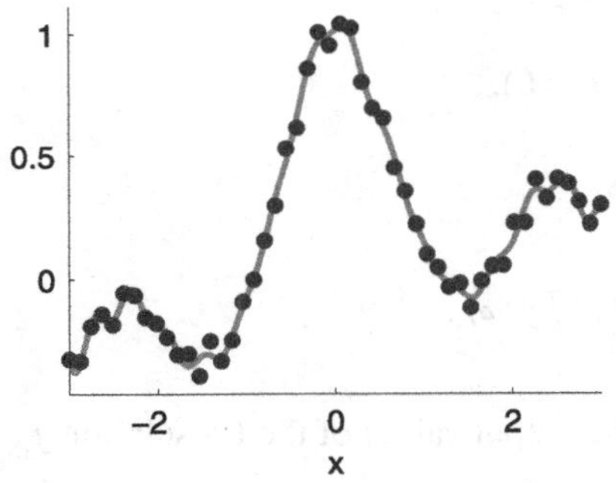

FIGURE 22.4

Example of LS regression with sinusoidal basis functions $\boldsymbol{\phi}(x) = (1, \sin\frac{x}{2}, \cos\frac{x}{2}, \sin\frac{2x}{2}, \cos\frac{2x}{2}, \ldots, \sin\frac{15x}{2}, \cos\frac{15x}{2})^\top$.

Thus, the LS solution for the kernel model can also be obtained in the same way, by replacing the design matrix $\mathbf{\Phi}$ with the *kernel matrix*:

$$\boldsymbol{K} = \begin{pmatrix} K(\boldsymbol{x}_1,\boldsymbol{x}_1) & \cdots & K(\boldsymbol{x}_1,\boldsymbol{x}_n) \\ \vdots & \ddots & \vdots \\ K(\boldsymbol{x}_n,\boldsymbol{x}_1) & \cdots & K(\boldsymbol{x}_n,\boldsymbol{x}_n) \end{pmatrix}.$$

22.3 PROPERTIES OF LS SOLUTION

Let us consider *singular value decomposition* (see Fig. 22.2) of design matrix $\boldsymbol{\Phi}$:

$$\boldsymbol{\Phi} = \sum_{k=1}^{\min(n,b)} \kappa_k \boldsymbol{\psi}_k \boldsymbol{\varphi}_k^\top,$$

where $\boldsymbol{\psi}_k$, $\boldsymbol{\varphi}_k$, and κ_k are a *left singular vector*, a *right singular vector*, and a *singular value* of $\boldsymbol{\Phi}$, respectively.

As explained in Fig. 22.1, the generalized inverse $\boldsymbol{\Phi}^\dagger$ of the matrix $\boldsymbol{\Phi}$ can be expressed as

$$\boldsymbol{\Phi}^\dagger = \sum_{k=1}^{\min(n,b)} \kappa_k^\dagger \boldsymbol{\varphi}_k \boldsymbol{\psi}_k^\top,$$

where $\kappa^\dagger$ is the generalized inverse of scalar κ:

$$\kappa^\dagger = \begin{cases} 1/\kappa & (\kappa \neq 0), \\ 0 & (\kappa = 0). \end{cases}$$

Then the LS solution $\widehat{\boldsymbol{\theta}}_{\mathrm{LS}}$ can be expressed as

$$\widehat{\boldsymbol{\theta}}_{\mathrm{LS}} = \sum_{k=1}^{\min(n,b)} \kappa_k^\dagger (\boldsymbol{\psi}_k^\top \boldsymbol{y}) \boldsymbol{\varphi}_k.$$

The n-dimensional vector consisting of the output values of the LS solution $f_{\widehat{\boldsymbol{\theta}}_{\mathrm{LS}}}$ at training input samples $\{\boldsymbol{x}_i\}_{i=1}^n$ is given by

$$(f_{\widehat{\boldsymbol{\theta}}_{\mathrm{LS}}}(\boldsymbol{x}_1), \ldots, f_{\widehat{\boldsymbol{\theta}}_{\mathrm{LS}}}(\boldsymbol{x}_n))^\top = \boldsymbol{\Phi}\widehat{\boldsymbol{\theta}}_{\mathrm{LS}} = \boldsymbol{\Phi}\boldsymbol{\Phi}^\dagger \boldsymbol{y}.$$

Since $(\boldsymbol{\Phi}\boldsymbol{\Phi}^\dagger)^2 = \boldsymbol{\Phi}\boldsymbol{\Phi}^\dagger$ and $(\boldsymbol{\Phi}\boldsymbol{\Phi}^\dagger)^\top = \boldsymbol{\Phi}\boldsymbol{\Phi}^\dagger$, $\boldsymbol{\Phi}\boldsymbol{\Phi}^\dagger$ is the *projection* matrix onto the range of $\boldsymbol{\Phi}$ and thus the LS solution actually projects the training output vector $\boldsymbol{y}$ onto the range of $\boldsymbol{\Phi}$.

When there exists a parameter $\boldsymbol{\theta}^*$ such that the true learning target function f is given by $f_{\boldsymbol{\theta}^*}$, the vector consisting of the output values of the true function f at training input samples $\{\boldsymbol{x}_i\}_{i=1}^n$ is given by

$$(f(\boldsymbol{x}_1), \ldots, f(\boldsymbol{x}_n))^\top = \boldsymbol{\Phi}\boldsymbol{\theta}^*.$$

This means that the true output vector belongs to the range of $\boldsymbol{\Phi}$, and thus the LS method tries to remove noise included in $\boldsymbol{y}$ by projecting it onto the range of $\boldsymbol{\Phi}$ (Fig. 22.5).

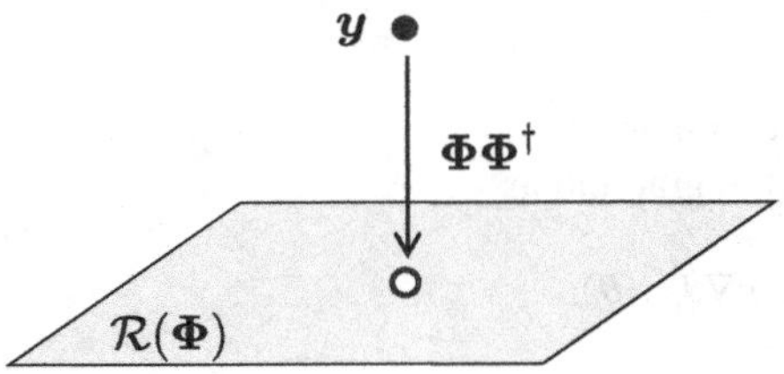

FIGURE 22.5

Geometric interpretation of LS method for linear-in-parameter model. Training output vector $\boldsymbol{y}$ is projected onto the range of $\boldsymbol{\Phi}$, denoted by $\mathcal{R}(\boldsymbol{\Phi})$, for denoising purposes.

When the expectation of noise is zero, the LS solution $\widehat{\boldsymbol{\theta}}_{\mathrm{LS}}$ is an *unbiased estimator* of the true parameter $\boldsymbol{\theta}^*$:

$$\mathbb{E}[\widehat{\boldsymbol{\theta}}_{\mathrm{LS}}] = \boldsymbol{\theta}^*,$$

where $\mathbb{E}$ denotes the expectation over the noise included in training output $\{y_i\}_{i=1}^n$. Even if the model is *misspecified* (see Section 17.1.2), i.e., $f \neq f_{\boldsymbol{\theta}}$ for any $\boldsymbol{\theta}$, $\mathbb{E}[\widehat{\boldsymbol{\theta}}_{\mathrm{LS}}]$ converges to the optimal parameter in the model, as the number of training samples, n, tends to infinity. This property is called *asymptotic unbiasedness* (see Section 13.2).

22.4 LEARNING ALGORITHM FOR LARGE-SCALE DATA

The size of design matrix $\boldsymbol{\Phi}$ is $n \times b$ and thus it cannot be stored in the memory space if the number of training samples, n, and the number of parameters, b, are large. In such a situation, the *stochastic gradient* algorithm introduced in Section 15.3 is useful, which updates parameter $\boldsymbol{\theta}$ to go down the gradient of the training squared error J_{LS}. For linear-in-parameter models, the training squared error J_{LS} is a *convex function* (see Fig. 8.3). Thus, the solution obtained by the stochastic gradient algorithm gives the global optimal solution. The algorithm of the stochastic gradient algorithm for linear-in-parameter models is summarized in Fig. 22.6.

In Fig. 22.7, a MATLAB code of stochastic gradient descent for LS regression with the Gaussian kernel model is provided:

$$f_{\boldsymbol{\theta}}(\boldsymbol{x}) = \sum_{j=1}^{n} \theta_j K(\boldsymbol{x}, \boldsymbol{x}_j),$$

where

$$K(\boldsymbol{x}, \boldsymbol{c}) = \exp\left(-\frac{\|\boldsymbol{x} - \boldsymbol{c}\|^2}{2h^2}\right).$$

1. Initialize the parameter $\boldsymbol{\theta}$.
2. Choose the ith training sample $(\boldsymbol{x}_i, y_i)$ randomly.
3. Update the parameter $\boldsymbol{\theta}$ to go down the gradient as

$$\boldsymbol{\theta} \longleftarrow \boldsymbol{\theta} - \varepsilon \nabla J_{\mathrm{LS}}^{(i)}(\boldsymbol{\theta}),$$

where ε is a small positive scalar called the *step size*, and $\nabla J_{\mathrm{LS}}^{(i)}$ is the gradient of the training squared error for the ith training sample:

$$\nabla J_{\mathrm{LS}}^{(i)}(\boldsymbol{\theta}) = -\boldsymbol{\phi}(\boldsymbol{x}_i)\big(y_i - f_{\boldsymbol{\theta}}(\boldsymbol{x}_i)\big).$$

4. Iterate 2–3 until convergence.

FIGURE 22.6

Algorithm of stochastic gradient descent for LS regression with a linear-in-parameter model.

For $n = 50$ training samples, the Gaussian bandwidth is set at $h = 0.3$. The behavior of the algorithm is illustrated in Fig. 22.8, where starting from a random initial value, a function close to the final solution is obtained after around 200 iterations. However, 11,556 iterations are required until convergence.

The convergence speed of the stochastic gradient algorithm depends on the step size (`e=0.1` in Fig. 22.7) and the convergence condition (`norm(t-t0)<0.000001` in Fig. 22.7). The algorithm may converge faster if these parameters are tuned. In particular, as for the step size, starting from a large value and gradually decreasing it would be appropriate, as discussed in Section 15.3. However, appropriately implementing this idea is not straightforward in practice.

22.5 LEARNING ALGORITHM FOR HIERARCHICAL MODEL

Stochastic gradient descent is a popular choice also for training hierarchical models such as neural networks introduced in Section 21.3. However, since the training squared error J_{LS} is not a convex function in hierarchical models, there exist multiple local optimal solutions in general and the gradient method may only find one of the local optimal solutions (Fig. 22.9). In practice, starting from different initial values, gradient descent is performed multiple times and the solution that gives the smallest training squared error is chosen as the most promising one.

For the sigmoidal neural network model introduced in Section 21.3,

$$f_{\boldsymbol{\theta}}(\boldsymbol{x}) = \sum_{j=1}^{b} \alpha_j \phi(\boldsymbol{x}; \boldsymbol{\beta}_j, \gamma_j),$$

```
n=50; x=linspace(-3,3,n)'; pix=pi*x;
y=sin(pix)./(pix)+0.1*x+0.05*randn(n,1);

hh=2*0.3^2; t=randn(n,1); e=0.1;
for o=1:n*1000
  i=ceil(rand*n);
  ki=exp(-(x-x(i)).^2/hh); t0=t+e*ki*(y(i)-ki'*t);
  if norm(t-t0)<0.000001, break, end
  t=t0;
end

N=1000; X=linspace(-3,3,N)';
K=exp(-(repmat(X.^2,1,n)+repmat(x.^2',N,1)-2*X*x')/hh);
F=K*t;
figure(1); clf; hold on; axis([-2.8 2.8 -0.5 1.2]);
plot(X,F,'g-'); plot(x,y,'bo');
```

FIGURE 22.7

MATLAB code of stochastic gradient descent for LS regression with the Gaussian kernel model.

where

$$\phi(\boldsymbol{x};\boldsymbol{\beta},\gamma) = \frac{1}{1+\exp\left(-\boldsymbol{x}^\top\boldsymbol{\beta}-\gamma\right)},$$

the gradient $\nabla_{\boldsymbol{\theta}} J_{\mathrm{LS}}^{(i)}$ can be computed efficiently as

$$\frac{\partial J_{\mathrm{LS}}^{(i)}}{\partial \alpha_j} = -z_{i,j} r_i,$$

$$\frac{\partial J_{\mathrm{LS}}^{(i)}}{\partial \beta_j^{(k)}} = -\alpha_j z_{i,j}\left(1-z_{i,j}\right) x_i^{(k)} r_i,$$

$$\frac{\partial J_{\mathrm{LS}}^{(i)}}{\partial \gamma_j} = -\alpha_j z_{i,j}\left(1-z_{i,j}\right) r_i,$$

where $\beta_j^{(k)}$ denotes the kth element of vector $\boldsymbol{\beta}_j$, $x_i^{(k)}$ denotes the kth element of vector $\boldsymbol{x}_i$, r_i denotes the residual for the ith training sample,

$$r_i = y_i - f_{\boldsymbol{\theta}}(\boldsymbol{x}_i),$$

and $z_{i,j}$ denotes the output of the jth basis function for the ith training sample:

$$z_{i,j} = \phi(\boldsymbol{x}_i;\boldsymbol{\beta}_j,\gamma_j).$$

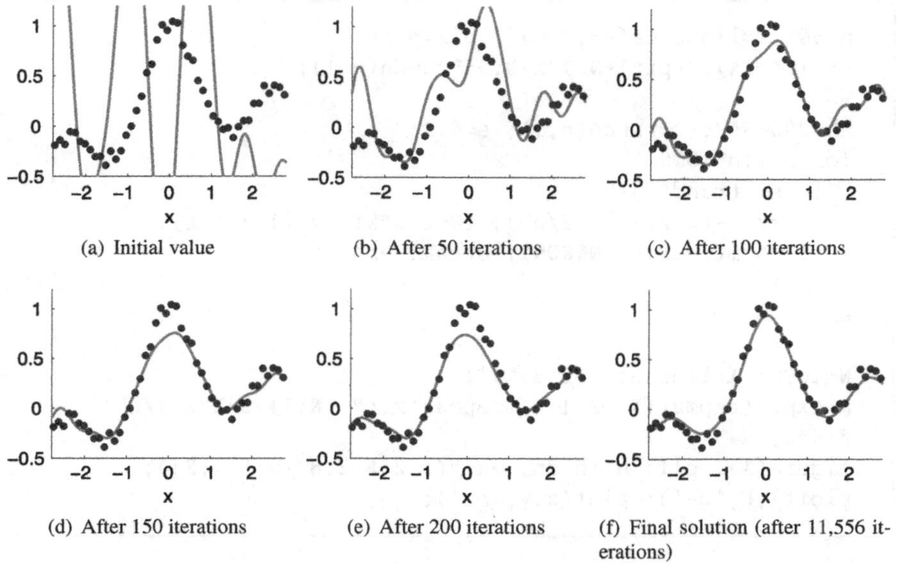
(a) Initial value
(b) After 50 iterations
(c) After 100 iterations
(d) After 150 iterations
(e) After 200 iterations
(f) Final solution (after 11,556 iterations)

FIGURE 22.8

Example of stochastic gradient descent for LS regression with the Gaussian kernel model. For $n = 50$ training samples, the Gaussian bandwidth is set at $h = 0.3$.

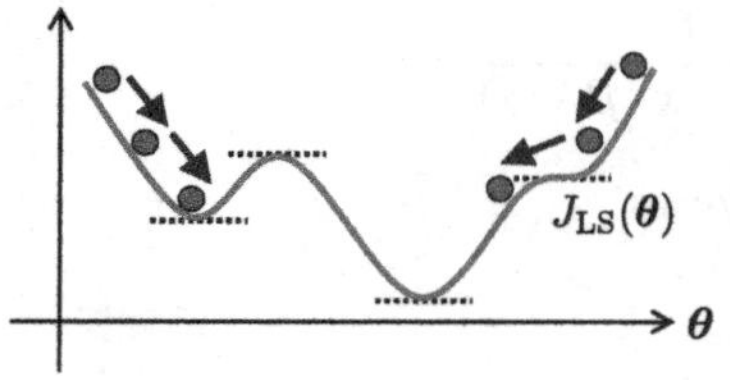

FIGURE 22.9

Gradient descent for nonlinear models. The training squared error J_{LS} is nonconvex and there exist multiple local optimal solutions in general.

Since the residual r_i is propagated backward when computing the gradient, the gradient method for neural networks is often referred to as *error back-propagation* [85].

If input $\boldsymbol{x}_i$ is augmented as

$$\widetilde{\boldsymbol{x}}_i = (\boldsymbol{x}_i^\top, 1)^\top \in \mathbb{R}^{d+1},$$

```
n=50; N=1000; x=linspace(-3,3,n)'; pix=pi*x;
y=sin(pix)./(pix)+0.1*x+0.05*randn(n,1);

x(:,2)=1; d=1; m=20; e=0.1; a=ones(m,1); b=randn(m,d+1);
for o=1:n*100000
  i=ceil(rand*n); z=1./(1+exp(-b*x(i,:)')); r=y(i)-a'*z;
  a0=a+e*z*r; b0=b+e*(a.*z.*(1-z)*x(i,:))*r;
  if norm(a-a0)+norm(b-b0)<0.00000001, break, end
  a=a0; b=b0;
end

X=linspace(-3,3,N)'; X(:,2)=1; Y=a'*(1./(1+exp(-b*X')));
figure(1); clf; hold on; axis([-2.8 2.8 -0.5 1.2]);
plot(X(:,1),Y,'g-'); plot(x(:,1),y,'bo');
```

FIGURE 22.10

MATLAB code for error back-propagation algorithm.

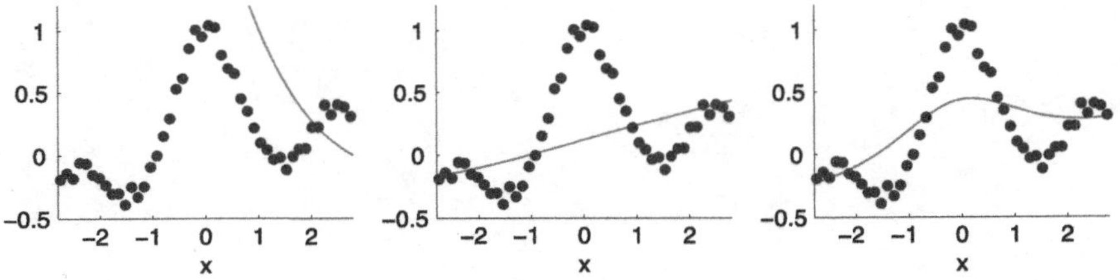

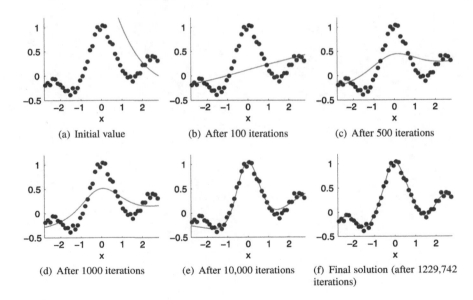

(a) Initial value (b) After 100 iterations (c) After 500 iterations

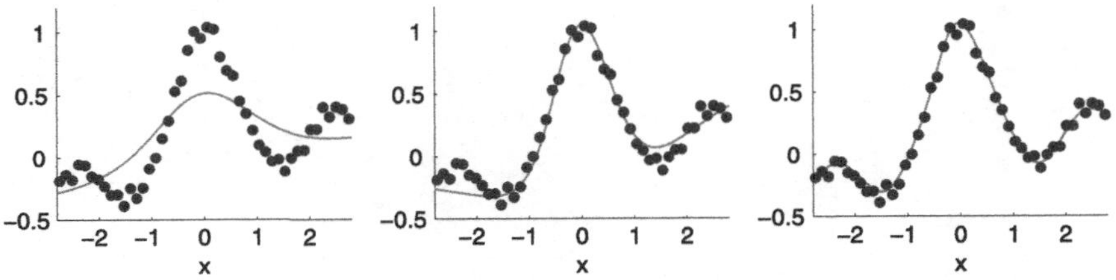

(d) After 1000 iterations (e) After 10,000 iterations (f) Final solution (after 1229,742 iterations)

FIGURE 22.11

Example of regression by error back-propagation algorithm.

the basis parameter,

$$\widetilde{\beta}_j = (\beta_j^\top, \gamma_j)^\top \in \mathbb{R}^{d+1},$$

can be updated together as

$$\frac{\partial J_{\mathrm{LS}}^{(i)}}{\partial \widetilde{\beta}_j^{(k)}} = -\alpha_j z_{i,j}\left(1 - z_{i,j}\right)\widetilde{x}_i^{(k)} r_i \quad \text{for } k = 1, \ldots, d+1.$$

A MATLAB code of the error back-propagation algorithm for a three-layered neural network is provided in Fig. 22.10, and its behavior is illustrated in Fig. 22.11. This shows that the solution fits the training samples well after many iterations.

Although the error back-propagation is applicable to any neural networks models, learning a neural network having many layers is often difficult in practice. Indeed, the error is propagated only to parameters close to the output layer, and parameters near the input layer are rarely updated. To efficiently learn a *deep neural network*, iteratively initializing each layer one by one based on unsupervised learning is shown to be useful [55]. Such a method will be explained in Chapter 36.

CHAPTER

CONSTRAINED LS REGRESSION

23

CHAPTER CONTENTS

The least squared method introduced in Chapter 22 forms the basis of various machine learning techniques. However, the naive LS method often yields *overfitting* to noisy training samples (Fig. 23.1(a)). This is caused by the fact that the model is too complicated compared with the number of available training samples. In this chapter, *constrained LS* methods are introduced for controlling the model complexity.

23.1 SUBSPACE-CONSTRAINED LS

In the LS method for linear-in-parameter model,

$$f_{\boldsymbol{\theta}}(\boldsymbol{x}) = \sum_{j=1}^{b} \theta_j \phi_j(\boldsymbol{x}) = \boldsymbol{\theta}^\top \boldsymbol{\phi}(\boldsymbol{x}),$$

parameters $\{\theta_j\}_{j=1}^{b}$ can be determined without any constraint, implying that the entire parameter space is used for learning (see Fig. 23.2(a)). In this section, the method of *subspace-constrained LSsubspace-constrained least squares* is introduced, which imposes a subspace constraint in the parameter space:

$$\min_{\boldsymbol{\theta}} J_{\mathrm{LS}}(\boldsymbol{\theta}) \quad \text{subject to } \boldsymbol{P}\boldsymbol{\theta} = \boldsymbol{\theta},$$

where $\boldsymbol{P}$ is a $b \times b$ projection matrix such that

$$\boldsymbol{P}^2 = \boldsymbol{P} \quad \text{and} \quad \boldsymbol{P}^\top = \boldsymbol{P}.$$

As illustrated in Fig. 23.2(b), with the constraint $\boldsymbol{P}\boldsymbol{\theta} = \boldsymbol{\theta}$, parameter $\boldsymbol{\theta}$ is confined in the range of $\boldsymbol{P}$.

The solution $\widehat{\boldsymbol{\theta}}$ of subspace-constrained LS can be obtained simply by replacing the design matrix $\boldsymbol{\Phi}$ with $\boldsymbol{\Phi}\boldsymbol{P}$:

$$\widehat{\boldsymbol{\theta}} = (\boldsymbol{\Phi}\boldsymbol{P})^\dagger \boldsymbol{y}.$$

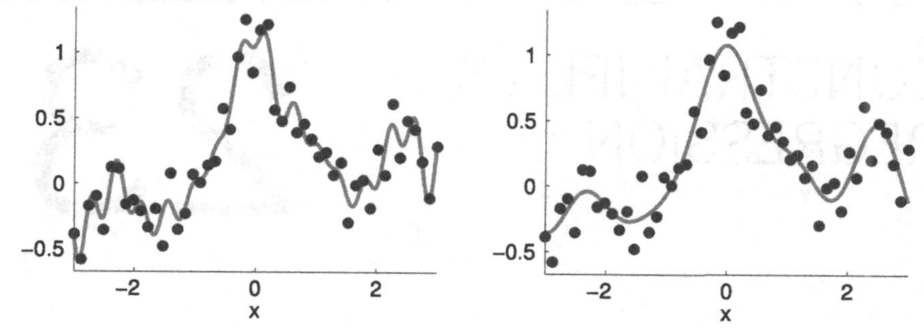

(a) Ordinary LS

(b) Subspace-constrained least squared

FIGURE 23.1

Examples of LS regression for linear-in-parameter model when the noise level in training output is high. Sinusoidal basis functions $\{1, \sin\frac{x}{2}, \cos\frac{x}{2}, \sin\frac{2x}{2}, \cos\frac{2x}{2}, \ldots, \sin\frac{15x}{2}, \cos\frac{15x}{2}\}$ are used in ordinary LS, while its subset $\{1, \sin\frac{x}{2}, \cos\frac{x}{2}, \sin\frac{2x}{2}, \cos\frac{2x}{2}, \ldots, \sin\frac{5x}{2}, \cos\frac{5x}{2}\}$ is used in the subspace-constrained LS method.

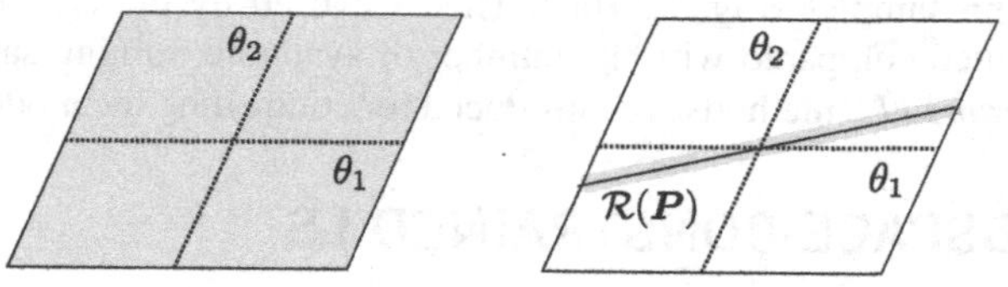

(a) Ordinary LS

(b) Subspace-constrained LS

FIGURE 23.2

Constraint in parameter space.

A MATLAB code for subspace-constrained LS is provided in Fig. 23.3, where a linear-in-parameter model with sinusoidal basis functions,

$$\left\{1, \sin\frac{x}{2}, \cos\frac{x}{2}, \sin\frac{2x}{2}, \cos\frac{2x}{2}, \ldots, \sin\frac{15x}{2}, \cos\frac{15x}{2}\right\},$$

is constrained to be confined in the subspace spanned by

$$\left\{1, \sin\frac{x}{2}, \cos\frac{x}{2}, \sin\frac{2x}{2}, \cos\frac{2x}{2}, \ldots, \sin\frac{5x}{2}, \cos\frac{5x}{2}\right\}.$$

The behavior of subspace-constrained LS is illustrated in Fig. 23.1(b), showing that the constraint effectively contributes to mitigating overfitting. Although the projection matrix $\boldsymbol{P}$ is manually determined in the above example, it may be determined in a data-dependent way, e.g., by PCA introduced in Section 35.2.1 or *multitask learning with the trace norm* introduced in Section 34.3.

```
n=50; N=1000; x=linspace(-3,3,n)'; X=linspace(-3,3,N)';
pix=pi*x; y=sin(pix)./(pix)+0.1*x+0.2*randn(n,1);

p(:,1)=ones(n,1); P(:,1)=ones(N,1);
for j=1:15
  p(:,2*j)=sin(j/2*x); p(:,2*j+1)=cos(j/2*x);
  P(:,2*j)=sin(j/2*X); P(:,2*j+1)=cos(j/2*X);
end
t1=p\y; F1=P*t1;
t2=(p*diag([ones(1,11) zeros(1,20)]))\y; F2=P*t2;

figure(1); clf; hold on; axis([-2.8 2.8 -0.8 1.2]);
plot(X,F1,'g-'); plot(X,F2,'r--'); plot(x,y,'bo');
legend('LS','Subspace-Constrained LS');
```

FIGURE 23.3

MATLAB code for subspace-constrained LS regression.

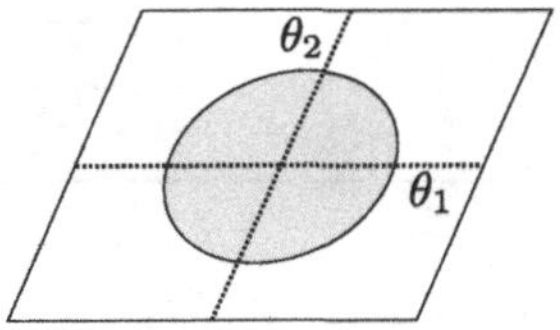

FIGURE 23.4

Parameter space in ℓ_2-constrained LS.

23.2 ℓ_2-CONSTRAINED LS

In the subspace-constrained LS method, the constraint is given by a projection matrix $\boldsymbol{P}$. However, since $\boldsymbol{P}$ has many degrees of freedom, it is not easy to handle in practice. In this section, an alternative approach called *ℓ_2-constrained LSℓ_2-constrained least squares* is introduced:

$$\min_{\boldsymbol{\theta}} J_{\mathrm{LS}}(\boldsymbol{\theta}) \quad \text{subject to } \|\boldsymbol{\theta}\|^2 \leq R^2, \tag{23.1}$$

where $R \geq 0$. As illustrated in Fig. 23.4, parameters are searched within an origin-centered hypersphere in the ℓ_2-constrained LS method, where R denotes the radius of the hypersphere.

Let us consider the constrained optimization problem,

$$\min_{t} f(t) \text{ subject to } g(t) \leq 0,$$

where $f : \mathbb{R}^d \to \mathbb{R}$ and $g : \mathbb{R}^d \to \mathbb{R}^p$ are the differentiable convex functions. Its *Lagrange dual problem* is given as

$$\max_{\lambda} \inf_{t} L(t, \lambda) \text{ subject to } \lambda \geq 0,$$

where

$$\lambda = (\lambda_1, \ldots, \lambda_p)^\top$$

is called the *Lagrange multipliers* and

$$L(t, \lambda) = f(t) + \lambda^\top g(t)$$

is called the *Lagrange function* (or simply the Lagrangian). The solution of the Lagrange dual problem for t agrees with the original constrained optimization problem.

FIGURE 23.5

Lagrange dual problem.

Lagrange duality (Fig. 23.5) yields that the solution of optimization problem (23.1) can be obtained by solving the following Lagrange dual problem:

$$\max_{\lambda} \min_{\theta} \left[J_{\text{LS}}(\theta) + \frac{\lambda}{2} \left(\|\theta\|^2 - R^2 \right) \right] \quad \text{subject to } \lambda \geq 0.$$

Although the Lagrange multiplier λ is determined based on the radius R in principle, λ may be directly specified in practice. Then the solution of ℓ_2-constrained LS $\widehat{\theta}$ is given by

$$\widehat{\theta} = \operatorname*{argmin}_{\theta} \left[J_{\text{LS}}(\theta) + \frac{\lambda}{2} \|\theta\|^2 \right]. \tag{23.2}$$

The first term in Eq. (23.2) measures the goodness of fit to training samples, while the second term $\frac{\lambda}{2}\|\theta\|^2$ measures the amount of overfitting.

Differentiating the objective function in Eq. (23.2) with respect to parameter θ and setting it to zero give the solution of ℓ_2-constrained LS analytically as

$$\widehat{\theta} = (\Phi^\top \Phi + \lambda I)^{-1} \Phi^\top y, \tag{23.3}$$

where $\boldsymbol{I}$ denotes the identity matrix. From Eq. (23.3), ℓ_2-constrained LS enhances the regularity of matrix $\boldsymbol{\Phi}^\top\boldsymbol{\Phi}$ by adding $\lambda\boldsymbol{I}$, which contributes to stabilizing the computation of its inverse. For this reason, ℓ_2-constrained LS is also called *ℓ_2-regularization learning*, the second term $\|\boldsymbol{\theta}\|^2$ in Eq. (23.2) is called a *regularizer*, and λ is called the *regularization parameter*. In statistics, ℓ_2-constrained LS is referred to as *ridge regression* [56].

Let us consider *singular value decomposition* (see Fig. 22.2) of design matrix $\boldsymbol{\Phi}$:

$$\boldsymbol{\Phi} = \sum_{k=1}^{\min(n,b)} \kappa_k \boldsymbol{\psi}_k \boldsymbol{\varphi}_k^\top,$$

where $\boldsymbol{\psi}_k$, $\boldsymbol{\phi}_k$, and κ_k are a *left singular vector*, a *right singular vector*, and a *singular value* of $\boldsymbol{\Phi}$, respectively. Then the solution of ℓ_2-constrained LS can be expressed as

$$\widehat{\boldsymbol{\theta}} = \sum_{k=1}^{\min(n,b)} \frac{\kappa_k}{\kappa_k^2 + \lambda} \boldsymbol{\psi}_k^\top \boldsymbol{y} \boldsymbol{\varphi}_k.$$

When $\lambda = 0$, ℓ_2-constrained LS is reduced to ordinary LS. When the design matrix $\boldsymbol{\Phi}$ is *ill-conditioned* in the sense that a very small singular value exists, the inverse of that singular value, κ_k/κ_k^2 $(= 1/\kappa_k)$, can be very large. Then noise included in training output vector $\boldsymbol{y}$ is magnified by the factor $1/\kappa_k$ in ordinary LS. On the other hand, in ℓ_2-constrained LS, κ_k^2 in the denominator is increased by adding a positive scalar λ, which prevents $\kappa_k/(\kappa_k^2 + \lambda)$ from being too large and thus overfitting can be mitigated.

A MATLAB code for ℓ_2-constrained LS is provided in Fig. 23.6, where the Gaussian kernel model is used:

$$f_{\boldsymbol{\theta}}(\boldsymbol{x}) = \sum_{j=1}^{n} \theta_j K(\boldsymbol{x}, \boldsymbol{x}_j),$$

where

$$K(\boldsymbol{x}, \boldsymbol{c}) = \exp\left(-\frac{\|\boldsymbol{x} - \boldsymbol{c}\|^2}{2h^2}\right).$$

The Gaussian bandwidth is set at $h = 0.3$, and the regularization parameter is set at $\lambda = 0.1$. The behavior of ℓ_2-constrained LS is illustrated in Fig. 23.7, showing that overfitting can be successfully avoided.

ℓ_2-constrained LS can be slightly generalized by using a $b \times b$ positive semidefinite matrix $\boldsymbol{G}$ as

$$\min_{\boldsymbol{\theta}} J_{\mathrm{LS}}(\boldsymbol{\theta}) \quad \text{subject to } \boldsymbol{\theta}^\top \boldsymbol{G} \boldsymbol{\theta} \le R^2,$$

which is called *generalized ℓ_2-constrained LS*. When matrix $\boldsymbol{G}$, called a *regularization matrix*, is positive and symmetric, $\boldsymbol{\theta}^\top \boldsymbol{G} \boldsymbol{\theta} \le R^2$ represents an ellipsoidal

```
n=50; N=1000; x=linspace(-3,3,n)'; X=linspace(-3,3,N)';
pix=pi*x; y=sin(pix)./(pix)+0.1*x+0.2*randn(n,1);

x2=x.^2; X2=X.^2; hh=2*0.3^2; l=0.1;
k=exp(-(repmat(x2,1,n)+repmat(x2',n,1)-2*x*x')/hh);
K=exp(-(repmat(X2,1,n)+repmat(x2',N,1)-2*X*x')/hh);
t1=k\y; F1=K*t1; t2=(k^2+l*eye(n))\(k*y); F2=K*t2;

figure(1); clf; hold on; axis([-2.8 2.8 -1 1.5]);
plot(X,F1,'g-'); plot(X,F2,'r--'); plot(x,y,'bo');
legend('LS','L2-Constrained LS');
```

FIGURE 23.6

MATLAB code of ℓ_2-constrained LS regression for Gaussian kernel model.

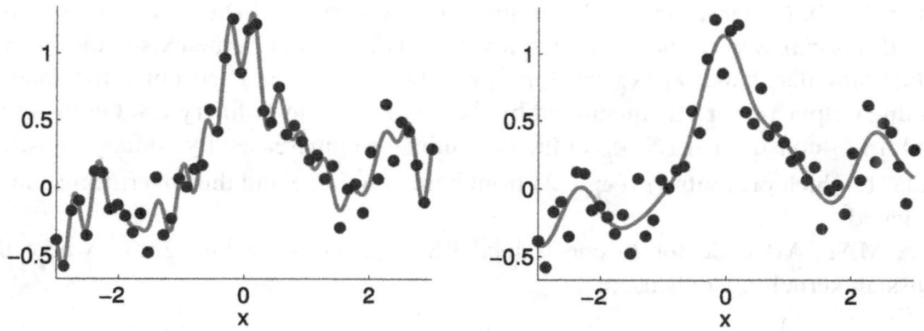

(a) Ordinary LS (b) ℓ_2-constrained LS

FIGURE 23.7

Example of ℓ_2-constrained LS regression for Gaussian kernel model. The Gaussian bandwidth is set at $h = 0.3$, and the regularization parameter is set at $\lambda = 0.1$.

constraint (Fig. 23.8). The solution of generalized ℓ_2-constrained LS can be obtained in the same way as ordinary ℓ_2-constrained LS by

$$\widehat{\theta} = (\Phi^\top \Phi + \lambda G)^{-1} \Phi^\top y.$$

23.3 MODEL SELECTION

In the previous sections, constrained LS was demonstrated to contribute to mitigating overfitting. However, the behavior of constrained LS depends on the choice of con-

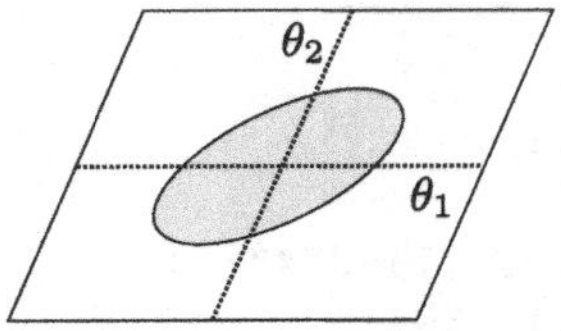

FIGURE 23.8

Parameter space in generalized ℓ_2-constrained
LS.

(a) $(h, \lambda) = (0.03, 0.0001)$ (b) $(h, \lambda) = (0.03, 0.1)$ (c) $(h, \lambda) = (0.03, 100)$

(d) $(h, \lambda) = (0.3, 0.0001)$ (e) $(h, \lambda) = (0.3, 0.1)$ (f) $(h, \lambda) = (0.3, 100)$

(g) $(h, \lambda) = (3, 0.0001)$ (h) $(h, \lambda) = (3, 0.1)$ (i) $(h, \lambda) = (3, 100)$

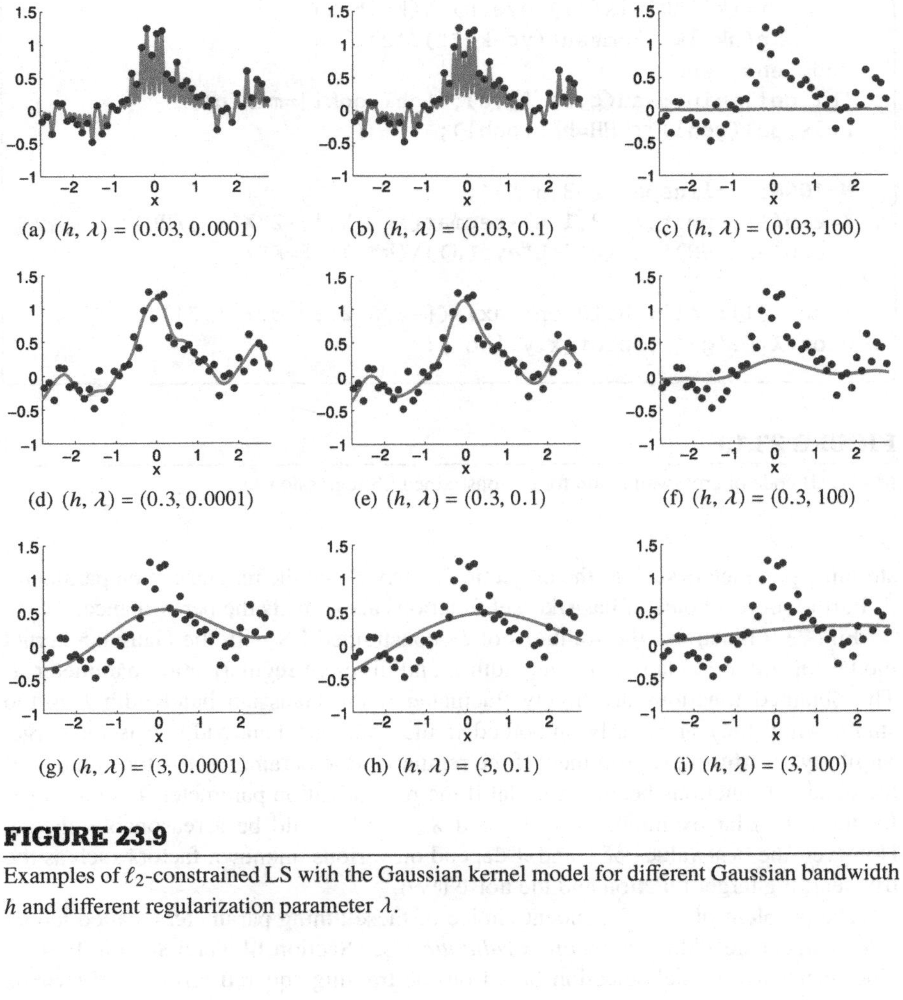

FIGURE 23.9

Examples of ℓ_2-constrained LS with the Gaussian kernel model for different Gaussian bandwidth
h and different regularization parameter λ.

```
n=50; x=linspace(-3,3,n)'; pix=pi*x;
y=sin(pix)./(pix)+0.1*x+0.2*randn(n,1);

x2=x.^2; xx=repmat(x2,1,n)+repmat(x2',n,1)-2*x*x';
hhs=2*[0.03 0.3 3].^2; ls=[0.0001 0.1 100];
m=5; u=mod(randperm(n),m)+1;
for hk=1:length(hhs)
  hh=hhs(hk); k=exp(-xx/hh);
  for i=1:m
    ki=k(u~=i,:); kc=k(u==i,:); yi=y(u~=i); yc=y(u==i);
    for lk=1:length(ls)
      t=(ki'*ki+ls(lk)*eye(n))\(ki'*yi);
      g(hk,lk,i)=mean((yc-kc*t).^2);
end, end, end
[gl,ggl]=min(mean(g,3),[],2); [ghl,gghl]=min(gl);
L=ls(ggl(gghl)); HH=hhs(gghl);

N=1000; X=linspace(-3,3,N)';
K=exp(-(repmat(X.^2,1,n)+repmat(x2',N,1)-2*X*x')/HH);
k=exp(-xx/HH); t=(k^2+L*eye(n))\(k*y); F=K*t;

figure(1); clf; hold on; axis([-2.8 2.8 -0.7 1.7]);
plot(X,F,'g-'); plot(x,y,'bo');
```

FIGURE 23.10

MATLAB code of cross validation for ℓ_2-constrained LS regression.

straining parameters such as the projection matrix $\boldsymbol{P}$ and the regularization parameter λ. Furthermore, choice of basis/kernel functions also affects the performance.

Fig. 23.9 illustrates the solutions of ℓ_2-constrained LS with the Gaussian kernel model for different Gaussian bandwidth h and different regularization parameter λ. The obtained functions are highly fluctuated if the Gaussian bandwidth h is too small, while they are overly smoothed if the Gaussian bandwidth h is too large. Similarly, overfitting is prominent if the regularization parameter λ is too small, and the obtained functions become too flat if the regularization parameter λ is too large. In this particular example, $h = 0.3$ and $\lambda = 0.1$ would be a reasonable choice. However, the best values of h and λ depend on various unknown factors such as the true learning target function and the noise level.

The problem of data-dependent choice of these tuning parameters, called *model selection*, can be addressed by *cross validation* (see Section 14.4 and Section 16.4.2). Note that naive model selection based on the training squared error simply yields

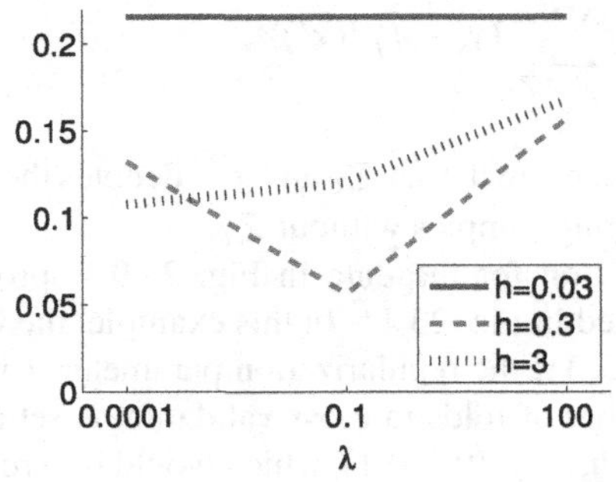

FIGURE 23.11

Example of cross validation for ℓ_2-constrained LS regression. The cross validation error for all Gaussian bandwidth h and regularization parameter λ is plotted, which is minimized at $(h, \lambda) = (0.3, 0.1)$. See Fig. 23.9 for learned functions.

For invertible and symmetric matrix $\boldsymbol{A}$, it holds that

$$(\boldsymbol{A} + \boldsymbol{b}\boldsymbol{b}^\top)^{-1} = \boldsymbol{A}^{-1} - \frac{\boldsymbol{A}^{-1}\boldsymbol{b}\boldsymbol{b}^\top\boldsymbol{A}^{-1}}{1 + \boldsymbol{b}^\top\boldsymbol{A}^{-1}\boldsymbol{b}}.$$

If $c - \boldsymbol{b}^\top\boldsymbol{A}^{-1}\boldsymbol{b} \neq 0$, it holds that

$$\begin{pmatrix} \boldsymbol{A} & \boldsymbol{b} \\ \boldsymbol{b}^\top & c \end{pmatrix}^{-1} = \begin{pmatrix} \boldsymbol{A}^{-1} + \alpha\boldsymbol{A}^{-1}\boldsymbol{b}\boldsymbol{b}^\top\boldsymbol{A}^{-1} & -\alpha\boldsymbol{A}^{-1}\boldsymbol{b} \\ -\alpha\boldsymbol{b}^\top\boldsymbol{A}^{-1} & \alpha \end{pmatrix},$$

where

$$\alpha = \frac{1}{c - \boldsymbol{b}^\top\boldsymbol{A}^{-1}\boldsymbol{b}}.$$

FIGURE 23.12

Matrix inversion lemma.

overfitting. For example, in Fig. 23.9, choosing the smallest h and λ minimizes the training squared error, which results in the heaviest overfitting. The algorithm of cross validation is exactly the same as the one described in Fig. 16.17, but the *squared loss function* is used to compute the validation error:

$$J_j^{(\ell)} = \frac{1}{|\mathcal{Z}_\ell|} \sum_{(\boldsymbol{x}',y')\in\mathcal{Z}_\ell} (y' - \widehat{f}_j^{(\ell)}(\boldsymbol{x}'))^2,$$

where $|\mathcal{Z}_\ell|$ denotes the number of elements in the set $\mathcal{Z}_\ell$ and $\widehat{f}_j^{(\ell)}$ denotes the function learned using model $\mathcal{M}_j$ from all training samples without $\mathcal{Z}_\ell$.

A MATLAB code of cross validation for the data in Fig. 23.9 is provided in Fig. 23.10, and its behavior is illustrated in Fig. 23.11. In this example, the Gaussian bandwidth h is chosen from $\{0.03, 0.3, 3\}$, the regularization parameter λ is chosen from $\{0.0001, 0.1, 100\}$, and the number of folds in cross validation is set at 5. The cross validation error is minimized at $(h, \lambda) = (0.3, 0.1)$, which would be a reasonable choice as illustrated in Fig. 23.9.

Cross validation when the number of folds is set at the number of training samples n, i.e., $n - 1$ samples are used for training and only the remaining single sample is used for validation, is called *leave-one-out cross validation*. Naive implementation of leave-one-out cross validation requires n repetitions of training and validation, which is computationally demanding when n is large. However, for ℓ_2-constrained LS, the score of leave-one-out cross validation can be computed analytically without repetition as follows [77]:

$$\frac{1}{n}\|\widetilde{\boldsymbol{H}}^{-1}\boldsymbol{H}\boldsymbol{y}\|^2, \tag{23.4}$$

where $\boldsymbol{H}$ is the $n \times n$ matrix defined as

$$\boldsymbol{H} = \boldsymbol{I} - \boldsymbol{\Phi}(\boldsymbol{\Phi}^\top\boldsymbol{\Phi} + \lambda\boldsymbol{I})^{-1}\boldsymbol{\Phi}^\top,$$

and $\widetilde{\boldsymbol{H}}$ is the diagonal matrix with diagonal elements the same as $\boldsymbol{H}$. In the derivation of Eq. (23.4), the *matrix inversion lemma* was utilized (see Fig. 23.12).

CHAPTER

SPARSE REGRESSION 24

CHAPTER CONTENTS

ℓ_2-constrained LS for linear-in-parameter model,

$$f_{\boldsymbol{\theta}}(\boldsymbol{x}) = \sum_{j=1}^{b} \theta_j \phi_j(\boldsymbol{x}),$$

is highly useful in practice. However, when the number of parameters, b, is very large, computing the output value $f_{\boldsymbol{\theta}}(\boldsymbol{x})$ for test sample $\boldsymbol{x}$ is time-consuming. In this chapter, a learning method that tends to produce a *sparse* solution is introduced. A sparse solution means that many of the parameters $\{\theta_j\}_{j=1}^{b}$ take exactly zero, and thus $f_{\boldsymbol{\theta}}(\boldsymbol{x})$ can be computed efficiently.

24.1 ℓ_1-CONSTRAINED LS

ℓ_2-constrained LS uses the ℓ_2-norm as a constraint, while *sparse learning* uses the ℓ_1-norm as a constraint:

$$\min_{\boldsymbol{\theta}} \frac{1}{2}\|\boldsymbol{y} - \boldsymbol{\Phi}\boldsymbol{\theta}\|^2 \text{ subject to } \|\boldsymbol{\theta}\|_1 \leq R^2, \tag{24.1}$$

where the ℓ_1-norm of vector $\boldsymbol{\theta} = (\theta_1, \ldots, \theta_b)^\top$ is defined as the absolute sum of all elements:

$$\|\boldsymbol{\theta}\|_1 = \sum_{j=1}^{b} |\theta_j|.$$

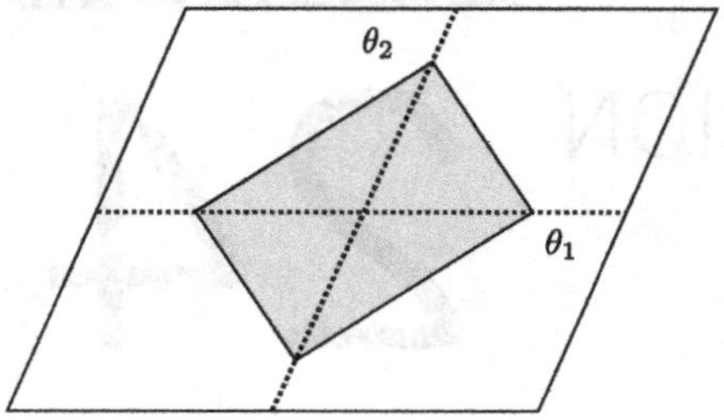

FIGURE 24.1

Parameter space in ℓ_1-constrained LS.

The region that $\|\theta\|_1 \leq R^2$ specifies is illustrated in Fig. 24.1, which has corners on the coordinate axes. This shape is called the ℓ_1-ball, and ℓ_1-constrained LS is also called *ℓ_1-regularization learning*.

Actually, these corners are the key to produce a sparse solution. Let us intuitively explain this idea using Fig. 24.2. For linear-in-parameter model,

$$f_{\theta}(x) = \sum_{j=1}^{b} \theta_j \phi_j(x) = \theta^{\top} \phi(x),$$

the training squared error is a convex quadratic function with respect to θ:

$$J_{\mathrm{LS}}(\theta) = \frac{1}{2}\|y - \Phi\theta\|^2.$$

Thus, in the parameter space, J_{LS} has ellipsoidal contours and its minimum is the LS solution $\hat{\theta}_{\mathrm{LS}}$. As illustrated in Fig. 24.2(a), the solution of ℓ_2-constrained LS is given as the point where the ellipsoidal contour and the ℓ_2-ball touch. Similarly, the solution of ℓ_1-constrained LS is given as the point where the ellipsoidal contour and the ℓ_1-ball touch. Since the ℓ_1-ball has corners on the coordinate axes, the solution tends to be at one of the corners, which is a sparse solution (see Fig. 24.2(b)).

In statistics, ℓ_1-constrained LS is referred to as *lasso regression* [110].

24.2 SOLVING ℓ_1-CONSTRAINED LS

Since the ℓ_1-norm is not differentiable at the origin, solving ℓ_1-constrained LS problem (24.1) is not as straightforward as ℓ_2-constrained LS. In this section, a general optimization technique called the *alternating direction method of multipliers* (ADMM) [20] is applied to ℓ_1-constrained LS and gives a simple yet practical algorithm. The algorithm of ADMM is summarized in Fig. 24.3.

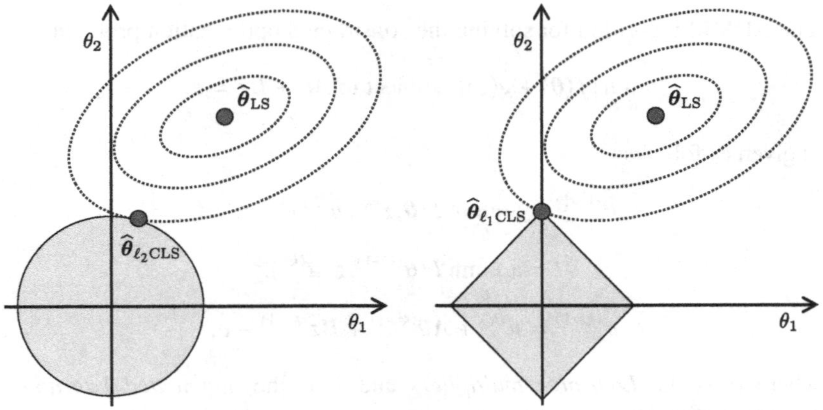

(a) ℓ_2-constrained LS (b) ℓ_1-constrained LS

FIGURE 24.2

The solution of ℓ_1-constrained LS tends to be on one of the coordinate axes, which is a sparse solution.

Let us express the optimization problem of ℓ_1-constrained LS, Eq. (24.1), as

$$\min_{\boldsymbol{\theta},\boldsymbol{z}} \left[\frac{1}{2}\|\boldsymbol{y} - \boldsymbol{\Phi\theta}\|^2 + \lambda\|\boldsymbol{z}\|_1\right] \quad \text{subject to } \boldsymbol{\theta} = \boldsymbol{z},$$

where $\lambda \geq 0$. The *augmented Lagrange function* for this optimization problem is given by

$$L(\boldsymbol{\theta},\boldsymbol{z},\boldsymbol{u}) = \frac{1}{2}\|\boldsymbol{y} - \boldsymbol{\Phi\theta}\|^2 + \lambda\|\boldsymbol{z}\|_1 + \boldsymbol{u}^\top(\boldsymbol{\theta} - \boldsymbol{z}) + \frac{1}{2}\|\boldsymbol{\theta} - \boldsymbol{z}\|^2,$$

where $\boldsymbol{u}$ is the *Lagrange multipliers*. Since

$$\frac{\partial L}{\partial \boldsymbol{\theta}} = -\boldsymbol{\Phi}^\top(\boldsymbol{y} - \boldsymbol{\Phi\theta}) + \boldsymbol{u} + \boldsymbol{\theta} - \boldsymbol{z},$$

setting this to zero yields the following update equation for $\boldsymbol{\theta}$:

$$\begin{aligned}\boldsymbol{\theta}^{(k+1)} &= \underset{\boldsymbol{\theta}}{\operatorname{argmin}}\, L(\boldsymbol{\theta},\boldsymbol{z}^{(k)},\boldsymbol{u}^{(k)}) \\ &= (\boldsymbol{\Phi}^\top\boldsymbol{\Phi} + \boldsymbol{I})^{-1}(\boldsymbol{\Phi}^\top\boldsymbol{y} + \boldsymbol{z}^{(k)} - \boldsymbol{u}^{(k)}).\end{aligned}$$

From

$$\begin{aligned}&\min_{z}\left[\lambda|z| + u(\theta - z) + \frac{1}{2}(\theta - z)^2\right] \\ &\quad = \max(0, \theta + u - \lambda) + \min(0, \theta + u + \lambda),\end{aligned}$$

The ADMM algorithm for solving the constrained optimization problem,

$$\min_{\boldsymbol{\theta},\boldsymbol{z}} [f(\boldsymbol{\theta}) + g(\boldsymbol{z})] \quad \text{subject to } \boldsymbol{A}\boldsymbol{\theta} + \boldsymbol{B}\boldsymbol{z} = \boldsymbol{c},$$

is given as follows:

$$\begin{aligned}\boldsymbol{\theta}^{(k+1)} &= \operatorname*{argmin}_{\boldsymbol{\theta}} L(\boldsymbol{\theta}, \boldsymbol{z}^{(k)}, \boldsymbol{u}^{(k)}),\\ \boldsymbol{z}^{(k+1)} &= \operatorname*{argmin}_{\boldsymbol{z}} L(\boldsymbol{\theta}^{(k+1)}, \boldsymbol{z}, \boldsymbol{u}^{(k)}),\\ \boldsymbol{u}^{(k+1)} &= \boldsymbol{u}^{(k)} + \boldsymbol{A}\boldsymbol{\theta}^{(k+1)} + \boldsymbol{B}\boldsymbol{z}^{(k+1)} - \boldsymbol{c},\end{aligned}$$

where $\boldsymbol{u}$ is the *Lagrange multipliers* and L is the *augmented Lagrange function* defined as

$$\begin{aligned}L(\boldsymbol{\theta}, \boldsymbol{z}, \boldsymbol{u}) = f(\boldsymbol{\theta}) + g(\boldsymbol{z}) &+ \boldsymbol{u}^\top(\boldsymbol{A}\boldsymbol{\theta} + \boldsymbol{B}\boldsymbol{z} - \boldsymbol{c})\\ &+ \frac{1}{2}\|\boldsymbol{A}\boldsymbol{\theta} + \boldsymbol{B}\boldsymbol{z} - \boldsymbol{c}\|^2.\end{aligned}$$

An advantage of ADMM is that no tuning parameter such as the step size in the gradient algorithm (see Fig. 22.6) is involved.

FIGURE 24.3

Alternating direction method of multipliers.

the update equation for $\boldsymbol{z}$ is given as

$$\begin{aligned}\boldsymbol{z}^{(k+1)} &= \operatorname*{argmin}_{\boldsymbol{z}} L(\boldsymbol{\theta}^{(k+1)}, \boldsymbol{z}, \boldsymbol{u}^{(k)})\\ &= \max(\boldsymbol{0}, \boldsymbol{\theta}^{(k+1)} + \boldsymbol{u}^{(k)} - \lambda\boldsymbol{1}) + \min(\boldsymbol{0}, \boldsymbol{\theta}^{(k+1)}\boldsymbol{u}^{(k)} + \lambda\boldsymbol{1}),\end{aligned}$$

where $\boldsymbol{1}$ is the vector with all ones. Finally, the update equation for $\boldsymbol{u}$ is given as follows (see Fig. 24.3):

$$\boldsymbol{u}^{(k+1)} = \boldsymbol{u}^{(k)} + \boldsymbol{\theta}^{(k+1)} - \boldsymbol{z}^{(k+1)}.$$

A MATLAB code of ℓ_1-constrained LS by ADMM for the Gaussian kernel model,

$$f_{\boldsymbol{\theta}}(\boldsymbol{x}) = \sum_{j=1}^{n} \theta_j \exp\left(-\frac{\|\boldsymbol{x} - \boldsymbol{x}_j\|^2}{2h^2}\right),$$

is provided in Fig. 24.4, and its behavior is illustrated in Fig. 24.5. The obtained solution is actually very similar to the one obtained by ℓ_2-constrained LS (see

```
n=50; x=linspace(-3,3,n)'; pix=pi*x;
y=sin(pix)./(pix)+0.1*x+0.2*randn(n,1);

hh=2*0.3^2; l=0.1; x2=x.^2;
k=exp(-(repmat(x2,1,n)+repmat(x2',n,1)-2*x*x')/hh);
ky=k*y; A=inv(k^2+eye(n)); t0=zeros(n,1); z=t0; u=t0;
for o=1:1000
 t=A*(ky+z-u); z=max(0,t+u-l)+min(0,t+u+l); u=u+t-z;
 if norm(t-t0)<0.0001, break, end
 t0=t;
end

N=1000; X=linspace(-3,3,N)';
K=exp(-(repmat(X.^2,1,n)+repmat(x2',N,1)-2*X*x')/hh); F=K*t;
figure(1); clf; hold on; axis([-2.8 2.8 -1 1.5]);
plot(X,F,'g-'); plot(x,y,'bo'); sum(abs(t)<0.001)
```

FIGURE 24.4

MATLAB code of ℓ_1-constrained LS by ADMM for Gaussian kernel model.

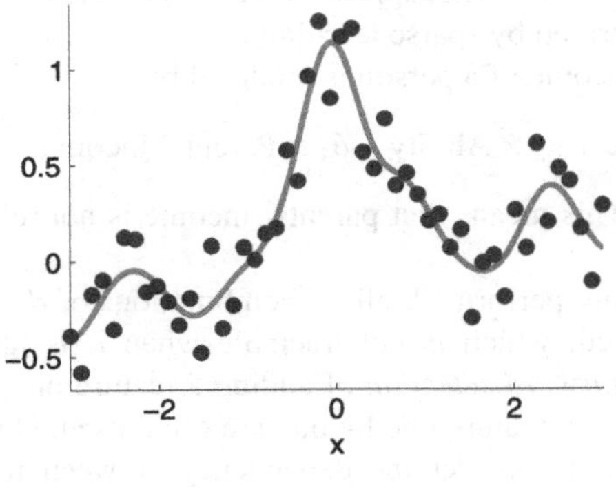

FIGURE 24.5

Example of ℓ_1-constrained LS for Gaussian kernel model. 38 out of 50 parameters are zero.

Fig. 23.7). However, 38 out of 50 parameters are zero in ℓ_1-constrained LS, while all 50 parameters are nonzero in ℓ_2-constrained LS.

When the *stochastic gradient* algorithm (see Fig. 22.6) is used to obtain the LS solution for linear-in-parameter model,

$$f_{\boldsymbol{\theta}}(\boldsymbol{x}) = \sum_{j=1}^{b} \theta_j \phi_j(\boldsymbol{x}) = \boldsymbol{\theta}^\top \boldsymbol{\phi}(\boldsymbol{x}),$$

parameter $\boldsymbol{\theta}$ is updated as

$$\boldsymbol{\theta} \longleftarrow \boldsymbol{\theta} + \varepsilon \boldsymbol{\phi}(\boldsymbol{x})\big(y - f_{\boldsymbol{\theta}}(\boldsymbol{x})\big),$$

where $\varepsilon > 0$ is the step size and $(\boldsymbol{x}, y)$ is a randomly chosen training sample. To obtain a sparse solution by the stochastic gradient algorithm, parameter $\boldsymbol{\theta} = (\theta_1, \ldots, \theta_b)^\top$ may be thresholded once in every several gradient updates as follows [66]:

$$\forall j = 1, \ldots, b, \quad \theta_j \longleftarrow \begin{cases} \max(0, \theta_j - \lambda\varepsilon) & (\theta_j > 0), \\ \min(0, \theta_j + \lambda\varepsilon) & (\theta_j \leq 0). \end{cases}$$

24.3 FEATURE SELECTION BY SPARSE LEARNING

Let us apply sparse learning to the linear-in-input model for $\boldsymbol{x} = (x^{(1)}, \ldots, x^{(d)})^\top$:

$$f_{\boldsymbol{\theta}}(\boldsymbol{x}) = \sum_{j=1}^{d} \theta_j x^{(j)} = \boldsymbol{\theta}^\top \boldsymbol{x}.$$

If $\theta_j = 0$, then the jth input variable $x^{(j)}$ does not appear in the final prediction model. Thus, *feature selection* can be performed by sparse learning.

For example, suppose that the income of a person is modeled by

$$\theta_1 \times \text{Education} + \theta_2 \times \text{Age} + \theta_3 \times \text{Ability} + \theta_4 \times \text{Parents' income},$$

and sparse learning gives $\theta_4 = 0$. This means that parents' income is not related to the income of the child.

If such feature selection is naively performed, all 2^d combinations of d features $x^{(1)}, \ldots, x^{(d)}$ need to be investigated, which is not tractable when d is large. In practice, greedy strategies such as *forward selection* of adding a feature one by one and *backward elimination* of deleting a feature one by one are often used. However, since such greedy approaches do not consider the dependency between features, they do not necessarily give good feature combinations. If sparse learning based on the ℓ_1-norm is used, dependency between features can be taken into account to some extent. However, feature selection by sparse learning is possible only for the *linear-in-input model*. If *linear-in-parameter models* are used, just a subset of basis functions is selected, which is different from feature selection because every basis function depends on all features in general.

24.4 VARIOUS EXTENSIONS

In this section, various extensions to ℓ_1-constrained LS are introduced.

24.4.1 GENERALIZED ℓ_1-CONSTRAINED LS

For some matrix $\boldsymbol{F}$, *generalized ℓ_1-constrained LS* is defined as

$$\min_{\boldsymbol{\theta}} \frac{1}{2}\|\boldsymbol{y} - \boldsymbol{\Phi}\boldsymbol{\theta}\|^2 \text{ subject to } \|\boldsymbol{F}\boldsymbol{\theta}\|_1 \leq R^2,$$

and its ADMM form is given by

$$\min_{\boldsymbol{\theta},\boldsymbol{z}} \left[\frac{1}{2}\|\boldsymbol{y} - \boldsymbol{\Phi}\boldsymbol{\theta}\|^2 + \lambda\|\boldsymbol{z}\|_1\right] \text{ subject to } \boldsymbol{F}\boldsymbol{\theta} = \boldsymbol{z}.$$

Then the ADMM update formulas are given by

$$\begin{aligned}
\boldsymbol{\theta}^{(k+1)} &= (\boldsymbol{\Phi}^\top\boldsymbol{\Phi} + \boldsymbol{F}^\top\boldsymbol{F})^{-1}(\boldsymbol{\Phi}^\top\boldsymbol{y} + \boldsymbol{F}^\top\boldsymbol{z}^{(k)} - \boldsymbol{F}^\top\boldsymbol{u}^{(k)}),\\
\boldsymbol{z}^{(k+1)} &= \max(\boldsymbol{0}, \boldsymbol{F}\boldsymbol{\theta}^{(k+1)} + \boldsymbol{u}^{(k)} - \lambda\boldsymbol{1}) + \min(\boldsymbol{0}, \boldsymbol{F}\boldsymbol{\theta}^{(k+1)}\boldsymbol{u}^{(k)} + \lambda\boldsymbol{1}),\\
\boldsymbol{u}^{(k+1)} &= \boldsymbol{u}^{(k)} + \boldsymbol{F}\boldsymbol{\theta}^{(k+1)} - \boldsymbol{z}^{(k+1)}.
\end{aligned}$$

A notable example of generalized ℓ_1-constrained LS in statistics is *fused lasso* [111]:

$$\min_{\boldsymbol{\theta}} \frac{1}{2}\|\boldsymbol{y} - \boldsymbol{\Phi}\boldsymbol{\theta}\|^2 \text{ subject to } \sum_{j=1}^{b-1} |\theta_{j+1} - \theta_j| \leq R^2,$$

which corresponds to

$$F_{j,j'} = \begin{cases} 1 & (j' = j+1),\\ -1 & (j' = j),\\ 0 & (\text{otherwise}). \end{cases}$$

When $\boldsymbol{\Phi}$ is the identity matrix, this problem is called the *total variation denoising* in the signal processing community.

24.4.2 ℓ_p-CONSTRAINED LS

Let us generalize ℓ_2-constrained LS and ℓ_1-constrained LS to ℓ_p-constrained LS for $p \geq 0$:

$$\|\boldsymbol{\theta}\|_p \leq R^2.$$

Here, the *ℓ_p-norm* of vector $\boldsymbol{\theta} = (\theta_1, \ldots, \theta_b)^\top$ is defined as

$$\|\boldsymbol{\theta}\|_p = \begin{cases} \sum_{j=1}^b \delta(\theta_j \neq 0) & (p = 0),\\ \left(\sum_{j=1}^b |\theta_j|^p\right)^{\frac{1}{p}} & (0 < p < \infty),\\ \max\left\{|\theta_1|, \ldots, |\theta_b|\right\} & (p = \infty), \end{cases}$$

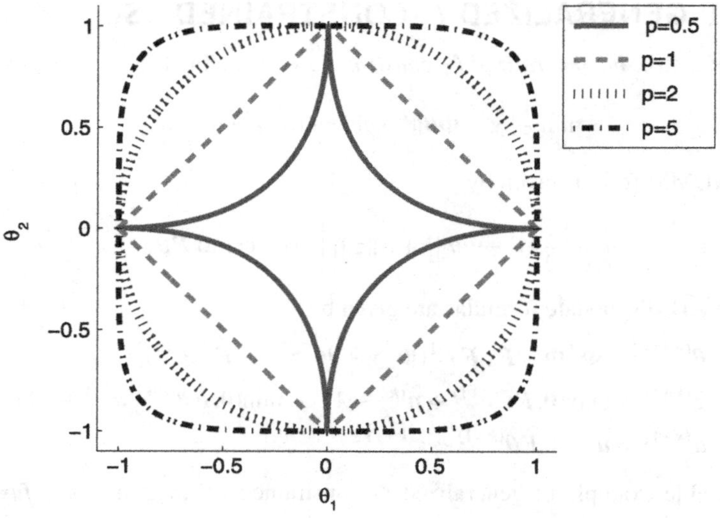

FIGURE 24.6

Unit ℓ_p-balls.

where

$$\delta(\theta_j \neq 0) = \begin{cases} 1 & (\theta_j \neq 0), \\ 0 & (\theta_j = 0). \end{cases}$$

Thus, the ℓ_0-norm gives the number of nonzero elements, and the ℓ_∞-norm gives the maximum element.

Fig. 24.6 shows the unit ℓ_p-balls (i.e. $\|\theta\|_p = 1$) for $p = 0.5$, 1, 2, and 5. The ℓ_p-ball has corners on the coordinate axes when $p \leq 1$, and the ℓ_p-ball has a convex shape when $p \geq 1$. As illustrated in Fig. 24.2, having corners on the coordinate axes is the key to produce a sparse solution. On the other hand, having a convex shape is essential to obtain the global optimal solution. Thus, $p = 1$ is the best choice that only allows sparsity induction under the convex formulation (see Fig. 24.7).

24.4.3 $\ell_1 + \ell_2$-CONSTRAINED LS

ℓ_1-constrained LS is a useful method in practice, but it has several drawbacks.

When the number of parameters, b, is larger than the number of training samples, n, the number of nonzero values in the solution of ℓ_1-constrained LS is at most n.

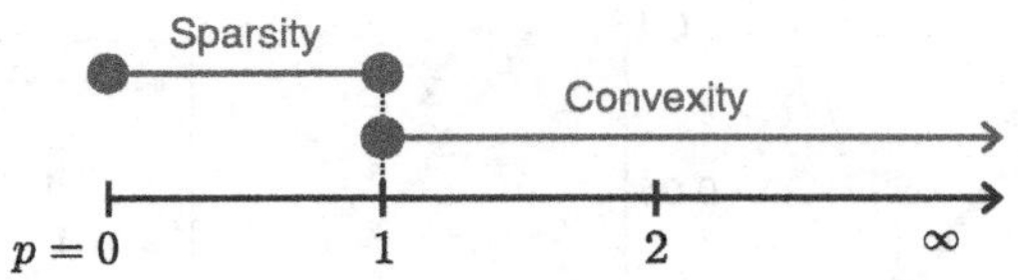

FIGURE 24.7

Properties of ℓ_p-constraint.

This is not a problem when a kernel model is used, because $b = n$:

$$f_{\boldsymbol{\theta}}(\boldsymbol{x}) = \sum_{j=1}^{n} \theta_j K(\boldsymbol{x}, \boldsymbol{x}_j).$$

However, this could be a critical limitation when feature selection is performed using the linear-in-input model (see Section 24.3),

$$f_{\boldsymbol{\theta}}(\boldsymbol{x}) = \sum_{j=1}^{d} \theta_j x^{(j)} = \boldsymbol{\theta}^\top \boldsymbol{x},$$

because only n features can be selected at most.

When several basis functions are similar to each other and having group structure, ℓ_1-constrained LS tends to choose only one of them and ignore the rest. This means that, when feature selection is performed using the linear-in-input model, only one feature is chosen from a group of correlated features. When kernel models are used, the kernel basis functions tend to have group structure, if training input samples $\{\boldsymbol{x}_i\}_{i=1}^n$ have cluster structure.

$\ell_1 + \ell_2$-constrained LS can overcome the above drawbacks of ℓ_1-constrained LS, which uses the sum of the ℓ_1-norm and the ℓ_2-norm as a constraint:

$$(1 - \tau)\|\boldsymbol{\theta}\|_1 + \tau\|\boldsymbol{\theta}\|^2 \leq R^2,$$

where $0 \leq \tau \leq 1$ controls the balance between the ℓ_1-norm and ℓ_2-norm constraints. The $(\ell_1 + \ell_2)$-constraint is reduced to the ℓ_1-constraint if $\tau = 0$, and the $(\ell_1 + \ell_2)$-constraint is reduced to the ℓ_2-constraint if $\tau = 1$. When $0 \leq \tau < 1$, the region $(1 - \tau)\|\boldsymbol{\theta}\|_1 + \tau\|\boldsymbol{\theta}\|^2 \leq R^2$ has corners on the coordinate axes.

Fig. 24.8 illustrates the unit $(\ell_1 + \ell_2)$-ball for balance parameter $\tau = 1/2$, which is similar to the unit $\ell_{1.4}$-ball. However, while the $\ell_{1.4}$-ball has no corner as the ℓ_2-ball, the $(\ell_1 + \ell_2)$-ball has corners. Thus, the $(\ell_1 + \ell_2)$-constraint tends to produce a sparse solution.

Let us express the optimization problem of $(\ell_1 + \ell_2)$-constrained LS as

$$\min_{\boldsymbol{\theta}, \boldsymbol{z}} \left[\frac{1}{2}\|\boldsymbol{y} - \boldsymbol{\Phi}\boldsymbol{\theta}\|^2 + \eta\|\boldsymbol{\theta}\|^2 + \lambda\|\boldsymbol{z}\|_1 \right] \text{ subject to } \boldsymbol{\theta} = \boldsymbol{z},$$

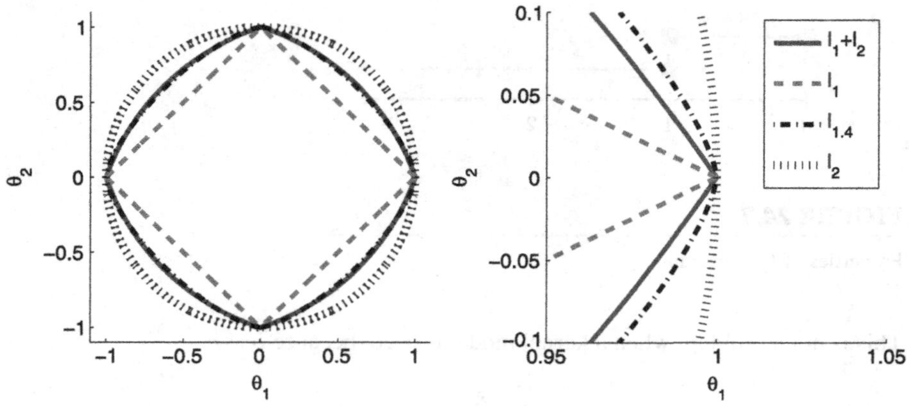

FIGURE 24.8

Unit $(\ell_1 + \ell_2)$-norm ball for balance parameter $\tau = 1/2$, which is similar to the unit $\ell_{1.4}$-ball. However, while the $\ell_{1.4}$-ball has no corner, the $(\ell_1 + \ell_2)$-ball has corners.

where $\lambda, \eta \geq 0$. Then the solution of $(\ell_1 + \ell_2)$-constrained LS can be obtained almost in the same way as ℓ_1-constrained LS by ADMM (see Section 24.2), but only the update equation for $\boldsymbol{\theta}$ is changed as

$$\boldsymbol{\theta}^{(k+1)} = (\boldsymbol{\Phi}^\top \boldsymbol{\Phi} + (\eta + 1)\boldsymbol{I})^{-1}(\boldsymbol{\Phi}^\top \boldsymbol{y} + \boldsymbol{z}^{(k)} - \boldsymbol{u}^{(k)}).$$

Even when the number of parameters is larger than the number of training samples, i.e. $b > n$, $(\ell_1 + \ell_2)$-constrained LS can produce a solution with more than n nonzero elements. Furthermore, $(\ell_1 + \ell_2)$-constrained LS tends to choose features in a groupwise manner, i.e. all features in the same group tend to be discarded simultaneously. However, in addition to the regularization parameter λ, the balance parameter τ needs to be tuned, which is cumbersome in practice.

In statistics, $(\ell_1 + \ell_2)$-constrained LS is referred to as *elastic-net regression* [124].

24.4.4 $\ell_{1,2}$-CONSTRAINED LS

Suppose that the parameter vector $\boldsymbol{\theta} = (\theta_1, \ldots, \theta_b)^\top$ has group structure as

$$\boldsymbol{\theta} = (\boldsymbol{\theta}^{(1)\top}, \ldots, \boldsymbol{\theta}^{(t)\top})^\top,$$

where $\boldsymbol{\theta}^{(j)} \in \mathbb{R}^{b_j}$ and $\sum_{j=1}^{t} b_j = b$. Then, LS with $\ell_{1,2}$-constraint,

$$\sum_{j=1}^{t} \|\boldsymbol{\theta}^{(j)}\| \leq R,$$

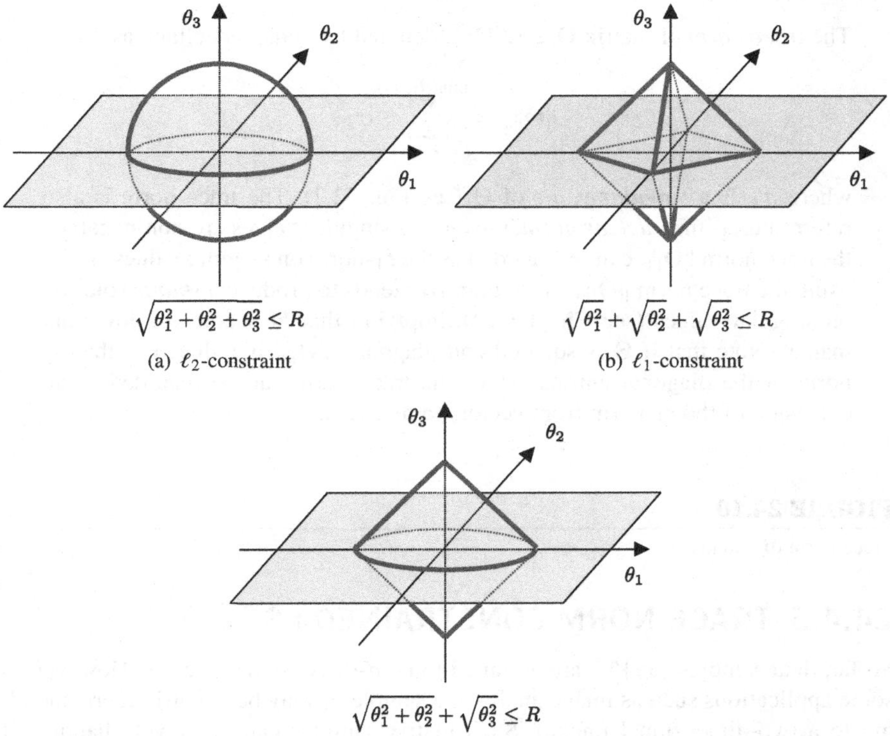

$\sqrt{\theta_1^2+\theta_2^2+\theta_3^2} \leq R$

(a) ℓ_2-constraint

$\sqrt{\theta_1^2}+\sqrt{\theta_2^2}+\sqrt{\theta_3^2} \leq R$

(b) ℓ_1-constraint

$\sqrt{\theta_1^2+\theta_2^2}+\sqrt{\theta_3^2} \leq R$

(c) $\ell_{1,2}$-constraint

FIGURE 24.9

Constraints in three-dimensional parameter space.

tends to give a groupwise sparse solution, i.e. some of parameter subvector $\boldsymbol{\theta}^{(j)}$ becomes zero. Such a learning method is called *$\ell_{1,2}$-constrained LS*.

The ℓ_2-constraint, ℓ_1-constraint, and $\ell_{1,2}$-constraint in three-dimensional parameter space are illustrated in Fig. 24.9. The $\ell_{1,2}$-region consists of the ℓ_2-part and the ℓ_1-part, and if the solution is on the top/bottom peak of the $\ell_{1,2}$-region, it will be sparse as $(\theta_1, \theta_2, \theta_3) = (0, 0, \pm R)$. On the other hand, if the solution is on the circle in the (θ_1, θ_2)-plane, the solution will be sparse as $(\theta_1, \theta_2, \theta_3) = (a, b, 0)$ for $a^2 + b^2 = R^2$.

In statistics, $\ell_{1,2}$-constrained LS is referred to as *group-lasso regression* [122]. The $\ell_{1,2}$-constraint plays an important role in advanced machine learning topics such as *multitask feature selection* (Section 34.3), *structural change detection* (Section 39.2.2), and *multiple kernel learning* [11].

The *trace norm* of matrix $\boldsymbol{\Theta} \in \mathbb{R}^{d_1 \times d_2}$, denoted by $\|\boldsymbol{\Theta}\|_{\mathrm{tr}}$, is defined as

$$\|\boldsymbol{\Theta}\|_{\mathrm{tr}} = \sum_{k=1}^{\min(d_1,d_2)} \sigma_k,$$

where σ_k is a *singular value* of $\boldsymbol{\Theta}$ (see Fig. 22.2). The trace norm is also referred to as the *nuclear norm*. Given that singular values are non-negative, the trace norm $\|\boldsymbol{\Theta}\|_{\mathrm{tr}}$ can be regarded as the ℓ_1-norm on singular values. Thus, using the trace norm $\|\boldsymbol{\Theta}\|_{\mathrm{tr}}$ as a regularizer tends to produce a *sparse* solution on singular values (see Chapter 24), implying that $\boldsymbol{\Theta}$ becomes a low-rank matrix. Note that if $\boldsymbol{\Theta}$ is squared and diagonal, $\|\boldsymbol{\Theta}\|_{\mathrm{tr}}$ is reduced to the ℓ_1-norm of the diagonal entries. Thus, the trace norm can be regarded as an extension of the ℓ_1-norm from vectors to matrices.

FIGURE 24.10

Trace norm of a matrix.

24.4.5 TRACE NORM CONSTRAINED LS

So far, data samples $\{\boldsymbol{x}_i\}_{i=1}^n$ are assumed to be d-dimensional vectors. However, in some applications such as image analysis, a sample $\boldsymbol{x}_i$ may be a matrix (corresponding to a two-dimensional image). Such matrix samples can be naively handled if they are vectorized (see Fig. 6.5), but then the two-dimensional structure of images is completely lost. Here, a regularization technique for matrix samples is introduced [112].

Suppose that matrix-input scalar-output paired samples $\{(\boldsymbol{X}_i, y_i)\}_{i=1}^n$ are given as training data, where $\boldsymbol{X}_i \in \mathbb{R}^{d_1 \times d_2}$ and $y_i \in \mathbb{R}$. Let us employ the *linear-in-input* model for parameter matrix $\boldsymbol{\Theta} \in \mathbb{R}^{d_1 \times d_2}$,

$$f_{\boldsymbol{\Theta}}(\boldsymbol{X}) = \mathrm{tr}\left(\boldsymbol{\Theta}^\top \boldsymbol{X}\right),$$

which is equivalent to the vectorized linear-in-input model $\mathrm{vec}(\boldsymbol{\Theta})^\top \mathrm{vec}(\boldsymbol{X})$. To utilize the two-dimensional structure, the *trace norm* $\|\boldsymbol{\Theta}\|_{\mathrm{tr}}$ is employed as a constraint (see Fig. 24.10):

$$\min_{\boldsymbol{\Theta} \in \mathbb{R}^{d_1 \times d_2}} \frac{1}{2} \sum_{i=1}^n \left(y_i - f_{\boldsymbol{\Theta}}(\boldsymbol{X}_i)\right)^2 \text{ subject to } \|\boldsymbol{\Theta}\|_{\mathrm{tr}} \leq R.$$

Note that this optimization problem is convex, and thus the global optimal solution can be easily obtained, e.g, by the *proximal gradient method* (see Fig. 34.8). Thanks to the trace norm constraint, the solution of $\boldsymbol{\Theta}$ tends to be a low-rank matrix.

CHAPTER

ROBUST REGRESSION 25

CHAPTER CONTENTS

Although LS is useful in various practical applications, it is sensitive to *outliers*, samples containing large noise. In this chapter, alternative learning methods that possess high robustness against outliers are introduced.

25.1 NONROBUSTNESS OF ℓ_2-LOSS MINIMIZATION

A LS solution for straight-line model $f_{\boldsymbol{\theta}}(x) = \theta_1 + \theta_2 x$ obtained from 10 training samples is illustrated in Fig. 25.1. If there is no outlier as in Fig. 25.1(a), a good solution can be obtained by LS. However, if there exists an outlier as in Fig. 25.1(b), the solution is strongly affected by the outlier.

When a large number of training samples are handled, it would be natural to consider that more or less outliers are included. In such a situation, LS is less reliable, as illustrated in Fig. 25.1. One way to cope with outliers is to remove them in advance, which will be discussed in Chapter 38. Another approach is to perform learning so that the solution is less sensitive to outliers. The ability to be less sensitive to outliers is referred to as *robustness*. In this chapter, robust learning methods are explored.

In the LS method, the goodness of fit to training samples is measured by the ℓ_2-loss function:

$$J_{\mathrm{LS}}(\boldsymbol{\theta}) = \frac{1}{2}\sum_{i=1}^{n} r_i^2,$$

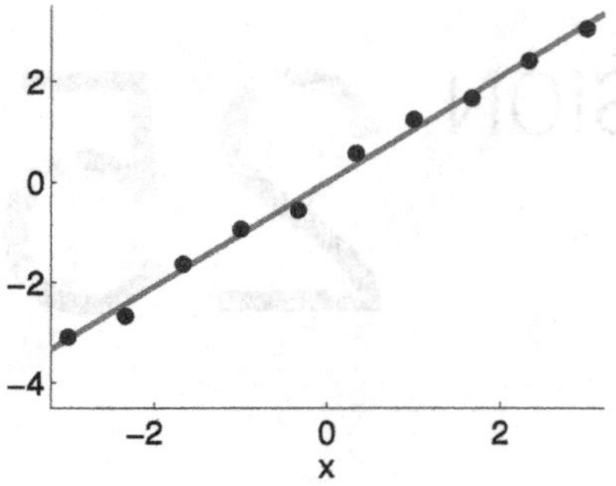

(a) Without outliers

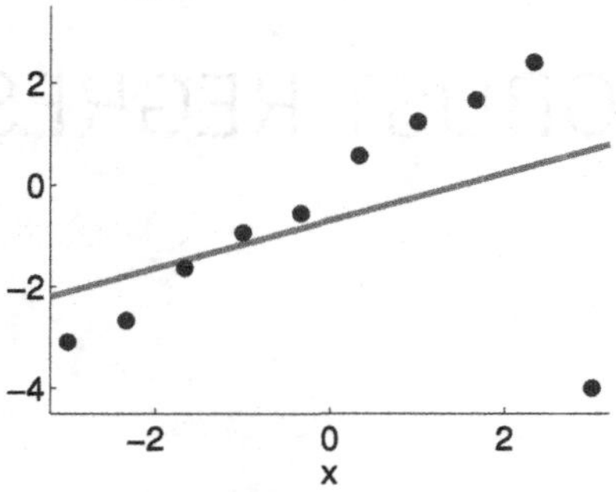

(b) With an outlier

FIGURE 25.1

LS solution for straight-line model $f_{\boldsymbol{\theta}}(x) = \theta_1 + \theta_2 x$, which is strongly affected by an outlier.

where r_i denotes the *residual* for the ith training sample $(\boldsymbol{x}_i, y_i)$:

$$r_i = y_i - f_{\boldsymbol{\theta}}(\boldsymbol{x}_i).$$

Since the ℓ_2-loss squares the error, large error tends to be magnified significantly. For example, in Fig. 25.1 the output of the rightmost training sample is moved from $y \approx 3$ to $y \approx -4$. Then the residual 7 has the "power" of $7^2 = 49$ to pull the regression function down, which causes significant change in the solution.

25.2 ℓ_1-LOSS MINIMIZATION

To cope with the nonrobustness of the LS method, let us employ the *ℓ_1-loss* or the *absolute loss* (Fig. 25.2):

$$\widehat{\boldsymbol{\theta}}_{\mathrm{LA}} = \underset{\boldsymbol{\theta}}{\operatorname{argmin}}\, J_{\mathrm{LA}}(\boldsymbol{\theta}), \quad \text{where } J_{\mathrm{LA}}(\boldsymbol{\theta}) = \sum_{i=1}^{n} |r_i|.$$

This learning method is called *ℓ_1-loss minimization* or *least absolute deviations regression*.

An example of ℓ_1-loss minimization is shown in Fig. 25.3, which is obtained in the same setup as in Fig. 25.1 (how to compute the solution of least absolute deviations will be explained later in Section 25.3). This shows that ℓ_1-loss minimization is much more robust against outliers than ℓ_2-loss minimization, and ℓ_1-loss minimization gives almost the same solution as ℓ_2-loss minimization if there are no outliers.

For a constant model $f_\theta(\boldsymbol{x}) = \theta$, the LS method gives the *mean* of training output samples $\{y_i\}_{i=1}^n$:

$$\widehat{\theta}_{\mathrm{LS}} = \underset{\theta}{\operatorname{argmin}} \sum_{i=1}^{n} (\theta - y_i)^2 = \mathrm{mean}\big(\{y_i\}_{i=1}^n\big).$$

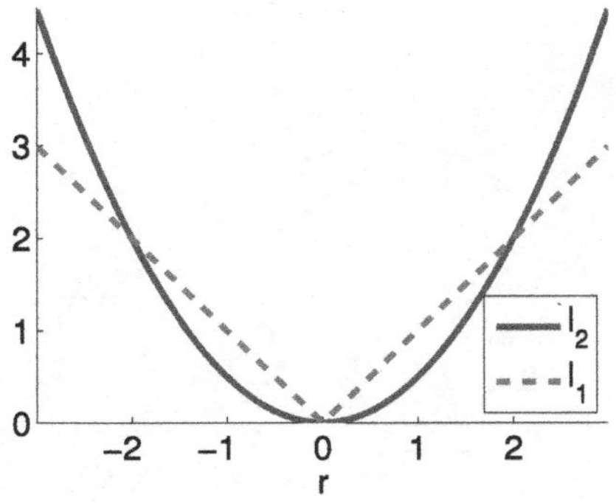

FIGURE 25.2

ℓ_2-loss and ℓ_1-loss. The ℓ_2-loss magnifies large residuals.

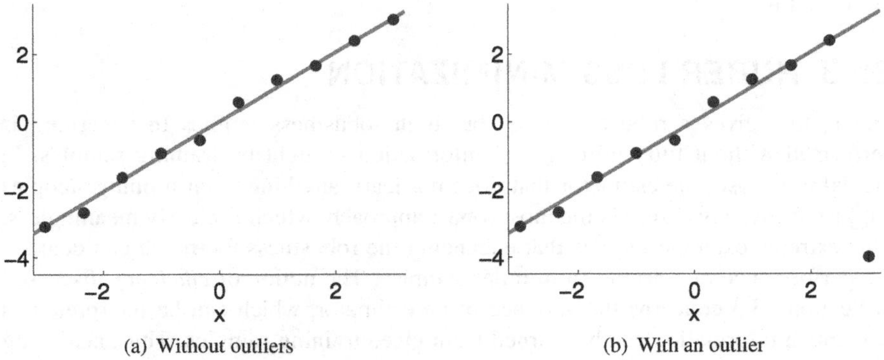

FIGURE 25.3

Solution of least absolute deviations for straight-line model $f_{\theta}(x) = \theta_1 + \theta_2 x$ for the same training samples as Fig. 25.1. Least absolute deviations give a much more robust solution than LS.

On the other hand, least absolute deviations gives the *median* (see Section 2.4.1) of the training output samples $\{y_i\}_{i=1}^n$:

$$\widehat{\theta}_{\mathrm{LA}} = \underset{\theta}{\operatorname{argmin}} \sum_{i=1}^{n} |\theta - y_i| = \operatorname{median}\Big(\{y_i\}_{i=1}^n\Big).$$

The mean of $\{y_i\}_{i=1}^n$ is affected if one of the sample values is changed, while the median is not affected by changing sample values as long as the order of samples is

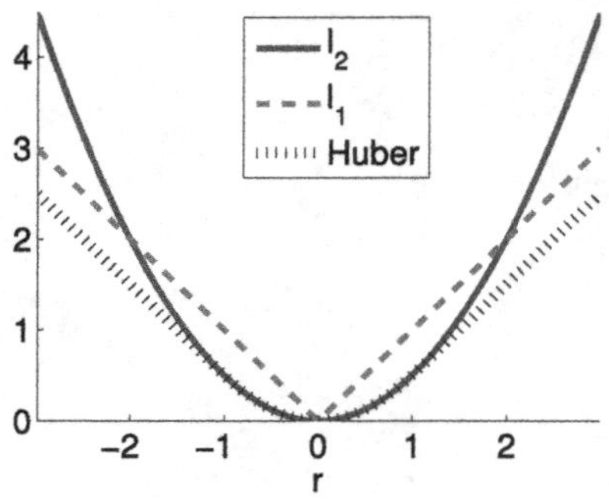

FIGURE 25.4

Huber loss, with threshold $\eta = 1$.

unchanged. The robustness of least absolute deviations is brought by such a property of the ℓ_1-loss.

25.3 HUBER LOSS MINIMIZATION

The ℓ_1-loss gives a robust solution, but high robustness implies that learning is performed without fully utilizing the information brought by training samples. In the extreme case, the estimator that does not learn anything from training samples but just always gives zero is the most robust approach, which is clearly meaningless. This extreme example implies that enhancing the robustness too much can degrade the performance of learning from inlier samples. The notion of *efficiency* discussed in Section 13.3 concerns the variance of an estimator, which can be interpreted as how much information can be learned from given training samples. Thus, achieving robustness and efficiency in a balanced way is crucial in practice. In this section, the *Huber loss* [58] is introduced, which can control the balance between robustness and efficiency.

25.3.1 DEFINITION

The Huber loss is defined with the ℓ_2-loss and the ℓ_1-loss as follows (Fig. 25.4):

$$\rho_{\text{Huber}}(r) = \begin{cases} r^2/2 & (|r| \leq \eta), \\ \eta|r| - \eta^2/2 & (|r| > \eta). \end{cases}$$

This means that the Huber loss is reduced to the ℓ_2-loss if the absolute residual $|r|$ is less than or equal to a threshold η (i.e., that sample may be an inlier), while it is reduced to the ℓ_1-loss if the absolute residual $|r|$ is larger than a threshold η (i.e., that sample may be an outlier). Note that the ℓ_1-loss is modified as $\eta|r| - \eta^2/2$ so that it is smoothly connected to the ℓ_2-loss $r^2/2$. The method of Huber loss minimization is given by

$$\min_{\theta} J(\theta), \quad \text{where } J(\theta) = \sum_{i=1}^{n} \rho_{\text{Huber}}(r_i). \tag{25.1}$$

25.3.2 STOCHASTIC GRADIENT ALGORITHM

Let us consider the linear-in-parameter model:

$$f_{\theta}(x) = \sum_{j=1}^{b} \theta_j \phi_j(x) = \theta^{\top} \phi(x).$$

Since the Huber loss is once differentiable as

$$\rho'_{\text{Huber}}(r) = \begin{cases} r & (|r| \leq \eta), \\ -\eta & (r < -\eta), \\ \eta & (r > \eta), \end{cases}$$

the solution of Huber loss minimization may be obtained by the *stochastic gradient* algorithm (see Section 15.3) as follows:

$$\theta \longleftarrow \theta + \varepsilon \frac{\partial \rho_{\text{Huber}}}{\partial \theta} \bigg|_{\theta = \widehat{\theta}},$$

where (x_i, y_i) is a randomly chosen training sample and $\varepsilon > 0$ is the step size.

25.3.3 ITERATIVELY REWEIGHTED LS

Another approach is *iteratively reweighted LS*, which considers the following quadratic upper bound of the absolute-value part of the Huber loss:

$$\eta|r| - \frac{\eta^2}{2} \leq \frac{\eta}{2c} r^2 + \frac{\eta c}{2} - \frac{\eta^2}{2} \quad \text{for } c > 0.$$

As illustrated in Fig. 25.5, this quadratic upper bound touches the Huber loss at $r = \pm c$.

Let us consider an iterative optimization procedure and construct an upper bound for $c_i = |\widetilde{r}_i|$:

$$\eta|r_i| - \frac{\eta^2}{2} \leq \frac{\eta}{2|\widetilde{r}_i|} r_i^2 + \frac{\eta|\widetilde{r}_i|}{2} - \frac{\eta^2}{2},$$

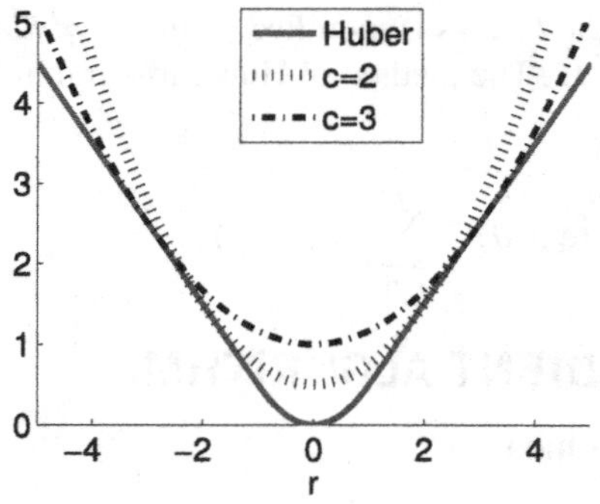

FIGURE 25.5

Quadratic upper bound $\frac{\eta r^2}{2c} + \frac{\eta c}{2} - \frac{\eta^2}{2}$ of Huber loss $\rho_{\text{Huber}}(r)$ for $c > 0$, which touches each other at $r = \pm c$.

where $\widetilde{r}_i$ denotes the residual of the current solution for the ith training sample $(\boldsymbol{x}_i, y_i)$. If the absolute-value part of the Huber loss is upper-bounded by this, Huber loss minimization problem Eq. (25.1) yields the following *weighted LS* problem:

$$\widehat{\boldsymbol{\theta}} = \underset{\boldsymbol{\theta}}{\text{argmin}}\, \widetilde{J}(\boldsymbol{\theta}), \quad \text{where } \widetilde{J}(\boldsymbol{\theta}) = \frac{1}{2}\sum_{i=1}^{n} \widetilde{w}_i r_i^2 + C,$$

where $C = \sum_{i:|\widetilde{r}_i|>\eta}(\eta|\widetilde{r}_i|/2 - \eta^2/2)$ is a constant that is independent of $\boldsymbol{\theta}$, and weight $\widetilde{w}_i$ is defined as follows (Fig. 25.6):

$$\widetilde{w}_i = \begin{cases} 1 & (|\widetilde{r}_i| \leq \eta), \\ \eta/|\widetilde{r}_i| & (|\widetilde{r}_i| > \eta). \end{cases}$$

As explained in Section 22.1, the solution of weighted LS is given analytically as

$$\widehat{\boldsymbol{\theta}} = (\boldsymbol{\Phi}^\top \widetilde{\boldsymbol{W}} \boldsymbol{\Phi})^{-1} \boldsymbol{\Phi}^\top \widetilde{\boldsymbol{W}} \boldsymbol{y}, \tag{25.2}$$

where $\widetilde{\boldsymbol{W}}$ denotes the diagonal matrix with diagonal elements $\widetilde{w}_1, \ldots, \widetilde{w}_n$.

Since the upper bound $\widetilde{J}$ touches the original objective function J at current solution $\boldsymbol{\theta} = \widetilde{\boldsymbol{\theta}}$,

$$J(\widetilde{\boldsymbol{\theta}}) = \widetilde{J}(\widetilde{\boldsymbol{\theta}})$$

holds. Since $\widehat{\boldsymbol{\theta}}$ is the minimizer of $\widetilde{J}$,

$$\widetilde{J}(\widetilde{\boldsymbol{\theta}}) \geq \widetilde{J}(\widehat{\boldsymbol{\theta}}).$$

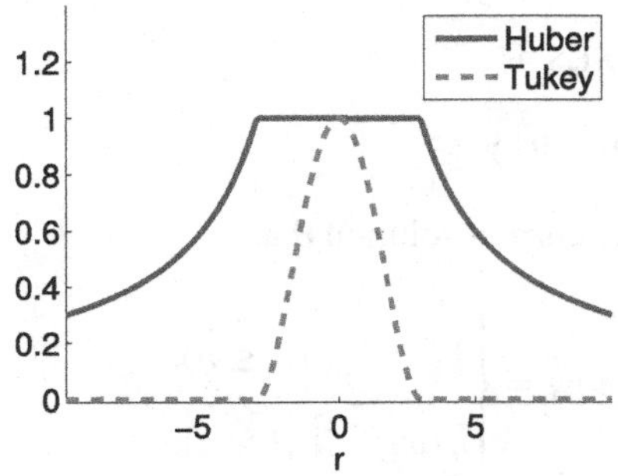

FIGURE 25.6

Weight functions for Huber loss minimization and Tukey loss minimization.

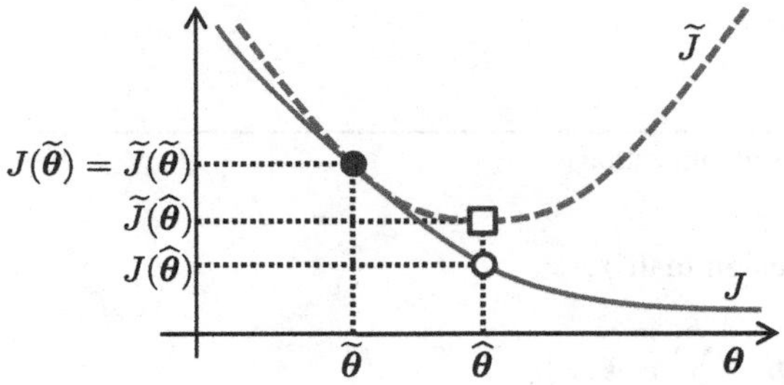

FIGURE 25.7

Updated solution $\widehat{\theta}$ is no worse than current solution $\widetilde{\theta}$.

Furthermore, $\widetilde{J}$ upper-bounds J and thus

$$\widetilde{J}(\widehat{\theta}) \geq J(\widehat{\theta})$$

holds. Summarizing the above relations,

$$J(\widetilde{\theta}) = \widetilde{J}(\widetilde{\theta}) \geq \widetilde{J}(\widehat{\theta}) \geq J(\widehat{\theta})$$

holds, meaning that the updated solution $\widehat{\theta}$ is no worse than the current solution $\widetilde{\theta}$ in terms of the objective value J (Fig. 25.7). By iterating this upper bound minimization, the solution of Huber loss minimization can be obtained. The pseudocode of this algorithm, called *iteratively reweighted LS*, is provided in Fig. 25.8. If *singular value*

1. Initialize parameter $\boldsymbol{\theta}$, e.g., by ordinary LS as

$$\boldsymbol{\theta} \longleftarrow (\boldsymbol{\Phi}^\top \boldsymbol{\Phi})^{-1} \boldsymbol{\Phi}^\top \boldsymbol{y}.$$

2. Compute the weight matrix $\boldsymbol{W}$ from the current solution $\boldsymbol{\theta}$ as

$$\boldsymbol{W} = \mathrm{diag}(w_1, \ldots, w_n), \quad \text{where } w_i = \begin{cases} 1 & (|r_i| \leq \eta), \\ \eta/|r_i| & (|r_i| > \eta), \end{cases}$$

and $r_i = y_i - f_{\boldsymbol{\theta}}(\boldsymbol{x}_i)$ is the residual.

3. Compute the solution $\boldsymbol{\theta}$ based on the weight matrix $\boldsymbol{W}$ as

$$\boldsymbol{\theta} \longleftarrow (\boldsymbol{\Phi}^\top \boldsymbol{W} \boldsymbol{\Phi})^{-1} \boldsymbol{\Phi}^\top \boldsymbol{W} \boldsymbol{y}.$$

4. Iterate 2–3 until convergence.

FIGURE 25.8

Iteratively reweighted LS for Huber loss minimization.

decomposition (see Fig. 22.2) of design matrix,

$$\boldsymbol{\Phi} = \sum_{j=1}^{b} \kappa_j \boldsymbol{\psi}_j \boldsymbol{\varphi}_j^\top,$$

is computed in advance, update of $\boldsymbol{\theta}$ can be performed more efficiently as

$$\boldsymbol{\theta} \longleftarrow \sum_{j=1}^{b} \frac{\boldsymbol{\psi}_j^\top \boldsymbol{W} \boldsymbol{y}}{\kappa_j \boldsymbol{\psi}_j^\top \boldsymbol{W} \boldsymbol{\psi}_j} \boldsymbol{\varphi}_j.$$

A MATLAB code of iteratively reweighted LS for Huber loss minimization is provided in Fig. 25.9, and its behavior is illustrated in Fig. 25.10. This shows that only two iterations give almost the final solution. The iteration converges after four iterations and a robust solution can be obtained.

When the threshold η is taken to be sufficiently small, the Huber loss may be regarded as a smooth approximation to the ℓ_1-loss. Thus, the solution of ℓ_1-loss minimization can be approximately obtained by the above iteratively weighted LS algorithm.

25.3.4 ℓ_1-CONSTRAINED HUBER LOSS MINIMIZATION

As explained in Chapter 24, the ℓ_1-constraint tends to give a sparse solution. Here, ℓ_1-constrained Huber loss minimization is considered:

```
n=10; N=1000; x=linspace(-3,3,n)'; X=linspace(-4,4,N)';
y=x+0.2*randn(n,1); y(n)=-4;

p(:,1)=ones(n,1); p(:,2)=x; t=p\y; e=1;
for o=1:1000
  r=abs(p*t-y); w=ones(n,1); w(r>e)=e./r(r>e);
  t0=(p'*(repmat(w,1,2).*p))\(p'*(w.*y));
  if norm(t-t0)<0.001, break, end
  t=t0;
end
P(:,1)=ones(N,1); P(:,2)=X; F=P*t;
figure(1); clf; hold on; axis([-4 4 -4.5 3.5]);
plot(X,F,'g-'); plot(x,y,'bo');
```

FIGURE 25.9

MATLAB code of iteratively reweighted LS for Huber loss minimization. Straight-line model $f_{\boldsymbol{\theta}}(x) = \theta_1 + \theta_2 x$ is used, with threshold $\eta = 1$.

$$\min_{\boldsymbol{\theta}} \sum_{i=1}^{n} \rho_{\text{Huber}}(r_i) \quad \text{subject to } \|\boldsymbol{\theta}\|_1 \leq R^2,$$

which produces a robust and sparse solution.

As shown above, the optimization problem of Huber loss minimization can be solved by iteratively reweighted LS, which was obtained by a quadratic upper bounding technique. Similarly, ℓ_1-regularized LS, whose solution was obtained by ADMM in Section 24.2, can also be solved by iteratively reweighted LS. More specifically, the absolute value of parameter θ can be upper-bounded by a quadratic function as

$$|\theta| \leq \frac{\theta^2}{2c} + \frac{c}{2} \quad \text{for } c > 0,$$

which touches the absolute value at $\theta = \pm c$ (see Fig. 25.11).

Let us consider an iterative optimization procedure and construct an upper bound for $c_j = \widetilde{\theta}_j$:

$$|\theta_j| \leq \frac{|\widetilde{\theta}_j|^{\dagger}}{2}\theta_j^2 + \frac{|\widetilde{\theta}_j|}{2},$$

where $\widetilde{\theta}_j$ is the current solution and † denotes the *generalized inverse* (see Fig. 22.1). If the ℓ_1-norm of parameter vector $\boldsymbol{\theta} = (\theta_1, \ldots, \theta_b)^{\top}$ is upper-bounded by this, the

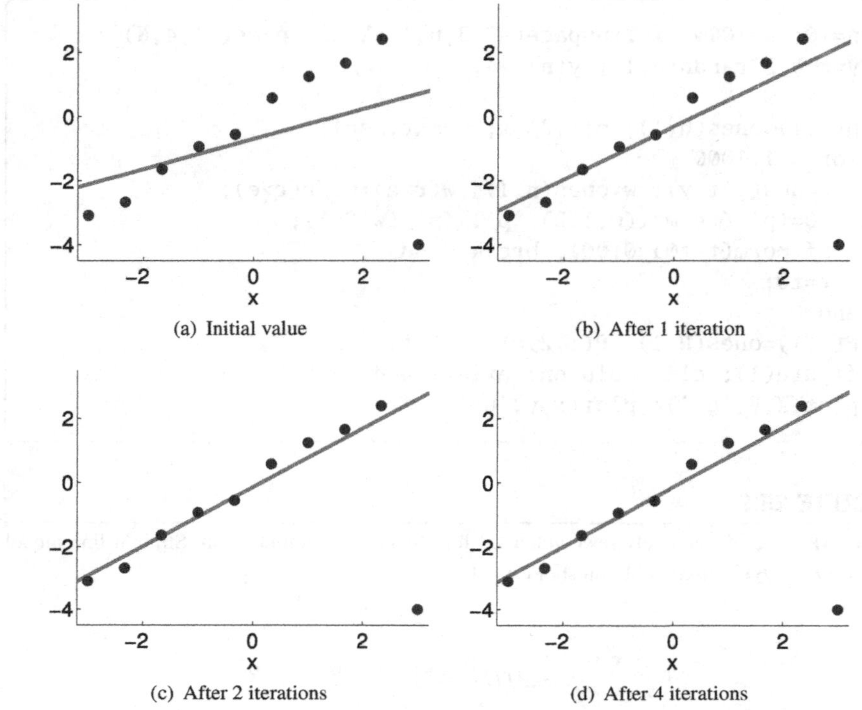

FIGURE 25.10

Examples of iteratively reweighted LS for Huber loss minimization. Straight-line model $f_{\boldsymbol{\theta}}(x) = \theta_1 + \theta_2 x$ is used, with threshold $\eta = 1$.

ℓ_1-regularized LS problem,

$$\min_{\boldsymbol{\theta}} \left[\frac{1}{2} \|\boldsymbol{y} - \boldsymbol{\Phi}\boldsymbol{\theta}\|^2 + \lambda \|\boldsymbol{\theta}\|_1 \right],$$

yields the following *generalized ℓ_2-regularized LS*:

$$\widehat{\boldsymbol{\theta}} = \underset{\boldsymbol{\theta}}{\text{argmin}}\, \widetilde{J}(\boldsymbol{\theta}), \quad \text{where } \widetilde{J}(\boldsymbol{\theta}) = J_{\text{LS}}(\boldsymbol{\theta}) + \frac{\lambda}{2} \boldsymbol{\theta}^\top \widetilde{\boldsymbol{\Theta}}^\dagger \boldsymbol{\theta} + C,$$

where $C = \sum_{j=1}^{b} |\widetilde{\theta}_j|/2$ is a constant that is independent of $\boldsymbol{\theta}$. As explained in Section 23.2, the solution of generalized ℓ_2-regularized LS is given analytically as

$$\widehat{\boldsymbol{\theta}} = (\boldsymbol{\Phi}^\top \boldsymbol{\Phi} + \lambda \widetilde{\boldsymbol{\Theta}}^\dagger)^{-1} \boldsymbol{\Phi}^\top \boldsymbol{y}.$$

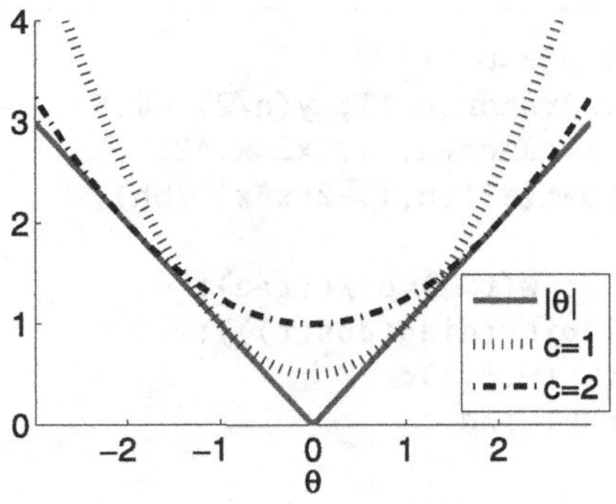

FIGURE 25.11

Quadratic upper bound $\frac{\theta^2}{2c} + \frac{c}{2}$ of absolute value $|\theta|$ for $c > 0$, which touches each other at $\theta = \pm c$.

1. Initialize parameter $\boldsymbol{\theta}$, e.g., randomly or by ordinary LS as
$$\boldsymbol{\theta} \longleftarrow \boldsymbol{\Phi}^{\dagger}\boldsymbol{y}.$$
2. Compute the weight matrix $\boldsymbol{W}$ and regularization matrix $\boldsymbol{\Theta}$ from the current solution $\boldsymbol{\theta}$ as
$$\boldsymbol{W} = \text{diag}(w_1, \ldots, w_n) \quad \text{and} \quad \boldsymbol{\Theta} = \text{diag}(|\theta_1|, \ldots, |\theta_b|),$$
where weight w_i is defined using residual $r_i = y_i - f_{\boldsymbol{\theta}}(\boldsymbol{x}_i)$ as
$$w_i = \begin{cases} 1 & (|r_i| \leq \eta), \\ \eta/|r_i| & (|r_i| > \eta). \end{cases}$$
3. Compute the solution $\boldsymbol{\theta}$ based on the weight matrix $\boldsymbol{W}$ regularization matrix $\boldsymbol{\Theta}$ as
$$\boldsymbol{\theta} \longleftarrow (\boldsymbol{\Phi}^{\top}\boldsymbol{W}\boldsymbol{\Phi} + \lambda\boldsymbol{\Theta}^{\dagger})^{-1}\boldsymbol{\Phi}^{\top}\boldsymbol{W}\boldsymbol{y}.$$
4. Iterate 2–3 until convergence.

FIGURE 25.12

Iteratively reweighted LS for ℓ_1-regularized Huber loss minimization.

```
n=50; x=linspace(-3,3,n)'; pix=pi*x;
y=sin(pix)./(pix)+0.1*x+0.2*randn(n,1); y(n/2)=-0.5;
hh=2*0.3^2; l=0.1; e=0.1; t=randn(n,1); x2=x.^2;
k=exp(-(repmat(x2,1,n)+repmat(x2',n,1)-2*x*x')/hh);
for o=1:1000
  r=abs(k*t-y); w=ones(n,1); w(r>e)=e./r(r>e);
  Z=k*(repmat(w,1,n).*k)+l*pinv(diag(abs(t)));
  t0=(Z+0.000001*eye(n))\(k*(w.*y));
  if norm(t-t0)<0.001, break, end
  t=t0;
end
N=1000; X=linspace(-3,3,N)';
K=exp(-(repmat(X.^2,1,n)+repmat(x2',N,1)-2*X*x')/hh); F=K*t;
figure(1); clf; hold on; axis([-2.8 2.8 -1 1.5]);
plot(X,F,'g-'); plot(x,y,'bo');
```

FIGURE 25.13

MATLAB code of iteratively reweighted LS for ℓ_1-regularized Huber loss minimization with Gaussian kernel model.

The iteratively reweighted LS algorithms for ℓ_1-regularized LS and Huber loss minimization can be combined to obtain a sparse and robust solution, as summarized in Fig. 25.12.

A MATLAB code of iteratively reweighted LS for ℓ_1-regularized Huber loss minimization is provided in Fig. 25.13, where the Gaussian kernel model,

$$f_{\boldsymbol{\theta}}(\boldsymbol{x}) = \sum_{j=1}^{n} \theta_j \exp\left(-\frac{\|\boldsymbol{x} - \boldsymbol{x}_j\|^2}{2h^2}\right),$$

is used and the Gaussian bandwidth is set at $h = 0.3$. Since plain LS is numerically unstable, it is stabilized by adding 10^{-6} to the diagonal elements of $\boldsymbol{\Phi}^\top \boldsymbol{W} \boldsymbol{\Phi}$ (more precisely, $\boldsymbol{K}\boldsymbol{W}\boldsymbol{K}$ since the Gaussian kernel model is used). The behavior of ℓ_1-regularized Huber loss minimization is illustrated in Fig. 25.14. Regardless of the presence or absence of the ℓ_1-regularizer, ℓ_2-loss minimization is strongly affected by the outlier at around $x = 0$, while Huber loss minimization can successfully suppress its influence. With the ℓ_1-regularizer, ℓ_2-loss minimization gives 38 zero parameters out of 50 parameters, and Huber loss minimization gives 36 zero parameters.

25.4 TUKEY LOSS MINIMIZATION

The Huber loss nicely controls the balance between efficiency and robustness by combining the ℓ_1-loss and the ℓ_2-loss. However, as long as the ℓ_1-loss is used, strong

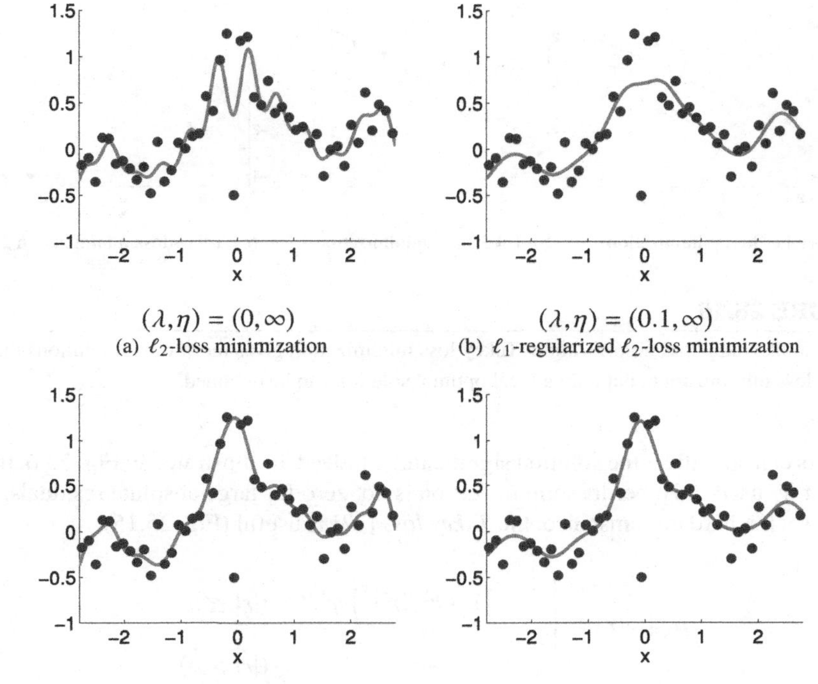

$(\lambda, \eta) = (0, \infty)$
(a) ℓ_2-loss minimization

$(\lambda, \eta) = (0.1, \infty)$
(b) ℓ_1-regularized ℓ_2-loss minimization

$(\lambda, \eta) = (0, 0.1)$
(c) Huber loss minimization

$(\lambda, \eta) = (0.1, 0.1)$
(d) ℓ_1-regularized Huber loss minimization

FIGURE 25.14

Example of ℓ_1-regularized Huber loss minimization with Gaussian kernel model.

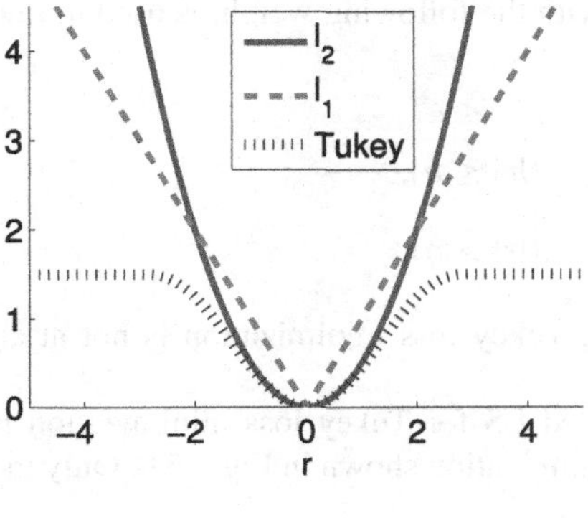

FIGURE 25.15

Tukey loss, with threshold $\eta = 3$.

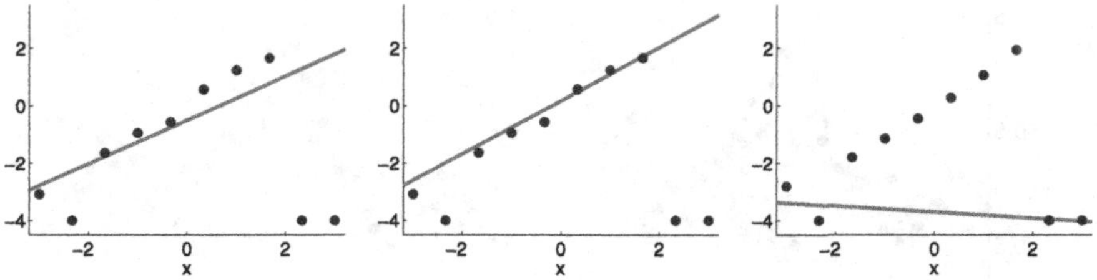

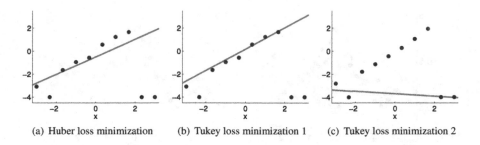

(a) Huber loss minimization (b) Tukey loss minimization 1 (c) Tukey loss minimization 2

FIGURE 25.16

Example of Tukey loss minimization. Tukey loss minimization gives more robust solutions than Huber loss minimization, but only a local optimal solution can be obtained.

outliers can still affect the solution significantly. Indeed, as illustrated in Fig. 25.6, the weight $\widetilde{w}_i$ used in Huber loss minimization is not zero for large absolute residuals.

In such a hard circumstance, the *Tukey loss* [73] is useful (Fig. 25.15):

$$\rho_{\text{Tukey}}(r) = \begin{cases} \left(1 - [1 - r^2/\eta^2]^3\right)\eta^2/6 & (|r| \leq \eta), \\ \eta^2/6 & (|r| > \eta). \end{cases}$$

The Tukey loss is constant $\eta^2/6$ if the absolute residual $|r|$ is larger than a threshold η. Thus, Tukey loss minimization is expected to be more robust than Huber loss minimization.

However, the Tukey loss is not a *convex function* (see Fig. 8.3). Thus, there may exist multiple local optimal solutions and finding the global one is not straightforward. In practice, *iteratively reweighted LS* with the following weight is used to find a local optimal solution (Fig. 25.6):

$$w = \begin{cases} (1 - r^2/\eta^2)^2 & (|r| \leq \eta), \\ 0 & (|r| > \eta). \end{cases}$$

Since the Tukey weight is zero for $|r| > \eta$, Tukey loss minimization is not at all influenced by strong outliers.

A MATLAB code of iteratively reweighted LS for Tukey loss minimization is essentially the same as that for Huber loss minimization shown in Fig. 25.9. Only the weight computation for the Huber loss,

```
w=ones(n,1); w(r>e)=e./r(r>e);
```

is replaced with that for the Tukey loss:

```
w=zeros(n,1); w(r<=e)=(1-r(r<=e).^2/e^2).^2;
```

The behavior of Tukey loss minimization is illustrated in Fig. 25.16, where the threshold is set at $\eta = 1$. In this example, Tukey loss minimization gives an even more robust solution than Huber loss minimization. However, since Tukey loss minimization is a nonconvex optimization problem, different solutions may be obtained if the initial solution is changed or training samples are slightly perturbed. Indeed, as illustrated in Fig. 25.16(c), slightly changing the noise included in training output samples gives another local optimal solution, which is significantly different from the original one in Fig. 25.16(b).

For general differentiable symmetric loss $\rho(r)$, iteratively reweighted LS with weight $w_i = \rho'(r_i)/r_i$ gives a (local) optimal solution, where $\rho'(r)$ denotes the derivative of $\rho(r)$ [57].

CHAPTER

LEAST SQUARES CLASSIFICATION

26

CHAPTER CONTENTS

In Chapter 22–Chapter 25, various regression techniques for estimating real-valued output were introduced. In this chapter, the classification problem of estimating category-valued output is tackled by the LS regression method introduced in Chapter 22. In the context of classification, input $\boldsymbol{x}$ is referred to as a *pattern* and output y is referred to as a *label*.

26.1 CLASSIFICATION BY LS REGRESSION

Let us consider a binary classification problem where the label y takes either $+1$ or -1. Then the classification problem can be regarded as approximating a binary-valued function $f(\boldsymbol{x}) \in \{+1, -1\}$ (Fig. 26.1). Such a function may be naively learned by the LS method introduced in Chapter 22:

$$\widehat{\boldsymbol{\theta}} = \underset{\boldsymbol{\theta}}{\operatorname{argmin}} \frac{1}{2} \sum_{i=1}^{n} \left(f_{\boldsymbol{\theta}}(\boldsymbol{x}_i) - y_i \right)^2.$$

Then, an estimate $\widehat{y}$ of the label y for a test pattern $\boldsymbol{x}$ is obtained by the sign of the learned function:

$$\widehat{y} = \begin{cases} +1 & (f_{\widehat{\boldsymbol{\theta}}}(\boldsymbol{x}) \geq 0), \\ -1 & (f_{\widehat{\boldsymbol{\theta}}}(\boldsymbol{x}) < 0). \end{cases} \tag{26.1}$$

Various extensions of the LS method introduced in Chapter 23, Chapter 24, and Chapter 25 may also be utilized for classification.

A MATLAB code of classification by ℓ_2-regularized LS for the Gaussian kernel model,

$$f_{\boldsymbol{\theta}}(\boldsymbol{x}) = \sum_{j=1}^{n} \theta_j \exp\left(-\frac{\|\boldsymbol{x} - \boldsymbol{x}_j\|^2}{2h^2} \right),$$

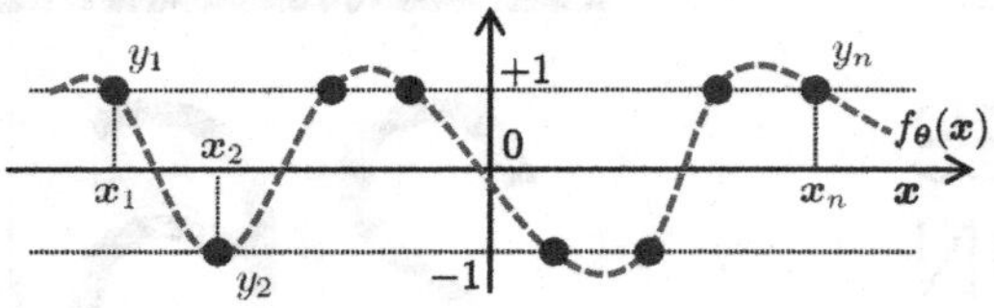

FIGURE 26.1

Binary classification as function approximation.

```
n=200; a=linspace(0,4*pi,n/2);
u=[a.*cos(a) (a+pi).*cos(a)]'+rand(n,1);
v=[a.*sin(a) (a+pi).*sin(a)]'+rand(n,1);
x=[u v]; y=[ones(1,n/2) -ones(1,n/2)]';

x2=sum(x.^2,2); hh=2*1^2; l=0.01;
k=exp(-(repmat(x2,1,n)+repmat(x2',n,1)-2*x*x')/hh);
t=(k^2+l*eye(n))\(k*y);

m=100; X=linspace(-15,15,m)'; X2=X.^2;
U=exp(-(repmat(u.^2,1,m)+repmat(X2',n,1)-2*u*X')/hh);
V=exp(-(repmat(v.^2,1,m)+repmat(X2',n,1)-2*v*X')/hh);
figure(1); clf; hold on;
contourf(X,X,sign(V'*(U.*repmat(t,1,m))));
plot(x(y==1,1),x(y==1,2),'bo');
plot(x(y==-1,1),x(y==-1,2),'rx');
colormap([1 0.7 1; 0.7 1 1]); axis([-15 15 -15 15]);
```

FIGURE 26.2

MATLAB code of classification by ℓ_2-regularized LS for Gaussian kernel model.

is provided in Fig. 26.2, and its behavior is illustrated in Fig. 26.3. This shows that entangled data can be successfully classified by ℓ_2-regularized LS.

Let us consider the linear-in-input model,

$$f_\theta(x) = \theta^\top x,$$

and assign $\{+1/n_+, -1/n_-\}$ to labels instead of $\{+1, -1\}$, where n_+ and n_- denote the number of positive and negative training samples, respectively. The LS solution in this setup is given by

$$\widehat{\theta}_{\mathrm{LS}} = \widehat{\Sigma}^{-1}(\widehat{\mu}_+ - \widehat{\mu}_-),$$

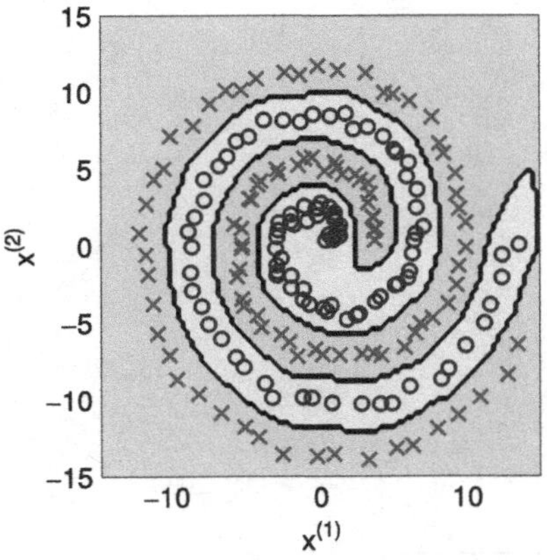

FIGURE 26.3

Example of classification by ℓ_2-regularized LS for Gaussian kernel model.

where

$$\widehat{\boldsymbol{\Sigma}} = \frac{1}{n}\sum_{i=1}^{n} \boldsymbol{x}_i \boldsymbol{x}_i^\top,\ \widehat{\boldsymbol{\mu}}_+ = \frac{1}{n_+}\sum_{i:y_i=+1/n_+}^{n} \boldsymbol{x}_i,\quad \text{and}\quad \widehat{\boldsymbol{\mu}}_- = \frac{1}{n_-}\sum_{i:y_i=-1/n_-}^{n} \boldsymbol{x}_i.$$

This solution actually agrees with FDA explained in Section 12.4, which was proved to achieve the optimal classification performance when samples in the positive and negative classes follow the Gaussian distributions with a common covariance matrix. Thus, LS classification for the linear-in-input model bridges the generative and discriminative approaches.

26.2 0/1-LOSS AND MARGIN

As shown in Eq. (26.1), classification is performed based on the sign of a learned function, not the value of the function itself. Thus, in classification, the 0/1-*loss* would be more natural than the ℓ_2-loss:

$$\frac{1}{2}\Big(1 - \text{sign}\left(f_{\boldsymbol{\theta}}(\boldsymbol{x})y\right)\Big).$$

The 0/1-loss can be equivalently expressed as

$$\delta\Big(\text{sign}\left(f_{\boldsymbol{\theta}}(\boldsymbol{x})\right) \neq y\Big) = \begin{cases} 1 & (\text{sign}\left(f_{\boldsymbol{\theta}}(\boldsymbol{x})\right) \neq y), \\ 0 & (\text{sign}\left(f_{\boldsymbol{\theta}}(\boldsymbol{x})\right) = y), \end{cases}$$

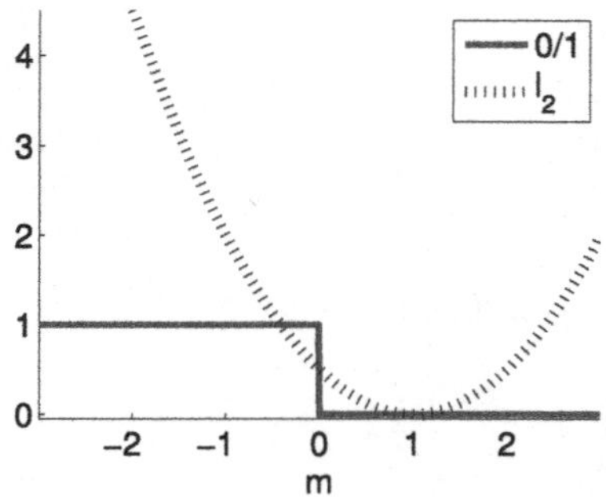

FIGURE 26.4

0/1-loss and ℓ_2-loss as functions of margin $m = f_{\theta}(x)y$.

which means that the 0/1-loss takes 1 if x is misclassified and 0 if x is correctly classified. Thus, the 0/1-loss counts the number of misclassified samples.

The 0/1-loss is plotted as a function of $m = f_{\theta}(x)y$ in Fig. 26.4, which takes 0 if $m > 0$ and 1 if $m \leq 0$. $m > 0$ means that $f_{\theta}(x)$ and y have the same sign, i.e., the sample x is classified correctly. On the other hand, $m \leq 0$ corresponds to misclassification of x. Thus, the value of the 0/1-loss does not depend on the magnitude of m, but only its sign. However, if m is decreased from a positive value, the 0/1-loss suddenly takes 1 if m goes below zero. Thus, having a larger m would be safer. For this reason, $m = f_{\theta}(x)y$ is called the *margin* for sample (x, y).

As illustrated in Fig. 26.4, the 0/1-loss is a binary-valued function and the derivative is zero everywhere, except at zero where the 0/1-loss is not differentiable. For this reason, 0/1-loss minimization for a rich enough model $f_{\theta}(x)$,

$$\min_{\theta} \frac{1}{2} \sum_{i=1}^{n} \left(1 - \mathrm{sign}\left(f_{\theta}(x_i)y_i\right)\right),$$

is essentially a discrete optimization problem of assigning either +1 or −1 to each training sample. This is a combinatorial problem with 2^n possibilities, and the number of combinations grows exponentially with respect to n. Therefore, it cannot be solved when n is not small. Even if a solution that vanishes the 0/1-loss can be found, the solution may not be unique due to the flatness of the 0/1-loss.

In regression, the ℓ_2-loss is commonly used as training and generalization error metrics. On the other hand, in classification, although it would be natural to use the 0/1-loss as a generalization error metric since it corresponds to the misclassification rate, the 0/1-loss cannot be directly used as a training error metric due to computational intractability. For this reason, a *surrogate loss* is usually employed for training a classifier. In the LS classification method, the ℓ_2-loss is used as a surrogate for the 0/1-loss.

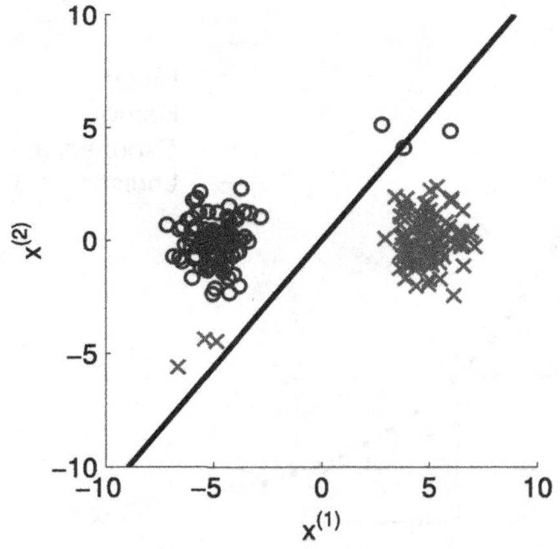

FIGURE 26.5

Example of ℓ_2-loss minimization for linear-in-input model. Since the ℓ_2-loss has a positive slope when $m > 1$, the obtained solution contains some classification error even though all samples can be correctly classified in principle.

When $y \in \{+1, -1\}$, $y^2 = 1$ and $1/y = y$ hold. Then the ℓ_2-loss, which is defined as the square of the residual $r = y - f_{\theta}(x)$, can be expressed in terms of the margin $m = f_{\theta}(x)y$ as

$$r^2 = \left(y - f_{\theta}(x)\right)^2 = y^2\left(1 - f_{\theta}(x)/y\right)^2 = \left(1 - f_{\theta}(x)y\right)^2$$
$$= \left(1 - m\right)^2.$$

The ℓ_2-loss is plotted as a function of margin m in Fig. 26.4. The ℓ_2-loss is positive when $m < 1$, and it has a negative slope as a function of m. Therefore, differently from the 0/1-loss, the ℓ_2-loss allows gradient-based learning to reduce the error. However, the ℓ_2-loss is positive also for $m > 1$ and it has a positive slope as a function of m. Thus, when $m > 1$, ℓ_2-loss minimization reduces the margin toward $m = 1$.

Since $m = 1$ still correctly classifies the sample, reducing the margin toward $m = 1$ when $m > 1$ does not cause any problem at a glance. However, if ℓ_2-loss minimization is applied to the data illustrated in Fig. 26.5, the obtained solution contains some classification error, even though the samples can be linearly separated in principle. This is actually caused by the fact that the ℓ_2-loss has a positive slope when $m > 1$. As a surrogate for the 0/1-loss, a monotone decreasing function of margin m would be reasonable.

In Fig. 26.6, various popular surrogate loss functions are plotted, which will be explained in the following chapters.

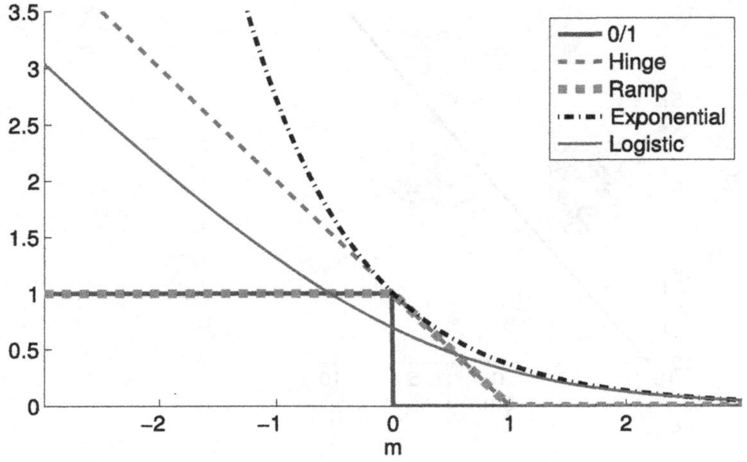

FIGURE 26.6

Popular surrogate loss functions.

26.3 MULTICLASS CLASSIFICATION

So far, binary classification was considered. However, in many practical applications of pattern recognition, the number of classes, c, may be more than two. For example, the number of classes is $c = 26$ in hand-written alphabet recognition. In this section, two methods to reduce a multiclass classification problem into a set of binary classification problems are explained. A method to directly solve the multiclass classification problems will be discussed in Chapter 27 and Chapter 28.

One way to reduce a multiclass classification problem into binary classification problems is the *one-versus-rest* method (Fig. 26.7), which considers c binary classification problems of one class versus the other classes. More specifically, for $y = 1, \ldots, c$, the yth binary classification problem assigns label $+1$ to samples in class y and -1 to samples in all other classes. Let $\widehat{f}_y(\boldsymbol{x})$ be a learned decision function for the yth binary classification problem. Then, test sample $\boldsymbol{x}$ is classified into class $\widehat{y}$ that gives the highest score:

$$\widehat{y} = \underset{y=1,\ldots,c}{\operatorname{argmax}}\, \widehat{f}_y(\boldsymbol{x}).$$

Another way to reduce a multiclass classification problem into binary classification problems is the *one-versus-one* method (Fig. 26.8), which considers $c(c-1)/2$ binary classification problems of one class versus another class. More specifically, for $y, y' = 1, \ldots, c$, the (y, y')th binary classification problem assigns label $+1$ to samples in class y and -1 to samples in class y'. Let $\widehat{f}_{y,y'}(\boldsymbol{x})$ be a learned decision function

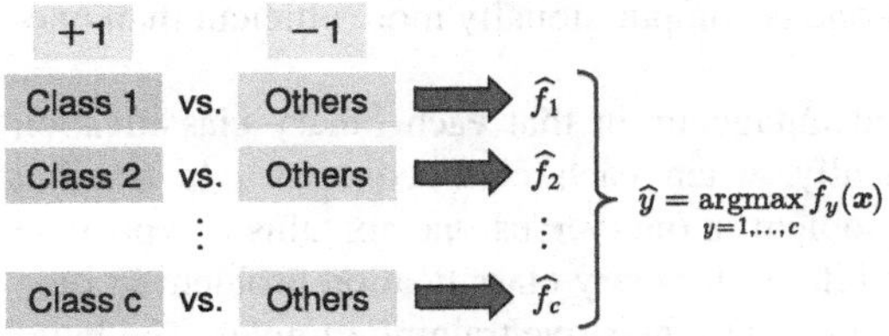

FIGURE 26.7

One-versus-rest reduction of multiclass classification problem.

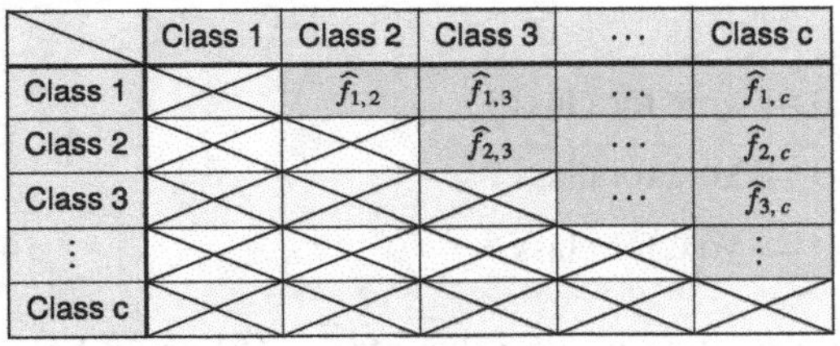
Class 1
Class 2
Class 3
Class c
$\widehat{f}_{1,2}$
$\widehat{f}_{1,3}$
$\widehat{f}_{1,c}$
$\widehat{f}_{2,3}$
$\widehat{f}_{2,c}$
$\widehat{f}_{3,c}$

FIGURE 26.8

One-versus-one reduction of multiclass classification problem.

for the (y, y')th binary classification problem. Then, test sample $\boldsymbol{x}$ is classified into class $\widehat{y}$ that gathers the highest votes:

$$\widehat{f}_{y,y'}(\boldsymbol{x}) \geq 0 \Rightarrow \text{Vote for class } y,$$
$$\widehat{f}_{y,y'}(\boldsymbol{x}) < 0 \Rightarrow \text{Vote for class } y'.$$

One-versus-rest consists only of c binary classification problems, while one-versus-one consists of $c(c-1)/2$ binary classification problems. Thus, one-versus-rest is more compact than one-versus-one. However, a binary classification problem in one-versus-one involves samples only in two classes, while that in one-versus-rest involves samples in all classes. Thus, each binary classification problem in one-versus-one may be solved more efficiently than that in one-versus-rest. Suppose that each class contains n/c training samples, and the computational complexity of classifier training is linear with respect to the number of training samples. Then the total computational complexity for one-versus-rest and one-versus-one is both $O(cn)$. However, if classifier training takes super-linear time, which is often the case in many classification algorithms, one-versus-one is computationally more efficient than one-versus-rest.

One-versus-one would also be advantageous in that each binary classification problem is balanced. More specifically, when each class contains n/c training samples, each binary classification problem in one-versus-one contains n/c positive and n/c negative training samples, while each binary classification problem in one-versus-rest contains n/c positive and $n(c-1)/c$ negative training samples. The latter situation is often referred to as *class imbalance*, and obtaining high classification accuracy in the imbalanced case is usually more challenging than the balanced case. Furthermore, each binary classification problem in one-versus-one would be simpler than one-versus-rest, because the "rest" class in one-versus-rest usually possesses multimodality, which makes classifier training harder.

On the other hand, one-versus-one has a potential limitation that voting can be tied: for example, when $c = 3$,

$$\begin{aligned}
\widehat{f}_{1,2}(\boldsymbol{x}) &\geq 0 \Rightarrow \text{Vote for class 1},\\
\widehat{f}_{2,3}(\boldsymbol{x}) &\geq 0 \Rightarrow \text{Vote for class 2},\\
\widehat{f}_{1,3}(\boldsymbol{x}) &< 0 \Rightarrow \text{Vote for class 3}.
\end{aligned}$$

In such a situation, weighted voting according to the value of $\widehat{f}_{y,y'}(\boldsymbol{x})$ could be a practical option. However, it is not clear what the best way to weight is.

As discussed above, both one-versus-rest and one-versus-one approaches have pros and cons. A method to directly solve multiclass classification problems will be introduced in Chapter 27 and Chapter 28. However, the direct method does not necessarily perform better than the reduction approaches, because multiclass classification problems are usually more complicated than binary classification problems. In practice, the best approach should be selected depending on the target problem and other constraints.

CHAPTER

SUPPORT VECTOR CLASSIFICATION

27

CHAPTER CONTENTS

In the previous chapter, LS regression was shown to be used also for classification. However, due to the nonmonotonicity of the ℓ_2-loss function, classification by ℓ_2-loss minimization is rather unnatural as a surrogate for the 0/1-loss. In this chapter, a classification algorithm called the *support vector machine* is introduced, which uses a more natural loss function called the *hinge loss*.

27.1 MAXIMUM MARGIN CLASSIFICATION

In this section, a classification algorithm called the *support vector machine* is introduced, which is based on the *margin maximization principle* [115].

27.1.1 HARD MARGIN SUPPORT VECTOR CLASSIFICATION

Let us consider a linear-in-input binary classifier:

$$f_{w,\gamma}(x) = w^\top x + \gamma, \tag{27.1}$$

where w and γ are the parameters of the model which correspond to the *normal vector* and the *intercept* of the decision boundary, respectively (Fig. 27.1). If w and

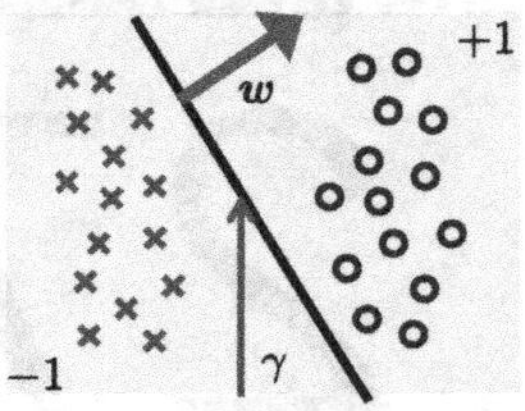

FIGURE 27.1

Linear-in-input binary classifier $f_{w,\gamma}(x) = w^\top x + \gamma$. w and γ are the normal vector and the intercept of the decision boundary, respectively.

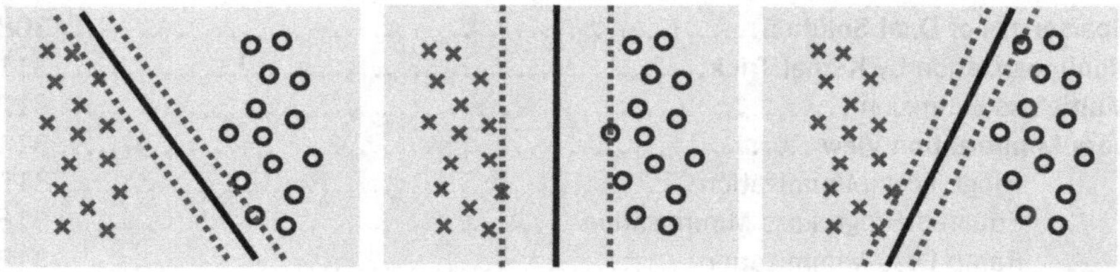

(a) Small margin (b) Large margin (c) Small margin

FIGURE 27.2

Decision boundaries that separate all training samples correctly.

γ are learned so that margins for all training samples are positive, i.e.

$$f_{w,\gamma}(x_i)y_i = (w^\top x_i + \gamma)y_i > 0, \quad \forall i = 1, \ldots, n,$$

all training samples $\{(x_i, y_i)\}_{i=1}^n$ can be correctly classified. Since the open set $(w^\top x_i + \gamma)y_i > 0$ is not easy to handle mathematically, let us convert this constraint to a closed set based on the fact that the scale of w and γ is arbitrary:

$$(w^\top x_i + \gamma)y_i \geq 1, \quad \forall i = 1, \ldots, n.$$

If there exists (w, γ) such that the above condition is fulfilled, the set of training samples $\{(x_i, y_i)\}_{i=1}^n$ is said to be *linearly separable*. For a linearly separable set of training samples, there usually exist infinitely many decision boundaries that correctly separate all training samples. Here, let us choose the one that separates all training samples with the *maximum margin*, where the *margin* for a set of training samples $\{(x_i, y_i)\}_{i=1}^n$ is defined as the minimum of the normalized margin

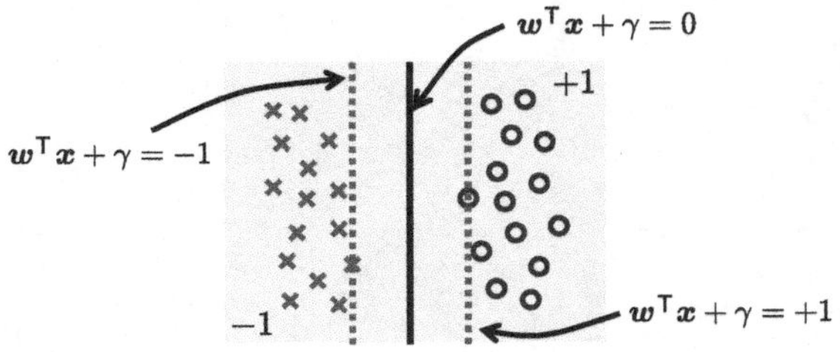

FIGURE 27.3

Decision boundary of hard margin support vector machine. It goes through the center of positive and negative training samples, $w^\top x_+ + \gamma = +1$ for some positive sample x_+ and $w^\top x_- + \gamma = -1$ for some negative sample x_-.

$m_i = (w^\top x_i + \gamma)y_i/\|w\|$ for $i = 1, \ldots, n$ (Fig. 27.2):

$$\min_{i=1,\ldots,n} \left\{ \frac{(w^\top x_i + \gamma)y_i}{\|w\|} \right\} = \frac{1}{\|w\|}.$$

Geometrically, the margin for a set of training samples corresponds to the half distance between two extreme decision boundaries $w^\top x + \gamma = +1$ and $w^\top x + \gamma = -1$ (Fig. 27.3). The classifier that maximizes the above margin (or equivalently minimizes the squared inverse margin) is called the *hard margin support vector machine* [18]:

$$\min_{w,\gamma} \frac{1}{2}\|w\|^2 \text{ subject to } (w^\top x_i + \gamma)y_i \geq 1, \quad \forall i = 1, \ldots, n.$$

27.1.2 SOFT MARGIN SUPPORT VECTOR CLASSIFICATION

The hard margin support vector machine requires linear separability, which may not always be satisfied in practice. The *soft margin support vector machine* [30] relaxes this requirement by allowing error $\xi = (\xi_1, \ldots, \xi_n)^\top$ for margins (Fig. 27.4):

$$\min_{w,\gamma,\xi} \left[\frac{1}{2}\|w\|^2 + C\sum_{i=1}^{n} \xi_i \right] \tag{27.2}$$
$$\text{subject to } (w^\top x_i + \gamma)y_i \geq 1 - \xi_i,\ \xi_i \geq 0, \quad \forall i = 1, \ldots, n,$$

where $C > 0$ is a tuning parameter that controls the margin errors. The margin error $\xi = (\xi_1, \ldots, \xi_n)^\top$ is also referred to as *slack variables* in optimization. Larger C

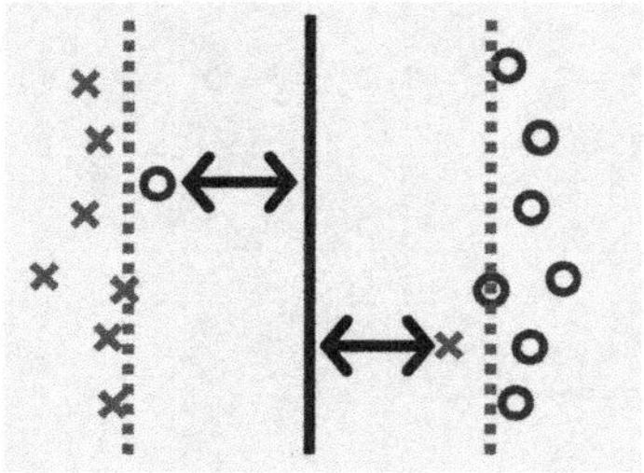

FIGURE 27.4

Soft margin support vector machine allows small margin errors.

makes the margin error $\sum_{i=1}^{n} \xi_i$ small and then soft margin support vector machine approaches hard margin support vector machine.

Below, the soft margin support vector machine may be merely called the support vector machine.

27.2 DUAL OPTIMIZATION OF SUPPORT VECTOR CLASSIFICATION

The optimization problem of support vector classification (27.2) takes the form of *quadratic programming* (Fig. 27.5), where the objective is a quadratic function and constraints are linear. Since quadratic programming has been extensively studied in the optimization community and various practical algorithms are available, which can be readily used for obtaining the solution of support vector classification.

However, although original linear-in-input model (27.1) contains only $d + 1$ parameters (i.e. $\boldsymbol{w} \in \mathbb{R}^d$ and $\gamma \in \mathbb{R}$), quadratic optimization problem (27.2) contains additional parameter $\boldsymbol{\xi} \in \mathbb{R}^n$. Thus, the total number of parameters to be optimized is $n + d + 1$, which does not scale well to large data sets.

The *Lagrange function* (Fig. 23.5) of optimization problem (27.2) is given by

$$L(\boldsymbol{w},\gamma,\boldsymbol{\xi},\boldsymbol{\alpha},\boldsymbol{\beta}) = \frac{1}{2}\|\boldsymbol{w}\|^2 + C\sum_{i=1}^{n} \xi_i - \sum_{i=1}^{n} \alpha_i\Big((\boldsymbol{w}^\top \boldsymbol{x}_i + \gamma)y_i - 1 + \xi_i\Big) - \sum_{i=1}^{n} \beta_i \xi_i.$$

Then Lagrange dual optimization problem is given by

$$\max_{\boldsymbol{\alpha},\boldsymbol{\beta}} \inf_{\boldsymbol{w},\gamma,\boldsymbol{\xi}} L(\boldsymbol{w},\gamma,\boldsymbol{\xi},\boldsymbol{\alpha},\boldsymbol{\beta}) \quad \text{subject to } \boldsymbol{\alpha} \geq \boldsymbol{0}, \ \boldsymbol{\beta} \geq \boldsymbol{0}.$$

Quadratic programming is the optimization problem defined for matrices $\boldsymbol{F}$ and $\boldsymbol{G}$ and vectors $\boldsymbol{f}$ and $\boldsymbol{g}$ as

$$\min_{\boldsymbol{\theta}} \left[\frac{1}{2}\boldsymbol{\theta}^\top \boldsymbol{F}\boldsymbol{\theta} + \boldsymbol{f}^\top\boldsymbol{\theta}\right] \quad \text{subject to } \boldsymbol{G}\boldsymbol{\theta} \leq \boldsymbol{g},$$

where vector inequality $\boldsymbol{G}\boldsymbol{\theta} \leq \boldsymbol{g}$ denotes the elementwise inequalities:

$$\begin{pmatrix} a \\ b \end{pmatrix} \leq \begin{pmatrix} c \\ d \end{pmatrix} \Longleftrightarrow \begin{cases} a \leq c, \\ b \leq d. \end{cases}$$

Matrix $\boldsymbol{F}$ is supposed to be positive, i.e. all eigenvalues (see Fig. 6.2) are positive. When $\boldsymbol{F}$ is *ill-conditioned* in the sense that a very small eigenvalue exists, a small positive constant may be added to the diagonal elements of $\boldsymbol{F}$ to improve numerical stability.

FIGURE 27.5

Quadratic programming.

The first-order optimality condition of $\inf_{\boldsymbol{w},\gamma,\boldsymbol{\xi}} L(\boldsymbol{w},\gamma,\boldsymbol{\xi},\boldsymbol{\alpha},\boldsymbol{\beta})$ yields

$$\frac{\partial L}{\partial \boldsymbol{w}} = 0 \implies \boldsymbol{w} = \sum_{i=1}^n \alpha_i y_i \boldsymbol{x}_i,$$
$$\frac{\partial L}{\partial \gamma} = 0 \implies \sum_{i=1}^n \alpha_i y_i = 0,$$
$$\frac{\partial L}{\partial \xi_i} = 0 \implies \alpha_i + \beta_i = C, \quad \forall i = 1,\ldots,n,$$

showing that $\boldsymbol{w}$ and $\boldsymbol{\beta}$ can be expressed by $\boldsymbol{\alpha}$. Furthermore, the condition $\alpha_i + \beta_i = C$ allows us to eliminate the margin error $\boldsymbol{\xi}$ from the Lagrange function.

Summarizing the above computations, the Lagrange dual optimization problem is simplified as

$$\widehat{\boldsymbol{\alpha}} = \underset{\boldsymbol{\alpha}}{\operatorname{argmax}} \left[\sum_{i=1}^n \alpha_i - \frac{1}{2}\sum_{i,j=1}^n \alpha_i\alpha_j y_i y_j \boldsymbol{x}_i^\top \boldsymbol{x}_j\right]$$
$$\text{subject to } \sum_{i=1}^n \alpha_i y_i = 0,\ 0 \leq \alpha_i \leq C \quad \text{for } i = 1,\ldots,n. \tag{27.3}$$

This is a quadratic programming problem which only contains n variables to be optimized, and therefore it may be solved more efficiently than original optimization problem (27.2). The solution $\widehat{\alpha}$ of Lagrange dual optimization problem (27.3) gives the solution of support vector classification as

$$\widehat{w} = \sum_{i=1}^{n} \widehat{\alpha}_i y_i \boldsymbol{x}_i.$$

The solution of intercept $\widehat{\gamma}$ can be obtained by using $\boldsymbol{x}_i$ such that $0 < \widehat{\alpha}_i < C$ as

$$\widehat{\gamma} = y_i - \sum_{j:\widehat{\alpha}_i > 0} \widehat{\alpha}_j y_j \boldsymbol{x}_i^\top \boldsymbol{x}_j. \tag{27.4}$$

The derivation of $\widehat{\gamma}$ above will be explained in Section 27.3. When the linear-in-input model without the intercept,

$$f_{w,\gamma}(\boldsymbol{x}) = \boldsymbol{w}^\top \boldsymbol{x},$$

is used, the constraint,

$$\sum_{i=1}^{n} \alpha_i y_i = 0,$$

can be removed from Lagrange dual optimization problem (27.3).

While LS classification could not perfectly classify the training samples plotted in Fig. 26.5, support vector classification can achieve perfect separation (Fig. 27.6). Among 200 dual parameters $\{\alpha_i\}_{i=1}^n$, 197 parameters take zero and only 3 parameters specified by the square in the plot take nonzero values. The reason for this sparsity induction will be explained in the next section.

The quadratic programming approach to obtaining the solution of support vector classification explained above is handy and useful to understand the properties of support vector classification. However, its scalability to large-scale data is limited and more advanced optimization techniques have been extensively studied, e.g. [25].

27.3 SPARSENESS OF DUAL SOLUTION

The ℓ_1-constraint was shown to induce a sparse solution in Chapter 24. On the other hand, as shown in Fig. 27.6, the dual solution $\widehat{\alpha}$ of support vector classification tends to be sparse even though the ℓ_1-constraint is not used.

To explain the sparsity induction mechanism of support vector classification, let us investigate the optimality conditions called the KKT conditions (Fig. 27.7). Dual variables and constraints satisfy the following conditions called *complementary slackness*:

$$\alpha_i(m_i - 1 + \xi_i) = 0 \quad \text{and} \quad \beta_i \xi_i = 0, \ \forall i = 1, \ldots, n,$$

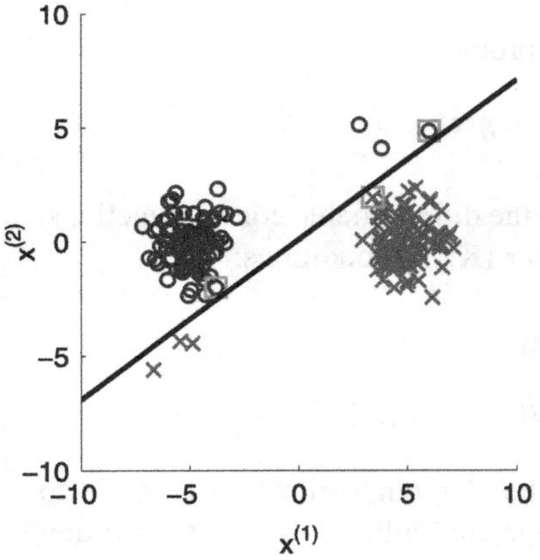

FIGURE 27.6

Example of linear support vector classification. Among 200 dual parameters $\{\alpha_i\}_{i=1}^n$, 197 parameters take zero and only 3 parameters specified by the square in the plot take nonzero values.

where $m_i = (\boldsymbol{w}^\top \boldsymbol{x}_i + \gamma)y_i$ denotes the margin for the ith training sample $(\boldsymbol{x}_i, y_i)$. Furthermore, $\frac{\partial L}{\partial \xi_i} = 0$ yields

$$\alpha_i + \beta_i = C.$$

Summarizing the above conditions, the dual variable α_i and margin m_i satisfy the following relations (Fig. 27.8):

- $\alpha_i = 0$ implies $m_i \geq 1$: the ith training sample $\boldsymbol{x}_i$ is on the margin border or inside the margin and is correctly classified.
- $0 < \alpha_i < C$ implies $m_i = 1$: $\boldsymbol{x}_i$ is on the margin border and correctly classified.
- $\alpha_i = C$ implies $m_i \leq 1$: $\boldsymbol{x}_i$ is on the margin border or outside the margin. If $\xi_i > 1$, $m_i < 0$ and then $\boldsymbol{x}_i$ is misclassified.
- $m_i > 1$ implies $\alpha_i = 0$: if $\boldsymbol{x}_i$ is inside the margin, $\alpha_i = 0$.
- $m_i < 1$ implies $\alpha_i = C$: if $\boldsymbol{x}_i$ is outside the margin, $\alpha_i = C$.

A training input sample $\boldsymbol{x}_i$ such that $\alpha_i > 0$ is called a *support vector*, because those points "support" the decision boundary. In other words, nonsupport vectors do not affect the classification boundary. Support vector $\boldsymbol{x}_i$ such that $0 < \alpha_i < C$ is on the margin border (see Fig. 27.8 again) and satisfy

$$m_i = (\boldsymbol{w}^\top \boldsymbol{x}_i + \gamma)y_i = 1.$$

The solution of constrained optimization problem,

$$\min_{t} f(t) \text{ subject to } g(t) \leq 0,$$

where $f : \mathbb{R}^d \rightarrow \mathbb{R}$ and $g : \mathbb{R}^d \rightarrow \mathbb{R}^p$ are the differentiable convex functions, satisfies the following *Karush-Kuhn-Tucker* (KKT) conditions:

$$\frac{\partial L}{\partial t} = \mathbf{0}, \quad g(t) \leq \mathbf{0}, \quad \lambda \geq \mathbf{0},$$
$$\text{and } \lambda_i g_i(t) = 0, \ \forall i = 1, \ldots, n,$$

where $L(t,\lambda) = f(t) + \lambda^\top g(t)$ is the the Lagrange function (Fig. 23.5) and $\lambda = (\lambda_1, \ldots, \lambda_p)^\top$ are the Lagrange multipliers. The last condition $\lambda_i g_i(t) = 0$ is called *complementary slackness* because at least either of λ_i or $g_i(t)$ is zero.

FIGURE 27.7

KKT optimality conditions.

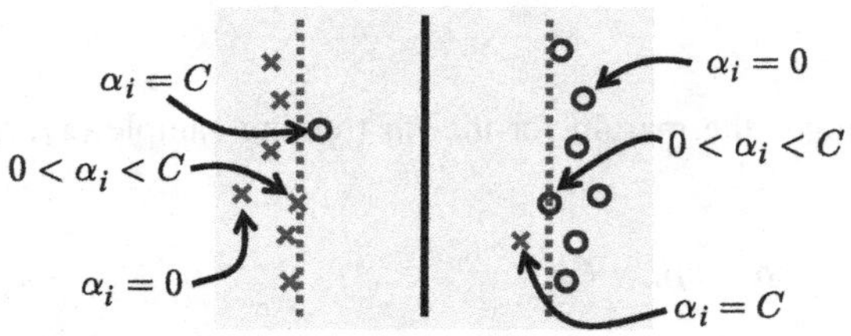

FIGURE 27.8

When $\alpha_i = 0$, x_i is inside the margin and correctly classified. When $0 < \alpha_i < C$, x_i is on the margin border (the dotted lines) and correctly classified. When $\alpha_i = C$, x_i is outside the margin, and if $\xi_i > 1$, $m_i < 0$ and thus x_i is misclassified.

This implies that the intercept γ satisfies

$$\gamma = 1/y_i - w^\top x_i = y_i - w^\top x_i, \ \forall i : 0 < \alpha_i < C,$$

from which the solution $\widehat{\gamma}$ in Eq. (27.4) is obtained.

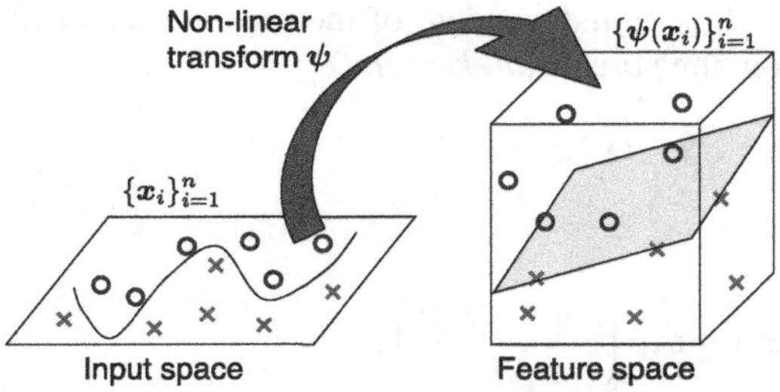

FIGURE 27.9

Nonlinearization of support vector machine by kernel trick.

27.4 NONLINEARIZATION BY KERNEL TRICK

In this section, support vector classification introduced for linear-in-input model (27.1) is extended to nonlinear models. More specifically, training input samples $\{x_i\}_{i=1}^n$ are transformed to a feature space using a nonlinear function ψ (Fig. 27.9). Then, for the transformed samples $\{\psi(x_i)\}_{i=1}^n$, the linear support vector machine is trained. A linear decision boundary obtained in the feature space is a nonlinear decision boundary obtained in the original input space.

If the dimensionality of the feature space is higher than the original input space, it is more likely that training samples are linearly separable. However, too high dimensionality causes large computation costs.

To cope with this problem, a technique called the *kernel trick* [30, 89] is useful. In the optimization problem of the linear support vector machine, Eq. (27.3), training input samples $\{x_i\}_{i=1}^n$ appear only in terms of their inner product:

$$x_i^\top x_j = \langle x_i, x_j \rangle.$$

Similarly, in the above nonlinear support vector machine, training input samples $\{x_i\}_{i=1}^n$ appear only in

$$\langle \psi(x_i), \psi(x_j) \rangle.$$

This means that the nonlinear support vector machine is trainable even if $\psi(x_i)$ and $\psi(x_j)$ are unknown, as long as their inner product $\langle \psi(x_i), \psi(x_j) \rangle$ is known.

The kernel trick is to directly specify the inner product by a kernel function $K(\cdot,\cdot)$ as

$$\langle \psi(x), \psi(x') \rangle = K(x, x').$$

This shows that, as long as the kernel function can be computed with a computation cost independent of the dimensionality of the feature space, the computation cost of training the support vector machine is independent of the dimensionality of the feature space. As a kernel function, the *polynomial kernel*,

$$K(\boldsymbol{x},\boldsymbol{x}') = (\boldsymbol{x}^\top \boldsymbol{x}' + a)^p,$$

and the *Gaussian kernel*,

$$K(\boldsymbol{x},\boldsymbol{x}') = \exp\left(-\frac{\|\boldsymbol{x}-\boldsymbol{x}'\|^2}{2h^2}\right),$$

are popular choices, where a is a positive real scalar, p is a positive integer, and h is a positive real scalar. It is known that the feature vector $\boldsymbol{\psi}$ that corresponds to the Gaussian kernel is actually *infinite dimensional*. Therefore, without the kernel trick, the solution cannot be computed explicitly.

A MATLAB code of support vector classification for the Gaussian kernel is provided in Fig. 27.10, and its behavior is illustrated in Fig. 27.11. This shows that nonlinearly entangled data can be successfully classified by support vector classification with the Gaussian kernel.

Since the kernel support vector machine only uses the kernel value $K(\boldsymbol{x},\boldsymbol{x}')$, as long as the kernel function is defined, nonvectorial pattern $\boldsymbol{x}$ such as sequences, trees, and graphs can be handled [46]. Note that the kernel trick can be applied not only to support vector machines but also to any algorithms that handle training samples only in terms of their inner products, for example, *clustering* (Chapter 37) and *dimensionality reduction* (Section 36.1).

27.5 MULTICLASS EXTENSION

As explained in Section 26.3, a multiclass classification problem can be reduced to a set of binary classification problems. In this section, a direct formulation of *multiclass support vector classification* [34] is provided.

Let us classify a test sample $\boldsymbol{x}$ into class $\widehat{y}$ as

$$\widehat{y} = \underset{y=1,\dots,c}{\text{argmax}}\ \boldsymbol{w}_y^\top \boldsymbol{x},$$

where $\boldsymbol{w}_y$ denotes the parameter vector for class y. In this situation, it is desirable that $\boldsymbol{w}_{\widehat{y}}^\top \boldsymbol{x}$ is larger than $\boldsymbol{w}_y^\top \boldsymbol{x}$ for $y \neq \widehat{y}$ with a *large margin*. Based on this idea, *hard margin* multiclass support vector classification for separable data is formulated as

$$\min_{\boldsymbol{w}_1,\dots,\boldsymbol{w}_c} \frac{1}{2}\sum_{y=1}^{c} \|\boldsymbol{w}_y\|^2$$
$$\text{subject to } \boldsymbol{w}_{y_i}^\top \boldsymbol{x}_i - \boldsymbol{w}_y^\top \boldsymbol{x}_i \geq 1, \quad \forall i = 1,\dots,n,\ \forall y \neq y_i.$$

Similarly to the binary case, the above formulation can be extended to *soft margin* multiclass support vector classification using the margin error $\boldsymbol{\xi} = (\xi_1,\dots,\xi_n)^\top$ as

```
n=200; a=linspace(0,4*pi,n/2);
u=[a.*cos(a) (a+pi).*cos(a)]'+rand(n,1);
v=[a.*sin(a) (a+pi).*sin(a)]'+rand(n,1);
x=[u v]; y=[ones(1,n/2) -ones(1,n/2)]';

x2=sum(x.^2,2); hh=2*1^2; l=0.01;
k=exp(-(repmat(x2,1,n)+repmat(x2',n,1)-2*x*x')/hh);
Q=(y*y').*k+0.01*eye(n); q=-ones(n,1);
H=[-eye(n); eye(n)]; h=[zeros(n,1); ones(n,1)/(2*l)];
a=quadprog(Q,q,H,h); t=y.*a;

m=100; X=linspace(-15,15,m)'; X2=X.^2;
U=exp(-(repmat(u.^2,1,m)+repmat(X2',n,1)-2*u*X')/hh);
V=exp(-(repmat(v.^2,1,m)+repmat(X2',n,1)-2*v*X')/hh);
figure(1); clf; hold on; axis([-15 15 -15 15]);
contourf(X,X,sign(V'*(U.*repmat(t,1,m))));
plot(x(y==1,1),x(y==1,2),'bo');
plot(x(y==-1,1),x(y==-1,2),'rx');
colormap([1 0.7 1; 0.7 1 1]);
```

FIGURE 27.10

MATLAB code of support vector classification for Gaussian kernel. quadprog.m included in Optimization Toolbox is required. Free alternatives to quadprog.m are available, e.g. from http://www.mathworks.com/matlabcentral/fileexchange/.

$$\min_{\boldsymbol{w}_1,\ldots,\boldsymbol{w}_c,\boldsymbol{\xi}} \left[\frac{1}{2}\sum_{y=1}^{c}\|\boldsymbol{w}_y\|^2 + C\sum_{i=1}^{n}\xi_i\right]$$
$$\text{subject to } \boldsymbol{w}_{y_i}^\top\boldsymbol{x}_i - \boldsymbol{w}_y^\top\boldsymbol{x}_i \geq t_{i,y} - \xi_i, \quad \forall i = 1,\ldots,n,\ \forall y = 1,\ldots,c,$$

where

$$t_{i,y} = \begin{cases} 0 & (y = y_i), \\ 1 & (y \neq y_i). \end{cases}$$

Note that the constraint $\boldsymbol{w}_{y_i}^\top\boldsymbol{x}_i - \boldsymbol{w}_y^\top\boldsymbol{x}_i \geq t_{i,y} - \xi_i$ is reduced to $\xi_i \geq 0$ when $y = y_i$.

The *Lagrange dual* (Fig. 23.5) of the above optimization problem is given by

$$\max_{\alpha_{1,1},\ldots,\alpha_{n,c}} \left[\sum_{i=1}^{n}\alpha_{i,y_i} - \frac{1}{2}\sum_{i,j=1}^{n}\sum_{y=1}^{c}\alpha_{i,y}\alpha_{j,y}\boldsymbol{x}_i^\top\boldsymbol{x}_j\right]$$

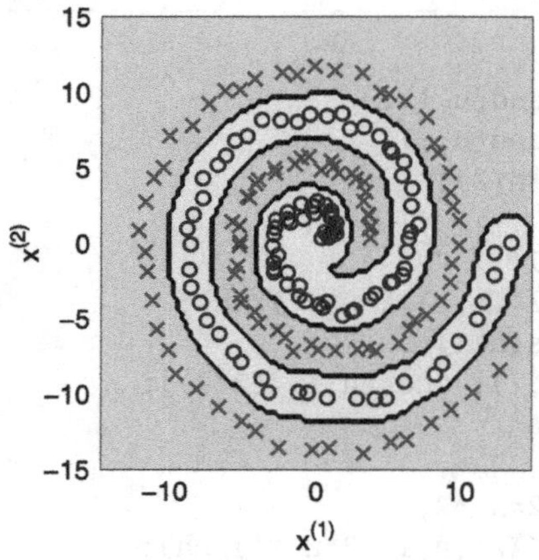

FIGURE 27.11

Example of support vector classification for Gaussian kernel.

$$\text{subject to} \sum_{y=1}^{c} \alpha_{i,y} = 0, \quad \forall i = 1, \ldots, n,$$

$$\alpha_{i,y} \leq C(1 - t_{i,y}), \quad \forall i = 1, \ldots, n, \ \forall y = 1, \ldots, c.$$

This is a quadratic programming problem and thus the solution $\widehat{\alpha}_{i,y}$ can be obtained by standard optimization software.

From the dual solution $\widehat{\alpha}_{i,y}$, the primal solution $\widehat{\boldsymbol{w}}_y$ can be obtained as

$$\widehat{\boldsymbol{w}}_y = \sum_{i=1}^{n} \widehat{\alpha}_{i,y} \boldsymbol{x}_i.$$

Then a test sample $\boldsymbol{x}$ is classified into class $\widehat{y}$ as

$$\widehat{y} = \underset{y=1,\ldots,c}{\operatorname{argmax}} \sum_{i=1}^{n} \widehat{\alpha}_{i,y} \boldsymbol{x}_i^\top \boldsymbol{x}.$$

Through this dual formulation, the *kernel trick* introduced in Section 27.4 can be applied to nonlinearize multiclass support vector classification by replacing $\boldsymbol{x}^\top \boldsymbol{x}'$ with $K(\boldsymbol{x}, \boldsymbol{x}')$.

27.6 LOSS MINIMIZATION VIEW

So far, support vector classification was derived based on margin maximization and the kernel trick. Although such a derivation is quite different from LS classification

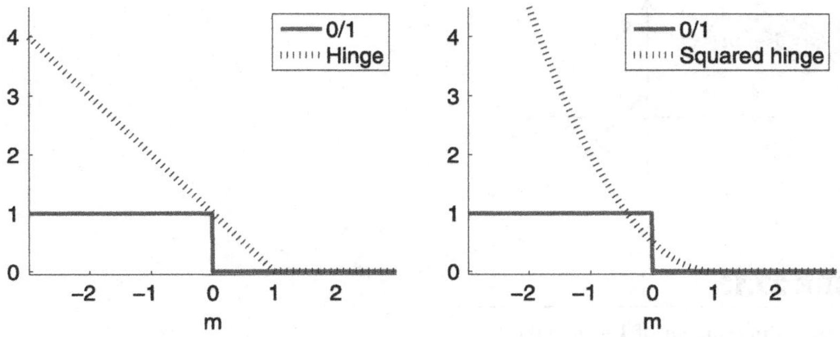

(a) Hinge loss (b) Squared hinge loss

FIGURE 27.12

Hinge loss and squared hinge loss.

introduced in Chapter 26, support vector classification can actually be regarded as an extension of LS classification. In this section, support vector classification is interpreted as loss minimization. Another view of support vector classification in L_1-distance approximation is given in Section 39.1.4.

27.6.1 HINGE LOSS MINIMIZATION

Let us consider the *hinge loss* as a surrogate for the 0/1-loss (Fig. 27.12(a)):

$$\max\left\{0, 1-m\right\} = \begin{cases} 1-m & (m \leq 1), \\ 0 & (m > 1), \end{cases}$$

where m denotes the margin $m = f_{\boldsymbol{\theta}}(\boldsymbol{x})y$. The hinge loss is zero for $m \geq 1$, which is the same as the 0/1-loss. On the other hand, for $m < 1$, the hinge loss takes a positive value $1 - m$ and its slope is negative.

For the kernel model with intercept γ,

$$f_{\boldsymbol{\theta},\gamma}(\boldsymbol{x}) = \sum_{j=1}^{n} \theta_j K(\boldsymbol{x}, \boldsymbol{x}_j) + \gamma,$$

let us consider hinge loss minimization with generalized ℓ_2-regularization:

$$\min_{\boldsymbol{\theta},\gamma} \left[\sum_{i=1}^{n} \max\left\{0, 1 - f_{\boldsymbol{\theta},\gamma}(\boldsymbol{x}_i)y_i\right\} + \frac{\lambda}{2}\boldsymbol{\theta}^\top \boldsymbol{K}\boldsymbol{\theta} \right],$$

where $K_{i,j} = K(\boldsymbol{x}_i, \boldsymbol{x}_j)$. As illustrated in Fig. 27.13, the hinge loss corresponds to the maximum of $1 - m$ and 0:

$$\max\{0, 1-m\} = \min_{\xi} \xi \quad \text{subject to } \xi \geq 1-m,\ \xi \geq 0.$$

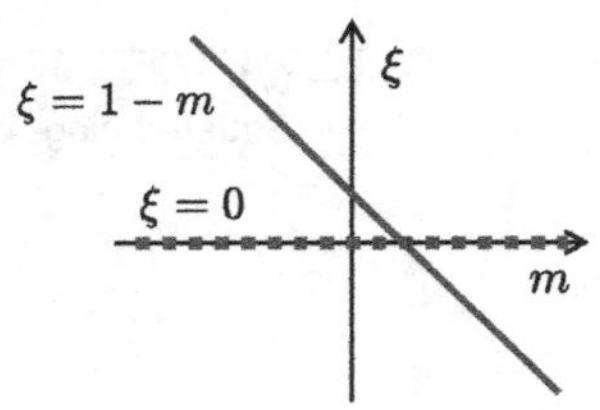

FIGURE 27.13

Hinge loss as maximizer of 1 − m and 0.

Then the optimization problem of regularized hinge loss minimization is expressed as

$$\min_{\theta,\gamma,\xi} \left[\sum_{i=1}^{n} \xi_i + \frac{\lambda}{2} \theta^\top K \theta \right]$$
$$\text{subject to } \xi_i \geq 1 - f_{\theta,\gamma}(x_i) y_i, \; \xi_i \geq 0, \quad \forall i = 1, \ldots, n. \tag{27.5}$$

Let us recall the primal optimization problem of kernel support vector classification (see (27.2)):

$$\min_{w,\gamma,\xi} \left[\frac{1}{2} \|w\|^2 + C \sum_{i=1}^{n} \xi_i \right]$$
$$\text{subject to } \xi_i \geq 1 - (w^\top \psi(x_i) + \gamma) y_i, \; \xi_i \geq 0, \quad \forall i = 1, \ldots, n.$$

Letting $C = 1/\lambda$ and $w = \sum_{j=1}^{n} \theta_j \psi(x_j)$ and using the kernel trick $\psi(x_i)^\top \psi(x_j) = K(x_i, x_j)$ in the above optimization problem show that support vector classification is actually equivalent to ℓ_2-regularized hinge loss minimization given by Eq. (27.5).

This loss minimization interpretation allows us to consider various extensions of support vector classification, which are described below.

27.6.2 SQUARED HINGE LOSS MINIMIZATION

For margin $m = f_\theta(x)y$, let us consider the *squared hinge loss*:

$$\frac{1}{2} \left(\max\left\{0, 1 - m\right\} \right)^2 = \begin{cases} \frac{1}{2}(1-m)^2 & (m \leq 1), \\ 0 & (m > 1), \end{cases}$$

which is differentiable everywhere (Fig. 27.12(b)). Then ℓ_2-regularized squared hinge loss minimization,

$$\min_{\theta} \left[\frac{1}{2} \sum_{i=1}^{n} \left(\max\left\{0, 1 - f_\theta(x_i) y_i\right\} \right)^2 + \frac{\lambda}{2} \|\theta\|^2 \right],$$

1. Initialize parameter $\boldsymbol{\theta}$, e.g. randomly or by ℓ_2-regularized LS as

$$\boldsymbol{\theta} \longleftarrow (\boldsymbol{\Phi}^\top \boldsymbol{\Phi} + \lambda \boldsymbol{I})^{-1} \boldsymbol{\Phi}^\top \boldsymbol{y}.$$

2. Compute the retarget matrix $\boldsymbol{V}$ from the current solution $\boldsymbol{\theta}$ as

$$\boldsymbol{V} = \mathrm{diag}\left(v_1, \ldots, v_n\right),$$

where

$$v_i = \boldsymbol{\theta}^\top \boldsymbol{\phi}(\boldsymbol{x}_i) y_i + \max(0, 1 - \boldsymbol{\theta}^\top \boldsymbol{\phi}(\boldsymbol{x}_i) y_i).$$

3. Compute the solution $\boldsymbol{\theta}$ based on the retarget matrix $\boldsymbol{V}$:

$$\boldsymbol{\theta} \longleftarrow (\boldsymbol{\Phi}^\top \boldsymbol{\Phi} + \lambda \boldsymbol{I})^{-1} \boldsymbol{\Phi}^\top \boldsymbol{V} \boldsymbol{y}.$$

4. Iterate 2–3 until convergence.

FIGURE 27.14

Iterative retargeted LS for ℓ_2-regularized squared hinge loss minimization.

may be solved by a simple stochastic gradient algorithm introduced in Section 15.3.

Another approach to obtaining the solution of ℓ_2-regularized squared hinge loss minimization is to put

$$v = m + \max\left\{0, 1 - m\right\}$$

and express the squared hinge loss as

$$\begin{aligned}\left(\max\left\{0, 1 - m\right\}\right)^2 &= (v - m)^2 \\ &= \left(v - f_{\boldsymbol{\theta}}(\boldsymbol{x}) y\right)^2 = \left(v y - f_{\boldsymbol{\theta}}(\boldsymbol{x})\right)^2,\end{aligned}$$

where $y = \pm 1$ is used. Since v is unknown, let us consider an iterative procedure and replace it with an estimate $\widehat{v}$ computed from the current solution. Then the solution of ℓ_2-regularized squared hinge loss minimization can be obtained by *iterative retargeted LS* (see Fig. 27.14).

A MATLAB code of iterative retargeted LS for ℓ_2-regularized squared hinge loss minimization is provided in Fig. 27.15, and its behavior is illustrated in Fig. 27.16. This shows that nonlinearly entangled data can be successfully classified almost in the same way as ordinary support vector classification with the Gaussian kernel (see Fig. 27.11).

```
n=200; a=linspace(0,4*pi,n/2);
u=[a.*cos(a) (a+pi).*cos(a)]'+rand(n,1);
v=[a.*sin(a) (a+pi).*sin(a)]'+rand(n,1);
x=[u v]; y=[ones(1,n/2) -ones(1,n/2)]';

x2=sum(x.^2,2); hh=2*1^2; l=1;
k=exp(-(repmat(x2,1,n)+repmat(x2',n,1)-2*x*x')/hh);
A=inv(k'*k+l*eye(n))*k'; t=rand(n,1);
for o=1:1000
  z=(k*t).*y; w=z+max(0,1-z); t0=A*(w.*y);
  if norm(t-t0)<0.001, break, end
  t=t0;
end

m=100; X=linspace(-15,15,m)'; X2=X.^2;
U=exp(-(repmat(u.^2,1,m)+repmat(X2',n,1)-2*u*X')/hh);
V=exp(-(repmat(v.^2,1,m)+repmat(X2',n,1)-2*v*X')/hh);
figure(1); clf; hold on; axis([-15 15 -15 15]);
contourf(X,X,sign(V'*(U.*repmat(t,1,m))));
plot(x(y==1,1),x(y==1,2),'bo');
plot(x(y==-1,1),x(y==-1,2),'rx');
colormap([1 0.7 1; 0.7 1 1]);
```

FIGURE 27.15

MATLAB code of iterative retargeted LS for ℓ_2-regularized squared hinge loss minimization.

27.6.3 RAMP LOSS MINIMIZATION

As illustrated in Fig. 27.12, the (squared) hinge loss is not upper-bounded. For this reason, learning with such an unbounded loss tends to be strongly affected by an *outlier* (Fig. 27.17(a)).

As discussed in Chapter 25, using a bounded loss can enhance the robustness against outliers. For margin $m = f_{\theta}(x)y$, let us consider the *ramp loss* (Fig. 27.18(a)):

$$\min\left\{1, \max\left(0, 1-m\right)\right\} = \begin{cases} 1 & (m < 0), \\ 1-m & (0 \le m \le 1), \\ 0 & (m > 1). \end{cases}$$

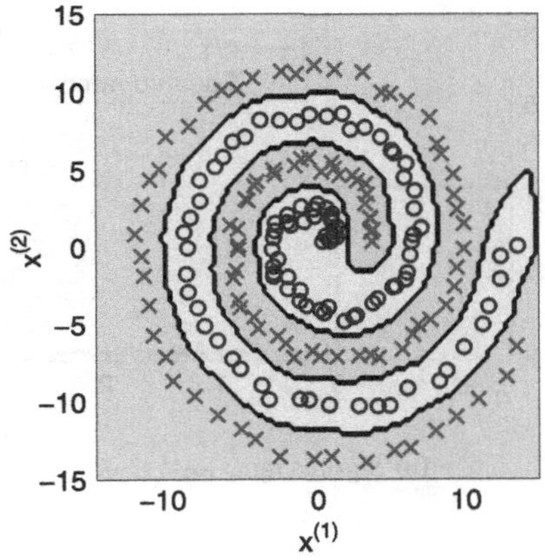

FIGURE 27.16

Example of ℓ_2-regularized squared hinge loss minimization.

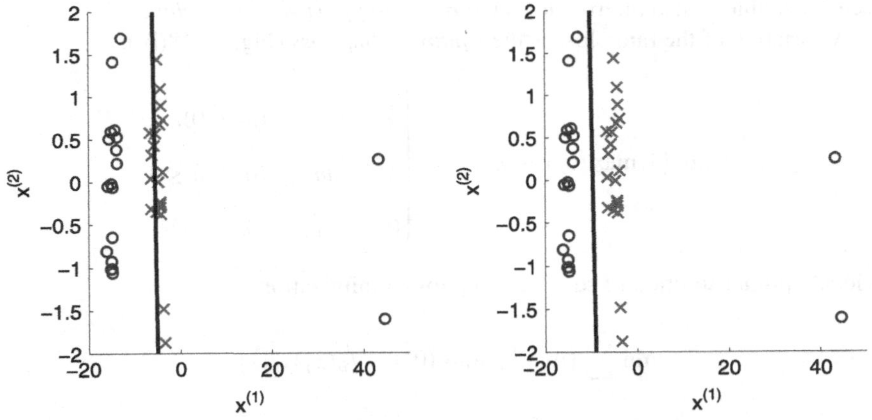

(a) Ordinary support vector classification

(b) Robust support vector classification

FIGURE 27.17

Examples of support vector classification with outliers.

Due to the boundedness of the ramp loss, ramp loss minimization,

$$\min_{\theta} \sum_{i=1}^{n} \min\left\{1, \max\left(0, 1 - f_{\theta}(\boldsymbol{x}_i)y_i\right)\right\},$$

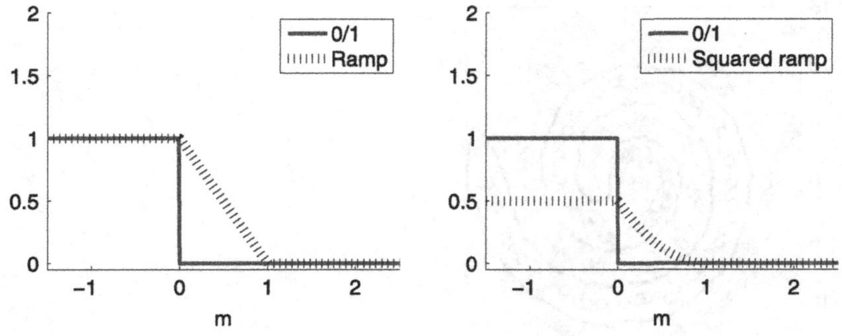

(a) Ramp loss (b) Squared ramp loss

FIGURE 27.18

Ramp loss and squared ramp loss.

is more robust against outliers than hinge loss minimization, as illustrated Fig. 27.17(b). However, since the ramp loss is a nonconvex function, obtaining the global optimal solution is not straightforward. A ramp loss variant of the support vector machine is also referred to as the *robust support vector machine*.

A variation of the ramp loss is the *squared ramp loss* (Fig. 27.18(b)):

$$\frac{1}{2}\left(\min\left\{1, \max\left(0, 1-m\right)\right\}\right)^2 = \begin{cases} \frac{1}{2} & (m < 0), \\ \frac{1}{2}(1-m)^2 & (0 \leq m \leq 1), \\ 0 & (m > 1). \end{cases}$$

A local optimal solution of squared ramp loss minimization,

$$\min_{\boldsymbol{\theta}} \sum_{i=1}^{n} \left(\min\left\{1, \max\left(0, 1 - f_{\boldsymbol{\theta}}(\boldsymbol{x}_i)y_i\right)\right\}\right)^2,$$

can be obtained by *iteratively retargeted LS* in the same as squared hinge loss minimization (see Section 27.6.2). The only difference is that the definition of v in iteratively retargeted LS is replaced with

$$v = m + \min\{1, \max(0, 1-m)\}.$$

CHAPTER

PROBABILISTIC CLASSIFICATION

28

CHAPTER CONTENTS

In Chapter 26 and Chapter 27, classification methods that *deterministically* assign a test pattern $\boldsymbol{x}$ into class y were discussed. In this chapter, methods of *probabilistic classification* are introduced, which choose the class probabilistically. More specifically, for a test pattern $\boldsymbol{x}$, the *class-posterior probability* $p(y|\boldsymbol{x})$ is learned in probabilistic classification. Then the class that maximizes the estimated class-posterior probability $\widehat{p}(y|\boldsymbol{x})$ is chosen:

$$\widehat{y} = \underset{y=1,\ldots,c}{\operatorname{argmax}}\, \widehat{p}(y|\boldsymbol{x}).$$

The value of $\widehat{p}(y|\boldsymbol{x})$ may be interpreted as the *reliability* of assigning label y to pattern $\boldsymbol{x}$, which allows the possibility of *rejecting* the pattern $\boldsymbol{x}$ if the reliability is low.

28.1 LOGISTIC REGRESSION

In this section, a standard probabilistic classifier called *logistic regression* is introduced.

28.1.1 LOGISTIC MODEL AND MLE

The *logistic model* parameterizes the class-posterior probability $p(y|\boldsymbol{x})$ in a log-linear form as

$$q(y|\boldsymbol{x};\boldsymbol{\theta}) = \frac{\exp\left(\sum_{j=1}^{b} \theta_j^{(y)} \phi_j(\boldsymbol{x})\right)}{\sum_{y'=1}^{c} \exp\left(\sum_{j=1}^{b} \theta_j^{(y')} \phi_j(\boldsymbol{x})\right)} = \frac{\exp\left(\boldsymbol{\theta}^{(y)\top} \boldsymbol{\phi}(\boldsymbol{x})\right)}{\sum_{y'=1}^{c} \exp\left(\boldsymbol{\theta}^{(y')\top} \boldsymbol{\phi}(\boldsymbol{x})\right)}, \tag{28.1}$$

where the exponential function in the numerator is for restricting the output to be positive and the summation in the denominator is for restricting the output summed

1. Initialize $\boldsymbol{\theta}$.
2. Choose the ith training sample randomly $(\boldsymbol{x}_i, y_i)$.
3. Update parameter $\boldsymbol{\theta} = (\boldsymbol{\theta}^{(1)\top}, \ldots, \boldsymbol{\theta}^{(c)\top})^\top$ to go up the gradient:

$$\boldsymbol{\theta}^{(y)} \longleftarrow \boldsymbol{\theta}^{(y)} + \varepsilon \nabla_y J_i(\boldsymbol{\theta}) \quad \text{for } y = 1, \ldots, c,$$

where $\varepsilon > 0$ is the step size, and $\nabla_y J_i$ is the gradient of $J_i(\boldsymbol{\theta}) = \log q(y_i | \boldsymbol{x}_i; \boldsymbol{\theta})$ with respect to $\boldsymbol{\theta}^{(y)}$:

$$\nabla_y J_i(\boldsymbol{\theta}) = -\frac{\exp\left(\boldsymbol{\theta}^{(y)\top}\boldsymbol{\phi}(\boldsymbol{x}_i)\right)\boldsymbol{\phi}(\boldsymbol{x}_i)}{\sum_{y'=1}^{c}\exp\left(\boldsymbol{\theta}^{(y')\top}\boldsymbol{\phi}(\boldsymbol{x}_i)\right)} + \begin{cases} \boldsymbol{\phi}(\boldsymbol{x}_i) & (y = y_i), \\ \boldsymbol{0} & (y \neq y_i). \end{cases}$$

4. Iterate 2 and 3 until convergence.

FIGURE 28.1

Stochastic gradient algorithm for logistic regression.

to one. Since the logistic model contains parameters $\{\theta_j^{(y)}\}_{j=1}^b$ for each class $y = 1, \ldots, c$, the entire parameter vector $\boldsymbol{\theta}$ is bc-dimensional:

$$\boldsymbol{\theta} = (\underbrace{\theta_1^{(1)}, \ldots, \theta_b^{(1)}}_{\text{Class 1}}, \ldots, \underbrace{\theta_1^{(c)}, \ldots, \theta_b^{(c)}}_{\text{Class } c})^\top.$$

Note that the number of parameters can be reduced from bc to $b(c-1)$ by using $\sum_{y=1}^{c} q(y|\boldsymbol{x}; \boldsymbol{\theta}) = 1$, but for simplicity the above bc-dimensional formulation is considered below.

The parameter $\boldsymbol{\theta}$ is learned by the *MLE* (Chapter 12). More specifically, $\boldsymbol{\theta}$ is determined so as to maximize the *log-likelihood*, i.e. the log-probability that the current training samples $\{(\boldsymbol{x}_i, y_i)\}_{i=1}^n$ are generated:

$$\max_{\boldsymbol{\theta}} \sum_{i=1}^{n} \log q(y_i | \boldsymbol{x}_i; \boldsymbol{\theta}).$$

Since the above objective function is differentiable with respect to $\boldsymbol{\theta}$, the solution may be obtained by a *stochastic gradient* algorithm (Section 15.3), as described in Fig. 28.1.

A MATLAB code of stochastic gradient ascent for logistic regression is provided in Fig. 28.2, where the log-Gaussian kernel model,

$$q(y|\boldsymbol{x}; \boldsymbol{\theta}) \propto \exp\left(\sum_{j=1}^{n} \theta_j \exp\left(-\frac{\|\boldsymbol{x} - \boldsymbol{x}_j\|^2}{2h^2}\right)\right),$$

```
n=90; c=3; y=ones(n/c,1)*[1:c]; y=y(:);
x=randn(n/c,c)+repmat(linspace(-3,3,c),n/c,1); x=x(:);

hh=2*1^2; t=randn(n,c);
for o=1:n*1000
  i=ceil(rand*n); yi=y(i); ki=exp(-(x-x(i)).^2/hh);
  ci=exp(ki'*t); t0=t-0.1*(ki*ci)/(1+sum(ci));
  t0(:,yi)=t0(:,yi)+0.1*ki;
  if norm(t-t0)<0.000001, break, end
  t=t0;
end

N=100; X=linspace(-5,5,N)';
K=exp(-(repmat(X.^2,1,n)+repmat(x.^2',N,1)-2*X*x')/hh);
figure(1); clf; hold on; axis([-5 5 -0.3 1.8]);
C=exp(K*t); C=C./repmat(sum(C,2),1,c);
plot(X,C(:,1),'b-'); plot(X,C(:,2),'r--');
plot(X,C(:,3),'g:');
plot(x(y==1),-0.1*ones(n/c,1),'bo');
plot(x(y==2),-0.2*ones(n/c,1),'rx');
plot(x(y==3),-0.1*ones(n/c,1),'gv');
legend('q(y=1|x)','q(y=2|x)','q(y=3|x)')
```

FIGURE 28.2

MATLAB code of stochastic gradient ascent for logistic regression.

is used for approximating the class-posterior probability. The behavior of Gaussian kernel logistic regression is illustrated in Fig. 28.3, showing that the class-posterior probability is well-approximated.

The log-likelihood of logistic regression for the ith training sample $(\boldsymbol{x}_i, y_i)$ is given by

$$\log q(y_i|\boldsymbol{x}_i;\boldsymbol{\theta}) = f_{y_i} - \log\left(\sum_{y=1}^{c}\exp(f_y)\right),$$

where $f_y = \boldsymbol{\theta}^{(y)\top}\boldsymbol{\phi}(\boldsymbol{x}_i)$. In numerical computation, the *log-sum-exp* term,

$$\log\left(\sum_{y=1}^{c}\exp(f_y)\right),$$

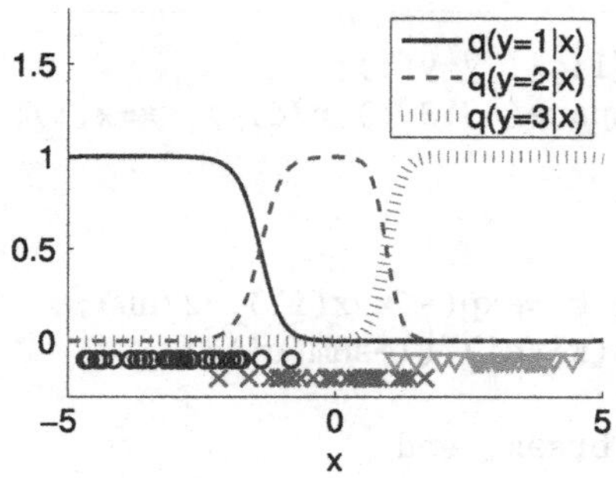

FIGURE 28.3

Example of stochastic gradient ascent for logistic regression.

is often cumbersome. When $f_y \gg 1$ for some y, $\exp(f_y)$ may cause overflow (e.g. $\exp(700) \approx 10^{300}$), resulting in $\log(\infty) = \infty$. When $f_y \ll -1$ for all $y = 1, \ldots, c$, $\exp(f_y)$ may cause underflow (e.g. $\exp(-700) \approx 10^{-300}$), resulting in $\log(0) = -\infty$. To cope with these numerical problems, the following expression of the log-sum-exp term is useful in practice:

$$\log\left(\sum_{y=1}^{c} \exp(f_y)\right) = \log\left(\exp(f_{y'})\left(\sum_{y=1}^{c} \exp\left(f_y - f_{y'}\right)\right)\right)$$
$$= f_{y'} + \log\left(\sum_{y=1}^{c} \exp\left(f_y - f_{y'}\right)\right),$$

where $y' = \mathrm{argmax}_{y=1,\ldots,c}\{\boldsymbol{\theta}^{(y)\top}\boldsymbol{\phi}(\boldsymbol{x}_i)\}$. With this expression, overflow can be avoided because $f_y - f_{y'} \leq 0$ for all $y = 1, \ldots, c$, and underflow does not cause $\log(0)$ because $f_y - f_{y'} = 0$ holds for $y = y'$.

28.1.2 LOSS MINIMIZATION VIEW

Let us focus on the binary classification problem for $y \in \{+1, -1\}$. Then the sum-to-one constraint,

$$q(y = +1|\boldsymbol{x};\boldsymbol{\theta}) + q(y = -1|\boldsymbol{x};\boldsymbol{\theta}) = 1,$$

implies that the number of parameters in the logistic model can be reduced from $2b$ to b:

$$q(y|\boldsymbol{x};\boldsymbol{\theta}) = \frac{1}{1 + \exp\left(-yf_{\boldsymbol{\theta}}(\boldsymbol{x})\right)},$$

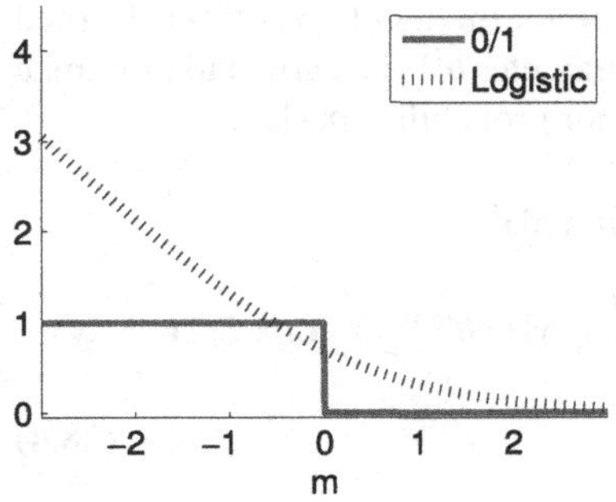

FIGURE 28.4

Logistic loss.

where $f_{\boldsymbol{\theta}}(\boldsymbol{x}) = \sum_{j=1}^{b} \theta_j \phi_j(\boldsymbol{x})$. Then the maximum log-likelihood criterion for this simplified model can be expressed as

$$\min_{\boldsymbol{\theta}} \sum_{i=1}^{n} \log\left(1 + \exp(-m_i)\right), \tag{28.2}$$

where $m_i = f_{\boldsymbol{\theta}}(\boldsymbol{x}_i) y_i$ is the *margin* for the ith training sample $(\boldsymbol{x}_i, y_i)$. Eq. (28.2) can be interpreted as *surrogate loss* minimization (see Section 26.2) with the *logistic loss* (Fig. 28.4):

$$\log(1 + \exp(-m)).$$

Thus, although logistic regression was derived in a very different framework from LS classification (Chapter 26) and support vector classification (Chapter 27), they can all be interpreted as surrogate loss minimization.

28.2 LS PROBABILISTIC CLASSIFICATION

In this section, an alternative to logistic regression based on LS fitting of the class-posterior probability is introduced.

Let us model the class-posterior probability $p(y|\boldsymbol{x})$ by the linear-in-parameter model:

$$q(y|\boldsymbol{x}; \boldsymbol{\theta}^{(y)}) = \sum_{j=1}^{b} \theta_j^{(y)} \phi_j(\boldsymbol{x}) = \boldsymbol{\theta}^{(y)\top} \boldsymbol{\phi}(\boldsymbol{x}). \tag{28.3}$$

Differently from logistic model (28.1), the above model for class y depends only on the parameter for class y, $\boldsymbol{\theta}^{(y)} = (\theta_1^{(y)}, \ldots, \theta_b^{(y)})^\top$, because the normalization term is not included.

In LS probabilistic classification, the class-posterior model $q(y|\boldsymbol{x};\boldsymbol{\theta}^{(y)})$ is learned for each class separately. More specifically, the parameter $\boldsymbol{\theta}^{(y)}$ is learned to minimize the expected squared error to the true class-posterior probability $p(y|\boldsymbol{x})$:

$$\begin{aligned} J_y(\boldsymbol{\theta}^{(y)}) &= \frac{1}{2}\int \left(q(y|\boldsymbol{x};\boldsymbol{\theta}^{(y)}) - p(y|\boldsymbol{x})\right)^2 p(\boldsymbol{x})\mathrm{d}\boldsymbol{x} \\ &= \frac{1}{2}\int q(y|\boldsymbol{x};\boldsymbol{\theta}^{(y)})^2 p(\boldsymbol{x})\mathrm{d}\boldsymbol{x} - \int q(y|\boldsymbol{x};\boldsymbol{\theta}^{(y)}) p(y|\boldsymbol{x})p(\boldsymbol{x})\mathrm{d}\boldsymbol{x} \\ &\quad + \frac{1}{2}\int p(y|\boldsymbol{x})^2 p(\boldsymbol{x})\mathrm{d}\boldsymbol{x}, \end{aligned} \tag{28.4}$$

where $p(\boldsymbol{x})$ denotes the marginal probability density of training input samples $\{\boldsymbol{x}_i\}_{i=1}^n$.

The second term in Eq. (28.4), $p(y|\boldsymbol{x})p(\boldsymbol{x})$, can be expressed as

$$p(y|\boldsymbol{x})p(\boldsymbol{x}) = p(\boldsymbol{x},y) = p(\boldsymbol{x}|y)p(y),$$

where $p(\boldsymbol{x}|y)$ denotes the *class-conditional probability density* for samples in class y, $\{\boldsymbol{x}_i\}_{i:y_i=y}$, and $p(y)$ denotes the *class-prior probability* for training output samples $\{y_i\}_{i=1}^n$. The first two terms in J_y,

$$\int q(y|\boldsymbol{x};\boldsymbol{\theta}^{(y)})^2 p(\boldsymbol{x})\mathrm{d}\boldsymbol{x} \quad \text{and} \quad \int q(y|\boldsymbol{x};\boldsymbol{\theta}^{(y)}) p(y)p(\boldsymbol{x}|y)\mathrm{d}\boldsymbol{x},$$

are the expectations over $p(\boldsymbol{x})$ and $p(\boldsymbol{x}|y)$, respectively, although they are not accessible. Here, the expectations are approximated by the sample averages:

$$\frac{1}{n}\sum_{i=1}^{n} q(y|\boldsymbol{x}_i;\boldsymbol{\theta}^{(y)})^2 \quad \text{and} \quad \frac{1}{n_y}\sum_{i:y_i=y} q(y|\boldsymbol{x}_i;\boldsymbol{\theta}^{(y)})p(y),$$

where n_y denotes the number of training samples in class y. Further approximating $p(y)$ by the sample ratio n_y/n, ignoring the third term in Eq. (28.4) because it is constant, and including the ℓ_2-regularizer yield the following training criterion:

$$\begin{aligned} \widehat{J}_y(\boldsymbol{\theta}^{(y)}) &= \frac{1}{2n}\sum_{i=1}^{n} q(y|\boldsymbol{x}_i;\boldsymbol{\theta}^{(y)})^2 - \frac{1}{n}\sum_{i:y_i=y} q(y|\boldsymbol{x}_i;\boldsymbol{\theta}^{(y)}) + \frac{\lambda}{2n}\|\boldsymbol{\theta}^{(y)}\|^2 \\ &= \frac{1}{2n}\boldsymbol{\theta}^{(y)\top}\boldsymbol{\Phi}^\top\boldsymbol{\Phi}\boldsymbol{\theta}^{(y)} - \frac{1}{n}\boldsymbol{\theta}^{(y)\top}\boldsymbol{\Phi}^\top\boldsymbol{\pi}^{(y)} + \frac{\lambda}{2n}\|\boldsymbol{\theta}^{(y)}\|^2, \end{aligned}$$

where $\boldsymbol{\pi}^{(y)} = (\pi_1^{(y)},\ldots,\pi_n^{(y)})^\top$ is defined as $\pi_i^{(y)} = \begin{cases} 1 & (y_i = y), \\ 0 & (y_i \neq y). \end{cases}$ Since $\widehat{J}_y(\boldsymbol{\theta}^{(y)})$ is a quadratic function, its minimizer can be obtained analytically by setting its derivative to zero:

$$\widehat{\boldsymbol{\theta}}^{(y)} = \left(\boldsymbol{\Phi}^\top\boldsymbol{\Phi} + \lambda\boldsymbol{I}\right)^{-1}\boldsymbol{\Phi}^\top\boldsymbol{\pi}^{(y)}.$$

```
n=90; c=3; y=ones(n/c,1)*[1:c]; y=y(:);
x=randn(n/c,c)+repmat(linspace(-3,3,c),n/c,1); x=x(:);

hh=2*1^2; x2=x.^2; l=0.1; N=100; X=linspace(-5,5,N)';
k=exp(-(repmat(x2,1,n)+repmat(x2',n,1)-2*x*x')/hh);
K=exp(-(repmat(X.^2,1,n)+repmat(x2',N,1)-2*X*x')/hh);
for yy=1:c
  yk=(y==yy); ky=k(:,yk);
  ty=(ky'*ky+l*eye(sum(yk)))\(ky'*yk);
  Kt(:,yy)=max(0,K(:,yk)*ty);
end
ph=Kt./repmat(sum(Kt,2),1,c);

figure(1); clf; hold on; axis([-5 5 -0.3 1.8]);
plot(X,ph(:,1),'b-'); plot(X,ph(:,2),'r--');
plot(X,ph(:,3),'g:');
plot(x(y==1),-0.1*ones(n/c,1),'bo');
plot(x(y==2),-0.2*ones(n/c,1),'rx');
plot(x(y==3),-0.1*ones(n/c,1),'gv');
legend('p(y=1|x)','p(y=2|x)','p(y=3|x)')
```

FIGURE 28.5

MATLAB code for LS probabilistic classification.

However, differently from logistic model (28.1), current linear-in-parameter model (28.3) can take a negative value and may not be summed to one. To cope with this problem, the solution is postprocessed as

$$\widehat{p}(y|\boldsymbol{x}) = \frac{\max(0, \widehat{\boldsymbol{\theta}}^{(y)\top}\boldsymbol{\phi}(\boldsymbol{x}))}{\sum_{y'=1}^{c}\max(0, \widehat{\boldsymbol{\theta}}^{(y')\top}\boldsymbol{\phi}(\boldsymbol{x}))}. \tag{28.5}$$

A MATLAB code of LS probabilistic classification for the Gaussian kernel model,

$$q(y|\boldsymbol{x};\boldsymbol{\theta}^{(y)}) = \sum_{j:y_j=y} \theta_j^{(y)} \exp\left(-\frac{\|\boldsymbol{x}-\boldsymbol{x}_j\|^2}{2h^2}\right),$$

is provided in Fig. 28.5, and its behavior is illustrated in Fig. 28.6. This shows that almost the same solution as logistic regression (see Fig. 28.3) can be obtained by LS probabilistic classification.

As illustrated above, LS probabilistic classification behaves similarly to logistic regression. In logistic regression, logistic model (28.1) containing bc parameters is

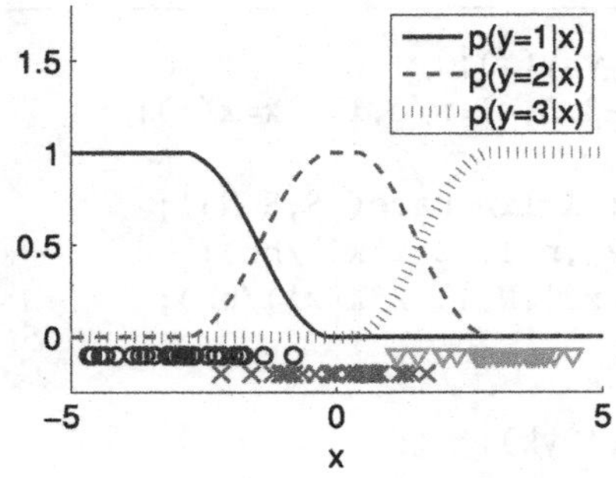

FIGURE 28.6

Example of LS probabilistic classification for the same data set as Fig. 28.3.

trained once for all classes. On the other hand, in LS probabilistic classification, linear-in-parameter model (28.3) containing only b parameters is trained for each class $y = 1, \dots, c$. When c is large, training small linear-in-parameter models containing only b parameters c times would be computationally more efficient than training a big logistic model containing bc parameters once. Furthermore, the solution of LS probabilistic classification can be obtained analytically. However, it involves *ad hoc* postprocessing, which may corrupt the solution when the number of training samples is small. Thus, using logistic regression would be more reliable when the number of training samples is small, while using LS probabilistic classification would be computationally more efficient when the number of training samples is large.

CHAPTER

STRUCTURED CLASSIFICATION

29

CHAPTER CONTENTS

When classifying a set of input patterns having certain data structure such as a sequence of letters and a parsing tree of a sentence, the classification performance is expected to be improved if such structural information is utilized. In this chapter, taking a sequence of letters as an example of such structure, methods of structured classification are introduced.

29.1 SEQUENCE CLASSIFICATION

Let us consider the problem of classifying a sequence of m patterns where each pattern belongs to one of the c classes. Let $\overline{\boldsymbol{x}}$ and $\overline{y}$ be sequences of m patterns and m classes, and let $\boldsymbol{x}^{(k)}$ and $y^{(k)}$ be their kth elements:

$$\overline{\boldsymbol{x}} = (\boldsymbol{x}^{(1)}, \ldots, \boldsymbol{x}^{(m)}),$$
$$\overline{y} = (y^{(1)}, \ldots, y^{(m)}).$$

For example, when recognizing a sequence of five hand-written digits (see Fig. 29.1), $m = 5$ and $c = 10$. If such a sequence is decomposed into five digits, ordinary classification algorithms introduced in the previous chapters can be used for classifying each digit. However, this decomposition approach loses useful contextual information that can drastically improve the classification accuracy, for example, the same number may rarely appear consecutively and some number appears frequently after a certain number. On the other hand, if the entire sequence of digits is directly recognized, the number of classes grows exponentially with respect to the length of

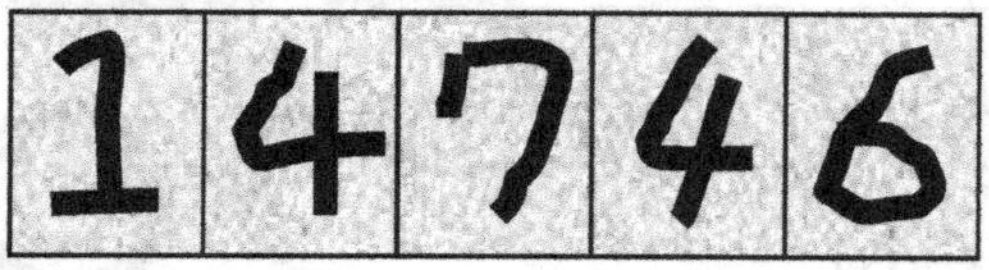

FIGURE 29.1

Classification of sequence of hand-written digits.

the sequence. In the example of Fig. 29.1, $(m, c) = (5, 10)$ and thus the number of classes is 10^5, which is hard to handle.

In this chapter, intermediate approaches are introduced, which take into account some contextual information with the computational complexity kept moderately. Note that, although only sequence data are considered, methods introduced in this chapter can be applied to more general structured data.

29.2 PROBABILISTIC CLASSIFICATION FOR SEQUENCES

In this section, the *logistic model* introduced in Section 28.1 is extended to the *conditional random field* [65] for sequence classification.

29.2.1 CONDITIONAL RANDOM FIELD

To introduce the conditional random field, let us first consider the logistic model applied to sequence classification, i.e. each pattern $\boldsymbol{x}^{(k)}$ is independently classified into one of the c classes (Fig. 29.2(a)):

$$q(y|\boldsymbol{x};\boldsymbol{\theta}) = \frac{\exp\left(\boldsymbol{\theta}_y^\top \boldsymbol{\phi}(\boldsymbol{x})\right)}{\sum_{y'=1}^{c} \exp\left(\boldsymbol{\theta}_{y'}^\top \boldsymbol{\phi}(\boldsymbol{x})\right)},$$

where $\boldsymbol{\phi}(\boldsymbol{x}) \in \mathbb{R}^b$ is the vector of basis functions for a single pattern $\boldsymbol{x}$, $\boldsymbol{\theta}_y \in \mathbb{R}^b$ is the parameter vector for class y, and

$$\boldsymbol{\theta} = (\boldsymbol{\theta}_1^\top, \ldots, \boldsymbol{\theta}_c^\top)^\top \in \mathbb{R}^{bc}$$

is the vector of all parameters. However, this decomposition approach cannot utilize the contextual information contained in sequence data.

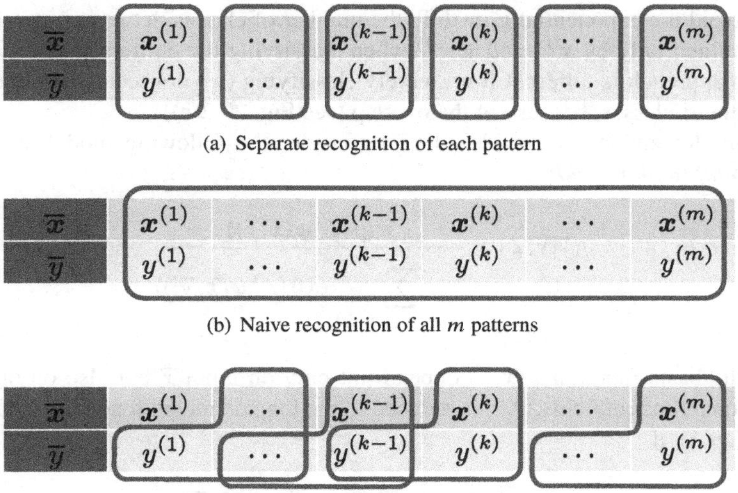
(a) Separate recognition of each pattern

(b) Naive recognition of all m patterns

(c) Efficient recognition of all m patterns

FIGURE 29.2

Sequence classification.

On the other hand, if the entire sequence is classified at once, a classification problem with $\overline{c} = c^m$ classes needs to be solved (Fig. 29.2(b)):

$$q(\overline{y}|\overline{x};\overline{\theta}) = \frac{\exp\left(\overline{\theta}_{\overline{y}}^{\top}\overline{\phi}(\overline{x})\right)}{\sum_{y^{(1)},\ldots,y^{(m)}=1}^{c}\exp\left(\overline{\theta}_{\overline{y}'}^{\top}\overline{\phi}(\overline{x})\right)}, \tag{29.1}$$

where

$$\overline{\phi}(\overline{x}) = (\phi(x^{(1)})^{\top},\ldots,\phi(x^{(m)})^{\top})^{\top} \in \mathbb{R}^{bm}$$

is the vector of basis functions for pattern sequence $\overline{x} = (x^{(1)},\ldots,x^{(m)})$,

$$\overline{\theta}_{\overline{y}} = (\theta_{\overline{y}}^{(1)\top},\ldots,\theta_{\overline{y}}^{(m)\top})^{\top} \in \mathbb{R}^{bm}$$

is the parameter vector for class $\overline{y} = (y^{(1)},\ldots,y^{(m)})$, and

$$\overline{\theta} = (\overline{\theta}_1{}^{\top},\ldots,\overline{\theta}_{\overline{c}}{}^{\top})^{\top} \in \mathbb{R}^{bm\overline{c}}$$

is the vector of all parameters. However, since the number of classes $\overline{c} = c^m$ grows exponentially with respect to the sequence length m, handling the above logistic model may be intractable.

To avoid the exponential growth in the number of classes, let us assume that $y^{(k)}$ is determined only by $\boldsymbol{x}^{(k)}$ and $y^{(k-1)}$ when classifying the entire sequence $\overline{\boldsymbol{x}}$. Note that this approach is different from merely classifying two consecutive patterns into one of the c^2 classes, because of the overlap (see Fig. 29.2(c)).

To implement the above idea, let us consider the following model, called the *conditional random field*:

$$q(\overline{y}|\overline{\boldsymbol{x}};\boldsymbol{\zeta}) = \frac{\exp\left(\boldsymbol{\zeta}^\top \boldsymbol{\varphi}(\overline{\boldsymbol{x}},\overline{y})\right)}{\sum_{y^{(1)},\ldots,y^{(m_i)}=1}^{c} \exp\left(\boldsymbol{\zeta}^\top \boldsymbol{\varphi}(\overline{\boldsymbol{x}},\overline{y}')\right)}, \tag{29.2}$$

where the basis function $\boldsymbol{\varphi}(\overline{\boldsymbol{x}},\overline{y})$ depend not only on input $\overline{\boldsymbol{x}}$ but also on output $\overline{y}$. Conditional random field (29.2) is reduced to the logistic model for sequence $\overline{\boldsymbol{x}}$ given by Eq. (29.1), if

$$\boldsymbol{\zeta} = \overline{\boldsymbol{\theta}} \quad \text{and} \quad \boldsymbol{\varphi}(\overline{\boldsymbol{x}},\overline{y}) = \boldsymbol{e}_{\overline{y}}^{(\overline{c})} \otimes \overline{\boldsymbol{\phi}}(\overline{\boldsymbol{x}}),$$

where $\boldsymbol{e}_{\overline{y}}^{(\overline{c})}$ is the $\overline{c}$-dimensional indicator vector of class $\overline{y}$ (i.e. the element corresponding class $\overline{y}$ is 1 and all other elements are 0), and $\otimes$ denotes the *Kronecker product* (see Fig. 6.5):

$$\begin{aligned} &\text{For } \boldsymbol{f} = (f_1,\ldots,f_u)^\top \in \mathbb{R}^u,\ \boldsymbol{g} = (g_1,\ldots,g_v)^\top \in \mathbb{R}^v, \\ &\boldsymbol{f} \otimes \boldsymbol{g} = (f_1g_1, f_1g_2,\ldots,f_1g_v, f_2g_1, f_2g_2,\ldots,f_2g_v,\ldots, \\ &\qquad\qquad f_ug_1, f_ug_2,\ldots,f_ug_v)^\top \in \mathbb{R}^{uv}. \end{aligned}$$

Thus, naively using conditional random field (29.2) is just an overly redundant expression of Eq. (29.1).

Let us simplify conditional random field (29.2) by taking into account the structure described in Fig. 29.2(c). More specifically, basis function $\boldsymbol{\varphi}(\overline{\boldsymbol{x}},\overline{y})$ is defined as the sum of basis functions depending only on two consecutive patterns:

$$\boldsymbol{\varphi}(\overline{\boldsymbol{x}},\overline{y}) = \sum_{k=1}^{m} \boldsymbol{\varphi}(\boldsymbol{x}^{(k)}, y^{(k)}, y^{(k-1)}), \tag{29.3}$$

where $y^{(0)} = y^{(1)}$. As two-pattern basis function $\boldsymbol{\varphi}(\boldsymbol{x}^{(k)}, y^{(k)}, y^{(k-1)})$, for example,

$$\boldsymbol{\varphi}(\boldsymbol{x}^{(k)}, y^{(k)}, y^{(k-1)}) = \begin{pmatrix} \boldsymbol{e}_{y^{(k)}}^{(c)} \otimes \boldsymbol{\phi}(\boldsymbol{x}^{(k)}) \\ \boldsymbol{e}_{y^{(k)}}^{(c)} \otimes \boldsymbol{e}_{y^{(k-1)}}^{(c)} \end{pmatrix} \in \mathbb{R}^{cb+c^2}, \tag{29.4}$$

may be used. Then the dimensionality of parameter vector $\boldsymbol{\zeta}$ is $cb + c^2$, which is independent of the sequence length m.

1. Initialize ζ.
2. Choose the ith training sample randomly $(\overline{\boldsymbol{x}}_i, \overline{y}_i)$.
3. Update parameter ζ to go up the gradient:

$$\zeta \longleftarrow \zeta + \varepsilon \left(\boldsymbol{\varphi}_i(\overline{\boldsymbol{x}}_i, \overline{y}_i) - \frac{\sum_{y^{(1)},\ldots,y^{(m_i)}=1}^{c} \exp\left(\zeta^\top \boldsymbol{\varphi}_i(\overline{\boldsymbol{x}}_i, \overline{y})\right) \boldsymbol{\varphi}_i(\overline{\boldsymbol{x}}_i, \overline{y})}{\sum_{y^{(1)},\ldots,y^{(m_i)}=1}^{c} \exp\left(\zeta^\top \boldsymbol{\varphi}_i(\overline{\boldsymbol{x}}_i, \overline{y})\right)} \right),$$

where $\varepsilon > 0$ is the step size and $\overline{y} = (y^{(1)}, \ldots, y^{(m_i)})$.
4. Iterate 2 and 3 until convergence.

FIGURE 29.3

Stochastic gradient algorithm for conditional random field.

29.2.2 MLE

Suppose that labeled sequences,

$$\left\{(\overline{\boldsymbol{x}}_i, \overline{y}_i) \,\middle|\, \overline{\boldsymbol{x}}_i = (\boldsymbol{x}_i^{(1)}, \ldots, \boldsymbol{x}_i^{(m_i)}),\ \overline{y}_i = (y_i^{(1)}, \ldots, y_i^{(m_i)})\right\}_{i=1}^{n},$$

are provided as training samples, where the length m_i of sequence $\overline{\boldsymbol{x}}_i$ can be different for each sample. With these training samples, the parameter ζ in conditional random field (29.2) is learned by *MLE* (see Chapter 12):

$$\max_{\zeta} \sum_{i=1}^{n} \log \frac{\exp\left(\zeta^\top \boldsymbol{\varphi}_i(\overline{\boldsymbol{x}}_i, \overline{y}_i)\right)}{\sum_{y^{(1)},\ldots,y^{(m_i)}=1}^{c} \exp\left(\zeta^\top \boldsymbol{\varphi}_i(\overline{\boldsymbol{x}}_i, \overline{y})\right)},$$

where

$$\boldsymbol{\varphi}_i(\overline{\boldsymbol{x}}_i, \overline{y}_i) = \sum_{k=1}^{m_i} \boldsymbol{\varphi}(\boldsymbol{x}_i^{(k)}, y_i^{(k)}, y_i^{(k-1)}).$$

The maximum likelihood solution may be obtained by a *stochastic gradient* algorithm (Section 15.3), as described in Fig. 29.3.

29.2.3 RECURSIVE COMPUTATION

However, the computational complexity of the second term of the gradient (see Fig. 29.3),

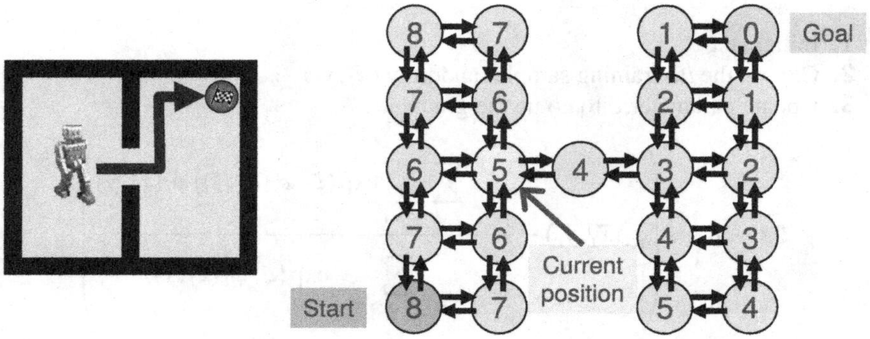

FIGURE 29.4

Dynamic programming, which solves a complex optimization problem by breaking it down into simpler subproblems recursively. When the number of steps to the goal is counted, dynamic programming trace back the steps from the goal. In this case, many subproblems of counting the number of steps from other positions are actually shared and thus dynamic programming can efficiently reuse the solutions to reduce the computation costs.

$$\frac{\sum_{y^{(1)},\ldots,y^{(m_i)}=1}^{c} \exp\left(\zeta^\top \varphi_i(\overline{x}_i,\overline{y})\right) \varphi_i(\overline{x}_i,\overline{y})}{\sum_{y^{(1)},\ldots,y^{(m_i)}=1}^{c} \exp\left(\zeta^\top \varphi_i(\overline{x}_i,\overline{y})\right)}, \tag{29.5}$$

grows exponentially with respect to the sequence length m_i. To cope with this problem, let us utilize the decomposition given by (29.3) based on *dynamic programming* (Fig. 29.4).

First, let us split the summation in the denominator of Eq. (29.5) into $(y^{(1)},\ldots, y^{(m_i-1)})$ and $y^{(m_i)}$:

$$\sum_{y^{(1)},\ldots,y^{(m_i)}=1}^{c} \exp\left(\sum_{k=1}^{m_i} \zeta^\top \varphi(x_i^{(k)}, y^{(k)}, y^{(k-1)})\right) = \sum_{y^{(m_i)}=1}^{c} A_{m_i}(y^{(m_i)}), \tag{29.6}$$

where

$$A_\tau(y) = \sum_{y^{(1)},\ldots,y^{(\tau-1)}=1}^{c} \exp\left(\sum_{k=1}^{\tau-1} \zeta^\top \varphi(x_i^{(k)}, y^{(k)}, y^{(k-1)}) + \zeta^\top \varphi(x_i^{(\tau)}, y, y^{(\tau-1)})\right).$$

$A_\tau(y^{(\tau)})$ can be expressed recursively using $A_{\tau-1}(y^{(\tau-1)})$ as

$$A_\tau(y^{(\tau)}) = \sum_{y^{(1)},\dots,y^{(\tau-1)}=1}^{c} \exp\left(\sum_{k=1}^{\tau} \zeta^\top \varphi(x_i^{(k)}, y^{(k)}, y^{(k-1)})\right)$$
$$= \sum_{y^{(\tau-1)}=1}^{c} A_{\tau-1}(y^{(\tau-1)}) \exp\left(\zeta^\top \varphi(x_i^{(\tau)}, y^{(\tau)}, y^{(\tau-1)})\right).$$

The computational complexity of naively computing $A_{m_i}(y^{(m_i)})$ is $O(c^{m_i})$, while that of computing $A_{m_i}(y^{(m_i)})$ using the above recursive expression from $A_1(y^{(1)})$ is only $O(c^2 m_i)$. Note that the complexity for computing $\zeta^\top \varphi(x, y, y')$ is not included here.

Next, let us split the summation in the numerator of Eq. (29.5) into $(y^{(1)}, \dots, y^{(k'-2)})$, $(y^{(k'-1)}, y^{(k')})$, and $(y^{(k'+1)}, \dots, y^{(m_i)})$:

$$\sum_{y^{(1)},\dots,y^{(m_i)}=1}^{c} \exp\left(\sum_{k=1}^{m_i} \zeta^\top \varphi(x_i^{(k)}, y^{(k)}, y^{(k-1)})\right)\left(\sum_{k'=1}^{m_i} \varphi(x_i^{(k')}, y^{(k')}, y^{(k'-1)})\right)$$
$$= \sum_{k'=1}^{m_i} \sum_{y^{(1)},\dots,y^{(k'-2)}=1}^{c} \sum_{y^{(k'-1)},y^{(k')}=1}^{c} \sum_{y^{(k'+1)},\dots,y^{(m_i)}=1}^{c}$$
$$\exp\left(\sum_{k=1}^{m_i} \zeta^\top \varphi(x_i^{(k)}, y^{(k)}, y^{(k-1)})\right) \varphi(x_i^{(k')}, y^{(k')}, y^{(k'-1)})$$
$$= \sum_{k'=1}^{m_i} \sum_{y^{(k'-1)},y^{(k')}=1}^{c} \varphi(x_i^{(k')}, y^{(k')}, y^{(k'-1)})$$
$$\times \exp\left(\zeta^\top \varphi(x_i^{(k')}, y^{(k')}, y^{(k'-1)})\right) A_{k'-1}(y^{(k'-1)}) B_{k'}(y^{(k')}), \tag{29.7}$$

where

$$B_\tau(y) = \sum_{y^{(\tau+1)},\dots,y^{(m_i)}=1}^{c} \exp\left(\sum_{k=\tau+2}^{m_i} \zeta^\top \varphi(x_i^{(k)}, y^{(k)}, y^{(k-1)})\right.$$
$$\left. + \zeta^\top \varphi(x_i^{(\tau+1)}, y^{(\tau+1)}, y)\right).$$

$B_\tau(y^{(\tau)})$ can be expressed recursively using $B_{\tau+1}(y^{(\tau+1)})$ as

$$B_\tau(y^{(\tau)}) = \sum_{y^{(\tau+1)},\dots,y^{(m_i)}=1}^{c} \exp\left(\sum_{k=\tau+1}^{m_i} \zeta^\top \varphi(x_i^{(k)}, y^{(k)}, y^{(k-1)})\right)$$
$$= \sum_{y^{(\tau+1)}=1}^{c} B_{\tau+1}(y^{(\tau+1)}) \exp\left(\zeta^\top \varphi(x_i^{(\tau+1)}, y^{(\tau+1)}, y^{(\tau)})\right).$$

The computational complexity of naively computing $B_1(y^{(1)})$ is $O(c^{m_i})$, while that of computing $B_1(y^{(1)})$ using the above recursive expression from $B_{m_i}(y^{(m_i)})$ is only $O(c^2 m_i)$. Note that the complexity for computing $\zeta^\top \varphi(x, y, y')$ is not included here.

Summarizing the above computations, stochastic gradient ascent for the conditional random field can be performed efficiently by computing $\{A_k(y^{(k)})\}_{k=1}^{m_i}$ and $\{B_k(y^{(k)})\}_{k=1}^{m_i}$ in advance using the recursive expressions.

29.2.4 PREDICTION FOR NEW SAMPLE

Trained conditional random field $q(\overline{y}|\overline{x}; \widehat{\zeta})$ allows us to predict the most probable label sequence $\overline{y} = (y^{(1)}, \ldots, y^{(m)})$ for a test pattern sequence $\overline{x} = (x^{(1)}, \ldots, x^{(m)})$ as

$$\underset{y^{(1)},\ldots,y^{(m)} \in \{1,\ldots,c\}}{\operatorname{argmax}} \quad q(\overline{y}|\overline{x}; \widehat{\zeta}).$$

However, the computational complexity for naive maximization over $y^{(1)}, \ldots, y^{(m)}$ grows exponentially with respect to the sequence length m. Here, dynamic programming is used again to speed up the maximization.

Let us simplify the maximization problem as

$$\underset{y^{(1)},\ldots,y^{(m)} \in \{1,\ldots,c\}}{\operatorname{argmax}} \frac{\exp\left(\widehat{\zeta}^\top \varphi(\overline{x}, \overline{y})\right)}{\sum_{\overline{y}=1}^{\overline{c}} \exp\left(\widehat{\zeta}^\top \varphi(\overline{x}, \overline{y})\right)} = \underset{y^{(1)},\ldots,y^{(m)} \in \{1,\ldots,c\}}{\operatorname{argmax}} \quad \widehat{\zeta}^\top \varphi(\overline{x}, \overline{y})$$

and decompose the maximization in the right-hand side into $(y^{(1)}, \ldots, y^{(m-1)})$ and $y^{(m)}$:

$$\max_{y^{(1)},\ldots,y^{(m)} \in \{1,\ldots,c\}} \widehat{\zeta}^\top \varphi(\overline{x}, \overline{y}) = \max_{y^{(m)} \in \{1,\ldots,c\}} P_m(y^{(m)}),$$

where

$$P_\tau(y) = \max_{y^{(1)},\ldots,y^{(\tau-1)} \in \{1,\ldots,c\}} \left[\sum_{k=1}^{\tau-1} \widehat{\zeta}^\top \varphi(x^{(k)}, y^{(k)}, y^{(k-1)}) + \widehat{\zeta}^\top \varphi(x^{(\tau)}, y, y^{(\tau-1)}) \right].$$

$P_\tau(y^{(\tau)})$ can be expressed by recursively using $P_{\tau-1}(y^{(\tau-1)})$ as

$$\begin{aligned}
P_\tau(y^{(\tau)}) &= \max_{y^{(1)},\ldots,y^{(\tau-1)} \in \{1,\ldots,c\}} \left[\sum_{k=1}^{\tau} \widehat{\zeta}^\top \varphi(x^{(k)}, y^{(k)}, y^{(k-1)}) \right] \\
&= \max_{y^{(\tau-1)} \in \{1,\ldots,c\}} \left[P_{\tau-1}(y^{(\tau-1)}) + \widehat{\zeta}^\top \varphi(x^{(\tau)}, y^{(\tau)}, y^{(\tau-1)}) \right].
\end{aligned}$$

The computational complexity of naively computing $P_m(y^{(m)})$ is $O(c^m)$, while that of computing $P_m(y^{(m)})$ using the above recursive expression from $P_1(y^{(1)})$ is only $O(c^2 m)$. Note that the complexity for computing $\zeta^\top \varphi(x, y, y')$ is not included here.

29.3 DETERMINISTIC CLASSIFICATION FOR SEQUENCES

The conditional random field models the class-posterior probability, which corresponds to a probabilistic classification method. On the other hand, in deterministic classification, only the class label is learned. In this section, based on the formulation of *multiclass support vector classification* given in Section 27.5, a deterministic sequence classification method is introduced. More specifically,

$$f_{\boldsymbol{w}_y}(\boldsymbol{x}) = \boldsymbol{w}_y^\top \boldsymbol{x}$$

is used as a score function for class y, and a test sample $\boldsymbol{x}$ is classified into class $\widehat{y}$ as

$$\widehat{y} = \underset{y=1,\ldots,c}{\operatorname{argmax}}\, f_{\boldsymbol{w}_y}(\boldsymbol{x}).$$

The parameter $\boldsymbol{w}_y$ is learned as

$$\begin{aligned}&\min_{\boldsymbol{w}_1,\ldots,\boldsymbol{w}_c,\boldsymbol{\xi}} \left[\frac{1}{2}\sum_{y=1}^{c}\|\boldsymbol{w}_y\|^2 + C\sum_{i=1}^{n}\xi_i\right]\\ &\text{subject to } \boldsymbol{w}_{y_i}^\top \boldsymbol{x}_i - \boldsymbol{w}_y^\top \boldsymbol{x}_i \geq t_{i,y} - \xi_i, \quad \forall i = 1,\ldots,n,\ \forall y = 1,\ldots,c,\end{aligned}$$

where

$$t_{i,y} = \begin{cases} 0 & (y = y_i),\\ 1 & (y \neq y_i).\end{cases}$$

Note that, when $y = y_i$, the constraint $\boldsymbol{w}_{y_i}^\top \boldsymbol{x}_i - \boldsymbol{w}_y^\top \boldsymbol{x}_i \geq t_{i,y} - \xi_i$ is reduced to $\xi_i \geq 0$.

Applying the feature expression of the conditional random field, $\boldsymbol{\zeta}^\top \boldsymbol{\varphi}(\overline{\boldsymbol{x}}, \overline{y})$, to the above optimization yields

$$\begin{aligned}&\min_{\boldsymbol{\zeta},\boldsymbol{\xi}} \left[\frac{1}{2}\|\boldsymbol{\zeta}\|^2 + C\sum_{i=1}^{n}\xi_i\right]\\ &\text{subject to } \boldsymbol{\zeta}^\top \Delta\boldsymbol{\varphi}_i(\overline{y}) \geq t_{i,\overline{y}} - \xi_i,\\ &\qquad \forall i = 1,\ldots,n,\ \forall \overline{y} = \underbrace{(1,\ldots,1)}_{m},\ldots,\underbrace{(c,\ldots,c)}_{m},\end{aligned}$$

where

$$\Delta\boldsymbol{\varphi}_i(\overline{y}) = \boldsymbol{\varphi}_i(\overline{\boldsymbol{x}}_i, \overline{y}_i) - \boldsymbol{\varphi}_i(\overline{\boldsymbol{x}}_i, \overline{y}).$$

The *Lagrange dual* (Fig. 23.5) of the above optimization problem is given by

$$\begin{aligned}&\max_{\{\alpha_{i,\overline{y}}\}_{i,\overline{y}}} J(\{\alpha_{i,\overline{y}}\}_{i,\overline{y}})\\ &\text{subject to } \sum_{y^{(1)},\ldots,y^{(m)}=1}^{c} \alpha_{i,\overline{y}} = C, \quad \forall i = 1,\ldots,n,\\ &\qquad \alpha_{i,\overline{y}} \geq 0, \quad \forall i = 1,\ldots,n,\ \forall y^{(1)},\ldots,y^{(m)} = 1,\ldots,c,\end{aligned} \tag{29.8}$$

where

$$J(\{\alpha_{i,\overline{y}}\}_{i,\overline{y}}) = \sum_{i=1}^{n} \sum_{y^{(1)},\ldots,y^{(m)}=1}^{c} \alpha_{i,\overline{y}} t_{i,\overline{y}} - \frac{1}{2} \sum_{i,i'=1}^{n} \sum_{y^{(1)},\ldots,y^{(m)}=1}^{c} \sum_{y'^{(1)},\ldots,y'^{(m)}=1}^{c} \alpha_{i,\overline{y}} \alpha_{i',\overline{y}'} \Delta\varphi_i(\overline{y})^\top \Delta\varphi_{i'}(\overline{y}').$$

This is a quadratic programming problem and thus the solution $\{\widehat{\alpha}_{i,\overline{y}}\}_{i,\overline{y}}$ can be obtained by standard optimization software in principle. However, the number of optimization variables is exponential with respect to the sequence length, m, and thus it is not computationally tractable.

Here, let us utilize the same simplification trick as the conditional random field:

$$\varphi(\overline{x}, \overline{y}) = \sum_{k=1}^{m} \varphi(x^{(k)}, y^{(k)}, y^{(k-1)}),$$

which yields

$$\Delta\varphi_i(\overline{y}) = \sum_{k=1}^{m_i} \Delta\varphi_i(x_i^{(k)}, y^{(k)}, y^{(k-1)}),$$

where

$$\Delta\varphi_i(x_i^{(k)}, y^{(k)}, y^{(k-1)}) = \varphi_i(x_i^{(k)}, y_i^{(k)}, y_i^{(k-1)}) - \varphi_i(x_i^{(k)}, y^{(k)}, y^{(k-1)}).$$

Then the objective function in optimization problem (29.8) can be expressed as

$$J(\{\alpha_{i,\overline{y}}\}_{i,\overline{y}}) = \sum_{i=1}^{n} \sum_{k=1}^{m} \sum_{y^{(k)}=1}^{c} \mu_i(y^{(k)}) t_{i,\overline{y}} - \frac{1}{2} \sum_{i,i'=1}^{n} \sum_{k,k'=1}^{m} \sum_{y^{(k-1)},y^{(k)}=1}^{c} \sum_{y^{(k'-1)},y^{(k')}=1}^{c} \Delta\varphi_i(x_i^{(k)}, y^{(k)}, y^{(k-1)})^\top \times \Delta\varphi_{i'}(x_{i'}^{(k')}, y^{(k')}, y^{(k'-1)}) \mu_i(y^{(k-1)}, y^{(k)}) \mu_i(y^{(k'-1)}, y^{(k')}),$$

where

$$\mu_i(y^{(k)}) = \sum_{y^{(1)},\ldots,y^{(k-1)},y^{(k+1)},\ldots,y^{(m_i)}=1}^{c} \alpha_{i,\overline{y}},$$

$$\mu_i(y^{(k-1)}, y^{(k)}) = \sum_{y^{(1)},\ldots,y^{(k-2)},y^{(k+1)},\ldots,y^{(m_i)}=1}^{c} \alpha_{i,\overline{y}}.$$

Thus, exponentially many variables $\{\alpha_{i,\overline{y}}\}_{i,\overline{y}}$ do not have to be optimized, but only optimizing $\{\mu_i(y^{(k)})\}_{k=1}^{m_i}$ and $\{\mu_i(y^{(k-1)}, y^{(k)})\}_{k=1}^{m_i}$ for $i = 1, \ldots, n$ is sufficient.

However, since $\{\mu_i(y^{(k)})\}_{k=1}^{m_i}$ and $\{\mu_i(y^{(k-1)}, y^{(k)})\}_{k=1}^{m_i}$ are related to each other, the additional constraint,

$$\sum_{y^{(k-1)}=1}^{c} \mu_i(y^{(k-1)}, y^{(k)}) = \mu_i(y^{(k)}),$$

is necessary, together with the original constraints:

$$\sum_{y^{(k-1)}=1}^{c} \mu_i(y^{(k)}) = C \quad \text{and} \quad \mu_i(y^{(k-1)}, y^{(k)}) \geq 0.$$

From the dual solution $\{\widehat{\mu}_i(y^{(k)})\}_{k=1}^{m_i}$ and $\{\widehat{\mu}_i(y^{(k-1)}, y^{(k)})\}_{k=1}^{m_i}$, the primal solution $\widehat{\zeta}$ can be obtained as

$$\widehat{\zeta} = \sum_{i=1}^{n} \sum_{k=1}^{m_i} \sum_{y^{(k-1)}, y^{(k)}=1}^{c} \widehat{\mu}_i(y^{(k-1)}, y^{(k)}) \Delta\varphi_i(\boldsymbol{x}_i^{(k)}, y^{(k)}, y^{(k-1)}).$$

Once the solution $\widehat{\zeta}$ is obtained, classification of a test sample $\overline{\boldsymbol{x}} = (\boldsymbol{x}^{(1)}, \ldots, \boldsymbol{x}^{(m)})$,

$$\underset{y^{(1)}, \ldots, y^{(m)} \in \{1, \ldots, c\}}{\text{argmax}} \quad \widehat{\zeta}^\top \varphi(\overline{\boldsymbol{x}}, \overline{y}),$$

can be performed exactly in the same way as that described in Section 29.2.4.

This method is referred to as the *structured support vector machine* [114].

PART

FURTHER TOPICS

5

Part 5 is devoted to introducing various advanced topics in machine learning.

In Chapter 30, methods of *ensemble learning* are introduced, which are aimed at combining multiple *weak* learning algorithms to produce a *strong* learning algorithm. In Chapter 31, methods of *online learning* are introduced, which provide computationally efficient means to learn from training data given sequentially. In Chapter 32, methods to estimate the *confidence of prediction* are introduced.

Then various techniques for improving the performance of supervised learning based on *side information* are discussed. In Chapter 33, the framework of *semisupervised learning* is discussed, which tries to make use of additional input-only samples. This chapter also includes methods of *transfer learning*, which are aimed at utilizing training data of other related learning tasks. In Chapter 34, methods of *multitask learning* are introduced, which solve multiple related learning tasks simultaneously by sharing common information.

In Chapter 35, methods of *dimensionality reduction* are introduced for extracting useful low-dimensional feature representations, covering linear supervised and unsupervised methods. Then Chapter 36 focuses on non-linear dimensionality reduction methods.

Finally, various unsupervised learning methods are covered. In Chapter 37, methods of *clustering* are introduced, which are aimed at grouping data samples based on their similarity. In Chapter 38, methods of *outlier detection* are introduced, which try to identify anomalous samples in a given data set. In Chapter 39, methods of *change detection* between data sets are introduced.

CHAPTER

ENSEMBLE LEARNING 30

CHAPTER CONTENTS

Ensemble learning is a framework of combining multiple *weak* learning algorithms to produce a *strong* learning algorithm. In this chapter, two types of ensemble learning approaches, *bagging* and *boosting*, are introduced (Fig. 30.1). Although the idea of ensemble learning can be applied to both regression and classification, only classification is considered in this chapter.

30.1 DECISION STUMP CLASSIFIER

As an example of a weak learning algorithm, let us consider a *decision stump* classifier, which is a depth-one version of *decision trees*. More specifically, a decision stump classifier randomly chooses one of the elements in the d-dimensional input vector $\boldsymbol{x} = (x^{(1)}, \ldots, x^{(d)})^\top$ and classification is performed by thresholding the chosen element. This means that the decision boundary is parallel to one of the coordinate axes (Fig. 30.2).

The decision stump may be a poor classifier in terms of the classification accuracy because of its low degree of freedom. Nevertheless, at least, it has advantage in computation costs because there only exist $n + 1$ solutions for n training samples. Indeed, the global optimal solution can be easily obtained by just sorting the n training samples along the chosen axis and find the best interval that minimizes the classification error.

A MATLAB code for decision stump classification is provided in Fig. 30.3, and its behavior is illustrated in Fig. 30.4. This shows that, as expected, decision stump classification performs poorly.

In the rest of this chapter, ensemble learning methods for improving the performance of this poor decision stump classifier are introduced.

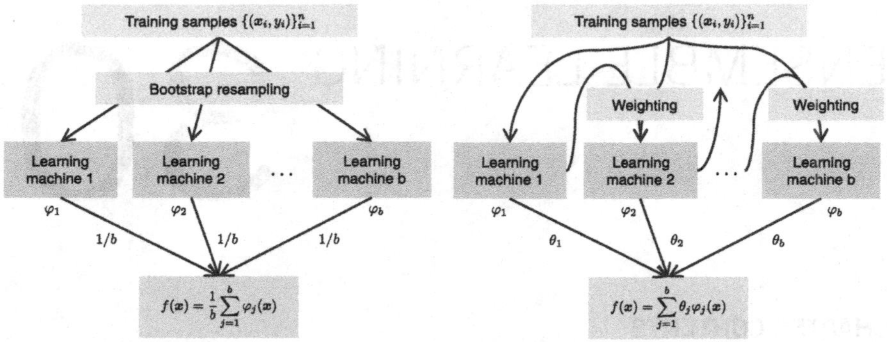

(a) Bagging (b) Boosting

FIGURE 30.1

Ensemble learning. Bagging trains weak learners in parallel, while boosting sequentially trains weak learners.

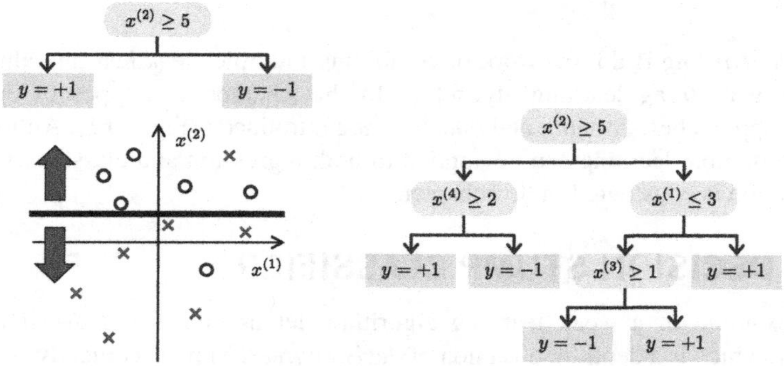

(a) Decision stump (b) Decision tree

FIGURE 30.2

Decision stump and decision tree classifiers. A decision stump is a depth-one version of a decision tree.

30.2 BAGGING

The name, *bagging*, stems from *bootstrap aggregation* [21]. The *bootstrap* method is a resampling technique to generate slightly different data sets from the original training data set (see Section 9.3.2), and bagging aggregates many classifiers trained with slightly different data sets (Fig. 30.1(a)). Since slightly different data sets give slightly different classifiers, averaging them is expected to give a more stable classifier. The algorithm of bagging is described in Fig. 30.5.

```
x=randn(50,2); y=2*(x(:,1)>x(:,2))-1;
X0=linspace(-3,3,50); [X(:,:,1) X(:,:,2)]=meshgrid(X0);

d=ceil(2*rand); [xs,xi]=sort(x(:,d));
el=cumsum(y(xi)); eu=cumsum(y(xi(end:-1:1)));
e=eu(end-1:-1:1)-el(1:end-1); [em,ei]=max(abs(e));
c=mean(xs(ei:ei+1)); s=sign(e(ei)); Y=sign(s*(X(:,:,d)-c));

figure(1); clf; hold on; axis([-3 3 -3 3]);
colormap([1 0.7 1; 0.7 1 1]); contourf(X0,X0,Y);
plot(x(y==1,1),x(y==1,2),'bo');
plot(x(y==-1,1),x(y==-1,2),'rx');
```

FIGURE 30.3

MATLAB code for decision stump classification.

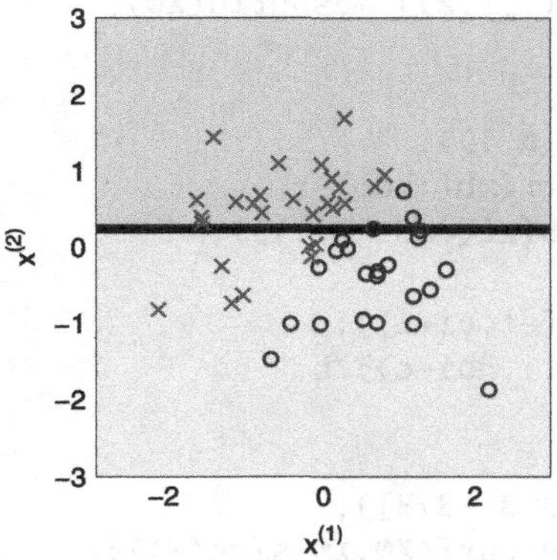

FIGURE 30.4

Example of decision stump classification.

A MATLAB code of bagging for decision stumps is provided in Fig. 30.6, and its behavior is illustrated in Fig. 30.7. This shows that ensemble learning by bagging improves the performance of a decision stump significantly. Note that each bootstrap training procedure is independent of each other, and thus bagging can be computed efficiently in parallel computing environments.

Bagging applied to decision trees is called the *random forest* [22], which is known to be a highly practical algorithm.

1. For $j = 1, \ldots, b$
   (a) Randomly choose n samples from $\{(\boldsymbol{x}_i, y_i)\}_{i=1}^n$ *with replacement* (see Fig. 3.3).
   (b) Train a classifier φ_j with the randomly resampled data set.
2. Output the average of $\{\varphi_j\}_{j=1}^b$ as the final solution f:

$$f(\boldsymbol{x}) \longleftarrow \frac{1}{b} \sum_{j=1}^{b} \varphi_j(\boldsymbol{x}).$$

FIGURE 30.5

Algorithm of bagging.

```
n=50; x=randn(n,2); y=2*(x(:,1)>x(:,2))-1;
b=5000; a=50; Y=zeros(a,a);
X0=linspace(-3,3,a); [X(:,:,1) X(:,:,2)]=meshgrid(X0);

for j=1:b
  db=ceil(2*rand); r=ceil(n*rand(n,1));
  xb=x(r,:); yb=y(r); [xs,xi]=sort(xb(:,db));
  el=cumsum(yb(xi)); eu=cumsum(yb(xi(end:-1:1)));
  e=eu(end-1:-1:1)-el(1:end-1);
  [em,ei]=max(abs(e)); c=mean(xs(ei:ei+1));
  s=sign(e(ei)); Y=Y+sign(s*(X(:,:,db)-c))/b;
end

figure(1); clf; hold on; axis([-3 3 -3 3]);
colormap([1 0.7 1; 0.7 1 1]); contourf(X0,X0,sign(Y));
plot(x(y==1,1),x(y==1,2),'bo');
plot(x(y==-1,1),x(y==-1,2),'rx');
```

FIGURE 30.6

MATLAB code of bagging for decision stumps.

30.3 BOOSTING

While bagging trained multiple weak learning machines in parallel, *boosting* [87] trains them in a sequential manner (Fig. 30.1(b)). In this section, a fundamental algorithm of boosting is introduced.

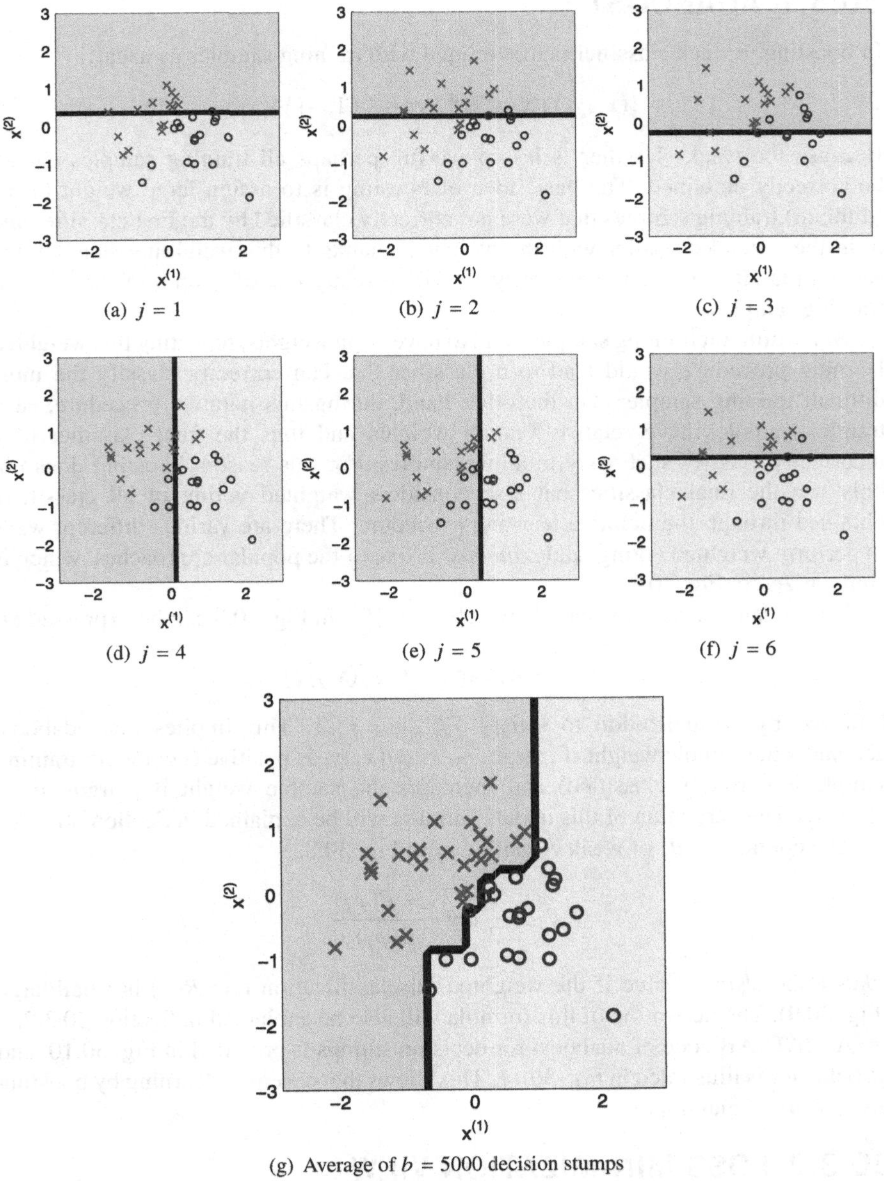

(a) $j = 1$ (b) $j = 2$ (c) $j = 3$

(d) $j = 4$ (e) $j = 5$ (f) $j = 6$

(g) Average of $b = 5000$ decision stumps

FIGURE 30.7

Example of bagging for decision stumps.

30.3.1 ADABOOST

In boosting, a weak classifier is first trained with training samples as usual:

$$\{(\boldsymbol{x}_i, y_i) \mid \boldsymbol{x}_i \in \mathbb{R}^d,\ y_i \in \{+1, -1\}\}_{i=1}^n.$$

Because the weak classifier is less powerful, perhaps all training samples cannot be correctly classified. The basic idea of boosting is to assign large weight to the (difficult) training samples that were not correctly classified by the first classifier and train the second classifier with the weights. Thanks to this weighting scheme, the second classifier trained in this way would correctly classify some of the difficult training samples.

Since difficult training samples tend to have large weights, repeating this weighted learning procedure would lead to a classifier that can correctly classify the most difficult training samples. On the other hand, during this iterative procedure, easy training samples have relatively small weights and thus the final classifier may incorrectly classify such easy training samples. For this reason, boosting does not only use the final classifier but also considers weighted voting of all classifiers obtained through the iterative learning procedure. There are various different ways to perform weighted voting, and *adaboost* is one of the popular approaches, which is summarized in Fig. 30.8.

The update formula for sample weights $\{w_i\}_{i=1}^n$ in Fig. 30.8 can be expressed as

$$w_i \longleftarrow w_i \exp\Big(- \theta_j \varphi_j(\boldsymbol{x}_i) y_i \Big),$$

followed by normalization to satisfy $\sum_{i=1}^n w_i = 1$. This implies that adaboost decreases the sample weight if margin $m_i = \varphi_j(\boldsymbol{x}_i) y_i$ is positive (i.e. the ith training sample is correctly classified) and increases the sample weight if margin m_i is negative. The derivation of this update formula will be explained in Section 30.3.2.

The confidence θ_j of weak classifier φ_j in Fig. 30.8,

$$\theta_j = \frac{1}{2} \log \frac{1 - R(\varphi_j)}{R(\varphi_j)},$$

takes a large/small value if the weighted misclassification rate $R(\varphi)$ is small/large (Fig. 30.9). The derivation of this formula will also be explained in Section 30.3.2.

A MATLAB code of adaboost for decision stumps is provided in Fig. 30.10, and its behavior is illustrated in Fig. 30.11. This shows that ensemble learning by boosting gives a strong classifier.

30.3.2 LOSS MINIMIZATION VIEW

The adaboost algorithm introduced above was derived as an ensemble learning method, which is quite different from the LS formulation explained in Chapter 26. However, adaboost can actually be interpreted as an extension of the LS method, and this interpretation allows us to derive, e.g. robust and probabilistic variations of adaboost.

1. Initialize sample weights $\{w_i\}_{i=1}^n$ for $\{(\boldsymbol{x}_i, y_i)\}_{i=1}^n$ to be uniform and the final strong classifier f to be zero:

$$w_1 = \cdots = w_n = 1/n \quad \text{and} \quad f \longleftarrow 0.$$

2. For $j = 1, \ldots, b$
   (a) Train a weak classifier φ_j with the current sample weights $\{w_i\}_{i=1}^n$ to minimize the weighted misclassification rate $R(\varphi)$:

$$\varphi_j = \underset{\varphi}{\operatorname{argmin}}\, R(\varphi), \quad \text{where} \quad R(\varphi) = \sum_{i=1}^n \frac{w_i}{2}\left(1 - \varphi(\boldsymbol{x}_i) y_i\right).$$

   (b) Set the confidence θ_j of weak classifier φ_j at

$$\theta_j = \frac{1}{2} \log \frac{1 - R(\varphi_j)}{R(\varphi_j)}.$$

   (c) Update the strong classifier f as

$$f \longleftarrow f + \theta_j \varphi_j.$$

   (d) Update the sample weights $\{w_i\}_{i=1}^n$ as

$$w_i \longleftarrow \frac{\exp\left(-f(\boldsymbol{x}_i) y_i\right)}{\sum_{i'=1}^n \exp\left(-f(\boldsymbol{x}_{i'}) y_{i'}\right)}, \quad \forall i = 1, \ldots, n.$$

FIGURE 30.8

Algorithm of adaboost.

As a surrogate loss to the 0/1-loss (see Section 26.2), let us consider the *exponential loss* (Fig. 30.12):

$$\exp(-m),$$

where $m = f_{\boldsymbol{\theta}}(\boldsymbol{x})y$ is the margin. For the linear-in-parameter model,

$$f_{\boldsymbol{\theta}}(\boldsymbol{x}) = \sum_{j=1}^b \theta_j \varphi_j(\boldsymbol{x}),$$

where $\{\varphi_j(\boldsymbol{x})\}_{j=1}^b$ are binary-valued basis functions that only takes either -1 or $+1$, let us consider exponential loss minimization:

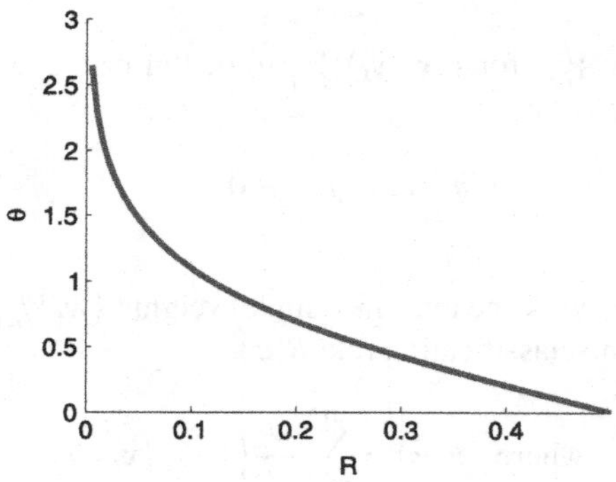

FIGURE 30.9

Confidence of classifier in adaboost. The confidence of classifier φ, denoted by θ, is determined based on the weighted misclassification rate R.

$$\min_{\theta} \sum_{i=1}^{n} \exp\Big(- f_{\theta}(\boldsymbol{x}_i) y_i \Big). \tag{30.1}$$

Suppose that not only the parameters $\{\theta_j\}_{j=1}^{b}$ but also the basis functions $\{\varphi_j\}_{j=1}^{b}$ can be learned in a sequential manner one by one. Let $\widetilde{f}$ be the learned function obtained so far and let θ and φ be the next parameter and basis function to be learned, respectively. Then Eq. (30.1) yields weighted exponential loss minimization:

$$\min_{\theta,\varphi} \sum_{i=1}^{n} \exp\Big(- \big\{\widetilde{f}(\boldsymbol{x}_i) + \theta\varphi(\boldsymbol{x}_i)\big\} y_i \Big) = \min_{\theta,\varphi} \sum_{i=1}^{n} \widetilde{w}_i \exp\Big(- \theta\varphi(\boldsymbol{x}_i) y_i \Big),$$

where the weight $\widetilde{w}_i$ is given by

$$\widetilde{w}_i = \exp\Big(- \widetilde{f}(\boldsymbol{x}_i) y_i \Big).$$

For the sake of simplicity, $\theta \geq 0$ is assumed below (when $\theta < 0$, the sign of φ is flipped to satisfy $\theta \geq 0$). Then the above sequential weighted exponential loss minimization can be rewritten as

$$\begin{aligned}
&\sum_{i=1}^{n} \widetilde{w}_i \exp\Big(- \theta\varphi(\boldsymbol{x}_i) y_i \Big) \\
&= \exp(-\theta) \sum_{i: y_i = \varphi(\boldsymbol{x}_i)} \widetilde{w}_i + \exp(\theta) \sum_{i: y_i \neq \varphi(\boldsymbol{x}_i)} \widetilde{w}_i \\
&= \Big\{\exp(\theta) - \exp(-\theta)\Big\} \sum_{i=1}^{n} \frac{\widetilde{w}_i}{2} \Big(1 - \varphi(\boldsymbol{x}_i) y_i \Big) + \exp(-\theta) \sum_{i=1}^{n} \widetilde{w}_i.
\end{aligned} \tag{30.2}$$

```
n=50; x=randn(n,2); y=2*(x(:,1)>x(:,2))-1; b=5000;
a=50; Y=zeros(a,a); yy=zeros(size(y)); w=ones(n,1)/n;
X0=linspace(-3,3,a); [X(:,:,1) X(:,:,2)]=meshgrid(X0);

for j=1:b
  wy=w.*y; d=ceil(2*rand); [xs,xi]=sort(x(:,d));
  el=cumsum(wy(xi)); eu=cumsum(wy(xi(end:-1:1)));
  e=eu(end-1:-1:1)-el(1:end-1);
  [em,ei]=max(abs(e)); c=mean(xs(ei:ei+1)); s=sign(e(ei));
  yh=sign(s*(x(:,d)-c)); R=w'*(1-yh.*y)/2;
  t=log((1-R)/R)/2; yy=yy+yh*t; w=exp(-yy.*y); w=w/sum(w);
  Y=Y+sign(s*(X(:,:,d)-c))*t;
end

figure(1); clf; hold on; axis([-3 3 -3 3]);
colormap([1 0.7 1; 0.7 1 1]); contourf(X0,X0,sign(Y));
plot(x(y==1,1),x(y==1,2),'bo');
plot(x(y==-1,1),x(y==-1,2),'rx');
```

FIGURE 30.10

MATLAB code of adaboost for decision stumps.

Thus, the minimizer $\widehat{\varphi}$ of Eq. (30.2) with respect to φ is given by weighted 0/1-loss minimization:

$$\widehat{\varphi} = \underset{\varphi}{\text{argmin}} \sum_{i=1}^{n} \frac{\widetilde{w}_i}{2} \Big(1 - \varphi(\boldsymbol{x}_i) y_i\Big).$$

The minimizer $\widehat{\theta}$ of Eq. (30.2) with respect to θ can be obtained by substituting $\widehat{\varphi}$ to φ in Eq. (30.2), differentiating it with respect to θ, and setting it at 0:

$$\Big\{\exp(\theta) + \exp(-\theta)\Big\} \sum_{i=1}^{n} \frac{\widetilde{w}_i}{2} \Big(1 - \widehat{\varphi}(\boldsymbol{x}_i) y_i\Big) - \exp(-\theta) \sum_{i=1}^{n} \widetilde{w}_i = 0.$$

Solving this for θ yields

$$\widehat{\theta} = \frac{1}{2} \log \frac{1 - \widehat{R}}{\widehat{R}},$$

where $\widehat{R}$ denotes the weighted 0/1-loss for $\widehat{\varphi}$:

$$\widehat{R} = \left\{\sum_{i=1}^{n} \frac{\widetilde{w}_i}{2} \Big(1 - \widehat{\varphi}(\boldsymbol{x}_i) y_i\Big)\right\} \Bigg/ \left\{\sum_{i'=1}^{n} \widetilde{w}_{i'}\right\}.$$

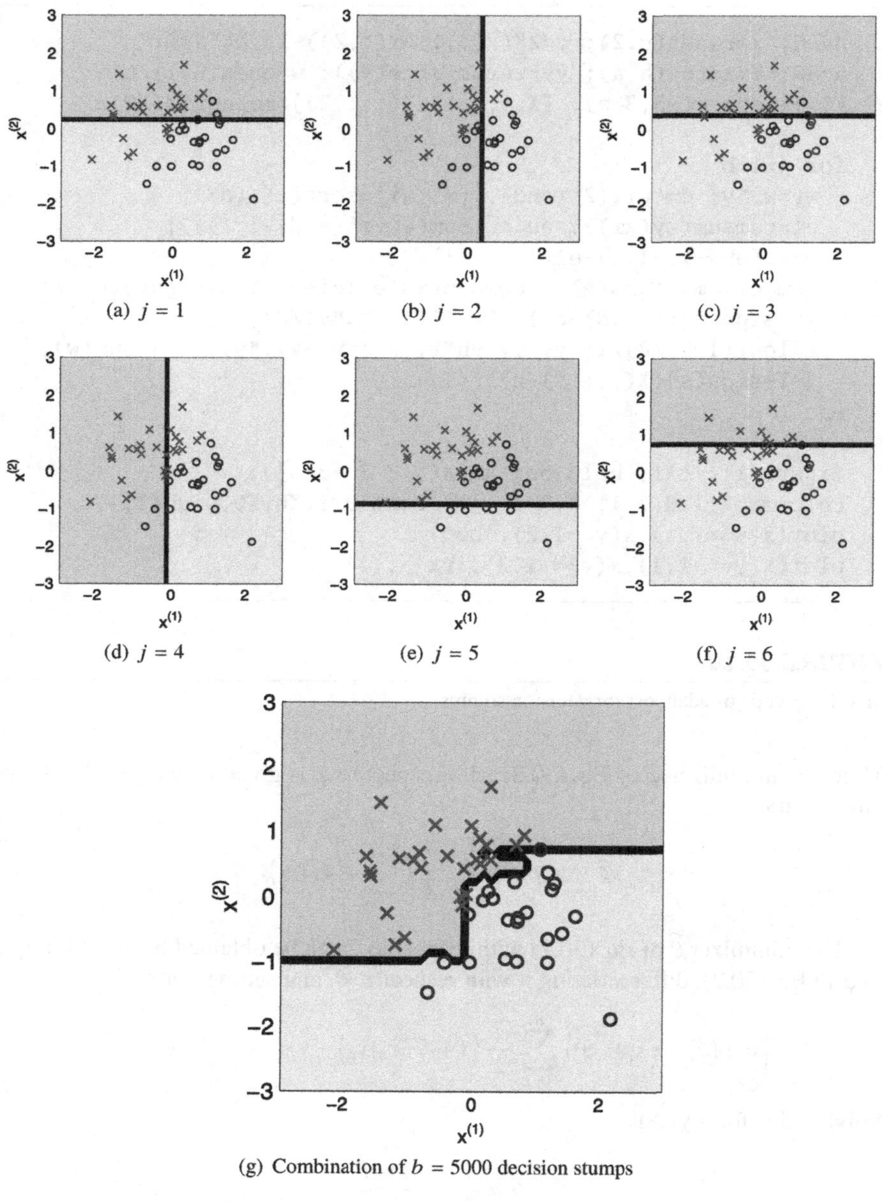

FIGURE 30.11

Example of adaboost for decision stumps.

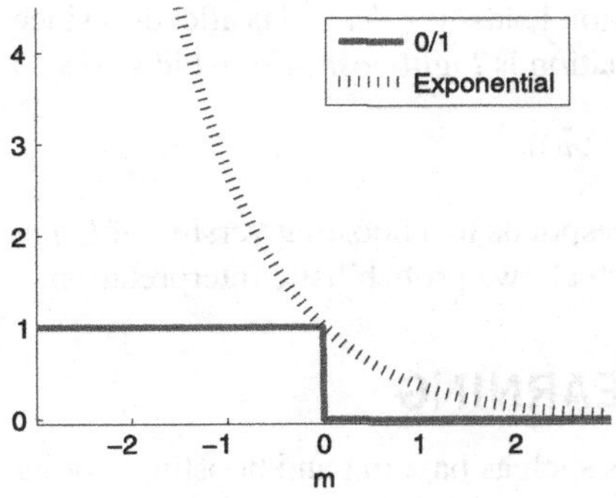

FIGURE 30.12

Exponential loss.

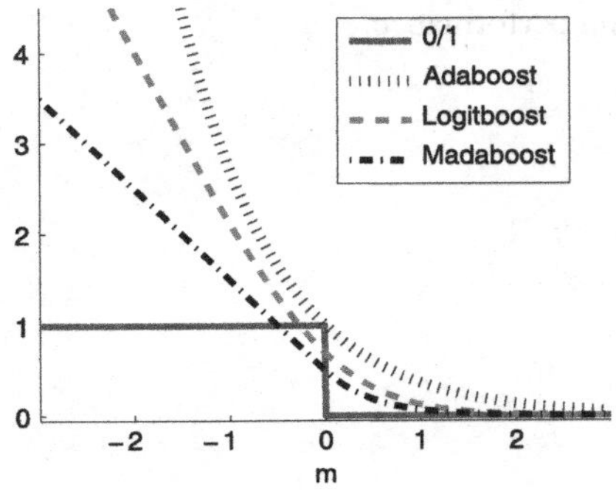

FIGURE 30.13

Loss functions for boosting.

Since $\widehat{\varphi}$ and $\widehat{\theta}$ obtained above are equivalent to the one used in adaboost (see Fig. 30.8), adaboost was shown to be equivalent to sequential weighted exponential loss minimization.

The above equivalence allows us to consider variations of adaboost for other loss functions. For example, a modification of adaboost, called *madaboost* [37], uses

$$\begin{cases} -m + 1/2 & (m \leq 0) \\ \exp(-2m)/2 & (m > 0) \end{cases}$$

as a loss function (Fig. 30.13). This loss function leads to *robust* classification since it increases only linearly. Another popular variation is *logitboost* [43], which uses

$$\log(1 + \exp(-2m))$$

as a loss function (Fig. 30.13). Logitboost corresponds to a boosting version of *logistic regression* introduced in Section 28.1, which allows probabilistic interpretation.

30.4 GENERAL ENSEMBLE LEARNING

Beyond systematic ensemble learning methods such as bagging and boosting, voting by various different learning algorithms is highly promising in practical applications. For example, the winning algorithms in real-world data analysis competitions such as *Netflix Prize*[1] and *KDD Cup 2013*[2] adopt such voting approaches. If additional validation data samples are available, voting weights may be optimized so that the prediction error for the validation samples is minimized.

Thus, beyond improving each learning algorithm, ensemble learning can be regarded as a final means to boost the prediction performance.

[1] http://www.netflixprize.com/.

[2] http://www.kdd.org/kddcup2013/content/kdd-cup-2013-workshop-0.

CHAPTER

ONLINE LEARNING

31

CHAPTER CONTENTS

The supervised learning methods explained so far handled all training samples $\{(\boldsymbol{x}_i, y_i)\}_{i=1}^n$ at the same time, which is called *batch learning*. On the other hand, when training samples are provided one by one in a sequential manner, it would be effective to perform *online learning*, i.e. a new training sample is incrementally learned upon the current learned result. Online learning is also useful when the number of training samples is so large that all training samples cannot be stored in memory. In this chapter, various online learning algorithms are introduced.

For simplicity, only the *linear-in-input model* is considered in this chapter:

$$f_{\boldsymbol{\theta}}(\boldsymbol{x}) = \boldsymbol{\theta}^\top \boldsymbol{x}.$$

However, all methods introduced in this section can be easily extended to $f_{\boldsymbol{\theta}}(\boldsymbol{x}) = \boldsymbol{\theta}^\top \boldsymbol{\phi}(\boldsymbol{x})$, *linear-in-parameter models* (see Section 21.1) with basis function $\boldsymbol{\phi}(\boldsymbol{x})$.

31.1 STOCHASTIC GRADIENT DESCENT

The *stochastic gradient* algorithm introduced in Section 22.4 is one of the typical online learning algorithms, which updates the parameter to reduce the loss for a new training sample.

More specifically, for the new training sample $(\boldsymbol{x}, y)$ and a loss function J, the parameter $\boldsymbol{\theta}$ is updated along the gradient of the loss ∇J as

$$\boldsymbol{\theta} \longleftarrow \boldsymbol{\theta} - \varepsilon \nabla J(\boldsymbol{\theta}),$$

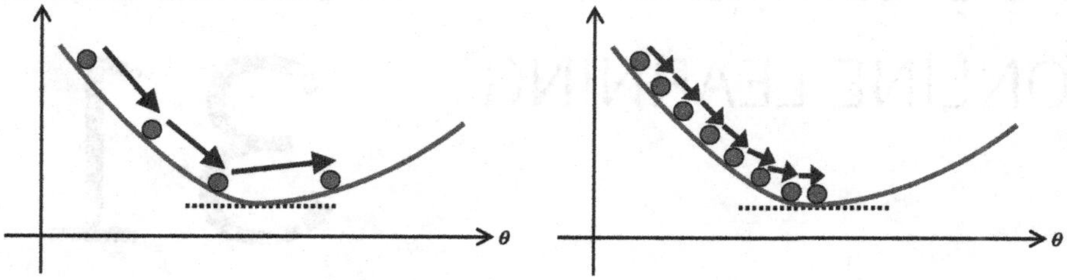

(a) Too large step size

(b) Too small step size

FIGURE 31.1

Choice of step size. Too large step size overshoots the optimal solution, while too small step size yields slow convergence.

where $\varepsilon > 0$ denotes the step size. See Fig. 22.6 for the detailed algorithm of the stochastic gradient method.

Although stochastic gradient is simple and easy to use, choice of step size ε is often cumbersome in practice, as illustrated in Fig. 31.1: too large step size overshoots the optimal solution, while too small step size yields slow convergence. Starting from a large ε and then reducing ε gradually, called *simulated annealing*, would be useful to control the step size. However, choice of initial ε and the decreasing factor of ε is not straightforward in practice.

For simple loss functions such as the squared loss,

$$J(\boldsymbol{\theta}) = \frac{1}{2}\left(y - \boldsymbol{\theta}^\top \boldsymbol{x}\right)^2,$$

it is actually possible to find the minimizer with respect to $\boldsymbol{\theta}$ analytically. Then step size selection no longer matters, which is highly appealing in practice. However, this often changes the parameter $\boldsymbol{\theta}$ drastically, implying that the knowledge acquired so far can be completely spoiled by addition of a single, possibly noisy training sample. Thus, in practice, it is important to choose the step size to be not too large.

31.2 PASSIVE-AGGRESSIVE LEARNING

In this section, a simple online learning method called *passive-aggressive learning* [32] is introduced, which better controls the step size.

More specifically, in passive-aggressive learning, the amount of change from the current solution $\widetilde{\boldsymbol{\theta}}$ is included as a penalty term:

$$\widehat{\boldsymbol{\theta}} = \underset{\boldsymbol{\theta}}{\operatorname{argmin}} \left[J(\boldsymbol{\theta}) + \frac{\lambda}{2}\|\boldsymbol{\theta} - \widetilde{\boldsymbol{\theta}}\|^2 \right], \tag{31.1}$$

where $\lambda > 0$ is the *passiveness* parameter. Minimizing the first term in Eq. (31.1) corresponds to aggressively updating the parameter, while minimizing the second term in Eq. (31.1) corresponds to passively keeping the previous solution. Below, passive-aggressive algorithms for classification and regression are introduced.

31.2.1 CLASSIFICATION

For classification, let us employ the *squared hinge loss* (see Fig. 27.12):

$$J(\boldsymbol{\theta}) = \frac{1}{2}\left(\max\left\{0, 1 - m\right\}\right)^2,$$

where $m = \boldsymbol{\theta}^\top \boldsymbol{x} y$ denotes the *margin*. Then the solution of passive-aggressive classification can be obtained analytically as follows.

First, the squared hinge loss can be expressed as

$$J(\boldsymbol{\theta}) = \min_{\xi} \frac{1}{2}\xi^2 \quad \text{subject to } \xi \geq 1 - m \text{ and } \xi \geq 0.$$

Note that $\xi \geq 0$ is actually automatically fulfilled because the objective function only contains ξ^2 and $\xi = 0$ is the solution when $1 - m \leq 0$. Then the objective function in Eq. (31.1) can be written as

$$\widehat{\boldsymbol{\theta}} = \underset{\boldsymbol{\theta}, \xi}{\operatorname{argmin}} \left[\frac{1}{2}\xi^2 + \frac{\lambda}{2}\|\boldsymbol{\theta} - \widetilde{\boldsymbol{\theta}}\|^2\right] \quad \text{subject to } \xi \geq 1 - m. \tag{31.2}$$

When $1 - m \leq 0$, $\xi = 0$ is the solution and thus $\widehat{\boldsymbol{\theta}} = \widetilde{\boldsymbol{\theta}}$ is obtained. This means that the new solution $\widehat{\boldsymbol{\theta}}$ is the same as the current solution $\widetilde{\boldsymbol{\theta}}$.

When $1 - m > 0$, let us introduce the *Lagrange multiplier* α to constrained optimization problem (31.2) and define the *Lagrange function* (see Fig. 23.5) as follows:

$$L(\boldsymbol{\theta}, \xi, \alpha) = \frac{1}{2}\xi^2 + \frac{\lambda}{2}\|\boldsymbol{\theta} - \widetilde{\boldsymbol{\theta}}\|^2 + \alpha(1 - m - \xi). \tag{31.3}$$

Then the KKT conditions (Fig. 27.7) yield

$$\frac{\partial L}{\partial \boldsymbol{\theta}} = \boldsymbol{0} \implies \boldsymbol{\theta} = \widetilde{\boldsymbol{\theta}} + \frac{\alpha y}{\lambda}\boldsymbol{x}, \tag{31.4}$$

$$\frac{\partial L}{\partial \xi} = 0 \implies \xi = \alpha.$$

Substituting these conditions into Lagrange function (31.3) and eliminating $\boldsymbol{\theta}$ and ξ yield

$$L(\alpha) = -\frac{\alpha^2}{2}\left(\frac{\|\boldsymbol{x}\|^2}{\lambda} + 1\right) + \alpha\left(1 - \widetilde{\boldsymbol{\theta}}^\top \boldsymbol{x} y\right).$$

Taking the derivative of $L(\alpha)$ and setting it at zero give the maximizer $\widehat{\alpha}$ analytically as

$$\widehat{\alpha} = \frac{1 - \widetilde{\boldsymbol{\theta}}^\top \boldsymbol{x} y}{\|\boldsymbol{x}\|^2/\lambda + 1}.$$

1. Initialize $\boldsymbol{\theta} \longleftarrow \mathbf{0}$.
2. Update the parameter $\boldsymbol{\theta}$ using a new training sample $(\boldsymbol{x}, y)$ as

$$\boldsymbol{\theta} \longleftarrow \boldsymbol{\theta} + \frac{y \max(0, 1 - \boldsymbol{\theta}^\top \boldsymbol{x} y)}{\|\boldsymbol{x}\|^2 + \lambda} \boldsymbol{x}.$$

3. Go to Step 2.

FIGURE 31.2

Algorithm of passive-aggressive classification.

Substituting this back into Eq. (31.4) gives

$$\widehat{\boldsymbol{\theta}} = \widetilde{\boldsymbol{\theta}} + \frac{(1 - \widetilde{\boldsymbol{\theta}}^\top \boldsymbol{x} y) y}{\|\boldsymbol{x}\|^2 + \lambda} \boldsymbol{x}.$$

Altogether, the final solution $\widehat{\boldsymbol{\theta}}$ is expressed as

$$\widehat{\boldsymbol{\theta}} = \widetilde{\boldsymbol{\theta}} + \frac{y \max(0, 1 - \widetilde{\boldsymbol{\theta}}^\top \boldsymbol{x} y)}{\|\boldsymbol{x}\|^2 + \lambda} \boldsymbol{x}.$$

The algorithm of passive-aggressive classification is summarized in Fig. 31.2.

A MATLAB code for passive-aggressive classification is provided in Fig. 31.3, and its behavior is illustrated in Fig. 31.4. This shows that, after three iterations, almost the same solution as the final one can be obtained that well separates positive and negative samples. In this implementation, the intercept of the linear model is expressed by augmenting input vector $\boldsymbol{x} = (x^{(1)}, \ldots, x^{(d)})$ as $(\boldsymbol{x}^\top, 1)^\top$:

$$f_{\boldsymbol{\theta}}(\boldsymbol{x}) = \boldsymbol{\theta}^\top (\boldsymbol{x}^\top, 1)^\top = \sum_{j=1}^{d} \theta_j x^{(j)} + \theta_{d+1}.$$

Note that, for the ordinary hinge loss (Fig. 27.12(a)),

$$J(\boldsymbol{\theta}) = \max\left\{0, 1 - m\right\},$$

the passive-aggressive algorithm still gives an analytic update formula:

$$\boldsymbol{\theta} \longleftarrow \boldsymbol{\theta} + y \min\left\{\frac{1}{\lambda}, \frac{\max(0, 1 - m)}{\|\boldsymbol{x}\|^2}\right\} \boldsymbol{x}.$$

31.2.2 REGRESSION

The idea of passive-aggressive learning can also be applied to regression. For residual $r = y - \boldsymbol{\theta}^\top \boldsymbol{x}$, let us employ the ℓ_2-loss and the ℓ_1-loss (Fig. 25.2):

$$J(\boldsymbol{\theta}) = \frac{1}{2} r^2 \quad \text{and} \quad J(\boldsymbol{\theta}) = |r|.$$

```
n=200; x=[randn(1,n/2)-5 randn(1,n/2)+5; 5*randn(1,n)]';
y=[ones(n/2,1);-ones(n/2,1)];
x(:,3)=1; p=randperm(n); x=x(p,:); y=y(p);
t=zeros(3,1); l=1;
for i=1:n
  xi=x(i,:)'; yi=y(i);
  t=t+yi*max(0,1-t'*xi*yi)/(xi'*xi+l)*xi;
end
figure(1); clf; hold on; axis([-10 10 -10 10]);
plot(x(y==1,1),x(y==1,2),'bo');
plot(x(y==-1,1),x(y==-1,2),'rx');
plot([-10 10],-(t(3)+[-10 10]*t(1))/t(2),'k-');
```

FIGURE 31.3

MATLAB code for passive-aggressive classification.

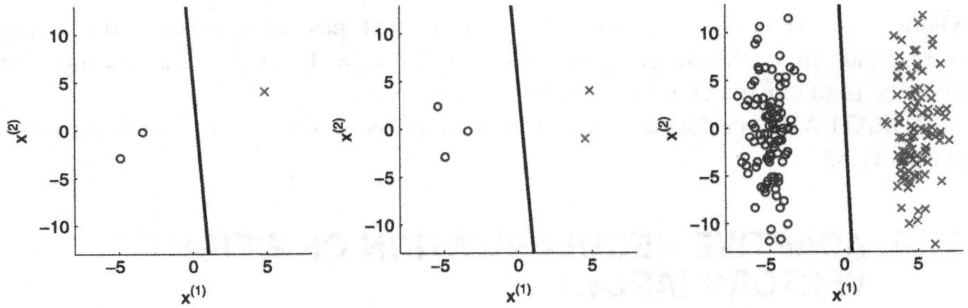

(a) After 3 iterations (b) After 5 iterations (c) After 200 iterations

FIGURE 31.4

Example of passive-aggressive classification.

Then, with a similar derivation to the classification case, the update formulas for passive-aggressive regression can be obtained analytically as

$$\boldsymbol{\theta} \longleftarrow \boldsymbol{\theta} + \frac{r}{\|\boldsymbol{x}\|^2 + \lambda}\boldsymbol{x} \quad \text{and} \quad \boldsymbol{\theta} \longleftarrow \boldsymbol{\theta} + \text{sign}\,(r) \min\left\{\frac{1}{\lambda}, \frac{|r|}{\|\boldsymbol{x}\|^2}\right\}\boldsymbol{x},$$

where sign (r) denotes the sign of r. As shown in Section 22.4, the stochastic gradient update rule for the ℓ_2-loss is given by

$$\boldsymbol{\theta} \longleftarrow \boldsymbol{\theta} + \varepsilon r \boldsymbol{x},$$

```
n=50; N=1000; x=linspace(-3,3,n)'; x=x(randperm(n));
pix=pi*x; y=sin(pix)./(pix)+0.1*x+0.05*randn(n,1);
hh=2*0.3^2; t=randn(n,1); l=1;
for i=1:n
  ki=exp(-(x-x(i)).^2/hh);
  t=t+(y(i)-t'*ki)/(ki'*ki+l)*ki;
end
X=linspace(-3,3,N)';
K=exp(-(repmat(X.^2,1,n)+repmat(x.^2',N,1)-2*X*x')/hh);
F=K*t;
figure(1); clf; hold on; axis([-2.8 2.8 -0.5 1.2]);
plot(X,F,'g-'); plot(x,y,'bo');
```

FIGURE 31.5

MATLAB code for passive-aggressive regression with the ℓ_2-loss.

where $\varepsilon > 0$ is the step size. This means that passive-aggressive regression corresponds to stochastic gradient with step size $\varepsilon = 1/(\|\boldsymbol{x}\|^2 + \lambda)$, and thus the step size is adaptively chosen based on the sample $\boldsymbol{x}$.

A MATLAB code for passive-aggressive regression with the ℓ_2-loss is provided in Fig. 31.5.

31.3 ADAPTIVE REGULARIZATION OF WEIGHT VECTORS (AROW)

Since the passive-aggressive algorithm uses unbounded loss functions, it suffers a lot from outliers. This problem can be mitigated if a bounded loss function, such as the *ramp loss* (see Section 27.6.3) and the *Tukey loss* (see Section 25.4), is used. However, bounded loss functions are nonconvex and thus optimization becomes cumbersome in practice. In this section, a robust online learning algorithm called AROW [33] is introduced, which utilizes the sequential nature of online learning.

31.3.1 UNCERTAINTY OF PARAMETERS

In AROW, parameter $\boldsymbol{\theta}$ is not point-estimated, but its distribution is learned to take into account its uncertainty.

As a distribution of parameters, let us consider the Gaussian distribution whose probability density function is given as follows (Fig. 6.1):

$$(2\pi)^{-d/2}\det(\boldsymbol{\Sigma})^{-1/2}\exp\left(-\frac{1}{2}(\boldsymbol{\theta}-\boldsymbol{\mu})^\top\boldsymbol{\Sigma}^{-1}(\boldsymbol{\theta}-\boldsymbol{\mu})\right),$$

where det(·) denotes the determinant, $\boldsymbol{\mu}$ denotes the expectation vector, and $\boldsymbol{\Sigma}$ denotes the variance-covariance matrix. Below, the Gaussian distribution with expectation vector $\boldsymbol{\mu}$ and variance-covariance matrix $\boldsymbol{\Sigma}$ is denoted by $N(\boldsymbol{\mu}, \boldsymbol{\Sigma})$.

AROW learns the expectation vector $\boldsymbol{\mu}$ and variance-covariance matrix $\boldsymbol{\Sigma}$ to minimize the following criterion:

$$J(\boldsymbol{\mu}) + \frac{1}{2}\boldsymbol{x}^\top \boldsymbol{\Sigma}\boldsymbol{x} + \lambda \mathrm{KL}\left(N(\boldsymbol{\mu}, \boldsymbol{\Sigma}) \middle\| N(\widetilde{\boldsymbol{\mu}}, \widetilde{\boldsymbol{\Sigma}})\right). \tag{31.5}$$

The first term $J(\boldsymbol{\mu})$ denotes the loss for a new training sample $(\boldsymbol{x}, y)$ when parameter $\boldsymbol{\theta} = \boldsymbol{\mu}$ is used for prediction. The second term $\frac{1}{2}\boldsymbol{x}^\top \boldsymbol{\Sigma}\boldsymbol{x}$ is the regularization term for variance-covariance matrix $\boldsymbol{\Sigma}$, which is adaptive to the training input vector $\boldsymbol{x}$. The third term $\lambda \mathrm{KL}(N(\boldsymbol{\mu}, \boldsymbol{\Sigma}) \| N(\widetilde{\boldsymbol{\mu}}, \widetilde{\boldsymbol{\Sigma}}))$ controls the amount of change from the current solution, where $\lambda > 0$ denotes the *passiveness* parameter, $\widetilde{\boldsymbol{\mu}}$ and $\widetilde{\boldsymbol{\Sigma}}$ are the current solutions for $\boldsymbol{\mu}$ and $\boldsymbol{\Sigma}$, and $\mathrm{KL}(p\|q)$ denotes the *KL divergence* (see Section 14.2) from density p to density q:

$$\mathrm{KL}(p\|q) = \int p(\boldsymbol{x}) \log \frac{p(\boldsymbol{x})}{q(\boldsymbol{x})} \mathrm{d}\boldsymbol{x}.$$

Passive-aggressive learning introduced in Section 31.2 used the Euclidean distance $\|\boldsymbol{\theta} - \widetilde{\boldsymbol{\theta}}\|^2$ for controlling the amount of change from the current solution $\widetilde{\boldsymbol{\theta}}$. On the other hand, AROW uses the KL divergence for taking into account the uncertainty of parameters, which is expected to contribute to better controlling the amount of change from the current solution.

The KL divergence for Gaussian distributions can be expressed *analytically* as

$$\begin{aligned}&\mathrm{KL}\left(N(\boldsymbol{\mu}, \boldsymbol{\Sigma}) \middle\| N(\widetilde{\boldsymbol{\mu}}, \widetilde{\boldsymbol{\Sigma}})\right)\\ &= \frac{1}{2}\left(\log \frac{\det(\widetilde{\boldsymbol{\Sigma}})}{\det(\boldsymbol{\Sigma})} + \mathrm{tr}(\widetilde{\boldsymbol{\Sigma}}^{-1}\boldsymbol{\Sigma}) + (\boldsymbol{\mu} - \widetilde{\boldsymbol{\mu}})^\top \widetilde{\boldsymbol{\Sigma}}^{-1}(\boldsymbol{\mu} - \widetilde{\boldsymbol{\mu}}) - d\right),\end{aligned}$$

where d denotes the dimensionality of input vector $\boldsymbol{x}$.

Below, specific AROW algorithms for classification and regression are introduced.

31.3.2 CLASSIFICATION

Let us use the squared hinge loss (see Fig. 27.12) for expectation vector $\boldsymbol{\mu}$:

$$J(\boldsymbol{\mu}) = \frac{1}{2}\left(\max\left\{0, 1 - \boldsymbol{\mu}^\top \boldsymbol{x} y\right\}\right)^2.$$

Then setting the derivative of Eq. (31.5) with respect to $\boldsymbol{\mu}$ at zero yields that the solution $\widehat{\boldsymbol{\mu}}$ satisfies

$$\widehat{\boldsymbol{\mu}} = \widetilde{\boldsymbol{\mu}} + y \max(0, 1 - \widetilde{\boldsymbol{\mu}}^\top \boldsymbol{x} y)\widetilde{\boldsymbol{\Sigma}}\boldsymbol{x}/\beta,$$

where $\beta = \boldsymbol{x}^\top \widetilde{\boldsymbol{\Sigma}}\boldsymbol{x} + \lambda$.

1. Initialize $\boldsymbol{\mu} \longleftarrow \mathbf{0}$ and $\boldsymbol{\Sigma} \longleftarrow \boldsymbol{I}$.
2. Update parameters $\boldsymbol{\mu}$ and $\boldsymbol{\Sigma}$ as follows, if the margin $m = \boldsymbol{\mu}^\top \boldsymbol{x} y$ for a new training sample $(\boldsymbol{x}, y)$ satisfies $m < 1$:

$$\boldsymbol{\mu} \longleftarrow \boldsymbol{\mu} + y \max(0, 1 - m)\boldsymbol{\Sigma}\boldsymbol{x}/\beta \quad \text{and} \quad \boldsymbol{\Sigma} \longleftarrow \boldsymbol{\Sigma} - \boldsymbol{\Sigma}\boldsymbol{x}\boldsymbol{x}^\top\boldsymbol{\Sigma}/\beta,$$

   where $\beta = \boldsymbol{x}^\top\boldsymbol{\Sigma}\boldsymbol{x} + \lambda$.
3. Go to Step 2.

FIGURE 31.6

Algorithm of AROW classification.

The matrix derivative formulas (see Fig. 12.3),

$$\frac{\partial}{\partial \boldsymbol{\Sigma}} \log(\det(\boldsymbol{\Sigma})) = \boldsymbol{\Sigma}^{-1} \quad \text{and} \quad \frac{\partial}{\partial \boldsymbol{\Sigma}} \mathrm{tr}\left(\widetilde{\boldsymbol{\Sigma}}^{-1}\boldsymbol{\Sigma}\right) = \widetilde{\boldsymbol{\Sigma}}^{-1},$$

allow us to compute the partial derivative of Eq. (31.5) with respect to $\boldsymbol{\Sigma}$, and setting it at zero gives the solution $\widehat{\boldsymbol{\Sigma}}$ analytically as

$$\widehat{\boldsymbol{\Sigma}}^{-1} = \widetilde{\boldsymbol{\Sigma}}^{-1} - \boldsymbol{x}\boldsymbol{x}^\top/\lambda.$$

Further applying the *matrix inversion lemma* (see Fig. 23.12) yields

$$\widehat{\boldsymbol{\Sigma}} = \widetilde{\boldsymbol{\Sigma}} - \widetilde{\boldsymbol{\Sigma}}\boldsymbol{x}\boldsymbol{x}^\top\widetilde{\boldsymbol{\Sigma}}/\beta.$$

This expression allows us to directly obtain the solution $\widehat{\boldsymbol{\Sigma}}$ without computing its inverse explicitly.

The algorithm of AROW classification is summarized in Fig. 31.6. When the dimensionality d of the input vector $\boldsymbol{x}$ is large, updating the $d \times d$ variance-covariance matrix $\boldsymbol{\Sigma}$ is computationally expensive. A practical approach to mitigating this problem is to only maintain the diagonal elements of $\boldsymbol{\Sigma}$ and regard all off-diagonal elements as zero, which corresponds to only considering the Gaussian distribution whose principal axes of the elliptic contour lines agree with the coordinate axes (see Fig. 6.1).

A MATLAB code for AROW classification is provided in Fig. 31.7, and its behavior is illustrated in Fig. 31.8. This shows that AROW suppresses the influence of outliers better than passive-aggressive learning.

31.3.3 REGRESSION

Let us use the squared loss for the expectation vector $\boldsymbol{\mu}$:

$$J(\boldsymbol{\mu}) = \frac{1}{2}\left(y - \boldsymbol{\mu}^\top\boldsymbol{x}\right)^2.$$

```
n=50; x=[randn(1,n/2)-15 randn(1,n/2)-5; randn(1,n)]';
y=[ones(n/2,1); -ones(n/2,1)]; x(1:2,1)=x(1:2,1)+10;
x(:,3)=1; p=randperm(n); x=x(p,:); y=y(p);
mu=zeros(3,1); S=eye(3); l=1;
for i=1:n
  xi=x(i,:)'; yi=y(i); z=S*xi; b=xi'*z+l; m=yi*mu'*xi;
  if m<1, mu=mu+yi*(1-m)*z/b; S=S-z*z'/b; end
end
figure(1); clf; hold on; axis([-20 0 -2 2]);
plot(x(y==1,1),x(y==1,2),'bo');
plot(x(y==-1,1),x(y==-1,2),'rx');
plot([-20 0],-(mu(3)+[-20 0]*mu(1))/mu(2),'k-');
```

FIGURE 31.7

MATLAB code for AROW classification.

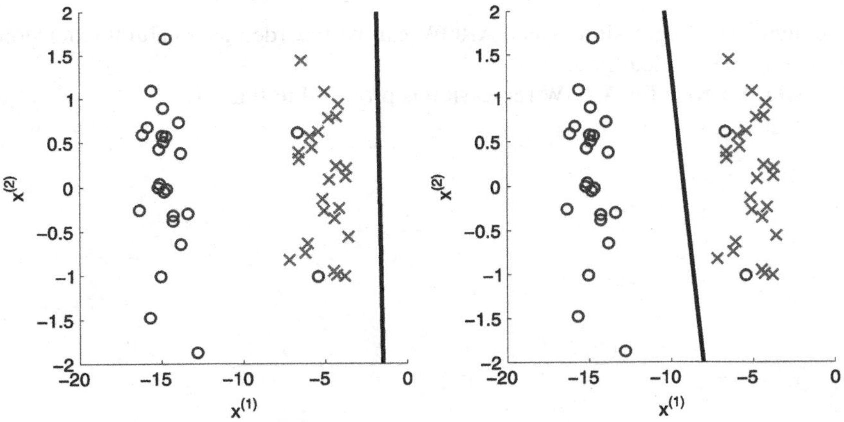

(a) Passive-aggressive (b) AROW

FIGURE 31.8

Examples of passive-aggressive and AROW classifications.

Then, with a similar derivation to the classification case, the update formula for AROW regression can be obtained analytically as

$$\boldsymbol{\mu} \longleftarrow \boldsymbol{\mu} + (y - \boldsymbol{\mu}^\top \boldsymbol{x})\boldsymbol{\Sigma}\boldsymbol{x}/\beta \quad \text{and} \quad \boldsymbol{\Sigma} \longleftarrow \boldsymbol{\Sigma} - \boldsymbol{\Sigma}\boldsymbol{x}\boldsymbol{x}^\top\boldsymbol{\Sigma}/\beta,$$

where $\beta = \boldsymbol{x}^\top\boldsymbol{\Sigma}\boldsymbol{x} + \lambda$. This method is also referred to as *recursive LS*. When $\boldsymbol{\Sigma}$ is fixed to the identity matrix and setting $\boldsymbol{\theta} = \boldsymbol{\mu}$, AROW regression is reduced to

```
n=50; N=1000; x=linspace(-3,3,n)'; X=linspace(-3,3,N)';
pix=pi*x; y=sin(pix)./(pix)+0.1*x+0.05*randn(n,1);
hh=2*0.3^2; m=randn(n,1); S=eye(n); l=1;
for j=1:100
  for i=1:n
    ki=exp(-(x-x(i)).^2/hh); Sk=S*ki;
    b=ki'*Sk+l; m=m+Sk*(y(i)-ki'*m)/b; S=S-Sk*Sk'/b;
end, end
K=exp(-(repmat(X.^2,1,n)+repmat(x.^2',N,1)-2*X*x')/hh);
F=K*m;
figure(1); clf; hold on; axis([-2.8 2.8 -0.5 1.2]);
plot(X,F,'g-'); plot(x,y,'bo');
```

FIGURE 31.9

MATLAB code for AROW regression.

passive-aggressive regression. Thus, AROW can be regarded as a natural extension of passive-aggressive learning.

A MATLAB code for AROW regression is provided in Fig. 31.9.

CHAPTER

CONFIDENCE OF PREDICTION

32

CHAPTER CONTENTS

By the supervised learning methods introduced in Part 4, output values for arbitrary input points can be predicted. In addition to predicting output values, however, it is useful to assess the *confidence* of prediction in some applications. The probabilistic classification methods introduced in Chapter 28 allow us to learn the confidence of classification, but its range of applications was limited to classification. In this chapter, more general methods for assessing the confidence of prediction are discussed.

32.1 PREDICTIVE VARIANCE FOR ℓ_2-REGULARIZED LS

Let us consider regression by *ℓ_2-regularized LS* for *linear-in-parameter models* (see Chapter 23). More specifically, from input-output paired samples $\{(\boldsymbol{x}_i, y_i)\}_{i=1}^n$, a predictor of output for input $\boldsymbol{x}$ is obtained as

$$\widehat{f}(\boldsymbol{x}) = \sum_{j=1}^{b} \widehat{\theta}_j \phi_j(\boldsymbol{x}) = \widehat{\boldsymbol{\theta}}^\top \boldsymbol{\phi}(\boldsymbol{x}),$$

where $\widehat{\boldsymbol{\theta}} = (\widehat{\theta}_1, \ldots, \widehat{\theta}_b)^\top$ is the vector of learned parameters given by

$$\widehat{\boldsymbol{\theta}} = (\boldsymbol{\Phi}^\top \boldsymbol{\Phi} + \lambda \boldsymbol{I})^{-1} \boldsymbol{\Phi}^\top \boldsymbol{y},$$

and $\boldsymbol{\phi}(\boldsymbol{x}) = (\phi_1(\boldsymbol{x}), \ldots, \phi_b(\boldsymbol{x}))^\top$ is the vector of basis functions. $\boldsymbol{\Phi}$ is the design matrix given by

$$\boldsymbol{\Phi} = \begin{pmatrix} \phi_1(\boldsymbol{x}_1) & \cdots & \phi_b(\boldsymbol{x}_1) \\ \vdots & \ddots & \vdots \\ \phi_1(\boldsymbol{x}_n) & \cdots & \phi_b(\boldsymbol{x}_n) \end{pmatrix},$$

$\lambda \geq 0$ is the regularization parameter, $\boldsymbol{I}$ is the identity matrix, and $\boldsymbol{y} = (y_1, \ldots, y_n)^\top$ is the training output vector.

Suppose that training output vector $\boldsymbol{y}$ is given by

$$\boldsymbol{y} = \boldsymbol{\Phi}\boldsymbol{\theta}^* + \boldsymbol{\epsilon}.$$

Here, $\boldsymbol{\theta}^*$ is the *true* parameter and $\boldsymbol{\epsilon}$ is a noise vector such that

$$\mathbb{E}[\boldsymbol{\epsilon}] = \boldsymbol{0} \quad \text{and} \quad \mathbb{E}[\boldsymbol{\epsilon}\boldsymbol{\epsilon}^\top] = \sigma^2 \boldsymbol{I},$$

where $\mathbb{E}$ denotes the expectation over $\boldsymbol{\epsilon}$ and σ^2 denotes the noise variance. Then the variance of prediction at a test input point $\boldsymbol{x}$,

$$V(\boldsymbol{x}) = \mathbb{E}\left[\left(\widehat{f}(\boldsymbol{x}) - \mathbb{E}[\widehat{f}(\boldsymbol{x})]\right)^2\right],$$

can be computed analytically as

$$\begin{aligned}
V(\boldsymbol{x}) &= \mathbb{E}\left[\left(\boldsymbol{\phi}(\boldsymbol{x})^\top \widehat{\boldsymbol{\theta}} - \boldsymbol{\phi}(\boldsymbol{x})^\top \mathbb{E}[\widehat{\boldsymbol{\theta}}]\right)^2\right] \\
&= \mathbb{E}\left[\left(\boldsymbol{\phi}(\boldsymbol{x})^\top ((\boldsymbol{\Phi}^\top\boldsymbol{\Phi} + \lambda\boldsymbol{I})^{-1}\boldsymbol{\Phi}^\top(\boldsymbol{\Phi}\boldsymbol{\theta}^* + \boldsymbol{\epsilon} - \boldsymbol{\Phi}\boldsymbol{\theta}^* - \mathbb{E}[\boldsymbol{\epsilon}]))\right)^2\right] \\
&= \mathbb{E}\left[\left(\boldsymbol{\phi}(\boldsymbol{x})^\top (\boldsymbol{\Phi}^\top\boldsymbol{\Phi} + \lambda\boldsymbol{I})^{-1}\boldsymbol{\Phi}^\top\boldsymbol{\epsilon}\right)^2\right] \\
&= \boldsymbol{\phi}(\boldsymbol{x})^\top (\boldsymbol{\Phi}^\top\boldsymbol{\Phi} + \lambda\boldsymbol{I})^{-1}\boldsymbol{\Phi}^\top \mathbb{E}[\boldsymbol{\epsilon}\boldsymbol{\epsilon}^\top]\boldsymbol{\Phi}(\boldsymbol{\Phi}^\top\boldsymbol{\Phi} + \lambda\boldsymbol{I})^{-1}\boldsymbol{\phi}(\boldsymbol{x}) \\
&= \sigma^2 \left\|\boldsymbol{\Phi}(\boldsymbol{\Phi}^\top\boldsymbol{\Phi} + \lambda\boldsymbol{I})^{-1}\boldsymbol{\phi}(\boldsymbol{x})\right\|^2.
\end{aligned}$$

The noise variance σ^2 may be estimated as

$$\widehat{\sigma}^2 = \frac{\|\boldsymbol{U}\boldsymbol{y}\|^2}{\mathrm{tr}(\boldsymbol{U})},$$

where

$$\boldsymbol{U} = \boldsymbol{I} - \boldsymbol{\Phi}(\boldsymbol{\Phi}^\top\boldsymbol{\Phi} + \lambda\boldsymbol{I})^{-1}\boldsymbol{\Phi}^\top.$$

$\mathrm{tr}(\boldsymbol{U})$ is often called the *effective number of parameters*. When $\lambda = 0$, the above $\widehat{\sigma}^2$ is reduced to the standard unbiased estimator of the noise variance, given by the sum of squared residuals divided by the number of parameters.

A MATLAB code for analytic computation of predictive variance is provided in Fig. 32.1, and its behavior is illustrated in Fig. 32.2. This shows that, depending on the density of training samples, confidence intervals are adaptively estimated.

Note that the above analytic computation technique can be extended to any estimator $\widehat{\boldsymbol{\theta}}$ that is linear with respect to $\boldsymbol{y}$. It can also be applied to correlated noise, as long as the variance-covariance matrix of the noise can be estimated.

```
n=50; x=linspace(-3,3,n)'; %x=randn(n,1);
N=1000; X=linspace(-3,3,N)';
pix=pi*x; y=sin(pix)./(pix)+0.1*x+0.2*randn(n,1);
x2=x.^2; xx=repmat(x2,1,n)+repmat(x2',n,1)-2*x*x';
hhs=2*[0.03 0.3 3].^2; ls=[0.0001 0.1 100];
m=5; u=mod(randperm(n),m)+1;
for hk=1:length(hhs)
  hh=hhs(hk); k=exp(-xx/hh);
  for i=1:m
    ki=k(u~=i,:); kc=k(u==i,:); yi=y(u~=i); yc=y(u==i);
    for lk=1:length(ls)
    t=(ki'*ki+ls(lk)*eye(n))\(ki'*yi);
    g(hk,lk,i)=mean((yc-kc*t).^2);
end, end, end
[gl,ggl]=min(mean(g,3),[],2); [ghl,gghl]=min(gl);
L=ls(ggl(gghl)); HH=hhs(gghl);
K=exp(-(repmat(X.^2,1,n)+repmat(x2',N,1)-2*X*x')/HH);
k=exp(-xx/HH); Q=inv(k^2+L*eye(n)); Qk=Q*k'; t=Qk*y; F=K*t;
V=sum((K*Q*K').^2,2)*sum((y-k'*t).^2)/(N-sum(sum(k.*Qk)));

figure(1); clf; hold on; axis([-2.8 2.8 -0.7 1.7]);
errorbar(X,F,sqrt(V),'y-'); plot(X,mean(F,2),'g-');
plot(x,y,'bo');
```

FIGURE 32.1

MATLAB code for analytic computation of predictive variance.

32.2 BOOTSTRAP CONFIDENCE ESTIMATION

The above analytic computation approach is useful to assess the confidence of prediction. However, its range of applications is rather limited due to the linearity assumption. Here, a more general approach based on the *bootstrap* technique (see Section 9.3.2) is introduced, which can be applied to *any* learning methods for assessing any statistics beyond the variance.

More specifically, from the original set of samples $\mathcal{D} = \{(\boldsymbol{x}_i, y_i)\}_{i=1}^n$, n pseudosamples $\mathcal{D}' = \{(\boldsymbol{x}'_i, y'_i)\}_{i=1}^n$ are generated by *sampling with replacement* (see Fig. 3.3). Then, from the bootstrap samples $\mathcal{D}' = \{(\boldsymbol{x}'_i, y'_i)\}_{i=1}^n$, a predictor $\widehat{f}(\boldsymbol{x})$ is computed. This resampling and prediction procedure is repeated many times and its variance (or any statistics such as *quantiles*) is computed.

A MATLAB code for bootstrap-based confidence estimation is provided in Fig. 32.3, and its behavior is illustrated in Fig. 32.4. This shows that similar results to Fig. 32.2 can be numerically obtained by bootstrapping.

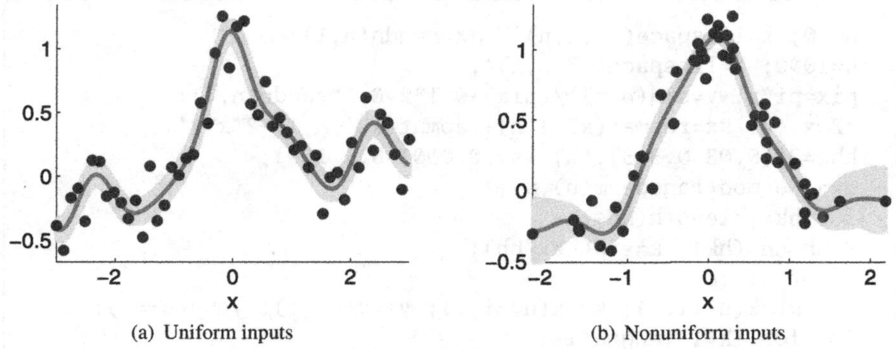

FIGURE 32.2

Examples of analytic computation of predictive variance. The shaded area indicates the confidence interval.

32.3 APPLICATIONS

In this section, useful applications of predictive confidence are described.

32.3.1 TIME-SERIES PREDICTION

Let us consider the problem of *time-series prediction*, i.e., given $\{y_i\}_{i=1}^t$, the objective is to predict y_{t+1} (Fig. 32.5).

Given that input variables $\{\boldsymbol{x}_i\}_{i=1}^t$ are not available in the current time-series prediction setup, a simple modeling approach is to use m previous output values $\{y_{t-i+1}\}_{i=1}^m$ to predict the next value y_{t+1} (Fig. 32.6):

$$\widehat{y}_{t+1} = \sum_{j=1}^{b} \theta_j \phi_j(y_{t-m+1}, \ldots, y_t).$$

This approach also allows multiple-step ahead prediction as

$$\widehat{y}_{t+2} = \sum_{j=1}^{b} \theta_j \phi_j(y_{t-m+2}, \ldots, y_t, \widehat{y}_{t+1}),$$

$$\widehat{y}_{t+3} = \sum_{j=1}^{b} \theta_j \phi_j(y_{t-m+3}, \ldots, y_t, \widehat{y}_{t+1}, \widehat{y}_{t+2}).$$

It is expected that prediction of distant future is more difficult, which may be assessed by the prediction variance.

A MATLAB code of time-series prediction by ℓ_2-regularized LS for the *linear-in-input model* (which is also referred to as the *autoregressive model* in the context

```
n=50; x0=linspace(-3,3,n)'; %x0=randn(n,1);
N=1000; X=linspace(-3,3,N)';
pix=pi*x0; y0=sin(pix)./(pix)+0.1*x0+0.2*randn(n,1);
for s=1:100
  r=ceil(n*rand(n,1));x=x0(r,:); y=y0(r);
  x2=x.^2; xx=repmat(x2,1,n)+repmat(x2',n,1)-2*x*x';
  hhs=2*[0.03 0.3 3].^2; ls=[0.0001 0.1 100];
  m=5; u=mod(randperm(n),m)+1;
  for hk=1:length(hhs)
    hh=hhs(hk); k=exp(-xx/hh);
    for i=1:m
      ki=k(u~=i,:); kc=k(u==i,:); yi=y(u~=i); yc=y(u==i);
      for lk=1:length(ls)
        t=(ki'*ki+ls(lk)*eye(n))\(ki'*yi);
        g(hk,lk,i)=mean((yc-kc*t).^2);
  end, end, end
  [gl,ggl]=min(mean(g,3),[],2); [ghl,gghl]=min(gl);
  L=ls(ggl(gghl)); HH=hhs(gghl);
  K=exp(-(repmat(X.^2,1,n)+repmat(x2',N,1)-2*X*x')/HH);
  k=exp(-xx/HH); t=(k^2+L*eye(n))\(k*y); F(:,s)=K*t;
end
figure(1); clf; hold on; axis([-2.8 2.8 -0.7 1.7]);
errorbar(X,mean(F,2),std(F,0,2),'y-');
plot(X,mean(F,2),'g-'); plot(x0,y0,'bo');
```

FIGURE 32.3

MATLAB code for bootstrap-based confidence estimation.

of time-series prediction),

$$\widehat{y}_{t+1} = \sum_{j=1}^{b} \theta_j y_{t-j+1},$$

is provided in Fig. 32.7. Its behavior illustrated in Fig. 32.8 shows that the confidence interval tends to grow as distant future is predicted.

32.3.2 TUNING PARAMETER OPTIMIZATION

The supervised learning methods introduced in Part 4 contain various tuning parameters such as the regularization parameter and the Gaussian bandwidth. Those tuning parameters may be optimized by *cross validation* with respect to the prediction error,

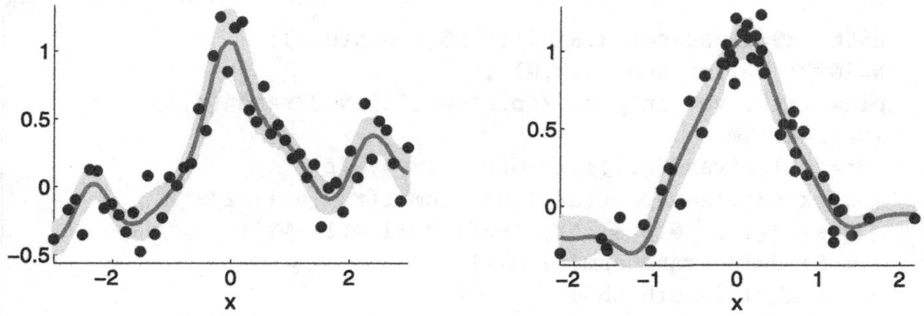

(a) Uniform inputs (b) Nonuniform inputs

FIGURE 32.4

Examples of bootstrap-based confidence estimation. The shaded area indicates the confidence interval.

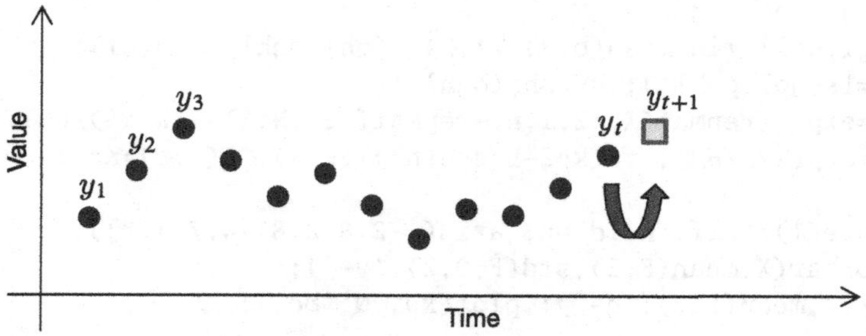

FIGURE 32.5

Problem of time-series prediction.

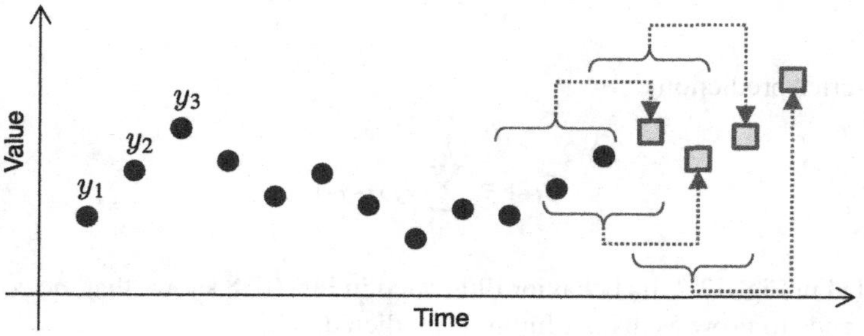

FIGURE 32.6

Time-series prediction from previous samples.

```
u=17; z(1:u+1,1)=1.2; n=200; d=20; N=n+130;
for t=u+1:u+N+d
  z(t+1)=0.9*z(t)+0.2*z(t-u)/(1+z(t-u)^10);
end
z=z(u+1:u+N+d); Y=z(d+1:end); x0=zeros(n+d-1,d);
%z=z+0.1*randn(N+d,1);
for i=1:d
  x0(i:n+d-1,i)=z(1:n+d-i);
end
x0=x0(d:end,:); y0=z(d+1:n+d);

B=100; v=zeros(N-n+d,B);
for s=1:B
  r=ceil(n*rand(n,1)); x=x0(r,:); y=y0(r);
  x2=sum(x.^2,2); xx=repmat(x2,1,n)+repmat(x2',n,1)-2*x*x';
  hhs=median(xx(:))*2*[0.5,0.2:1.5].^2; ls=[0.01 0.1 1 10];
  m=5; u=mod(randperm(n),m)+1;
  for hk=1:length(hhs)
    hh=hhs(hk); k=exp(-xx/hh);
    for i=1:m
      ki=k(u~=i,:); kc=k(u==i,:); yi=y(u~=i); yc=y(u==i);
      for lk=1:length(ls)
        t=(ki'*ki+ls(lk)*eye(n))\(ki'*yi);
        g(hk,lk,i)=mean((yc-kc*t).^2);
  end, end, end
  [gl,ggl]=min(mean(g,3),[],2); [ghl,gghl]=min(gl);
  L=ls(ggl(gghl)); HH=hhs(gghl); k=exp(-xx/HH);
  t=(k^2+L*eye(n))\(k*y); v(:,s)=[y0(n-d+1:n); zeros(N-n,1)];
  for i=1:N-n
    X=fliplr(v(i:d+i-1,s)');
    K=exp(-(repmat(sum(X.^2),1,n)+x2'-2*X*x')/HH);
    v(d+i,s)=K*t;
end, end

figure(1); clf; hold on; a=mean(v(d+1:end,:),2);
errorbar([n+1:N],a,std(v(d+1:end,:),0,2),'y-');
plot([1:N],Y,'r--'); plot([n+1:N],a,'g-');
plot([1:n],y0,'ko'); legend('','True','Estimated','Sample',4)
```

FIGURE 32.7

MATLAB code for time-series prediction by ℓ_2-regularized LS.

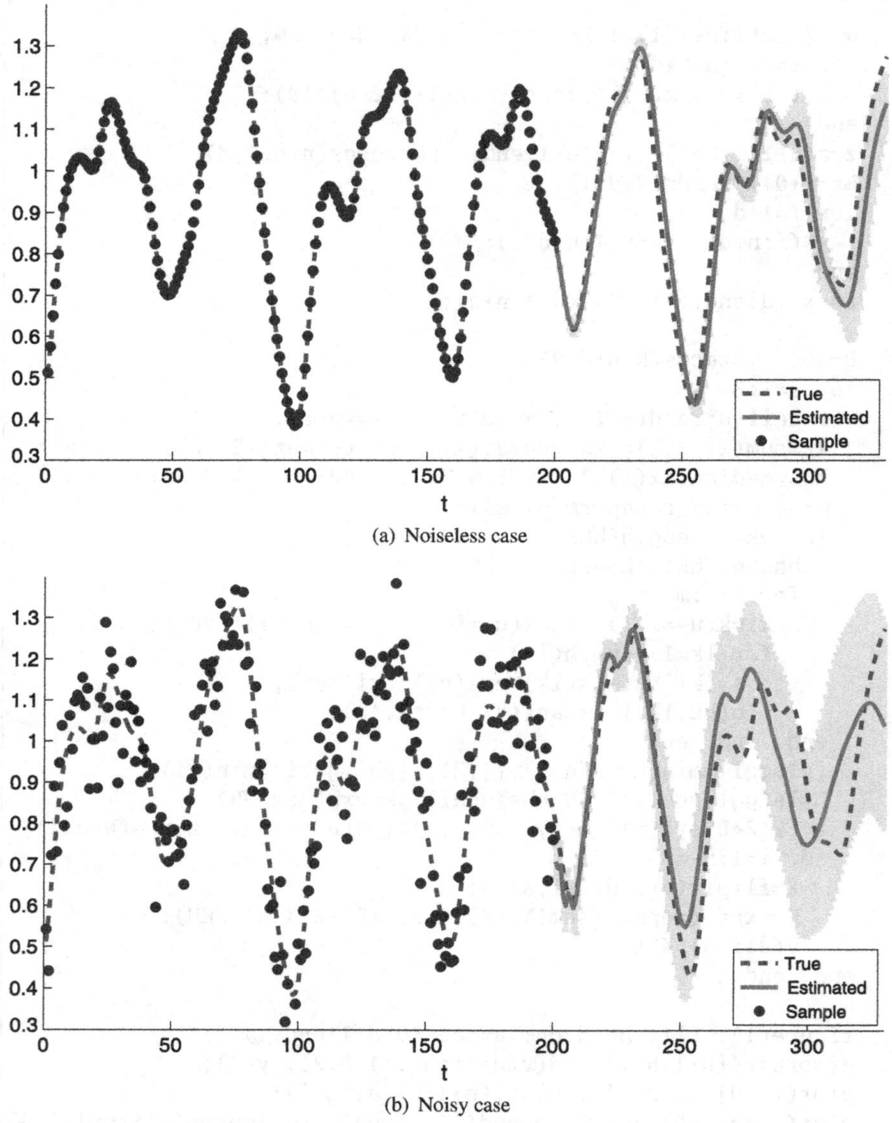

FIGURE 32.8

Examples of time-series prediction by ℓ_2-regularized LS. The shaded areas indicate the confidence intervals.

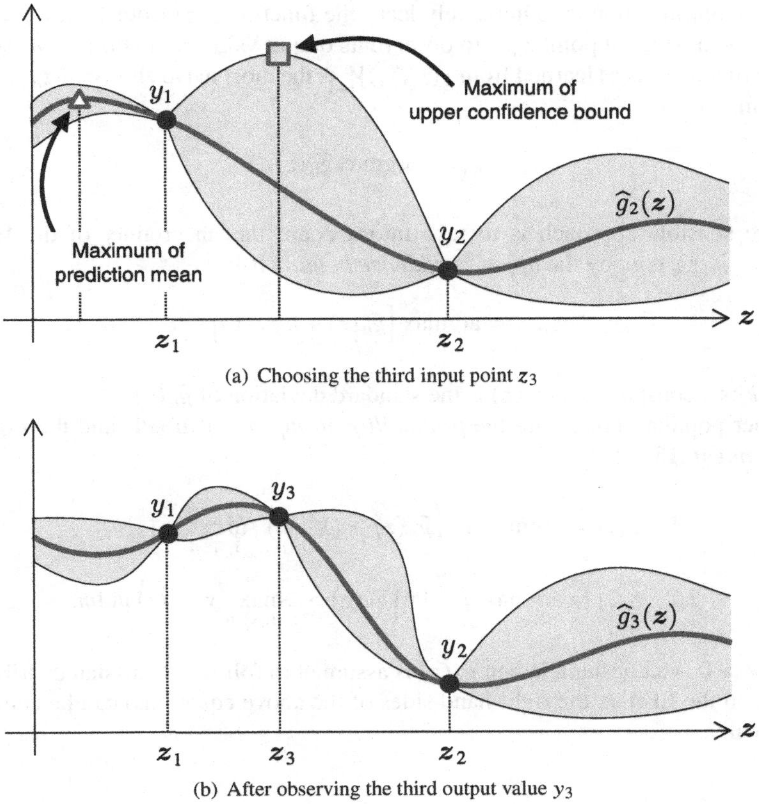

FIGURE 32.9

Bayesian optimization. The shaded areas indicate the confidence intervals.

as discussed in Section 23.3. However, naive grid search over tuning parameters is computationally expensive when multiple tuning parameters are optimized.

To cope with this problem, *Bayesian optimization* [12] has attracted a great deal of attention recently. Its basic idea is to find the maximizer of an unknown function $g(z)$:

$$z^* = \underset{z}{\text{argmax}}\, g(z).$$

In the current setup, the function g corresponds the prediction performance (i.e., the negative of the prediction error), and z corresponds to the tuning parameters.

If cross validation is performed for some values z_i, corresponding output values $y_i = g(z_i)$ can be observed. Then the unknown function $g(z)$ can be learned from those obtained values $\{(z_i, y_i)\}_{i=1}^n$ by any regression method. The basic idea of

Bayesian optimization is to iteratively learn the function $g(z)$ from $\{(z_i, y_i)\}_{i=1}^n$ and choose the next input point z_{n+1} to observe its output value y_{n+1} (Fig. 32.9(a)).

For function $\widehat{g}_n(z)$ learned from $\{(z_i, y_i)\}_{i=1}^n$, the most naive choice of z_{n+1} would be the maximizer:

$$z_{n+1} = \underset{z}{\operatorname{argmax}}\, \widehat{g}_n(z).$$

A more sensible approach is to take into account the uncertainty of the learned function $\widehat{g}_n(z)$, e.g., by the *upper confidence bound* [10]:

$$z_{n+1} = \underset{z}{\operatorname{argmax}} \left[\widehat{g}_n(z) + k\sigma_n(z)\right],$$

where k is a constant and $\sigma_n(z)$ is the standard deviation of $\widehat{g}_n(z)$.

Other popular choices are the *probability of improvement* (PI) and the *expected improvement* (EI):

$$\text{PI:} \quad z_{n+1} = \underset{z}{\operatorname{argmax}}\, \Pr\left(\widehat{g}_n(z) \geq (1+k) \max_{i=1,\ldots,n} y_i\right),$$

$$\text{EI:} \quad z_{n+1} = \underset{z}{\operatorname{argmax}} \int_0^\infty \Pr\left(\widehat{g}_n(z) > \max_{i=1,\ldots,n} y_i + m\right) m \mathrm{d}m,$$

where $k > 0$ is a constant. When $\widehat{g}_n(z)$ is assumed to follow a Gaussian distribution, the PI and the EI (i.e., the right-hand sides of the above equations) can be computed analytically.

CHAPTER

SEMISUPERVISED LEARNING

33

CHAPTER CONTENTS

Supervised learning is performed based on input-output paired training samples $\{(\boldsymbol{x}_i, y_i)\}_{i=1}^n$. However, gathering many input-output paired samples is often expensive in practical applications. On the other hand, *input-only* samples $\{\boldsymbol{x}_i\}_{i=n+1}^{n+n'}$ can be easily collected abundantly. For example, in web page classification, class label y_i (such as "sport", "computer", and "politics") should be *manually* given after carefully investigating web page $\boldsymbol{x}_i$, which requires a huge amount of human labor. On the other hand, input-only web pages $\{\boldsymbol{x}_i\}_{i=n+1}^{n+n'}$ can be automatically collected by crawlers.

In this chapter, methods of *semisupervised learning* [28, 101] are introduced, which utilize input-only samples $\{\boldsymbol{x}_i\}_{i=n+1}^{n+n'}$ in addition to input-output paired samples $\{(\boldsymbol{x}_i, y_i)\}_{i=1}^n$.

33.1 MANIFOLD REGULARIZATION

In this section, a method of semisupervised learning based on *manifold regularization* is introduced.

33.1.1 MANIFOLD STRUCTURE BROUGHT BY INPUT SAMPLES

Supervised learning from input-output paired samples $\{(\boldsymbol{x}_i, y_i)\}_{i=1}^n$ can be regarded as estimating the conditional density $p(y|\boldsymbol{x})$. On the other hand, unsupervised

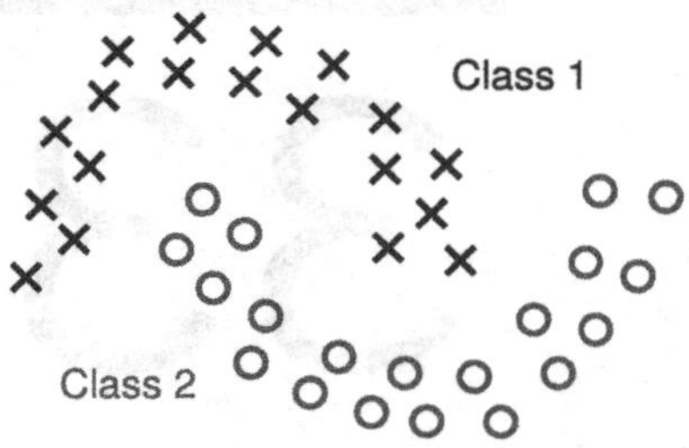

FIGURE 33.1

Semisupervised classification. Samples in the same cluster are assumed to belong to the same class.

learning from input-only samples $\{x_i\}_{i=n+1}^{n+n'}$ can be regarded as estimating the marginal density $p(x)$. Thus, without any assumption, input-only samples $\{x_i\}_{i=n+1}^{n+n'}$ do not help improve the estimation of the conditional density $p(y|x)$. For this reason, semisupervised learning imposes a certain assumption between $p(x)$ and $p(y|x)$ and utilizes the estimation of $p(x)$ to improve the accuracy of estimating $p(y|x)$.

Below, a semisupervised learning method based on a *manifold* assumption is introduced. Mathematically, a manifold is a topological space that can be locally approximated by Euclidean space. On the other hand, a manifold is just regarded as a local region in the context of semisupervised learning. More specifically, the manifold assumption means that input samples appear only on manifolds and output values change *smoothly* on the manifolds. In the case of classification, this means that samples in the same cluster belong to the same class (Fig. 33.1).

The Gaussian kernel model introduced in Section 21.2 actually utilizes this manifold assumption (see Fig. 21.5):

$$f_\theta(x) = \sum_{j=1}^{n} \theta_j K(x, x_j), \quad K(x, c) = \exp\left(-\frac{\|x - c\|^2}{2h^2}\right).$$

That is, by locating smooth Gaussian functions on input samples $\{x_i\}_{i=1}^{n}$, a smooth function over the input manifold can be learned. In semisupervised learning, the above model may be augmented to locate Gaussian kernels also on input-only samples $\{x_i\}_{i=n+1}^{n+n'}$:

$$f_\theta(x) = \sum_{j=1}^{n+n'} \theta_j K(x, x_j). \tag{33.1}$$

The parameters in model (33.1) are learned so that output at input samples, $\{f_\theta(x_i)\}_{i=1}^{n+n'}$, is similar to each other. For example, in the case of ℓ_2-regularized

LS, the optimization problem is given as

$$\min_{\theta}\left[\frac{1}{2}\sum_{i=1}^{n}\left(f_{\theta}(\boldsymbol{x}_i)-y_i\right)^2+\frac{\lambda}{2}\|\boldsymbol{\theta}\|^2\right.$$
$$\left.+\frac{\nu}{4}\sum_{i,i'=1}^{n+n'}W_{i,i'}\left(f_{\theta}(\boldsymbol{x}_i)-f_{\theta}(\boldsymbol{x}_{i'})\right)^2\right], \quad (33.2)$$

where the first and second terms correspond to ℓ_2-regularized LS and the third term is called the *Laplacian regularizer*, following the terminology in *spectral graph theory* [29]. $\nu \geq 0$ is the regularization parameter for semisupervised learning that controls the smoothness on the manifolds. $W_{i,i'} \geq 0$ denotes the similarity between $\boldsymbol{x}_i$ and $\boldsymbol{x}_{i'}$, which takes a large value if $\boldsymbol{x}_i$ and $\boldsymbol{x}_{i'}$ are similar, and a small value if $\boldsymbol{x}_i$ and $\boldsymbol{x}_{i'}$ are dissimilar. Popular choices of the similarity measure are described in Fig. 35.8. Below the similarity matrix $\boldsymbol{W}$ is assumed to be symmetric (i.e. $W_{i,i'} = W_{i',i}$).

33.1.2 COMPUTING THE SOLUTION

Here, how to compute the solution of Laplacian-regularized LS is explained. Let $\boldsymbol{D}$ be the diagonal matrix whose diagonal elements are given by the row-sums of matrix $\boldsymbol{W}$:

$$\boldsymbol{D} = \mathrm{diag}\left(\sum_{i=1}^{n+n'}W_{1,i},\ldots,\sum_{i=1}^{n+n'}W_{n+n',i}\right),$$

and let $\boldsymbol{L} = \boldsymbol{D} - \boldsymbol{W}$. Then the third term in Eq. (33.2) can be rewritten as

$$\sum_{i,i'=1}^{n+n'}W_{i,i'}\left(f_{\theta}(\boldsymbol{x}_i)-f_{\theta}(\boldsymbol{x}_{i'})\right)^2$$
$$=\sum_{i=1}^{n+n'}D_{i,i}f_{\theta}(\boldsymbol{x}_i)^2-2\sum_{i,i'=1}^{n+n'}W_{i,i'}f_{\theta}(\boldsymbol{x}_i)f_{\theta}(\boldsymbol{x}_{i'})+\sum_{i'=1}^{n+n'}D_{i',i'}f_{\theta}(\boldsymbol{x}_i)^2$$
$$=2\sum_{i,i'=1}^{n+n'}L_{i,i'}f_{\theta}(\boldsymbol{x}_i)f_{\theta}(\boldsymbol{x}_{i'}).$$

Thus, Eq. (33.2) for kernel model (33.1) can be reduced to the generalized ℓ_2-regularized LS as follows:

$$\min_{\theta}\left[\frac{1}{2}\sum_{i=1}^{n}\left(\sum_{j=1}^{n+n'}\theta_jK(\boldsymbol{x}_i,\boldsymbol{x}_j)-y_i\right)^2+\frac{\lambda}{2}\sum_{j=1}^{n+n'}\theta_j^2\right.$$
$$\left.+\frac{\nu}{2}\sum_{j,j'=1}^{n+n'}\theta_j\theta_{j'}\sum_{i,i'=1}^{n+n'}L_{i,i'}K(\boldsymbol{x}_i,\boldsymbol{x}_j)K(\boldsymbol{x}_{i'},\boldsymbol{x}_{j'})\right].$$

This can be compactly rewritten as

$$\min_{\boldsymbol{\theta}} \left[\frac{1}{2} \|\boldsymbol{K\theta} - \boldsymbol{y}\|^2 + \frac{\lambda}{2} \|\boldsymbol{\theta}\|^2 + \frac{\nu}{2} \boldsymbol{\theta}^\top \boldsymbol{KLK\theta} \right],$$

where

$$\boldsymbol{K} = \begin{pmatrix} K(\boldsymbol{x}_1, \boldsymbol{x}_1) & \cdots & K(\boldsymbol{x}_1, \boldsymbol{x}_{n+n'}) \\ \vdots & \ddots & \vdots \\ K(\boldsymbol{x}_{n+n'}, \boldsymbol{x}_1) & \cdots & K(\boldsymbol{x}_{n+n'}, \boldsymbol{x}_{n+n'}) \end{pmatrix},$$
$$\boldsymbol{\theta} = (\theta_1, \ldots, \theta_n, \theta_{n+1}, \ldots, \theta_{n+n'})^\top,$$
$$\boldsymbol{y} = (y_1, \ldots, y_n, \underbrace{0, \ldots, 0}_{n'})^\top.$$

Note that the vector $\boldsymbol{y}$ defined above contains n' zeros since $\{y_i\}_{i=n+1}^{n+n'}$ are not available. The solution $\widehat{\boldsymbol{\theta}}$ can be obtained analytically as

$$\widehat{\boldsymbol{\theta}} = (\boldsymbol{K}^2 + \lambda \boldsymbol{I} + \nu \boldsymbol{KLK})^{-1} \boldsymbol{K} \boldsymbol{y}.$$

A MATLAB code for Laplacian-regularized LS is provided in Fig. 33.2, and its behavior is illustrated in Fig. 33.3. This is an extremely difficult classification problem where only two labeled training samples are available. Laplacian-regularized LS gives a decision boundary that separates two point clouds, which is appropriate when the cluster assumption holds. On the other hand, ordinary LS gives a decision boundary that goes through the middle of the two labeled training samples, which is not reasonable under the cluster assumption.

In the above explanation, Laplacian regularization was applied to LS. However, the idea of Laplacian regularization is generic and it can be combined with various regression and classification techniques introduced in Part 4.

33.2 COVARIATE SHIFT ADAPTATION

The Laplace regularization method introduced above utilized the manifold structure of input samples. In this section, a semisupervised learning method called *covariate shift* adaptation is introduced [101], which explicitly takes into account the probability distributions of $\{(\boldsymbol{x}_i, y_i)\}_{i=1}^n$ and $\{\boldsymbol{x}'_{i'}\}_{i'=1}^{n'}$. A covariate is another name for an input variable, and covariate shift refers to the situation where $\{\boldsymbol{x}_i\}_{i=1}^n$ and $\{\boldsymbol{x}'_{i'}\}_{i'=1}^{n'}$ follow different probability distributions, but the conditional density $p(y|\boldsymbol{x})$ is unchanged.

33.2.1 IMPORTANCE WEIGHTED LEARNING

An example of covariate shift in regression is illustrated in Fig. 33.4, where the learning target function $f(x)$ is unchanged. However, $\{x_i\}_{i=1}^n$ are distributed

```
n=200; a=linspace(0,pi,n/2);
u=-10*[cos(a)+0.5 cos(a)-0.5]'+randn(n,1);
v=10*[sin(a) -sin(a)]'+randn(n,1);
x=[u v]; y=zeros(n,1); y(1)=1; y(n)=-1;
x2=sum(x.^2,2); hh=2*1^2;
k=exp(-(repmat(x2,1,n)+repmat(x2',n,1)-2*x*x')/hh); w=k;
t=(k^2+1*eye(n)+10*k*(diag(sum(w))-w)*k)\(k*y);

m=100; X=linspace(-20,20,m)'; X2=X.^2;
U=exp(-(repmat(u.^2,1,m)+repmat(X2',n,1)-2*u*X')/hh);
V=exp(-(repmat(v.^2,1,m)+repmat(X2',n,1)-2*v*X')/hh);
figure(1); clf; hold on; axis([-20 20 -20 20]);
colormap([1 0.7 1; 0.7 1 1]);
contourf(X,X,sign(V'*(U.*repmat(t,1,m))));
plot(x(y==1,1),x(y==1,2),'bo');
plot(x(y==-1,1),x(y==-1,2),'rx');
plot(x(y==0,1),x(y==0,2),'k.');
```

FIGURE 33.2

MATLAB code for Laplacian-regularized LS.

around $x = 1$, while $\{x'_{i'}\}_{i'=1}^{n'}$ are distributed around $x = 2$. Thus, this is a (weak) *extrapolation* problem.

Fig. 33.6(a) illustrates the function obtained by fitting the straight line model,

$$f_{\boldsymbol{\theta}}(x) = \theta_1 + \theta_2 x, \tag{33.3}$$

to samples $\{(\boldsymbol{x}_i, y_i)\}_{i=1}^{n}$ by ordinary LS:

$$\min_{\boldsymbol{\theta}} \frac{1}{2} \sum_{i=1}^{n} \left(f_{\boldsymbol{\theta}}(\boldsymbol{x}_i) - y_i\right)^2.$$

This shows that, although the samples $\{(\boldsymbol{x}_i, y_i)\}_{i=1}^{n}$ are appropriately fitted by LS, prediction of output values at $\{\boldsymbol{x}'_{i'}\}_{i'=1}^{n'}$ is poor.

In such a covariate shift situation, it intuitively seems that only using samples $\{(\boldsymbol{x}_i, y_i)\}_{i=1}^{n}$ *near* $\{\boldsymbol{x}'_{i'}\}_{i'=1}^{n'}$ can provide good prediction of output values at $\{\boldsymbol{x}'_{i'}\}_{i'=1}^{n'}$. This intuitive idea can be more formally realized by *importance weighting*. More specifically, *importance weighted LS* is given by

$$\min_{\boldsymbol{\theta}} \frac{1}{2} \sum_{i=1}^{n} w(\boldsymbol{x}_i)\left(f_{\boldsymbol{\theta}}(\boldsymbol{x}_i) - y_i\right)^2,$$

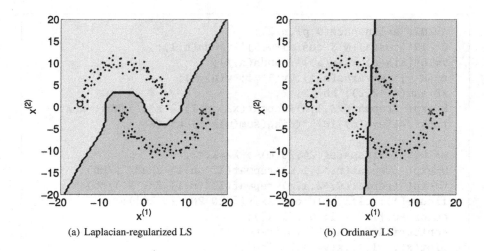

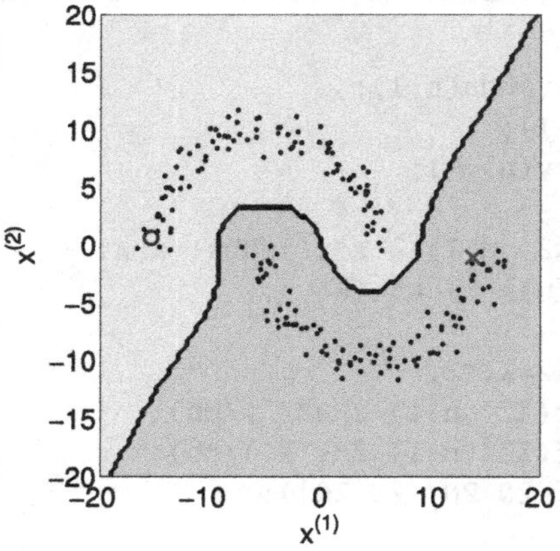

(a) Laplacian-regularized LS (b) Ordinary LS

FIGURE 33.3

Examples of Laplacian-regularized LS compared with ordinary LS. Dots denote unlabeled training samples.

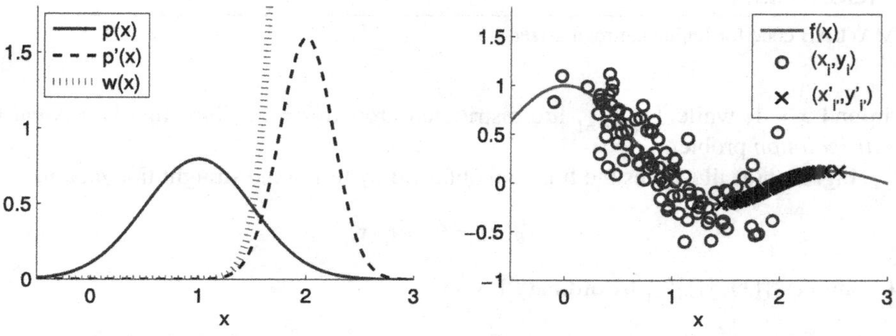

(a) Input densities and importance (b) Learning target function and training samples

FIGURE 33.4

Covariate shift in regression. Input distributions change, but the input-output relation is unchanged.

where $w(\boldsymbol{x})$ is called the importance function which is defined as the ratio of probability density $p'(\boldsymbol{x})$ for $\{\boldsymbol{x}'_{i'}\}_{i'=1}^{n'}$ and probability density $p(\boldsymbol{x})$ for $\{\boldsymbol{x}_i\}_{i=1}^{n}$:

$$w(\boldsymbol{x}) = \frac{p'(\boldsymbol{x})}{p(\boldsymbol{x})}.$$

```
n=100; u=randn(n,1)/4+2; x=randn(n,1)/2+1;
w=2*exp(-8*(x-2).^2+2*(x-1).^2); %w=ones(n,1);

y=sin(pi*x)./(pi*x)+0.1*randn(n,1);
x(:,2)=1; t=(x'*(repmat(w,1,2).*x))\(x'*(w.*y));
X=linspace(-1,3,100); Y=sin(pi*X)./(pi*X);
u(:,2)=1; v=u*t;
figure(1); clf; hold on;
plot(x(:,1),y,'bo'); plot(X,Y,'r-'); plot(u(:,1),v,'kx');
```

FIGURE 33.5

MATLAB code for importance weighted LS.

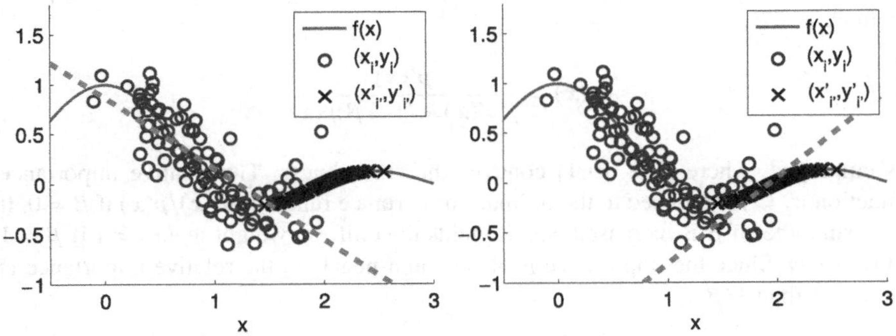

FIGURE 33.6

Example of LS learning under covariate shift. The dashed lines denote learned functions.

This importance weighted learning is based on the idea of *importance sampling* (see Section 19.2), which approximates the expectation with respect to $p'(\boldsymbol{x})$ by importance weighted average with respect to $p(\boldsymbol{x})$:

$$\int g(\boldsymbol{x})p'(\boldsymbol{x})\mathrm{d}\boldsymbol{x} = \int g(\boldsymbol{x})\frac{p'(\boldsymbol{x})}{p(\boldsymbol{x})}p(\boldsymbol{x})\mathrm{d}\boldsymbol{x} \approx \frac{1}{n}\sum_{i=1}^{n} g(\boldsymbol{x}_i)w(\boldsymbol{x}_i).$$

A MATLAB code of importance weighted LS for straight line model (33.3) is provided in Fig. 33.5, and its behavior is illustrated in Fig. 33.6(b). This shows that importance weighting contributes to improving the accuracy of predicting output values at $\{\boldsymbol{x}'_{i'}\}_{i'=1}^{n'}$.

In the above example, importance weighting was applied to LS. However, the idea of importance weighting is generic and it can be combined with various regression and classification techniques introduced in Part 4.

33.2.2 RELATIVE IMPORTANCE WEIGHTED LEARNING

As illustrated in Fig. 33.6(b), prediction performance under covariate shift can be improved by importance weighted learning.

Fig. 33.4(a) illustrates the importance function $w(x)$, which is monotone increasing as x grows. This means that, among all training samples $\{(x_i, y_i)\}_{i=1}^n$, only a few of them with large x have large importance values and the importance values of others are negligibly small. Thus, importance weighted learning is actually performed only from a few training samples, which can be unstable in practice.

This instability is caused by the fact that the importance function $w(\boldsymbol{x})$ takes large values for some $\boldsymbol{x}$. Thus, the instability problem is expected to be mitigated if a smooth variant of the importance function, such as the *relative importance* [121] defined as

$$w_\beta(\boldsymbol{x}) = \frac{p'(\boldsymbol{x})}{\beta p'(\boldsymbol{x}) + (1-\beta)p(\boldsymbol{x})},$$

is employed, where $\beta \in [0,1]$ controls the smoothness. The relative importance function $w_\beta(\boldsymbol{x})$ is reduced to the ordinary importance function $p'(\boldsymbol{x})/p(\boldsymbol{x})$ if $\beta = 0$. It gets smoother if β is increased, and it yields the uniform weight $w_\beta(\boldsymbol{x}) = 1$ if $\beta = 1$ (Fig. 33.7). Since the importance is always non-negative, the relative importance is no larger than $1/\beta$:

$$w_\beta(\boldsymbol{x}) = \frac{1}{\beta + (1-\beta)\frac{p(\boldsymbol{x})}{p'(\boldsymbol{x})}} \leq \frac{1}{\beta}.$$

33.2.3 IMPORTANCE WEIGHTED CROSS VALIDATION

The performance of relative importance weighted learning depends on the choice of smoothness β. Also, choice of model $f_{\boldsymbol{\theta}}(\boldsymbol{x})$ and regularization parameters significantly affects the final performance.

In Section 23.3, a model selection method based on *cross validation* was introduced, which estimates the prediction error for test samples. Actually, the validity of cross validation is ensured only when $\{\boldsymbol{x}_i\}_{i=1}^n$ and $\{\boldsymbol{x}'_{i'}\}_{i'=1}^{n'}$ follow the same probability distributions. Thus, performing cross validation under covariate shift can produce undesired solutions. Under covariate shift, a variant called *importance weighted cross validation* [102] is useful. The algorithm of importance weighted cross validation is summarized in Fig. 33.8, which is essentially the same as ordinary cross validation, but the validation error is computed with importance weights.

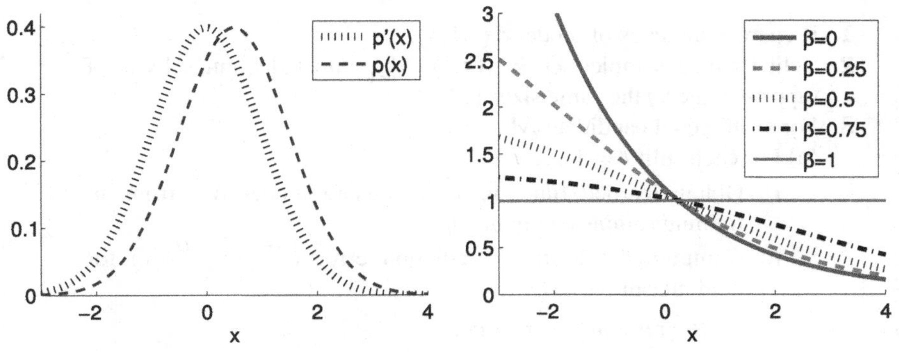

(a) Probability densities (b) Relative importance

FIGURE 33.7

Relative importance when $p'(x)$ is the Gaussian density with expectation 0 and variance 1 and $p(x)$ is the Gaussian density with expectation 0.5 and variance 1.

33.2.4 IMPORTANCE ESTIMATION

In (relative) importance weighted learning and cross validation, importance weights are needed. However, the importance function is usually unknown and only samples $\{\boldsymbol{x}_i\}_{i=1}^n$ and $\{\boldsymbol{x}'_{i'}\}_{i'=1}^{n'}$ are available. A naive way to estimate the importance function $w(\boldsymbol{x}) = p'(\boldsymbol{x})/p(\boldsymbol{x})$ is to separately estimate density $p'(\boldsymbol{x})$ from $\{\boldsymbol{x}'_{i'}\}_{i'=1}^{n'}$ and $p(\boldsymbol{x})$ from $\{\boldsymbol{x}_i\}_{i=1}^n$ and then compute the ratio of estimated densities. However, such a two-step approach is unreliable because division by an estimated density can magnify the estimation error significantly. Here, a direct importance estimator without going through density estimation is introduced.

Let us model the relative importance function $w_\beta(\boldsymbol{x})$ by a linear-in-parameter model:

$$w_{\boldsymbol{\alpha}}(\boldsymbol{x}) = \sum_{j=1}^{b} \alpha_j \psi_j(\boldsymbol{x}) = \boldsymbol{\alpha}^\top \boldsymbol{\psi}(\boldsymbol{x}),$$

where $\boldsymbol{\alpha} = (\alpha_1, \ldots, \alpha_b)^\top$ is the parameter vector and $\boldsymbol{\psi}(\boldsymbol{x}) = (\psi_1(\boldsymbol{x}), \ldots, \psi_b(\boldsymbol{x}))^\top$ is the vector of basis functions. The parameter $\boldsymbol{\alpha}$ is learned so that the following $J(\boldsymbol{\alpha})$ is minimized:

$$\begin{aligned} J(\boldsymbol{\alpha}) &= \frac{1}{2} \int \Big(w_{\boldsymbol{\alpha}}(\boldsymbol{x}) - w_\beta(\boldsymbol{x})\Big)^2 \Big(\beta p'(\boldsymbol{x}) + (1-\beta) p(\boldsymbol{x})\Big) \mathrm{d}\boldsymbol{x} \\ &= \frac{1}{2} \int \boldsymbol{\alpha}^\top \boldsymbol{\psi}(\boldsymbol{x}) \boldsymbol{\psi}(\boldsymbol{x})^\top \boldsymbol{\alpha} \Big(\beta p'(\boldsymbol{x}) + (1-\beta) p(\boldsymbol{x})\Big) \mathrm{d}\boldsymbol{x} \\ &\quad - \int \boldsymbol{\alpha}^\top \boldsymbol{\psi}(\boldsymbol{x}) p'(\boldsymbol{x}) \mathrm{d}\boldsymbol{x} + C, \end{aligned}$$

1. Prepare candidates of models: $\{\mathcal{M}_j\}_j$.
2. Split training samples $\mathcal{D} = \{(\boldsymbol{x}_i, y_i)\}_{i=1}^n$ into t disjoint subsets of (approximately) the same size: $\{\mathcal{D}_\ell\}_{\ell=1}^t$.
3. For each model candidate $\mathcal{M}_j$
   (a) For each split $\ell = 1, \ldots, t$
      i. Obtain learned function $\widehat{f}_j^{(\ell)}(\boldsymbol{x})$ using model $\mathcal{M}_j$ from all training samples without $\mathcal{D}_\ell$.
      ii. Compute the average prediction error $\widehat{G}_j^{(\ell)}$ of $\widehat{f}_j^{(\ell)}(\boldsymbol{x})$ for holdout samples $\mathcal{D}_\ell$.
         - Regression (squared loss):
$$\widehat{G}_j^{(\ell)} = \frac{1}{|\mathcal{D}_\ell|} \sum_{(\boldsymbol{x},y)\in\mathcal{D}_\ell} w(\boldsymbol{x})\left(y - \widehat{f}_j^{(\ell)}(\boldsymbol{x})\right)^2,$$
           where $|\mathcal{D}_\ell|$ denotes the number of elements in the set $\mathcal{D}_\ell$.
         - Classification (0/1-loss):
$$\widehat{G}_j^{(\ell)} = \frac{1}{|\mathcal{D}_\ell|} \sum_{(\boldsymbol{x},y)\in\mathcal{D}_\ell} \frac{w(\boldsymbol{x})}{2}\left(1 - \operatorname{sign}\left(\widehat{f}_j^{(\ell)}(\boldsymbol{x})y\right)\right).$$
   (b) Compute the average prediction error $\widehat{G}_j$ over all t splits:
$$\widehat{G}_j = \frac{1}{t}\sum_{\ell=1}^t \widehat{G}_j^{(\ell)}.$$
4. Choose the model $\mathcal{M}_{\widehat{j}}$ that minimizes the average prediction error:
$$\widehat{j} = \underset{j}{\operatorname{argmin}}\ \widehat{G}_j.$$
5. Obtain the final function approximator using chosen model $\mathcal{M}_{\widehat{j}}$ from all training samples $\{(\boldsymbol{x}_i, y_i)\}_{i=1}^n$.

FIGURE 33.8

Algorithm of importance weighted cross validation.

where $C = \frac{1}{2}\int w_\beta(\boldsymbol{x})p'(\boldsymbol{x})\mathrm{d}\boldsymbol{x}$ is independent of α and thus can be ignored. Approximating the expectations by sample averages and adding the ℓ_2-regularizer yield the following optimization problem:

$$\min_{\alpha} \left[\frac{1}{2} \alpha^\top \widehat{G}_\beta \alpha - \alpha^\top \widehat{h} + \frac{\lambda}{2} \|\alpha\|^2 \right],$$

where

$$\widehat{G}_\beta = \frac{\beta}{n'} \sum_{i'=1}^{n'} \psi(x'_{i'}) \psi(x'_{i'})^\top + \frac{1-\beta}{n} \sum_{i=1}^{n} \psi(x_i) \psi(x_i)^\top,$$

$$\widehat{h} = \frac{1}{n'} \sum_{i'=1}^{n'} \psi(x'_{i'}).$$

The minimizer $\widehat{\alpha}$ can be obtained analytically as

$$\widehat{\alpha} = \left(\widehat{G} + \lambda I \right)^{-1} \widehat{h}.$$

This method is called *LS relative density ratio estimation* [121]. The regularization parameter λ and parameters included in basis functions ψ can be optimized by cross validation with respect to the squared error J.

A MATLAB code of LS relative density ratio estimation for the Gaussian kernel model,

$$w_\alpha(x) = \sum_{j=1}^{n} \alpha_j \exp\left(-\frac{\|x - x_j\|^2}{2h^2} \right),$$

is provided in Fig. 33.9, and its behavior is illustrated in Fig. 33.10. This shows that the true relative importance function is nicely estimated.

33.3 CLASS-BALANCE CHANGE ADAPTATION

In the previous section, semisupervised learning methods for covariate shift adaptation were introduced, which can be naturally applied in regression. On the other hand, in classification, *class-balance change* is a natural situation, where the class-prior probabilities differ in $\{(x_i, y_i)\}_{i=1}^{n}$ and $\{x'_{i'}\}_{i'=1}^{n'}$ (see Fig. 33.11), but the class-conditional probability $p(x|y)$ remains unchanged. In this section, adaptation methods for class-balance change are introduced.

33.3.1 CLASS-BALANCE WEIGHTED LEARNING

The bias caused by class-balance change can be canceled by *class-balance weighted learning*.

More specifically, the class-prior ratio $p'(y)/p(y)$ is used as a weighting factor, where $p(y)$ and $p'(y)$ are the class-prior probabilities for $\{x_i\}_{i=1}^{n}$ and $\{x'_{i'}\}_{i'=1}^{n'}$, respectively. For example, in the case of LS, the learning criterion is given by

$$\min_{\theta} \frac{1}{2} \sum_{i=1}^{n} \frac{p'(y_i)}{p(y_i)} \left(f_\theta(x_i) - y_i \right)^2,$$

```
n=300; x=randn(n,1); y=randn(n,1)+0.5;
hhs=2*[1 5 10].^2; ls=10.^[-3 -2 -1]; m=5; b=0.5;
x2=x.^2; xx=repmat(x2,1,n)+repmat(x2',n,1)-2*x*x';
y2=y.^2; yx=repmat(y2,1,n)+repmat(x2',n,1)-2*y*x';
u=mod(randperm(n),m)+1; v=mod(randperm(n),m)+1;

for hk=1:length(hhs)
  hh=hhs(hk); k=exp(-xx/hh); r=exp(-yx/hh);
  for i=1:m
    ki=k(u~=i,:); ri=r(v~=i,:); h=mean(ki)';
    kc=k(u==i,:); rj=r(v==i,:);
    G=b*ki'*ki/sum(u~=i)+(1-b)*ri'*ri/sum(v~=i);
    for lk=1:length(ls)
      l=ls(lk); a=(G+l*eye(n))\h; kca=kc*a;
      g(hk,lk,i)=b*mean(kca.^2)+(1-b)*mean((rj*a).^2);
      g(hk,lk,i)=g(hk,lk,i)/2-mean(kca);
end, end, end
[gl,ggl]=min(mean(g,3),[],2); [ghl,gghl]=min(gl);
L=ls(ggl(gghl)); HH=hhs(gghl);
k=exp(-xx/HH); r=exp(-yx/HH);
s=r*((b*k'*k/n+(1-b)*r'*r/n+L*eye(n))\(mean(k)'));
figure(1); clf; hold on; plot(y,s,'rx');
```

FIGURE 33.9

MATLAB code for LS relative density ratio estimation for Gaussian kernel model.

which is called *class-balance weighted LS*. Beyond LS, this class-balance change adaptation technique can be applied to various classification methods introduced in Part 4.

Model selection of class-balance weighted learning may be carried out by *class-balance weighted cross validation*, which is essentially the same as *importance weighted cross validation* introduced in Section 33.2.3, but $p'(y_i)/p(y_i)$ is used as a weight.

33.3.2 CLASS-BALANCE ESTIMATION

To use class-balance weighted learning, the class-prior probabilities $p(y)$ and $p'(y)$ are needed, which are often unknown in practice. For labeled samples $\{(\boldsymbol{x}_i, y_i)\}_{i=1}^n$, estimating the class prior $p(y)$ is straightforward by n_y/n, where n_y denotes the number of samples in class y. However, $p'(y)$ cannot be estimated naively because of

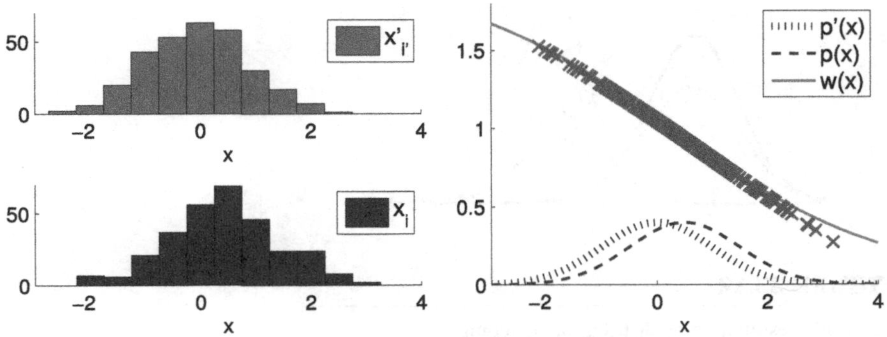

(a) Input samples (b) Relative importance ($\beta = 0.5$)

FIGURE 33.10

Example of LS relative density ratio estimation. ×'s in the right plot show estimated relative importance values at $\{x_i\}_{i=1}^n$.

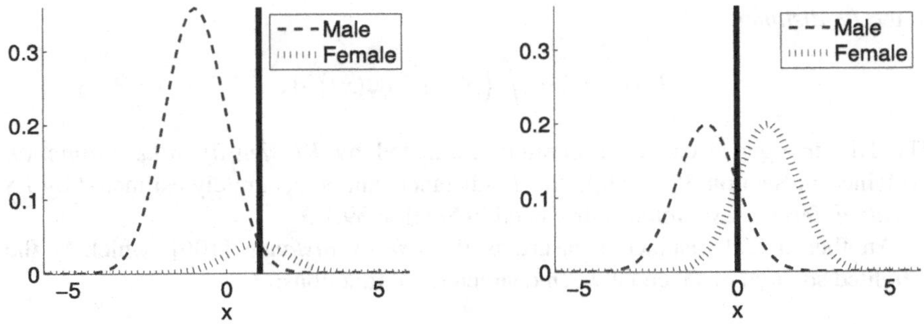

(a) Labeled samples $\{(x_i, y_i)\}_{i=1}^n$ (b) Unlabeled samples $\{x'_{i'}\}_{i'=1}^{n'}$

FIGURE 33.11

Class-balance change, which affects the decision boundary.

lack of output values $\{y'_{i'}\}_{i'=1}^{n'}$ for $\{x'_{i'}\}_{i'=1}^{n'}$. Below, a practical estimator of $p'(y)$ is introduced [38].

The basic idea of estimating $p'(y)$ is to fit a mixture $q_\pi(x)$ of classwise input densities $p(x|y)$ to $p'(x)$ (Fig. 33.12):

$$q_\pi(x) = \sum_{y=1}^{c} \pi_y p(x|y),$$

where c denotes the number of classes and the coefficient π_y corresponds to $p'(y)$.

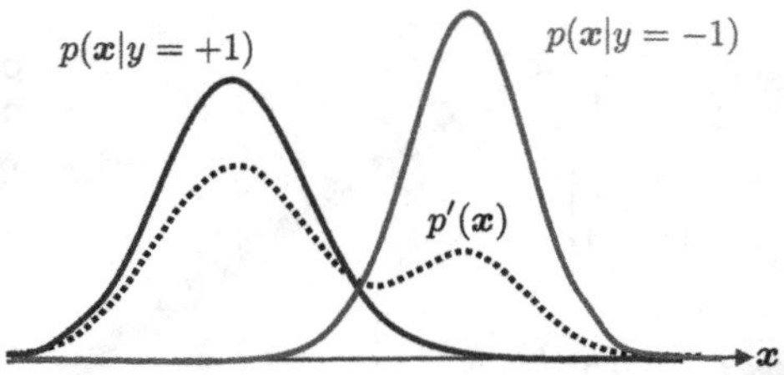

FIGURE 33.12

Class-prior estimation by distribution matching.

Matching of q_π to p' can be performed, for example, under the *KL divergence* (see Section 14.2),

$$\mathrm{KL}(p'\|q_\pi) = \int p'(\boldsymbol{x}) \log \frac{p'(\boldsymbol{x})}{q_\pi(\boldsymbol{x})} \mathrm{d}\boldsymbol{x},$$

or the *L_2-distance*:

$$L_2(p', q_\pi) = \int \left(p'(\boldsymbol{x}) - q_\pi(\boldsymbol{x})\right)^2 \mathrm{d}\boldsymbol{x}.$$

The KL divergence can be accurately estimated by *KL density ratio estimation* explained in Section 38.3, while the L_2-distance can be accurately estimated by *LS density difference estimation* introduced in Section 39.1.3.

Another useful distance measure is the *energy distance* [106], which is the weighted squared distance between characteristic functions:

$$D_{\mathrm{E}}(p', q_\pi) = \int_{\mathbb{R}^d} \|\varphi_{p'}(\boldsymbol{t}) - \varphi_{q_\pi}(\boldsymbol{t})\|^2 \left(\frac{\pi^{\frac{d+1}{2}}}{\Gamma\left(\frac{d+1}{2}\right)} \|\boldsymbol{t}\|^{d+1} \right)^{-1} \mathrm{d}\boldsymbol{t},$$

where $\|\cdot\|$ denotes the Euclidean norm, φ_p denotes the *characteristic function* (see Section 2.4.3) of p, $\Gamma(\cdot)$ is the *gamma function* (see Section 4.3), and d denotes the dimensionality of $\boldsymbol{x}$. Thanks to the careful design of the weight function, the energy distance can be equivalently expressed as

$$\begin{aligned} D_{\mathrm{E}}(p', q_\pi) &= 2\mathbb{E}_{\boldsymbol{x}' \sim p', \boldsymbol{x} \sim q_\pi} \|\boldsymbol{x}' - \boldsymbol{x}\| - \mathbb{E}_{\boldsymbol{x}', \widetilde{\boldsymbol{x}}' \sim p'} \|\boldsymbol{x}' - \widetilde{\boldsymbol{x}}'\| - \mathbb{E}_{\boldsymbol{x}, \widetilde{\boldsymbol{x}} \sim q_\pi} \|\boldsymbol{x} - \widetilde{\boldsymbol{x}}\| \\ &= 2\boldsymbol{\pi}^\top \boldsymbol{b} - \boldsymbol{\pi}^\top \boldsymbol{A} \boldsymbol{\pi} + C, \end{aligned}$$

where $\mathbb{E}_{\boldsymbol{x}' \sim p'}$ denotes the expectation with respect to $\boldsymbol{x}'$ following density p' and C is a constant independent of $\boldsymbol{\pi}$. $\boldsymbol{A}$ is the $c \times c$ symmetric matrix and $\boldsymbol{b}$ is the c-dimensional vector defined as

$$A_{y,\widetilde{y}} = \mathbb{E}_{\boldsymbol{x} \sim p(\boldsymbol{x}|y), \widetilde{\boldsymbol{x}} \sim p(\boldsymbol{x}|\widetilde{y})} \|\boldsymbol{x} - \widetilde{\boldsymbol{x}}\|,$$

```
x=[[randn(90,1)-2; randn(10,1)+2] 2*randn(100,1)];
x(:,3)=1; y=[ones(90,1); 2*ones(10,1)]; n=length(y);
X=[[randn(10,1)-2; randn(90,1)+2] 2*randn(100,1)];
X(:,3)=1; Y=[ones(10,1); 2*ones(90,1)]; N=length(Y);
x2=sum(x.^2,2); X2=sum(X.^2,2);
xx=sqrt(repmat(x2,1,n)+repmat(x2',n,1)-2*x*x');
xX=sqrt(repmat(x2,1,N)+repmat(X2',n,1)-2*x*X');
for i=1:2
  s(i)=sum(y==i)/n; b(i)=mean(mean(xX(y==i,:)));
  for j=1:2
    A(i,j)=mean(mean(xx(y==i,y==j)));
end, end
v=(A(1,2)-A(2,2)-b(1)+b(2))/(2*A(1,2)-A(1,1)-A(2,2));
v=min(1,max(0,v)); v(2)=1-v; w=v(y)./s(y); z=2*y-3;
u=x\z; t=(x'*(repmat(w',1,size(x,2)).*x))\(x'*(w'.*z));
figure(1); clf; hold on
plot([-5 5],-(u(3)+[-5 5]*u(1))/u(2),'k--');
plot([-5 5],-(t(3)+[-5 5]*t(1))/t(2),'g-');
plot(X(Y==1,1),X(Y==1,2),'bo');
plot(X(Y==2,1),X(Y==2,2),'rx');
legend('Unweighted','Weighted'); axis([-5 5 -10 10])
```

FIGURE 33.13

MATLAB code for class-balance weighted LS.

$$b_y = \mathbb{E}_{x' \sim p', x \sim p(x|y)} \|x' - x\|.$$

Although $D_E(p', q_{\pi})$ is a concave function with respect to $\pi = (\pi_1, \ldots, \pi_c)^\top$, it is a convex function with respect to $\pi_1, \ldots, \pi_{c-1}$ for $\pi_c = 1 - \sum_{y=1}^{c-1} \pi_y$ and thus its minimizer can be easily obtained [61]. For example, when $c = 2$ with $\pi_1 = \pi$ and $\pi_2 = 1 - \pi$, $D_E(p', q_{\pi})$ can be expressed as a function of π up to a constant as

$$J(\pi) = a\pi^2 - 2b\pi,$$

where

$$a = 2A_{1,2} - A_{1,1} - A_{2,2} = D_E(p(x|y=1), p(x|y=2)) \geq 0,$$
$$b = A_{1,2} - A_{2,2} - b_1 + b_2.$$

Since $J(\pi)$ is convex with respect to π, its minimizer is given analytically as $\min(1, \max(0, b/a))$. Note that $A_{y,\widetilde{y}}$ and b_y can be empirically approximated as

$$\widehat{A}_{y,\widetilde{y}} = \frac{1}{n_y n_{\widetilde{y}}} \sum_{i:y_i=y} \sum_{\widetilde{i}:y_{\widetilde{i}}=\widetilde{y}} \|x_i - x_{\widetilde{i}}\|,$$

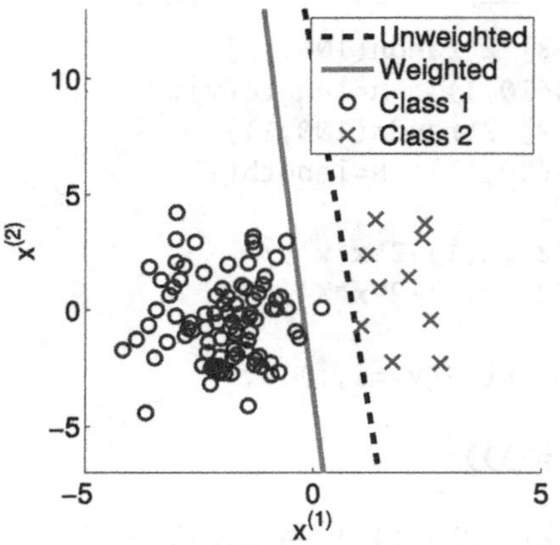

(a) $\{x_i\}_{i=1}^{n}$ with $(n_1, n_2) = (90, 10)$

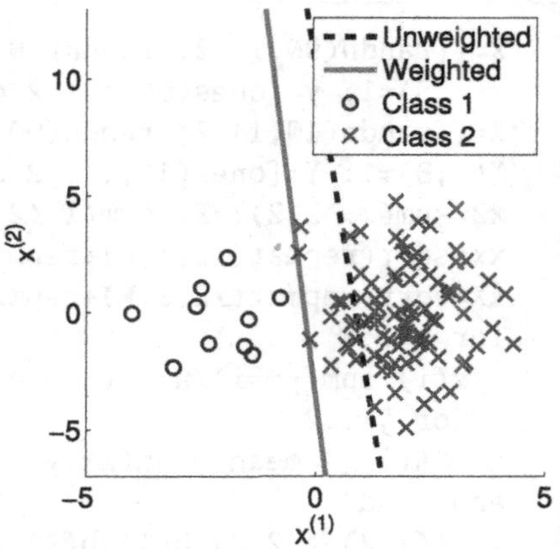

(b) $\{x'_{i'}\}_{i'=1}^{n'}$ with $(n'_1, n'_2) = (10, 90)$

FIGURE 33.14

Example of class-balance weighted LS. The test class priors are estimated as $\widehat{p}'(y = 1) = 0.18$ and $\widehat{p}'(y = 2) = 0.82$, which are used as weights in class-balance weighted LS.

$$\widehat{b}_y = \frac{1}{n'n_y} \sum_{i'=1}^{n'} \sum_{i:y_i=y} \|x'_{i'} - x_i\|.$$

A MATLAB code for class-balance weighted LS is provided in Fig. 33.13, and its behavior is illustrated in Fig. 33.14. This shows that the class prior can be estimated reasonably well and class-balance weighted learning contributes to improving the classification accuracy for test input points $\{x'_{i'}\}_{i'=1}^{n'}$.

CHAPTER

MULTITASK LEARNING 34

CHAPTER CONTENTS

When solving multiple related learning tasks, solving them together simultaneously by sharing information is expected to give a better solution than solving them independently. This is the basic idea of *multitask learning* [24]. In this chapter, practically useful multitask learning methods are introduced.

Let us consider T tasks indexed by $t = 1, \ldots, T$ and assume that each input-output paired training sample $(\boldsymbol{x}_i, y_i)$ is accompanied with task index t_i:

$$\{(\boldsymbol{x}_i, y_i, t_i) \mid t_i \in \{1, \ldots, T\}\}_{i=1}^{n}.$$

34.1 TASK SIMILARITY REGULARIZATION

In this section, a multitask learning method with *task similarity regularization* [40] is introduced.

34.1.1 FORMULATION

For the tth learning task, let us employ a linear-in-parameter model,

$$\sum_{j=1}^{b} \theta_{t,j} \phi_j(\boldsymbol{x}) = \boldsymbol{\theta}_t^{\top} \boldsymbol{\phi}(\boldsymbol{x}),$$

where $\boldsymbol{\theta}_t = (\theta_{t,1},\ldots,\theta_{t,b})^\top$ is the parameter vector for the tth task and the basis functions $\boldsymbol{\phi}(\boldsymbol{x})$ are *common* to all tasks.

The idea of task similarity regularization is to impose the parameters $\boldsymbol{\theta}_1,\ldots,\boldsymbol{\theta}_T$ to take similar values and learn all parameters,

$$\boldsymbol{\theta} = (\boldsymbol{\theta}_1^\top,\ldots,\boldsymbol{\theta}_T^\top)^\top \in \mathbb{R}^{bT},$$

simultaneously. Let us employ the *ℓ_2-regularized LS* (see Section 23.2) and learn $\boldsymbol{\theta}$ so that the following $J(\boldsymbol{\theta})$ is minimized:

$$J(\boldsymbol{\theta}) = \frac{1}{2}\sum_{i=1}^{n}\left(y_i - \boldsymbol{\phi}(\boldsymbol{x}_i)^\top\boldsymbol{\theta}_{t_i}\right)^2 + \frac{1}{2}\sum_{t=1}^{T}\lambda_t\|\boldsymbol{\theta}_t\|^2 + \frac{1}{4}\sum_{t,t'=1}^{T}\gamma_{t,t'}\|\boldsymbol{\theta}_t - \boldsymbol{\theta}_{t'}\|^2,$$

where $\lambda_t \geq 0$ is the ℓ_2-regularization parameter for the tth task, and $\gamma_{t,t'} \geq 0$ is the similarity between the tth task and the t'th task. If $\gamma_{t,t'} = 0$ for all $t,t' = 1,\ldots,T$, the third term in $J(\boldsymbol{\theta})$ disappears. Then there is no interaction between tasks and thus this corresponds to merely learning T tasks separately. On the other hand, if $\gamma_{t,t'} > 0$, $\boldsymbol{\theta}_t$ and $\boldsymbol{\theta}_{t'}$ are imposed to be close to each other and thus information can be implicitly shared between the tth task and the t'th task. If all $\gamma_{t,t'}$ are large enough, all solutions are imposed to be the same and thus a single solution that is common to all tasks is obtained from all training samples $\{(\boldsymbol{x}_i, y_i)\}_{i=1}^n$.

34.1.2 ANALYTIC SOLUTION

$J(\boldsymbol{\theta})$ can be compactly expressed as

$$J(\boldsymbol{\theta}) = \frac{1}{2}\|\boldsymbol{y} - \boldsymbol{\Psi}\boldsymbol{\theta}\|^2 + \frac{1}{2}\boldsymbol{\theta}^\top(\boldsymbol{C}\otimes\boldsymbol{I}_b)\boldsymbol{\theta}, \tag{34.1}$$

where

$$\boldsymbol{y} = (y_1,\ldots,y_n)^\top \in \mathbb{R}^n,$$
$$\boldsymbol{\Psi} = (\boldsymbol{\psi}_{t_1}(\boldsymbol{x}_1),\ldots,\boldsymbol{\psi}_{t_n}(\boldsymbol{x}_n))^\top \in \mathbb{R}^{n\times bT},$$
$$\boldsymbol{\psi}_t(\boldsymbol{x}) = \left(\boldsymbol{0}_{b(t-1)}^\top, \boldsymbol{\phi}(\boldsymbol{x})^\top, \boldsymbol{0}_{b(T-t)}^\top\right)^\top \in \mathbb{R}^{bT}.$$

$\boldsymbol{C}$ is the $T\times T$ matrix defined as

$$C_{t,t'} = \begin{cases} \lambda_t + \sum_{t''=1}^{T}\gamma_{t,t''} - \gamma_{t,t} & (t = t'), \\ -\gamma_{t,t'} & (t \neq t'), \end{cases}$$

and $\otimes$ denotes the *Kronecker product*, i.e. for $\boldsymbol{E} \in \mathbb{R}^{m\times n}$ and $\boldsymbol{F} \in \mathbb{R}^{p\times q}$,

$$\boldsymbol{E}\otimes\boldsymbol{F} = \begin{pmatrix} E_{1,1}\boldsymbol{F} & \cdots & E_{1,n}\boldsymbol{F} \\ \vdots & \ddots & \vdots \\ E_{m,1}\boldsymbol{F} & \cdots & E_{m,n}\boldsymbol{F} \end{pmatrix} \in \mathbb{R}^{mp\times nq}.$$

Then the minimizer $\widehat{\boldsymbol{\theta}}$ of $J(\boldsymbol{\theta})$ satisfies

$$(\boldsymbol{\Psi}^\top \boldsymbol{\Psi} + \boldsymbol{C} \otimes \boldsymbol{I}_b)\boldsymbol{\theta} = \boldsymbol{\Psi}^\top \boldsymbol{y}, \tag{34.2}$$

and $\widehat{\boldsymbol{\theta}}$ can be obtained analytically as

$$\widehat{\boldsymbol{\theta}} = \left(\boldsymbol{\Psi}^\top \boldsymbol{\Psi} + \boldsymbol{C} \otimes \boldsymbol{I}_b\right)^{-1} \boldsymbol{\Psi}^\top \boldsymbol{y}. \tag{34.3}$$

34.1.3 EFFICIENT COMPUTATION FOR MANY TASKS

The size of matrix $(\boldsymbol{\Psi}^\top \boldsymbol{\Psi} + \boldsymbol{C} \otimes \boldsymbol{I}_b)$ is $bT \times bT$, and thus directly computing the solution by the above analytic form is not tractable if the number of tasks, T, is large. However, since the rank of matrix $\boldsymbol{\Psi}^\top \boldsymbol{\Psi}$ is at most n, the solution can be computed efficiently if $n < bT$.

More specifically, the solution $\widehat{\boldsymbol{\theta}}_t^\top \boldsymbol{\phi}(\boldsymbol{x})$ for the tth task can be expressed as

$$\widehat{\boldsymbol{\theta}}_t^\top \boldsymbol{\phi}(\boldsymbol{x}) = \widehat{\boldsymbol{\theta}}^\top \boldsymbol{\psi}_t(\boldsymbol{x}) = \boldsymbol{y}^\top \boldsymbol{A}^{-1} \boldsymbol{b}_t. \tag{34.4}$$

Here, $\boldsymbol{A}$ and $\boldsymbol{b}_t$ are the $n \times n$ matrix and n-dimensional vector defined as

$$\begin{aligned}
A_{i,i'} &= [\boldsymbol{\Psi}(\boldsymbol{C}^{-1} \otimes \boldsymbol{I}_b)\boldsymbol{\Psi}^\top + \boldsymbol{I}_n]_{i,i'} \\
&= [\boldsymbol{C}^{-1}]_{t_i,t_{i'}} \boldsymbol{\phi}(\boldsymbol{x}_i)^\top \boldsymbol{\phi}(\boldsymbol{x}_{i'}) + \begin{cases} 1 & (i = i'), \\ 0 & (i \neq i'), \end{cases} \\
b_{t,i} &= [\boldsymbol{\Psi}(\boldsymbol{C}^{-1} \otimes \boldsymbol{I}_b)\boldsymbol{\psi}_t(\boldsymbol{x})]_i \\
&= [\boldsymbol{C}^{-1}]_{t,t_i} \boldsymbol{\phi}(\boldsymbol{x}_i)^\top \boldsymbol{\phi}(\boldsymbol{x}),
\end{aligned}$$

where $[\boldsymbol{C}^{-1}]_{t,t'}$ denotes the (t,t')th element of matrix $\boldsymbol{C}^{-1}$. Since the size of matrix $\boldsymbol{A}$ and the dimensionality of vector $\boldsymbol{b}_t$ are independent of the number of tasks, T, Eq. (34.4) would be computationally more efficient than Eq. (34.3) if $n < bT$. Moreover, since $\boldsymbol{A}^{-1}$ is independent of task index t, it needs to be computed only once across all tasks. Note that the following formulas regarding the Kronecker product were utilized in the derivation of Eq. (34.4):

$$\begin{aligned}
\boldsymbol{\Psi}\left(\boldsymbol{\Psi}^\top \boldsymbol{\Psi} + \boldsymbol{C} \otimes \boldsymbol{I}_b\right)^{-1} &= (\boldsymbol{\Psi}(\boldsymbol{C} \otimes \boldsymbol{I}_b)^{-1}\boldsymbol{\Psi}^\top + \boldsymbol{I}_n)^{-1}\boldsymbol{\Psi}(\boldsymbol{C} \otimes \boldsymbol{I}_b)^{-1}, \\
(\boldsymbol{C} \otimes \boldsymbol{I}_b)^{-1} &= \boldsymbol{C}^{-1} \otimes \boldsymbol{I}_b.
\end{aligned}$$

A MATLAB code for multitask LS is provided in Fig. 34.1, and its behavior is illustrated in Fig. 34.2. This shows that multitask classification outperforms single-task classification.

In the above multitask method, the task similarity $\gamma_{t,t'}$ was assumed to be known. When $\gamma_{t,t'}$ is unknown, it may be alternately learned as described in Fig. 34.3.

```
n=2; T=6; y=[ones(n/2,T); -ones(n/2,T)];
x=[randn(2,n,T); ones(1,n,T)]; r(1,1,:)=pi*[1:T]/T/10;
c=repmat(cos(r),[1 n/2]); x(1,:,:)=x(1,:,:)+[c -c];
s=repmat(sin(r),[1 n/2]); x(2,:,:)=x(2,:,:)+[s -s];
Ci=inv(-ones(T,T)+diag(T*ones(T,1)+0.01));
a=repmat([1:T],[n 1]); a=a(:); m=20; X=linspace(-4,4,m);
b=repmat(X,[m 1]); bt=b'; XX=[b(:)'; bt(:)'; ones(1,m^2)];
yAi=y(:)'*inv(Ci(a,a).*(x(:,:)'*x(:,:))+eye(n*T));

figure(1); clf; colormap([1 0.7 1; 0.7 1 1]);
for k=1:T
%  Y=yAi*(repmat(Ci(a,k),[1 3]).*x(:,:)')*XX;
  q=x(:,:,k); Y=((q*q'+0.01*eye(3))\(q*y(:,k)))'*XX;
  subplot(2,3,k); hold on; contourf(X,X,reshape(Y,m,m));
  plot(x(1,y(:,k)==1,k),x(2,y(:,k)==1,k),'bo');
  plot(x(1,y(:,k)==-1,k),x(2,y(:,k)==-1,k),'rx');
  plot(99*sin(r(k))*[1 -1],99*cos(r(k))*[-1 1],'k--');
  axis([-4 4 -4 4]);
end
```

FIGURE 34.1

MATLAB code for multitask LS.

34.2 MULTIDIMENSIONAL FUNCTION LEARNING

In this section, the problem of *multidimensional function learning* is discussed, which can be regarded as a special case of multitask learning.

34.2.1 FORMULATION

Let us consider the problem of learning a T-dimensional function,

$$f(x) = (f_1(x), \ldots, f_T(x))^\top,$$

from input-output paired training samples:

$$\{(x_i, y_i) \mid x_i \in \mathbb{R}^d,\ y_i = (y_i^{(1)}, \ldots, y_i^{(T)})^\top \in \mathbb{R}^T\}_{i=1}^n.$$

If each dimension of output y is regarded as a task, the problem of multidimensional function learning can be regarded as multitask learning. The difference is that multidimensional function learning shares input points across all tasks, while input points are generally different in multitask learning (Fig. 34.4). Thus, the number of training samples in multidimensional function learning is actually nT in the context of multitask learning.

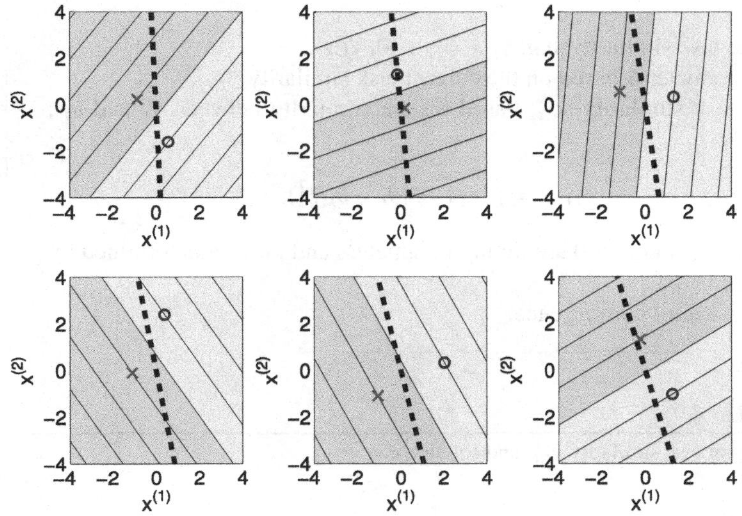

(a) Single-task classification

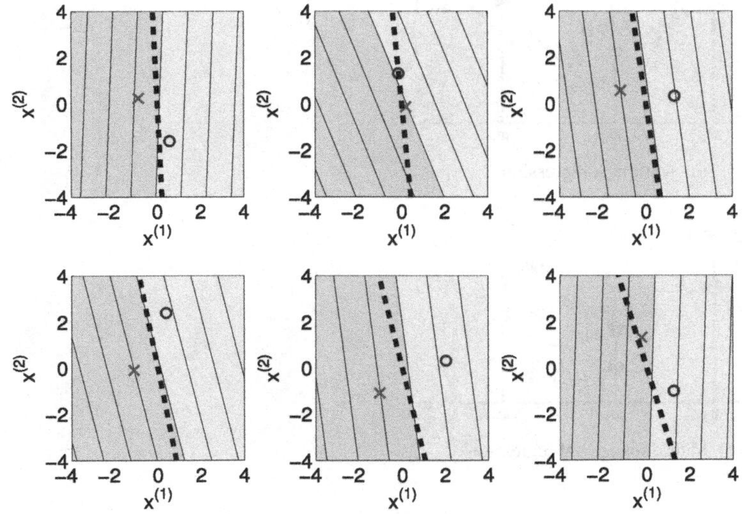

(b) Multitask classification

FIGURE 34.2

Examples of multitask LS. The dashed lines denote true decision boundaries and the contour lines denote learned results.

Multidimensional function learning for classification is specifically called *multilabel classification*, which can be regarded as a generalization of *multiclass*

1. Initialize task similarity, e.g. $\gamma_{t,t'} = \gamma > 0, \forall t, t'$.
2. Learn parameter $\boldsymbol{\theta}$ based on the current task similarity $\gamma_{t,t'}$.
3. Update task similarity $\gamma_{t,t'}$ based on the similarity between $\boldsymbol{\theta}_t$ and $\boldsymbol{\theta}_{t'}$, e.g.

$$\gamma_{t,t'} = \eta \exp(-\kappa \|\boldsymbol{\theta}_t - \boldsymbol{\theta}_{t'}\|^2),$$

   where $\eta \geq 0$ and $\kappa \geq 0$ are tuning parameters and may be determined by cross validation.
4. Iterate 2–3 until convergence.

FIGURE 34.3

Alternate learning of task similarity $\gamma_{t,t'}$ and solution $\boldsymbol{\theta}$.

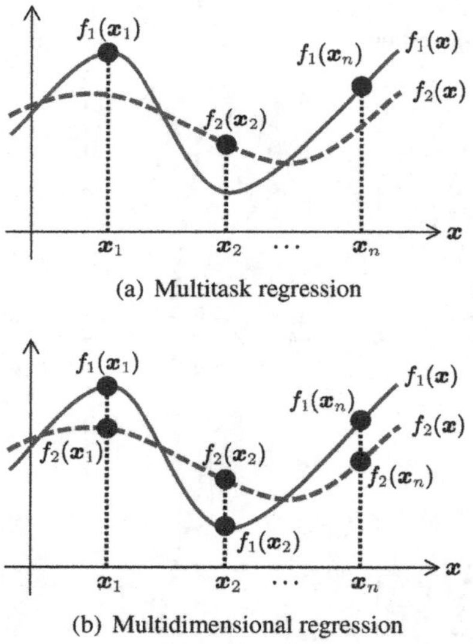

FIGURE 34.4

Multidimensional function learning.

classification (see Section 26.3). A pattern $\boldsymbol{x}$ belongs to one of the classes exclusively in multiclass classification, while $\boldsymbol{x}$ can belong to multiple classes simultaneously in multilabel classification. For example, an image can belong to the classes of "human," "dog," and "building" at the same time in image recognition, and a sound clip can belong to the classes of "conversation," "noise," and "background music" at the same time in audio recognition. In multilabel classification, co-occurrence of multiple classes can be utilized by multitask learning. For example, an image containing a dog often contains a human at the same time.

However, in multidimensional function learning, the number of parameters is bT and the *actual* number of samples is nT, both of which are proportional to the number of tasks, T. Thus, both the solutions explained in Section 34.1.2 and Section 34.1.3 are not computationally efficient in multidimensional function learning.

34.2.2 EFFICIENT ANALYTIC SOLUTION

Let us arrange the parameter vectors $\boldsymbol{\theta}_1,\ldots,\boldsymbol{\theta}_T$, the output vectors $\boldsymbol{y}_1,\ldots,\boldsymbol{y}_n$, and the basis vectors $\boldsymbol{\phi}(\boldsymbol{x}_1),\ldots,\boldsymbol{\phi}(\boldsymbol{x}_n)$ into matrices as

$$\boldsymbol{\Theta} = (\boldsymbol{\theta}_1,\ldots,\boldsymbol{\theta}_T) \in \mathbb{R}^{b\times T},$$
$$\boldsymbol{Y} = (\boldsymbol{y}_1,\ldots,\boldsymbol{y}_n)^\top \in \mathbb{R}^{n\times T},$$
$$\boldsymbol{\Phi} = (\boldsymbol{\phi}(\boldsymbol{x}_1),\ldots,\boldsymbol{\phi}(\boldsymbol{x}_n))^\top \in \mathbb{R}^{n\times b}.$$

Then, Eq. (34.2), which multitask solutions should satisfy, can be rewritten as

$$\boldsymbol{\Phi}^\top\boldsymbol{\Phi}\boldsymbol{\Theta} + \boldsymbol{\Theta}\boldsymbol{C} = \boldsymbol{\Phi}^\top\boldsymbol{Y}.$$

This form is known as the *continuous Sylvester equation*, which often arises in control theory [94], and can be solved efficiently, as explained in Fig. 34.5.

A MATLAB code for multidimensional regression is provided in Fig. 34.6, and its behavior is illustrated in Fig. 34.7. This shows that multitask regression outperforms single-task regression.

34.3 MATRIX REGULARIZATION

The multitask learning method introduced in Section 34.1 explicitly used task similarity $\gamma_{t,t'}$ to control the amount of information sharing. In this section, another multitask learning approach is introduced, which does not involve task similarity.

34.3.1 PARAMETER MATRIX REGULARIZATION

The basic idea is to share information across multiple tasks by regularizing the *parameter matrix*:

$$\boldsymbol{\Theta} = (\boldsymbol{\theta}_1,\ldots,\boldsymbol{\theta}_T) \in \mathbb{R}^{b\times T}.$$

More specifically, for a squared loss function, the multitask learning criterion to be minimized with respect to $\boldsymbol{\Theta}$ is given by

$$\frac{1}{2}\sum_{i=1}^{n}\left(y_i - \boldsymbol{\phi}(\boldsymbol{x}_i)^\top\boldsymbol{\theta}_{t_i}\right)^2 + \lambda R(\boldsymbol{\Theta}),$$

where $R(\boldsymbol{\Theta})$ denotes some regularization functional for $\boldsymbol{\Theta}$.

The *continuous Sylvester equation* for some matrices $\boldsymbol{A} \in \mathbb{R}^{b \times b}$, $\boldsymbol{B} \in \mathbb{R}^{T \times T}$, and $\boldsymbol{Z} \in \mathbb{R}^{b \times T}$ with respect to $\boldsymbol{\Theta} \in \mathbb{R}^{b \times T}$ is given by

$$\boldsymbol{A}\boldsymbol{\Theta} + \boldsymbol{\Theta}\boldsymbol{B} = \boldsymbol{Z}.$$

The use of the *Kronecker product* and the *vectorization operator* (see Fig. 6.5) allows us to rewrite the above equation as

$$(\boldsymbol{I}_T \otimes \boldsymbol{A} + \boldsymbol{B} \otimes \boldsymbol{I}_b)\mathrm{vec}(\boldsymbol{\Theta}) = \mathrm{vec}(\boldsymbol{Z}),$$

where $\boldsymbol{I}_T$ denotes the $T \times T$ identity matrix. This shows that the continuous Sylvester equation with respect to a $b \times T$ matrix can be seen as a linear equation with respect to a bT-dimensional vector, which is computationally expensive to solve naively. However, the continuous Sylvester equation can be solved more efficiently. For example, let $u_1, \ldots, u_b$ and $\boldsymbol{u}_1, \ldots, \boldsymbol{u}_b$ be *eigenvalues* and *eigenvectors* of $\boldsymbol{A}$ (see Fig. 6.2), and let $v_1, \ldots, v_T$ and $\boldsymbol{v}_1, \ldots, \boldsymbol{v}_T$ be eigenvalues and eigenvectors of $\boldsymbol{B}$, respectively. Then, when $u_j + v_t \neq 0$ for all $j = 1, \ldots, b$ and $t = 1, \ldots, T$, the solution $\widehat{\boldsymbol{\Theta}}$ of the above continuous Sylvester equation with respect to $\boldsymbol{\Theta}$ can be obtained analytically as

$$\widehat{\boldsymbol{\Theta}} = (\boldsymbol{u}_1, \ldots, \boldsymbol{u}_b)\boldsymbol{Q}(\boldsymbol{v}_1, \ldots, \boldsymbol{v}_T)^\top,$$

where $\boldsymbol{Q}$ is the $b \times T$ matrix defined as

$$Q_{j,t} = \frac{\boldsymbol{u}_j^\top \boldsymbol{Z} \boldsymbol{v}_t}{u_j + v_t}.$$

FIGURE 34.5

Continuous Sylvester equation.

If the squared *Frobenius norm* $\|\boldsymbol{\Theta}\|_{\mathrm{Frob}}^2$ is used for regularization, no information is shared across different tasks since this is equivalent to the sum of ℓ_2-norms of parameter vectors $\boldsymbol{\theta}_t$:

$$\|\boldsymbol{\Theta}\|_{\mathrm{Frob}}^2 = \sum_{t=1}^{T}\sum_{j=1}^{b} \Theta_{t,j}^2 = \sum_{t=1}^{T} \|\boldsymbol{\theta}_t\|^2.$$

Thus, a more intricate norm should be used for matrix regularization.

For example, the *trace norm* (see Fig. 24.10) tends to produce a *low-rank* solution:

```
n=30; x=linspace(-3,3,n)'; pix=pi*x; T=3;
y=repmat(sin(pix)./(pix)+0.1*x,1,T)+0.1*repmat([1:T],n,1);
y=y+0.5*randn(n,T); N=1000; X=linspace(-3,3,N)'; piX=pi*X;
Y=repmat(sin(piX)./(piX)+0.1*X,1,T)+0.1*repmat([1:T],N,1);
G=10*ones(T,T); %G=zeros(T,T);
l=0.1; C=l*eye(T)+diag(sum(G))-G; hh=1; x2=x.^2;
k=exp(-(repmat(x2,1,n)+repmat(x2',n,1)-2*x*x')/hh);
K=exp(-(repmat(X.^2,1,n)+repmat(x2',N,1)-2*X*x')/hh);

[U,u]=eig(k'*k); [V,v]=eig(C);
Q=U'*k'*y*V./(repmat(diag(u),1,T)+repmat(diag(v)',n,1));
S=U*Q*V'; F=K*S;

figure(1); clf; hold on; axis([-inf inf -inf inf])
plot(X,Y,'r-'); plot(X,F,'g--'); plot(x,y,'bo');
```

FIGURE 34.6

MATLAB code for multidimensional regression.

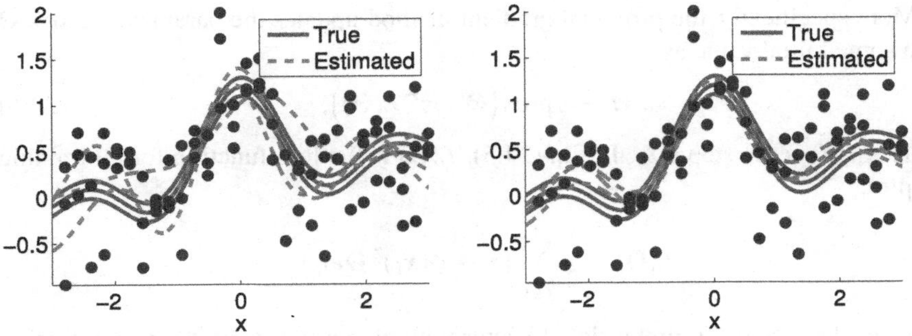

(a) Single-task regression (b) Multitask regression

FIGURE 34.7

Examples of multidimensional regression.

$$\|\boldsymbol{\Theta}\|_{\mathrm{tr}} = \sum_{k=1}^{\min(b,T)} \sigma_k,$$

where σ_k is a *singular value* of $\boldsymbol{\Theta}$ (see Fig. 22.2). Since a low-rank solution in multitask learning confines parameters of each task in a common subspace, information can be shared across different tasks [6]. In Section 23.1, the method of *subspace-constrained LS* was introduced, which confines the LS solution in a given subspace. Multitask learning based on the trace norm can be regarded as automatically learning the subspace with the help of other learning tasks.

Another possibility of matrix regularization for multitask learning is to use the $\ell_{2,1}$-norm $\|\boldsymbol{\Theta}\|_{2,1}$:

$$\|\boldsymbol{\Theta}\|_{2,1} = \sum_{t=1}^{T} \sqrt{\sum_{j=1}^{b} \Theta_{t,j}^2},$$

which tends to give a *row-wise* sparse solution (see Section 24.4.4). This means that $\theta_{1,j}, \ldots, \theta_{T,j}$ tend to simultaneously vanish for several $j \in \{1, \ldots, b\}$, and thus *feature selection* can be performed across different tasks, if linear basis function $\boldsymbol{\phi}(\boldsymbol{x}) = \boldsymbol{x}$ is used [8] (see also Section 24.3).

Below, a practical optimization method for multitask LS with trace norm regularization is introduced.

34.3.2 PROXIMAL GRADIENT FOR TRACE NORM REGULARIZATION

Since the trace norm $\|\boldsymbol{\Theta}\|_{\mathrm{tr}}$ is a convex function, the global optimal solution to multitask LS with trace norm regularization can be obtained, for example, by the *proximal gradient method* described in Fig. 34.8.

More specifically, the proximal gradient method updates the parameter matrix $\boldsymbol{\Theta}$ from some initial value as

$$\boldsymbol{\Theta} \longleftarrow \mathrm{prox}\Big(\boldsymbol{\Theta} - \varepsilon \nabla L(\boldsymbol{\Theta})\Big), \tag{34.5}$$

where $\varepsilon > 0$ is the step size. In Eq. (34.5), $L(\boldsymbol{\Theta})$ is the loss function for all training samples:

$$L(\boldsymbol{\Theta}) = \frac{1}{2}\sum_{i=1}^{n}\Big(y_i - \boldsymbol{\phi}(\boldsymbol{x}_i)^\top \boldsymbol{\Theta}\boldsymbol{e}_{t_i}\Big)^2,$$

where $\boldsymbol{e}_t$ denotes the T-dimensional vector with all zeros but the tth element being one, i.e. $\boldsymbol{\Theta}\boldsymbol{e}_t = \boldsymbol{\theta}_t$. $\nabla L(\boldsymbol{\Theta})$ is the gradient of L given by

$$\nabla L(\boldsymbol{\Theta}) = \sum_{i=1}^{n}\Big(\boldsymbol{\phi}(\boldsymbol{x}_i)^\top \boldsymbol{\Theta}\boldsymbol{e}_{t_i} - y_i\Big)\boldsymbol{\phi}(\boldsymbol{x}_i)\boldsymbol{e}_{t_i}^\top,$$

where the following matrix derivative formula is used:

$$\frac{\partial \boldsymbol{\phi}(\boldsymbol{x}_i)^\top \boldsymbol{\Theta}\boldsymbol{e}_{t_i}}{\partial \boldsymbol{\Theta}} = \boldsymbol{\phi}(\boldsymbol{x}_i)\boldsymbol{e}_{t_i}^\top.$$

The *proximal operator* $\mathrm{prox}(\boldsymbol{\Theta})$ in Eq. (34.5) is given by

$$\mathrm{prox}(\boldsymbol{\Theta}) = \underset{\boldsymbol{U}}{\mathrm{argmin}}\left(\|\boldsymbol{U}\|_{\mathrm{tr}} + \frac{1}{2\varepsilon\lambda}\|\boldsymbol{U} - \boldsymbol{\Theta}\|_{\mathrm{Frob}}^2\right),$$

Let us consider the following optimization problem:

$$\min_{\boldsymbol{\theta}} L(\boldsymbol{\theta}) + R(\boldsymbol{\theta}),$$

where $L(\boldsymbol{\theta})$ is a convex differentiable function and $R(\boldsymbol{\theta})$ is a closed convex function. The *proximal gradient method* finds the minimizer of $L(\boldsymbol{\theta}) + R(\boldsymbol{\theta})$ by updating $\boldsymbol{\theta}$ from some initial value as

$$\boldsymbol{\theta} \longleftarrow \operatorname*{argmin}_{\boldsymbol{u}} \left(R(\boldsymbol{u}) + L(\boldsymbol{\theta}) + \nabla L(\boldsymbol{\theta})^\top (\boldsymbol{u} - \boldsymbol{\theta}) + \frac{1}{2\varepsilon}\|\boldsymbol{u} - \boldsymbol{\theta}\|^2 \right),$$

where $\nabla L(\boldsymbol{\theta})$ is the gradient of L with respect to $\boldsymbol{\theta}$ and $\varepsilon > 0$ corresponds to the step size. If the above minimization with respect to $\boldsymbol{u}$ can be solved efficiently (e.g. analytically), the proximal gradient method is computationally efficient. The above update rule can be equivalently expressed as

$$\boldsymbol{\theta} \longleftarrow \operatorname{prox}_{\varepsilon R}\big(\boldsymbol{\theta} - \varepsilon \nabla L(\boldsymbol{\theta})\big),$$

where $\operatorname{prox}_R(\boldsymbol{\theta})$ is called the *proximal operator* defined as

$$\operatorname{prox}_R(\boldsymbol{\theta}) = \operatorname*{argmin}_{\boldsymbol{u}} \left(R(\boldsymbol{u}) + \frac{1}{2}\|\boldsymbol{u} - \boldsymbol{\theta}\|^2 \right).$$

This implies that the proximal gradient method can be regarded as a generalization of the projected gradient method. Indeed, when $R(\boldsymbol{\theta})$ is the *indicator function* for some set $\mathcal{S}$,

$$R(\boldsymbol{\theta}) = \begin{cases} 0 & (\boldsymbol{\theta} \in \mathcal{S}), \\ \infty & (\boldsymbol{\theta} \notin \mathcal{S}), \end{cases}$$

the proximal operator is reduced to an ordinary projection operator and thus the proximal gradient method is a projected gradient method.

FIGURE 34.8

Proximal gradient method.

where $\|\boldsymbol{\Theta}\|_{\text{Frob}}$ denotes the *Frobenius norm*:

$$\|\boldsymbol{\Theta}\|_{\text{Frob}} = \sqrt{\sum_{k_1,k_2=1}^{b_1,b_2} \Theta_{k_1,k_2}^2} = \sqrt{\operatorname{tr}\left(\boldsymbol{\Theta}\boldsymbol{\Theta}^\top\right)}.$$

```
n=2; T=6; y=[ones(n/2,T); -ones(n/2,T)];
x=[randn(2,n,T); ones(1,n,T)]; r(1,1,:)=pi*[1:T]/T/10;
c=repmat(cos(r),[1 n/2]); x(1,:,:)=x(1,:,:)+[c -c];
s=repmat(sin(r),[1 n/2]); x(2,:,:)=x(2,:,:)+[s -s];
t0=randn(3,T); e=0.1; l=4;
for o=1:1000
  for k=1:T
    gt(:,k)=x(:,:,k)*(x(:,:,k)'*t0(:,k)-y(:,k));
  end
  [U,S,V]=svd(t0-e*gt,'econ');
  S=diag(max(0,diag(S)-e*l)); t=U*S*V';
  if norm(t-t0)<0.001, break, end
  t0=t;
end

figure(1); clf; colormap([1 0.7 1; 0.7 1 1]);
m=20; X=linspace(-4,4,m); b=repmat(X,[m 1]); bt=b';
for k=1:T
  Y=t(:,k)'*[b(:)'; bt(:)'; ones(1,m^2)];
  subplot(2,3,k); hold on; contourf(X,X,reshape(Y,m,m));
  plot(x(1,y(:,k)==1,k),x(2,y(:,k)==1,k),'bo');
  plot(x(1,y(:,k)==-1,k),x(2,y(:,k)==-1,k),'rx');
  plot(99*sin(r(k))*[1 -1],99*cos(r(k))*[-1 1],'k--');
  axis([-4 4 -4 4]);
end
```

FIGURE 34.9

MATLAB code for multitask learning with trace norm regularization.

A notable fact is that the above proximal operator can be analytically expressed as follows [78]:

$$\text{prox}(\boldsymbol{\Theta}) = \sum_{k=1}^{\min(b_1,b_2)} \max(0, \sigma_k - \varepsilon\lambda)\boldsymbol{\psi}_k\boldsymbol{\phi}_k^\top,$$

where $\boldsymbol{\psi}_k$, $\boldsymbol{\phi}_k$, and σ_k are a left singular vector, a right singular vector, and a singular value of $\boldsymbol{\Theta}$, respectively:

$$\boldsymbol{\Theta} = \sum_{k=1}^{\min(b_1,b_2)} \sigma_k\boldsymbol{\psi}_k\boldsymbol{\phi}_k^\top.$$

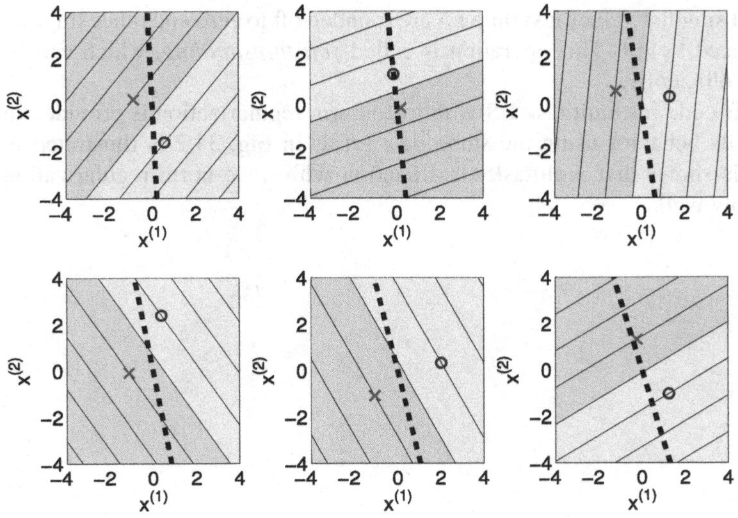

(a) Single-task classification

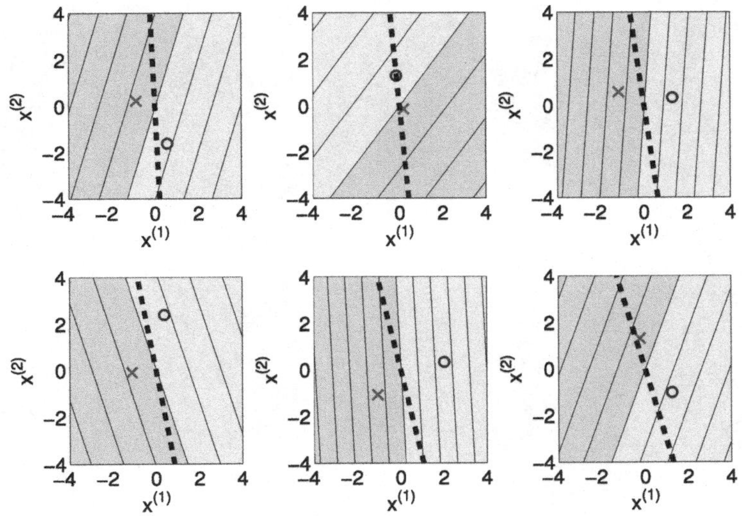

(b) Multitask classification

FIGURE 34.10

Examples of multitask LS with trace norm regularization. The data set is the same as Fig. 34.2. The dashed lines denote true decision boundaries and the contour lines denote learned results.

This means that singular values less than $\varepsilon\lambda$ are rounded off to zero and other singular values are reduced by $\varepsilon\lambda$. This operation is called *soft thresholding*, which can be computed very efficiently.

A MATLAB code for multitask LS with trace norm regularization is provided in Fig. 34.9, and its behavior using the same data set as in Fig. 34.2 is illustrated in Fig. 34.10. This shows that multitask classification with trace norm regularization works reasonably well.

CHAPTER

LINEAR DIMENSIONALITY REDUCTION

35

CHAPTER CONTENTS

Handling high-dimensional data is often cumbersome in practical data analysis. In this chapter, supervised and unsupervised methods of *linear dimensionality reduction* are introduced for reducing the dimensionality of data while preserving intrinsic information contained in the data.

35.1 CURSE OF DIMENSIONALITY

As the dimensionality of input $\boldsymbol{x}$ grows, any learning problem significantly gets harder and harder. For example, let us sample 5 points from $[0, 1]$ at regular intervals. Collecting samples in the same way in d-dimensional space requires 5^d points (Fig. 35.1(a)), which grows exponentially with respect to d. Since collecting 5^d points when d is large is not possible in practice (e.g. $5^{100} \approx 10^{70}$), samples are always scarce in high-dimensional problems.

Another trouble in high-dimensional problems is that our geometric intuition can be misleading. For example, let us consider the *inscribed hypersphere* of the unit hypercube in d-dimensional space (Fig. 35.1(b)). When $d = 1$, the volume of the inscribed hypersphere is 1, which is the same as the hypercube. When d is increased to 2 and 3, the volume of the hypercube is still 1, but the volume of the inscribed

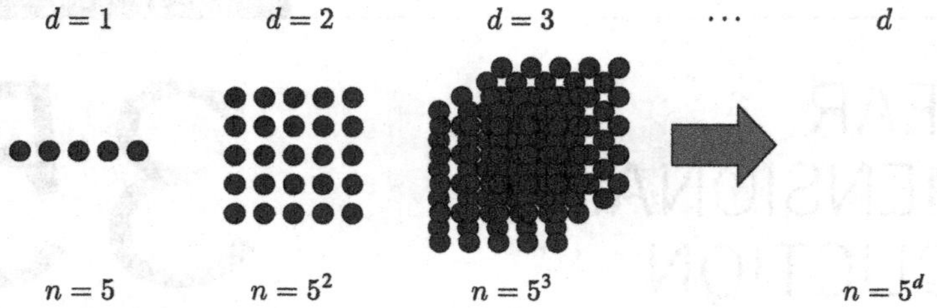

(a) Grid sampling in high-dimensional space. The required number of samples grows exponentially with respect to the dimensionality. Thus, samples are always scarce in high-dimensional space

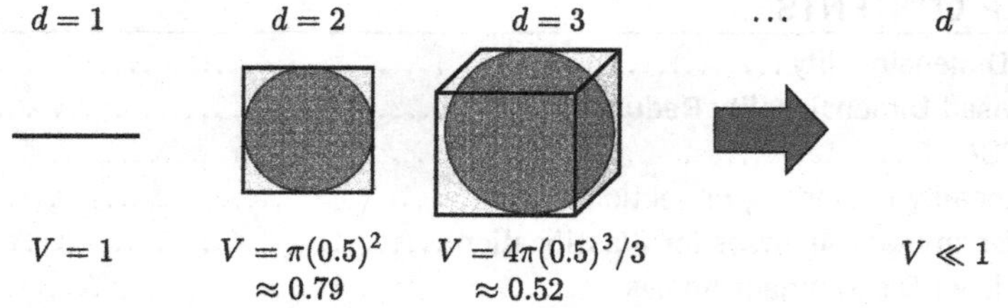

(b) Volume of inscribed hypersphere in high-dimensional space. While the volume of the unit hypercube is always 1, the volume of the inscribed hypersphere tends to 0 as the dimensionality grows. Thus, the inscribed hypersphere can be ignored in high-dimensional space

FIGURE 35.1

Curse of dimensionality.

hypersphere is decreased to approximately 0.79 and 0.52, respectively. Our geometric intuition is that, even though the inscribed hypersphere is smaller than the hypercube, the inscribed hypersphere is not extremely small. However, as d is further increased, the volume of the hypercube is still 1, but the volume of the inscribed hypersphere tends to 0. This means that, when d is large, the inscribed hypersphere is almost negligible.

As illustrated above, handling high-dimensional data is cumbersome in practice, which is often referred to as the *curse of dimensionality*. In the following sections, various methods of dimensionality reduction are introduced.

Linear dimensionality reduction transforms the original samples $\{\boldsymbol{x}_i\}_{i=1}^n$ to low-dimensional expressions $\{\boldsymbol{z}_i\}_{i=1}^n$ by a linear transformation $\boldsymbol{T} \in \mathbb{R}^{m \times d}$ (Fig. 35.2):

$$\boldsymbol{z}_i = \boldsymbol{T}\boldsymbol{x}_i,$$

$\boldsymbol{T}$ is called an *embedding matrix*. This chapter is devoted to introducing various linear dimensionality reduction methods. Nonlinear methods will be covered in Chapter 36.

In the rest of this chapter, training input samples $\{\boldsymbol{x}_i\}_{i=1}^n$ are assumed to be *centralized* as Fig. 35.3:

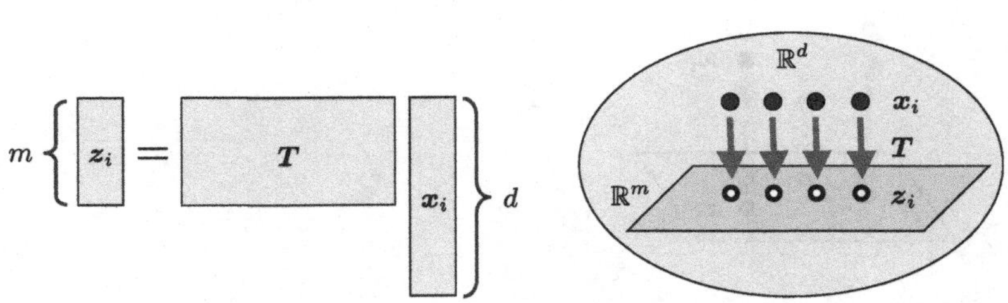

(a) Dimensionality reduction with matrix T (b) Projection onto a linear subspace

FIGURE 35.2

Linear dimensionality reduction. Transformation by a fat matrix T corresponds to projection onto a subspace.

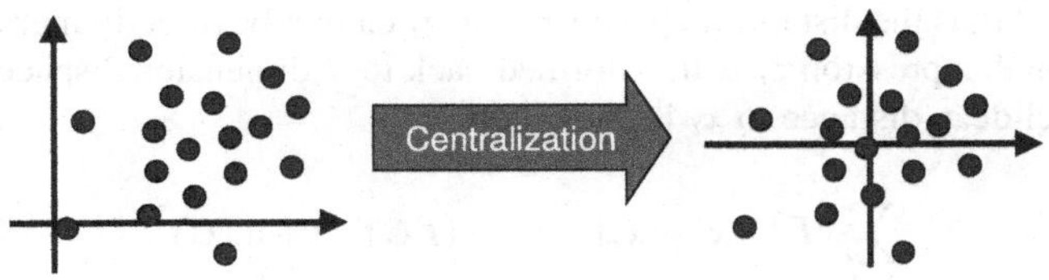

FIGURE 35.3

Data centering.

$$x_i \longleftarrow x_i - \frac{1}{n}\sum_{i'=1}^{n} x_{i'}.$$

35.2 UNSUPERVISED DIMENSIONALITY REDUCTION

This section covers *unsupervised dimensionality reduction*, which is aimed at decreasing the dimensionality of input samples $\{x_i\}_{i=1}^n$ without losing their intrinsic information.

35.2.1 PCA

First, a classical unsupervised linear dimensionality reduction method called PCA [59] is introduced.

PCA tries to keep the position of original samples when the dimensionality is reduced (Fig. 35.4). More specifically, under the constraint that z_i is an orthogonal projection of x_i, embedding matrix T that keeps z_i as close to x_i as possible is found. The constraint that z_i is an orthogonal projection of x_i is equivalent to the fact that embedding matrix T satisfies $TT^\top = I_m$, where I_m is the $m \times m$ identity matrix.

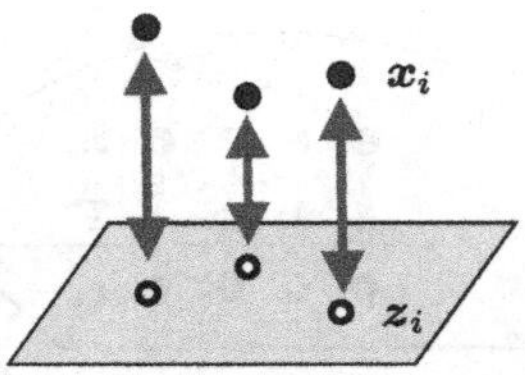

FIGURE 35.4

PCA, which tries to keep the position of original samples when the dimensionality is reduced.

However, as illustrated in Fig. 35.2(b), the dimensionality of $\boldsymbol{x}_i$ is different from that of $\boldsymbol{z}_i$, and thus the distance between $\boldsymbol{x}_i$ and $\boldsymbol{z}_i$ cannot be directly measured. Here, m-dimensional expression $\boldsymbol{z}_i$ is transformed back to d-dimensional space by $\boldsymbol{T}^\top$ and then the Euclidean distance to $\boldsymbol{x}_i$ is computed:

$$\sum_{i=1}^{n} \|\boldsymbol{T}^\top \boldsymbol{T}\boldsymbol{x}_i - \boldsymbol{x}_i\|^2 = -\mathrm{tr}\left(\boldsymbol{T}\boldsymbol{C}\boldsymbol{T}^\top\right) + \mathrm{tr}\left(\boldsymbol{C}\right),$$

where $\boldsymbol{C}$ is the *total scatter matrix*:

$$\boldsymbol{C} = \sum_{i=1}^{n} \boldsymbol{x}_i \boldsymbol{x}_i^\top.$$

Summarizing the above discussion, the optimization problem of PCA is given by

$$\max_{\boldsymbol{T} \in \mathbb{R}^{m \times d}} \mathrm{tr}\left(\boldsymbol{T}\boldsymbol{C}\boldsymbol{T}^\top\right) \text{ subject to } \boldsymbol{T}\boldsymbol{T}^\top = \boldsymbol{I}_m.$$

Although this is a nonconvex optimization problem, a global optimal solution is given analytically by

$$\boldsymbol{T} = (\boldsymbol{\xi}_1, \ldots, \boldsymbol{\xi}_m)^\top,$$

where $\boldsymbol{\xi}_1, \ldots, \boldsymbol{\xi}_d$ are eigenvectors of matrix $\boldsymbol{C}$ associated with eigenvalues $\lambda_1 \geq \cdots \geq \lambda_d \geq 0$ (Fig. 6.2):

$$\boldsymbol{C}\boldsymbol{\xi} = \lambda \boldsymbol{\xi}.$$

Note that other global optimal solutions are given by

$$\boldsymbol{T} = \boldsymbol{G}(\boldsymbol{\xi}_1, \ldots, \boldsymbol{\xi}_m)^\top,$$

where $\boldsymbol{G}$ is any *orthogonal matrix* such that $\boldsymbol{G}^{-1} = \boldsymbol{G}^\top$.

The embedding matrix of PCA is an orthogonal projection onto the subspace spanned by eigenvectors associated with large eigenvalues. In other words, by

```
n=100; x=[2*randn(n,1) randn(n,1)];
%x=[2*randn(n,1) 2*round(rand(n,1))-1+randn(n,1)/3];
x=x-repmat(mean(x),[n,1]);
[t,v]=eigs(x'*x,1);

figure(1); clf; hold on; axis([-6 6 -6 6]);
plot(x(:,1),x(:,2),'rx')
plot(9*[-t(1) t(1)],9*[-t(2) t(2)]);
```

FIGURE 35.5

MATLAB code for PCA.

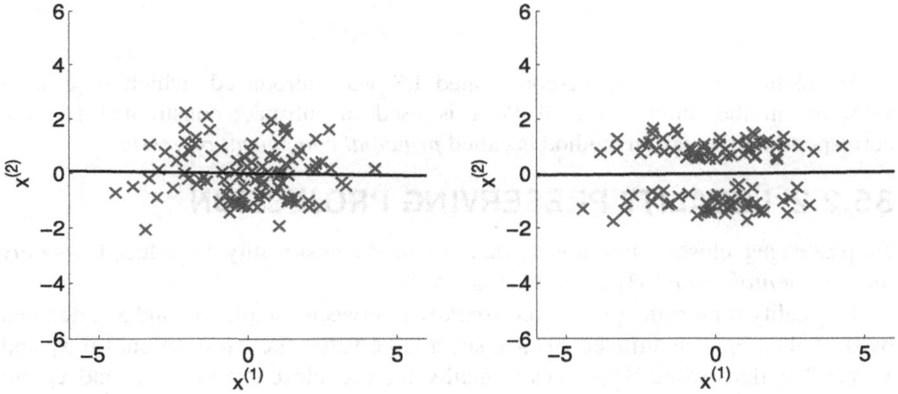

FIGURE 35.6

Example of PCA. The solid line denotes the one-dimensional embedding subspace found by PCA.

removing eigenvectors associated with small eigenvalues, the gap from the original samples is kept minimum. Note that PCA makes data samples *uncorrelated*, i.e. the sample variance-covariance matrix of $\{z_i\}_{i=1}^n$ is *diagonal*:

$$\sum_{i=1}^{n} z_i z_i^\top = \mathrm{diag}\left(\lambda_1, \ldots, \lambda_m\right).$$

A MATLAB code for PCA is provided in Fig. 35.5, and its behavior is illustrated in Fig. 35.6. This shows that samples are embedded into a one-dimensional subspace so that they do not change a lot. However, PCA does not necessarily maintain useful structure of data such as *clusters*.

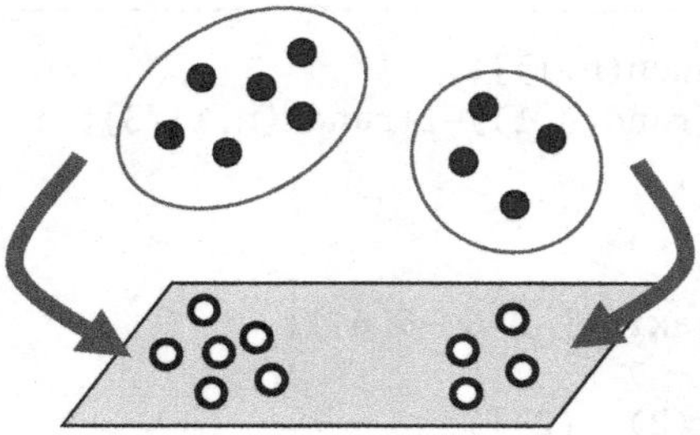

FIGURE 35.7

Locality preserving projection, which tries to keep the cluster structure of original samples when the dimensionality is reduced.

In Section 23.1, subspace-constrained LS was introduced, which requires a subspace in the input space. If PCA is used in subspace-constrained LS, the corresponding regression method is called *principal component regression*.

35.2.2 LOCALITY PRESERVING PROJECTION

To preserving cluster structure of data when dimensionality is reduced, *locality preserving projection* [53] is useful (Fig. 35.7).

In locality preserving projection, *similarity* between samples $\boldsymbol{x}_i$ and $\boldsymbol{x}_{i'}$, denoted by $0 \leq W_{i,i'} \leq 1$, is utilized. $W_{i,i'}$ takes a large value (i.e. close to one) if $\boldsymbol{x}_i$ and $\boldsymbol{x}_{i'}$ are "similar," while $W_{i,i'}$ takes a small value (i.e. close to zero) if $\boldsymbol{x}_i$ and $\boldsymbol{x}_{i'}$ are "dissimilar." The similarity is assumed to be symmetric, i.e. $W_{i,i'} = W_{i',i}$. Popular choices of the similarity measure are summarized in Fig. 35.8.

Locality preserving projection determines embedding matrix $\boldsymbol{T}$ so that sample pairs with high similarity are close to each other in the embedding space, i.e. the following criterion is minimized:

$$\frac{1}{2}\sum_{i,i'=1}^{n} W_{i,i'}\|\boldsymbol{T}\boldsymbol{x}_i - \boldsymbol{T}\boldsymbol{x}_{i'}\|^2. \tag{35.1}$$

However, this minimization problem has a trivial solution $\boldsymbol{T} = \boldsymbol{O}$, which is meaningless. To avoid such a degenerated solution, an appropriate constraint such as

$$\boldsymbol{T}\boldsymbol{X}\boldsymbol{D}\boldsymbol{X}^\top\boldsymbol{T}^\top = \boldsymbol{I}_m$$

is imposed, where

$$\boldsymbol{X} = (\boldsymbol{x}_1, \ldots, \boldsymbol{x}_n) \in \mathbb{R}^{d\times n},$$

- Gaussian similarity:

$$W_{i,i'} = \exp\left(-\frac{\|\boldsymbol{x}_i - \boldsymbol{x}_{i'}\|^2}{2t^2}\right),$$

where $t > 0$ is a tuning parameter that controls the decay of Gaussian tails.

- k-nearest-neighbor similarity:

$$W_{i,i'} = \begin{cases} 1 & (\boldsymbol{x}_i \in \mathcal{N}_k(\boldsymbol{x}_{i'}) \text{ or } \boldsymbol{x}_{i'} \in \mathcal{N}_k(\boldsymbol{x}_i)), \\ 0 & (\text{otherwise}), \end{cases}$$

where $\mathcal{N}_k(\boldsymbol{x})$ denotes the set of k-nearest-neighbor samples of $\boldsymbol{x}$ in $\{\boldsymbol{x}_i\}_{i=1}^n$, and $k \in \{1,\ldots,n\}$ is a tuning parameter that controls the locality. Note that k-nearest-neighbor similarity produces a *sparse* similarity matrix $\boldsymbol{W}$, which is often advantageous in practice.

- Local scaling similarity [123]:

$$W_{i,i'} = \exp\left(-\frac{\|\boldsymbol{x}_i - \boldsymbol{x}_{i'}\|^2}{2t_i t_{i'}}\right),$$

where t_i is the local scaling defined by

$$t_i = \|\boldsymbol{x}_i - \boldsymbol{x}_i^{(k)}\|,$$

and $\boldsymbol{x}_i^{(k)}$ denotes the kth nearest neighbor of $\boldsymbol{x}_i$ in $\{\boldsymbol{x}_i\}_{i=1}^n$. It is also practical to combine k-nearest-neighbor similarity and local scaling similarity.

FIGURE 35.8

Popular choices of similarity measure.

$$D_{i,i'} = \begin{cases} \sum_{i''=1}^{n} W_{i,i''} & (i = i'), \\ 0 & (i \neq i'). \end{cases}$$

For

$$\boldsymbol{L} = \boldsymbol{D} - \boldsymbol{W}, \tag{35.2}$$

```
n=100; x=[2*randn(n,1) randn(n,1)];
%x=[2*randn(n,1) 2*round(rand(n,1))-1+randn(n,1)/3];
x=x-repmat(mean(x),[n,1]); x2=sum(x.^2,2);
W=exp(-(repmat(x2,1,n)+repmat(x2',n,1)-2*x*x'));
D=diag(sum(W,2)); L=D-W; z=x'*D*x; z=(z+z')/2;
[t,v]=eigs(x'*L*x,z,1,'sm');

figure(1); clf; hold on; axis([-6 6 -6 6]);
plot(x(:,1),x(:,2),'rx')
plot(9*[-t(1) t(1)],9*[-t(2) t(2)]);
```

FIGURE 35.9

MATLAB code for locality preserving projection.

Eq. (35.1) can be compactly expressed as $\mathrm{tr}(\boldsymbol{T}\boldsymbol{X}\boldsymbol{L}\boldsymbol{X}^\top\boldsymbol{T}^\top)$. The matrix $\boldsymbol{L}$ is called the *graph Laplacian matrix*, which plays an important role in *spectral graph theory* (see Section 33.1).

Summarizing the above discussion, the optimization problem of locality preserving projection is given as

$$\min_{\boldsymbol{T}\in\mathbb{R}^{m\times d}} \mathrm{tr}\left(\boldsymbol{T}\boldsymbol{X}\boldsymbol{L}\boldsymbol{X}^\top\boldsymbol{T}^\top\right) \quad \text{subject to } \boldsymbol{T}\boldsymbol{X}\boldsymbol{D}\boldsymbol{X}^\top\boldsymbol{T}^\top = \boldsymbol{I}_m.$$

Although this is a nonconvex optimization problem, a global optimal solution is given analytically by

$$\boldsymbol{T} = (\boldsymbol{\xi}_d, \boldsymbol{\xi}_{d-1} \ldots, \boldsymbol{\xi}_{d-m+1})^\top,$$

where $\boldsymbol{\xi}_1, \ldots, \boldsymbol{\xi}_d$ are generalized eigenvectors of $(\boldsymbol{X}\boldsymbol{L}\boldsymbol{X}^\top, \boldsymbol{X}\boldsymbol{D}\boldsymbol{X}^\top)$ associated with generalized eigenvalues $\lambda_1 \geq \ldots \geq \lambda_d \geq 0$ (Fig. 6.2):

$$\boldsymbol{X}\boldsymbol{L}\boldsymbol{X}^\top\boldsymbol{\xi} = \lambda\boldsymbol{X}\boldsymbol{D}\boldsymbol{X}^\top\boldsymbol{\xi}.$$

Thus, the embedding matrix of locality preserving projection is given by the minor generalized eigenvectors of $(\boldsymbol{X}\boldsymbol{L}\boldsymbol{X}^\top, \boldsymbol{X}\boldsymbol{D}\boldsymbol{X}^\top)$.

A MATLAB code for locality preserving projection is provided in Fig. 35.9, and its behavior is illustrated in Fig. 35.10. This shows that cluster structure is nicely preserved.

35.3 LINEAR DISCRIMINANT ANALYSES FOR CLASSIFICATION

In this section, methods of *supervised linear dimensionality reduction* based on input-output paired training samples $\{(\boldsymbol{x}_i, y_i)\}_{i=1}^n$ for classification (i.e. $y \in \{1, \ldots, c\}$) are introduced.

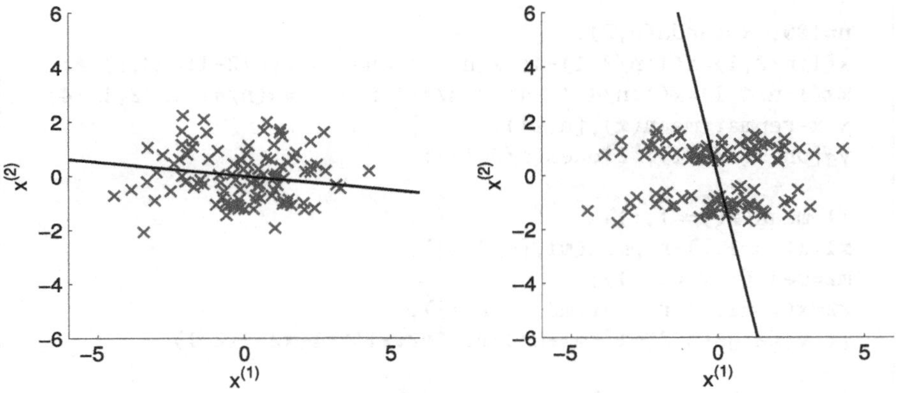

FIGURE 35.10

Example of locality preserving projection. The solid line denotes the one-dimensional embedding subspace found by locality preserving projection.

35.3.1 FISHER DISCRIMINANT ANALYSIS

First, let us introduce a classical supervised linear dimensionality reduction method called *Fisher discriminant analysis* [42].

The basic idea of Fisher discriminant analysis is to find a transformation matrix $\boldsymbol{T}$ so that sample pairs in the same class get closer to each other and sample pairs in different classes are far apart. More specifically, let $\boldsymbol{S}^{(\mathrm{w})}$ be the *within-class scatter matrix* and $\boldsymbol{S}^{(\mathrm{b})}$ be the *between-class scatter matrix* defined as

$$\boldsymbol{S}^{(\mathrm{w})} = \sum_{y=1}^{c} \sum_{i:y_i=y} (\boldsymbol{x}_i - \boldsymbol{\mu}_y)(\boldsymbol{x}_i - \boldsymbol{\mu}_y)^\top \in \mathbb{R}^{d\times d}, \tag{35.3}$$

$$\boldsymbol{S}^{(\mathrm{b})} = \sum_{y=1}^{c} n_y \boldsymbol{\mu}_y \boldsymbol{\mu}_y^\top \in \mathbb{R}^{d\times d}, \tag{35.4}$$

where $\sum_{i:y_i=y}$ denotes the summation over i such that $y_i = y$. $\boldsymbol{\mu}_y$ denotes the mean of training samples in class y:

$$\boldsymbol{\mu}_y = \frac{1}{n_y} \sum_{i:y_i=y} \boldsymbol{x}_i,$$

where n_y denotes the number of training samples in class y. Then the optimization problem of Fisher discriminant analysis is given as

$$\max_{\boldsymbol{T}\in\mathbb{R}^{m\times d}} \mathrm{tr}\Big((\boldsymbol{T}\boldsymbol{S}^{(\mathrm{w})}\boldsymbol{T}^\top)^{-1}\boldsymbol{T}\boldsymbol{S}^{(\mathrm{b})}\boldsymbol{T}^\top\Big),$$

where $\boldsymbol{T}\boldsymbol{S}^{(\mathrm{w})}\boldsymbol{T}^\top$ denotes the within-class scatter matrix after dimensionality reduction and $\boldsymbol{T}\boldsymbol{S}^{(\mathrm{b})}\boldsymbol{T}^\top$ denotes the between-class scatter matrix after dimensionality reduction.

```
n=100; x=randn(n,2);
x(1:n/2,1)=x(1:n/2,1)-4; x(n/2+1:end,1)=x(n/2+1:end,1)+4;
%x(1:n/4,1)=x(1:n/4,1)-4; x(n/4+1:n/2,1)=x(n/4+1:n/2,1)+4;
x=x-repmat(mean(x),[n,1]);
y=[ones(n/2,1); 2*ones(n/2,1)];

m1=mean(x(y==1,:));
x1=x(y==1,:)-repmat(m1,[n/2,1]);
m2=mean(x(y==2,:));
x2=x(y==2,:)-repmat(m2,[n/2,1]);
[t,v]=eigs(n/2*m1'*m1+n/2*m2'*m2,x1'*x1+x2'*x2,1);

figure(1); clf; hold on; axis([-8 8 -6 6]);
plot(x(y==1,1),x(y==1,2),'bo')
plot(x(y==2,1),x(y==2,2),'rx')
plot(99*[-t(1) t(1)],99*[-t(2) t(2)], 'k-')
```

FIGURE 35.11

MATLAB code for Fisher discriminant analysis.

Thus, Fisher discriminant analysis finds a transformation matrix $\boldsymbol{T}$ that decreases the within-class scatter and increases the between-class scatter.

Let $\lambda_1 \geq \cdots \geq \lambda_d \geq 0$ and $\boldsymbol{\xi}_1, \ldots, \boldsymbol{\xi}_d$ be the *generalized eigenvalues* and *generalized eigenvectors* of $(\boldsymbol{S}^{(\mathrm{b})}, \boldsymbol{S}^{(\mathrm{w})})$:

$$\boldsymbol{S}^{(\mathrm{b})}\boldsymbol{\xi} = \lambda \boldsymbol{S}^{(\mathrm{w})}\boldsymbol{\xi}.$$

Then the solution $\widehat{\boldsymbol{T}}$ of Fisher discriminant analysis is given analytically as

$$\widehat{\boldsymbol{T}} = (\boldsymbol{\xi}_1, \ldots, \boldsymbol{\xi}_m)^{\top}.$$

A MATLAB code for Fisher discriminant analysis is provided in Fig. 35.11, and its behavior is illustrated in Fig. 35.12. In these examples, two-dimensional samples are projected onto one-dimensional subspaces, and a subspace that nicely separates samples in two classes can be found in Fig. 35.12(a). However, in Fig. 35.12(b), samples in two classes are mixed up due to the within-class *multimodality* in class "o."

35.3.2 LOCAL FISHER DISCRIMINANT ANALYSIS

As illustrated in Fig. 35.12, Fisher discriminant analysis can give an undesired solution if within-class multimodality exists. Another limitation of Fisher discriminant analysis is that the between-class scatter matrix $\boldsymbol{S}^{(\mathrm{b})}$ has rank at most $c - 1$.

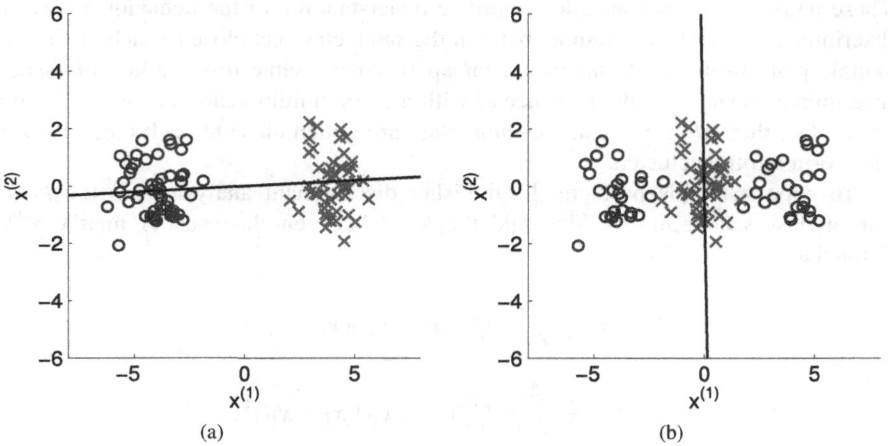

FIGURE 35.12

Examples of Fisher discriminant analysis. The solid lines denote the found subspaces to which training samples are projected.

This means that generalized eigenvectors $\xi_1, \ldots, \xi_d$ used for computing the solution make sense only up to the $(c-1)$th generalized eigenvector. Thus, in practice, the reduced dimensionality m should be less than the number of classes, c. This is highly restrictive when c is small. Indeed, in binary classification where $c = 2$, m should be 1. To overcome these limitations, *local Fisher discriminant analysis* [98] introduced here.

The within-class scatter matrix $\boldsymbol{S}^{(\mathrm{w})}$ and the between-class scatter matrix $\boldsymbol{S}^{(\mathrm{b})}$ used in Fisher discriminant analysis defined by Eqs (35.3) and (35.4) can be expressed as

$$\boldsymbol{S}^{(\mathrm{w})} = \frac{1}{2}\sum_{i,i'=1}^{n} Q^{(\mathrm{w})}_{i,i'}(\boldsymbol{x}_i - \boldsymbol{x}_{i'})(\boldsymbol{x}_i - \boldsymbol{x}_{i'})^\top,$$

$$\boldsymbol{S}^{(\mathrm{b})} = \frac{1}{2}\sum_{i,i'=1}^{n} Q^{(\mathrm{b})}_{i,i'}(\boldsymbol{x}_i - \boldsymbol{x}_{i'})(\boldsymbol{x}_i - \boldsymbol{x}_{i'})^\top,$$

where

$$Q^{(\mathrm{w})}_{i,i'} = \begin{cases} 1/n_y > 0 & (y_i = y_{i'} = y), \\ 0 & (y_i \neq y_{i'}), \end{cases}$$

$$Q^{(\mathrm{b})}_{i,i'} = \begin{cases} 1/n - 1/n_y < 0 & (y_i = y_{i'} = y), \\ 1/n > 0 & (y_i \neq y_{i'}). \end{cases}$$

These pairwise expressions allow intuitive understanding of the behavior of Fisher discriminant analysis, i.e. sample pairs in the same class get close to each other and sample pairs in different classes are far apart. At the same time, failure of Fisher discriminant analysis in the presence of within-class multimodality can be explained by the fact that *all* samples in the same class are gotten close to each other even if they form multiple clusters.

To cope with this problem, local Fisher discriminant analysis uses the *local* within-class scatter matrix $\boldsymbol{S}^{(\mathrm{lw})}$ and the *local* between-class scatter matrix $\boldsymbol{S}^{(\mathrm{lb})}$ defined as

$$\boldsymbol{S}^{(\mathrm{lw})} = \frac{1}{2}\sum_{i,i'=1}^{n} Q_{i,i'}^{(\mathrm{lw})}(\boldsymbol{x}_i - \boldsymbol{x}_{i'})(\boldsymbol{x}_i - \boldsymbol{x}_{i'})^\top,$$

$$\boldsymbol{S}^{(\mathrm{lb})} = \frac{1}{2}\sum_{i,i'=1}^{n} Q_{i,i'}^{(\mathrm{lb})}(\boldsymbol{x}_i - \boldsymbol{x}_{i'})(\boldsymbol{x}_i - \boldsymbol{x}_{i'})^\top,$$

where

$$Q_{i,i'}^{(\mathrm{lw})} = \begin{cases} W_{i,i'}/n_y & (y_i = y_{i'} = y), \\ 0 & (y_i \neq y_{i'}), \end{cases}$$

$$Q_{i,i'}^{(\mathrm{lb})} = \begin{cases} W_{i,i'}(1/n - 1/n_y) & (y_i = y_{i'} = y), \\ 1/n & (y_i \neq y_{i'}). \end{cases}$$

$0 \leq W_{i,i'} \leq 1$ denotes a similarity between sample $\boldsymbol{x}_i$ and sample $\boldsymbol{x}_{i'}$ (see Fig. 35.8). In the above local scatter matrices, similarity $W_{i,i'}$ is applied to sample pairs in the same class, which mitigates faraway sample pairs in the same class to be gotten close to each other strongly and thus within-class cluster structure tends to be preserved. Local Fisher discriminant analysis can be interpreted as applying *locality preserving projection* introduced in Section 35.2.2 in a classwise manner on top of Fisher discriminant analysis. Note that similarity $W_{i,i'}$ is *not* applied to sample pairs in different classes, since they should be faraway even if they belong to different clusters.

The optimization problem of local Fisher discriminant analysis is given by

$$\max_{\boldsymbol{T}\in\mathbb{R}^{m\times d}} \mathrm{tr}\Big((\boldsymbol{T}\boldsymbol{S}^{(\mathrm{lw})}\boldsymbol{T}^\top)^{-1}\boldsymbol{T}\boldsymbol{S}^{(\mathrm{lb})}\boldsymbol{T}^\top\Big),$$

which has exactly the same form as the original Fisher discriminant analysis. Thus, the solution $\widehat{\boldsymbol{T}}$ of local Fisher discriminant analysis can be obtained analytically in the same way by

$$\widehat{\boldsymbol{T}} = (\boldsymbol{\xi}_1, \ldots, \boldsymbol{\xi}_m)^\top,$$

```
n=100; x=randn(n,2);
x(1:n/2,1)=x(1:n/2,1)-4; x(n/2+1:end,1)=x(n/2+1:end,1)+4;
%x(1:n/4,1)=x(1:n/4,1)-4; x(n/4+1:n/2,1)=x(n/4+1:n/2,1)+4;
x=x-repmat(mean(x),[n,1]); y=[ones(n/2,1); 2*ones(n/2,1)];

Sw=zeros(2,2); Sb=zeros(2,2);
for j=1:2
  p=x(y==j,:); p1=sum(p); p2=sum(p.^2,2); nj=sum(y==j);
  W=exp(-(repmat(p2,1,nj)+repmat(p2',nj,1)-2*p*p'));
  G=p'*(repmat(sum(W,2),[1 2]).*p)-p'*W*p;
  Sb=Sb+G/n+p'*p*(1-nj/n)+p1'*p1/n; Sw=Sw+G/nj;
end
[t,v]=eigs((Sb+Sb')/2,(Sw+Sw')/2,1);

figure(1); clf; hold on; axis([-8 8 -6 6]);
plot(x(y==1,1),x(y==1,2),'bo')
plot(x(y==2,1),x(y==2,2),'rx')
plot(99*[-t(1) t(1)],99*[-t(2) t(2)], 'k-')
```

FIGURE 35.13

MATLAB code for local Fisher discriminant analysis.

where $\boldsymbol{\xi}_1, \ldots, \boldsymbol{\xi}_d$ are generalized eigenvectors associated with generalized eigenvalues $\lambda_1 \geq \cdots \geq \lambda_d \geq 0$ of $(\boldsymbol{S}^{(\mathrm{lb})}, \boldsymbol{S}^{(\mathrm{lw})})$:

$$\boldsymbol{S}^{(\mathrm{lb})}\boldsymbol{\xi} = \lambda \boldsymbol{S}^{(\mathrm{lw})}\boldsymbol{\xi}.$$

Another important advantage of local Fisher discriminant analysis is that the low-rank problem of $\boldsymbol{S}^{(\mathrm{b})}$ can also be avoided. More specifically, since $\boldsymbol{S}^{(\mathrm{lb})}$ contains the similarity $W_{i,i'}$, $\boldsymbol{S}^{(\mathrm{lb})}$ usually has full rank. Then all generalized eigenvectors $\boldsymbol{\xi}_1, \ldots, \boldsymbol{\xi}_d$ make sense and thus the reduced dimensionality m can be arbitrarily large.

A MATLAB code for local Fisher discriminant analysis is provided in Fig. 35.13, and its behavior is illustrated in Fig. 35.14. This shows that local Fisher discriminant analysis performs well even in the presence of within-class multimodality.

35.3.3 SEMISUPERVISED LOCAL FISHER DISCRIMINANT ANALYSIS

Supervised dimensionality reduction tends to *overfit* if the number of training samples is small. Here, *semisupervised local Fisher discriminant analysis* [100] is introduced, which utilizes unlabeled samples $\{\boldsymbol{x}_i\}_{i=n+1}^{n+n'}$ in addition to ordinary labeled training samples $\{(\boldsymbol{x}_i, y_i)\}_{i=1}^{n}$ to mitigate overfitting (Section 33.1).

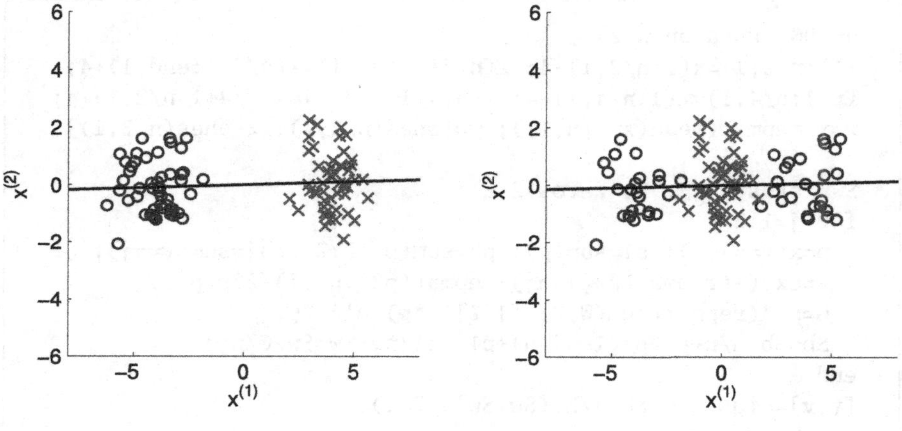

FIGURE 35.14

Examples of local Fisher discriminant analysis for the same data sets as Fig. 35.12. The solid lines denote the found subspaces to which training samples are projected.

The basic idea of semisupervised local Fisher discriminant analysis is to combine (supervised) local Fisher discriminant analysis with (unsupervised) PCA. As explained in Section 35.2.1, the solution of PCA is given by the leading eigenvectors of the *total scatter matrix*:

$$S^{(t)} = \sum_{i=1}^{n+n'} (x_i - \mu^{(t)})(x_i - \mu^{(t)})^\top,$$

where $\mu^{(t)}$ denotes the mean of all (i.e. labeled and unlabeled) input samples $\{x_i\}_{i=1}^{n+n'}$:

$$\mu^{(t)} = \frac{1}{n+n'} \sum_{i=1}^{n+n'} x_i.$$

Note that $\mu^{(t)}$ is not necessarily zero even though input of labeled samples, $\{x_i\}_{i=1}^{n}$, is assumed to be centralized through this chapter (Fig. 35.3). Similarly, the solution of local Fisher discriminant analysis is given by solving a generalized eigenvalue problem, as shown in Section 35.3.2. Based on these facts, the basic idea of semisupervised local Fisher discriminant analysis is to combine these eigenvalue problems.

More specifically, instead of the local scatter matrices used in local Fisher discriminant analysis, semisupervised local Fisher discriminant analysis uses the following *semisupervised* local scatter matrices:

$$S^{(\mathrm{slw})} = (1-\beta)S^{(\mathrm{lw})} + \beta S^{(t)},$$

$$S^{(\mathrm{slb})} = (1-\beta)S^{(\mathrm{lb})} + \beta I,$$

where I denotes the identity matrix and $\beta \in [0,1]$ controls the balance between labeled and unlabeled samples. Based on these scatter matrices, the optimization problem of semisupervised local Fisher discriminant analysis is given as

$$\max_{T \in \mathbb{R}^{m\times d}} \mathrm{tr}\Big((TS^{(\mathrm{slw})}T^\top)^{-1}TS^{(\mathrm{slb})}T^\top\Big),$$

which has exactly the same form as the original Fisher discriminant analysis. Thus, the solution $\widehat{T}$ of semisupervised local Fisher discriminant analysis can be obtained analytically in the same way by

$$\widehat{T} = (\xi_1,\ldots,\xi_m)^\top,$$

where $\xi_1,\ldots,\xi_d$ are generalized eigenvectors associated with generalized eigenvalues $\lambda_1 \geq \cdots \geq \lambda_d \geq 0$ of $(S^{(\mathrm{slb})}, S^{(\mathrm{slw})})$:

$$S^{(\mathrm{slb})}\xi = \lambda S^{(\mathrm{slw})}\xi.$$

This solution is reduced to (supervised) local Fisher discriminant analysis if $\beta = 0$ and is reduced to (unsupervised) PCA if $\beta = 1$. For $0 < \beta < 1$, semisupervised local Fisher discriminant analysis is expected to give a solution that bridges the two extreme cases.

A MATLAB code for semisupervised local Fisher discriminant analysis is provided in Fig. 35.15, and its behavior is illustrated in Fig. 35.16. This shows that semisupervised local Fisher discriminant analysis can successfully avoid overfitting.

35.4 SUFFICIENT DIMENSIONALITY REDUCTION FOR REGRESSION

The discriminant analysis methods introduced in the previous section explicitly used the class labels and thus cannot be applied to regression. Here, another supervised dimensionality reduction method called *sufficient dimensionality reduction* [67] is introduced, which can also be applied to regression.

35.4.1 INFORMATION THEORETIC FORMULATION

The basic idea of sufficient dimensionality reduction is to find a transformation matrix T that makes x *conditionally independent* of output y given projection $z = Tx$:

$$p(x, y|z) = p(x|z)p(y|z).$$

This means that z contains all information about output y.

Such a transformation matrix T is characterized as the maximizer of *mutual information* (MI) [92]:

$$\mathrm{MI} = \iint p(z,y)\log\frac{p(z,y)}{p(z)p(y)}\mathrm{d}z\mathrm{d}y.$$

```
n=2; m=200; x=0.1*randn(n,2); b=0.9; %b=0.001; b=1;
x(:,1)=x(:,1)+[repmat(3,[n/2,1]); repmat(-3,[n/2,1])];
%x(1:n/2,2)=x(1:n/2,2)+repmat(5,[n/2,1]);
xx=randn(m,2).*repmat([1 2],[m 1]);
xx(:,1)=xx(:,1)+[repmat(-3,[m/2,1]); repmat(3,[m/2,1])];
%x(:,2)=x(:,2)*1.7; xx(:,2)=xx(:,2)*1.7;
mu=mean([x;xx]); x=x-repmat(mu,[n,1]);
xx=xx-repmat(mu,[m,1]); y=[ones(n/2,1); 2*ones(n/2,1)];

x2=sum(x.^2,2); Qlb=zeros(n,n); Qlw=zeros(n,n);
W=exp(-(repmat(x2,1,n)+repmat(x2',n,1)-2*x*x'));
for j=1:2
  Wy=W.*((y==j)/2*(y==j)'); Qlw=Qlw+Wy/sum(y==j);
  Qlb=Qlb+Wy*(1/n-1/sum(y==j))+(y==j)/n/2*(y~=j)';
end
Srlb=(1-b)*x'*(diag(sum(Qlb))-Qlb)*x+b*cov([x; xx],1);
Srlw=(1-b)*x'*(diag(sum(Qlw))-Qlw)*x+b*eye(2);
[t,v]=eigs((Srlb+Srlb')/2,(Srlw+Srlw')/2,1);

figure(1); clf; hold on; axis([-6 6 -6 6]);
plot(xx(:,1),xx(:,2),'k.');
plot(x(y==1,1),x(y==1,2),'bo');
plot(x(y==2,1),x(y==2,2),'rx');
plot(99*[-t(1) t(1)],99*[-t(2) t(2)], 'k-');
```

FIGURE 35.15

MATLAB code for semisupervised local Fisher discriminant analysis.

MI is actually the *KL divergence* (see Section 14.2) from the joint probability density $p(z,y)$ to the product of marginals $p(z)p(y)$. Therefore, it is always non-negative and is zero if and only if $p(z,y) = p(z)p(y)$, i.e. z and y are *statistically independent* (see Section 5.6). Maximizing MI with respect to $\boldsymbol{T}$ implies maximizing *statistical dependency* between z and y, which is intuitively understandable as supervised dimensionality reduction.

The value of MI can be approximated by the *KL density ratio estimator* described in Section 38.3. However, because the log function and the density ratio $\frac{p(z,y)}{p(z)p(y)}$ are included in MI, it tends to be sensitive to outliers. Here, let us use a variant of MI based on the *L_2-distance* called the *quadratic mutual information* (QMI) [113]:

$$\mathrm{QMI} = \frac{1}{2}\iint f(z,y)^2 \mathrm{d}z\mathrm{d}y,$$

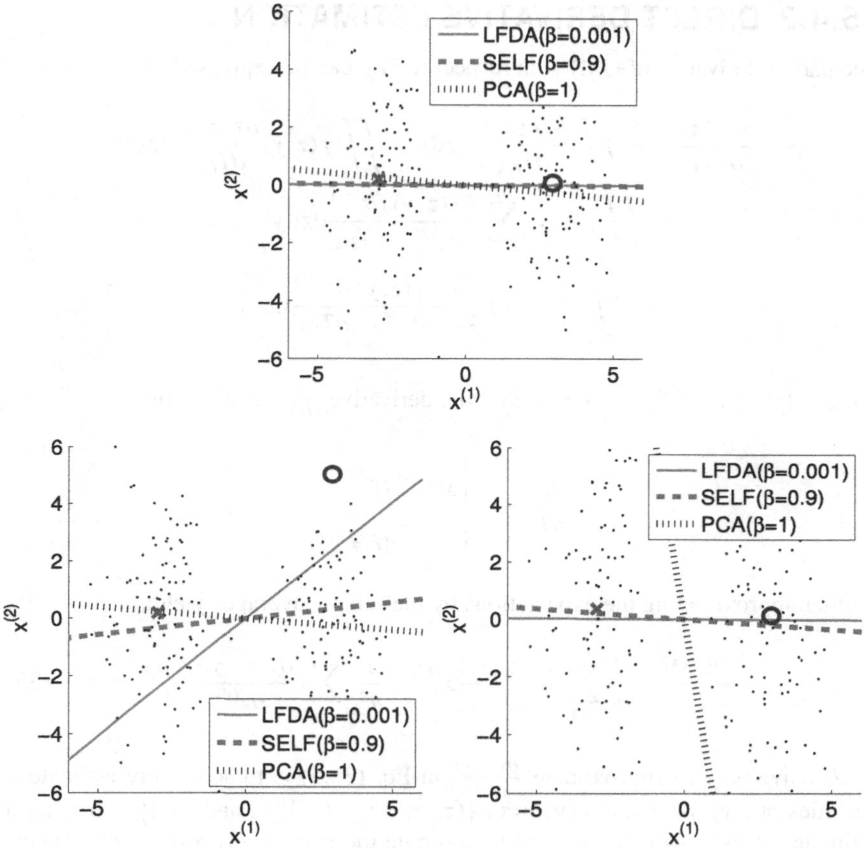

FIGURE 35.16

Examples of semisupervised local Fisher discriminant analysis. Lines denote the found subspaces to which training samples are projected. "LFDA" stands for local Fisher discriminant analysis, "SELF" stands for semisupervised LFDA, and "PCA" stands for principal component analysis.

where $f(z, y)$ is the density difference function defined as

$$f(z, y) = p(z, y) - p(z)p(y).$$

Then a maximizer of QMI with respect to transformation matrix $\boldsymbol{T}$ may be obtained by *gradient ascent*:

$$\boldsymbol{T} \longleftarrow \boldsymbol{T} + \varepsilon \frac{\partial \mathrm{QMI}}{\partial \boldsymbol{T}},$$

where $\varepsilon > 0$ is the step size.

35.4.2 DIRECT DERIVATIVE ESTIMATION

The partial derivative of QMI with respect to $T_{\ell,k}$ can be expressed as

$$\begin{aligned}\frac{\partial \mathrm{QMI}}{\partial T_{\ell,k}} &= \frac{1}{2}\iint \frac{\partial f(z,y)^2}{\partial T_{\ell,k}}\mathrm{d}z\mathrm{d}y = \iint f(z,y)\frac{\partial f(z,y)}{\partial T_{\ell,k}}\mathrm{d}z\mathrm{d}y \\ &= \iint p(z,y)\sum_{\ell'=1}^{m}\frac{\partial f(z,y)}{\partial z^{(\ell')}}\frac{\partial z^{(\ell')}}{\partial T_{\ell,k}}\mathrm{d}z\mathrm{d}y \\ &\quad - \iint p(z)p(y)\sum_{\ell'=1}^{m}\frac{\partial f(z,y)}{\partial z^{(\ell')}}\frac{\partial z^{(\ell')}}{\partial T_{\ell,k}}\mathrm{d}z\mathrm{d}y.\end{aligned}$$

For $\boldsymbol{z} = (z^{(1)},\ldots,z^{(m)})^\top = \boldsymbol{T}\boldsymbol{x}$, the partial derivative $\frac{\partial z^{(\ell')}}{\partial T_{\ell,k}}$ is given by

$$\frac{\partial z^{(\ell')}}{\partial T_{\ell,k}} = \begin{cases} x^{(k)} & (\ell = \ell'), \\ 0 & (\ell \neq \ell'). \end{cases}$$

Further approximating the expectations by the sample averages yields

$$\frac{\partial \mathrm{QMI}}{\partial T_{\ell,k}} \approx \frac{1}{n}\sum_{i=1}^{n}\frac{\partial f(z_i,y_i)}{\partial z^{(\ell)}}x_i^{(k)} - \frac{1}{n^2}\sum_{i,i'=1}^{n}\frac{\partial f(z_i,y_{i'})}{\partial z^{(\ell)}}x_i^{(k)}. \tag{35.5}$$

A naive way to approximate $\frac{\partial f(z,y)}{\partial z^{(\ell)}}$ in Eq. (35.5) is to separately estimate the densities $p(z,y)$, $p(z)$, and $p(y)$ from $\{(z_i,y_i)\}_{i=1}^n$, $\{z_i\}_{i=1}^n$, and $\{y_i\}_{i=1}^n$, to plug the estimated densities in $f(z,y)$, and to compute the partial derivative with respect to $z^{(\ell)}$. However, density estimation is a hard statistical estimation problem and thus such a plug-in derivative estimator may not be reliable. Below, a direct estimator of $\frac{\partial f(z,y)}{\partial z^{(\ell)}}$ without density estimation is introduced [108].

Let us employ the following Gaussian kernel model for approximating $\frac{\partial f(z,y)}{\partial z^{(\ell)}}$:

$$g_\alpha(z,y) = \sum_{j=1}^{n}\alpha_j \exp\left(-\frac{\|z-z_j\|^2 + (y-y_j)^2}{2h^2}\right).$$

The parameter vector $\alpha = (\alpha_1,\ldots,\alpha_n)^\top$ is learned so that the following squared error is minimized:

$$\begin{aligned}J(\alpha) &= \iint \left(g_\alpha(z,y) - \frac{\partial f(z,y)}{\partial z^{(\ell)}}\right)^2 \mathrm{d}z\mathrm{d}y \\ &= \iint g_\alpha(z,y)^2\mathrm{d}z\mathrm{d}y - 2\iint g_\alpha(z,y)\frac{\partial f(z,y)}{\partial z^{(\ell)}}\mathrm{d}z\mathrm{d}y + C,\end{aligned}$$

```
n=500; x=(rand(n,2)*2-1)*10; y=sin(x(:,1)/10*pi);
y2=y.^2; yy=repmat(y2,1,n)+repmat(y2',n,1)-2*y*y';
e=10; h=1; r=exp(-yy/(2*h)); rr=sum(r)'/(n^2);
t0=randn(2,1); t0=t0/norm(t0); c=pi*h;
for o=1:10000
  z=x*t0; z2=z.^2; zz=repmat(z2,1,n)+repmat(z2',n,1)-2*z*z';
  k=exp(-zz/(2*h)); kz=k.*(repmat(z',[n 1])-repmat(z,[1 n]));
  U=c*exp(-(zz+yy)/(4*h)); v=mean(kz.*r/h,2)-sum(kz,2).*rr/h;
  a=(U+0.1*eye(n))\v; g=(k.*r)*x/n-(k*x).*repmat(rr,[1 2]);
  t=t0+e*g'*a; t=t/norm(t);
  if norm(t-t0)<0.00001, break, end
  t0=t;
end

figure(1); clf; hold on; axis([-10 10 -10 10]); colormap gray
scatter3(x(:,1),x(:,2),y,100,y,'filled'); colorbar;
plot(99*[-t(1) t(1)],99*[-t(2) t(2)],'k-');
```

FIGURE 35.17

MATLAB code for supervised dimensionality reduction based on QMI.

where $C = \iint \left(\frac{\partial f(z,y)}{\partial z^{(\ell)}}\right)^2 \mathrm{d}z\mathrm{d}y$ is a constant independent of parameter α and thus is ignored. Suppose that $g_\alpha(z,y)f(z,y) \to 0$ as z and y tend to $\pm\infty$. Then *integration by parts* (4.4) yields

$$\iint g_\alpha(z,y)\frac{\partial f(z,y)}{\partial z^{(\ell)}}\mathrm{d}z\mathrm{d}y = -\iint \frac{\partial g_\alpha(z,y)}{\partial z^{(\ell)}} f(z,y)\mathrm{d}z\mathrm{d}y.$$

Plugging this into $J(\alpha)$, approximating the expectations by the sample averages, and adding the ℓ_2-regularizer result in the following optimization problem:

$$\min_{\alpha} \left[\alpha^\top \boldsymbol{U}\alpha - 2\alpha^\top \widehat{v}_\ell + \lambda\|\alpha\|^2\right],$$

where $\lambda \geq 0$ is the regularization parameter, and $\boldsymbol{U}$ and $\widehat{v}_\ell$ are the $n \times n$ matrix and the n-dimensional vector defined as

$$U_{j,j'} = (\sqrt{\pi}h)^{m+1} \exp\left(-\frac{\|z_j - z_{j'}\|^2 + (y_j - y_{j'})^2}{4h^2}\right),$$

$$\widehat{v}_{\ell,j} = \frac{1}{nh^2}\sum_{i=1}^{n} \exp\left(-\frac{\|z_i - z_j\|^2 + (y_i - y_j)^2}{2h^2}\right)(z_i^{(\ell)} - z_j^{(\ell)})$$

$$- \frac{1}{n^2h^2}\sum_{i,i'=1}^{n} \exp\left(-\frac{\|z_i - z_j\|^2 + (y_{i'} - y_j)^2}{2h^2}\right)(z_i^{(\ell)} - z_j^{(\ell)}).$$

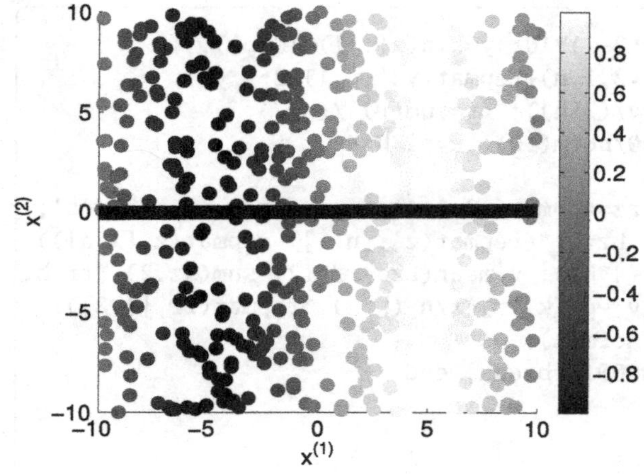

FIGURE 35.18

Example of supervised dimensionality reduction based on QMI. The solid line denotes the found subspace to which training samples are projected.

The above minimizer can be obtained analytically as

$$\widehat{\alpha}_\ell = (U + \lambda I)^{-1} \widehat{v}_\ell.$$

Then, Eq. (35.5) yields

$$\frac{\partial \text{QMI}}{\partial T_{\ell,k}} \approx \frac{1}{n} \sum_{i=1}^{n} g_{\widehat{\alpha}_\ell}(z_i, y_i) x_i^{(k)} - \frac{1}{n^2} \sum_{i,i'=1}^{n} g_{\widehat{\alpha}_\ell}(z_i, y_{i'}) x_i^{(k)}.$$

Tuning parameters such as the regularization parameter λ and the Gaussian bandwidth h can be optimized by cross validation with respect to the squared error J (or the misclassification error if a classifier is applied after dimensionality reduction).

A MATLAB code for QMI-based supervised dimensionality reduction is provided in Fig. 35.17, and its behavior is illustrated in Fig. 35.18. This shows that a subspace in the input space that strongly depends on output is obtained.

If QMI between $\boldsymbol{x}$ and its projection $\boldsymbol{z} = \boldsymbol{T}\boldsymbol{x}$ is considered, the QMI-based dimensionality reduction method can actually be applied in unsupervised dimensionality reduction. A MATLAB code for unsupervised dimensionality reduction based on QMI is provided in Fig. 35.19, and its behavior is illustrated in Fig. 35.20. This shows that similar results to locality preserving projection are obtained *without* explicitly using sample similarity.

Note that the sufficient dimensionality reduction method introduced above can also be applied to classification.

```
n=100; x=[2*randn(n,1) randn(n,1)];
%x=[2*randn(n,1) 2*round(rand(n,1))-1+randn(n,1)/3];
x=x-repmat(mean(x),[n,1]); x2=sum(x.^2,2);
yy=repmat(x2,1,n)+repmat(x2',n,1)-2*x*x';

e=10; h=1; l=exp(-yy/(2*h)); ll=sum(l)'/(n^2);
t0=randn(2,1); t0=t0/norm(t0); c=pi*h;
for o=1:10000
  z=x*t0; z2=z.^2; zz=repmat(z2,1,n)+repmat(z2',n,1)-2*z*z';
  k=exp(-zz/(2*h)); kz=k.*(repmat(z',[n 1])-repmat(z,[1 n]));
  U=c*exp(-(zz+yy)/(4*h)); v=mean(kz.*l/h,2)-sum(kz,2).*ll/h;
  a=(U+0.1*eye(n))\v; g=(k.*l)*x/n-(k*x).*repmat(ll,[1 2]);
  t=t0+e*g'*a; t=t/norm(t);
  if norm(t-t0)<0.00001, break, end
  t0=t;
end

figure(1); clf; hold on; axis([-6 6 -6 6]);
plot(x(:,1),x(:,2),'rx','LineWidth',2,'MarkerSize',12);
plot(9*[-t(1) t(1)],9*[-t(2) t(2)],'k-');
```

FIGURE 35.19

MATLAB code for unsupervised dimensionality reduction based on QMI.

35.5 MATRIX IMPUTATION

Let us consider a matrix $\boldsymbol{X} \in \mathbb{R}^{d_1 \times d_2}$ with *missing entries*, i.e. X_{k_1,k_2} is observed only for $(k_1,k_2) \in \mathcal{K}$ and X_{k_1,k_2} is missing for all $(k_1,k_2) \notin \mathcal{K}$. The objective of *matrix imputation* is to fill the missing entries from observed entries. Such an imputation problem arises, for example, in *recommender systems* [83], where entry X_{k_1,k_2} corresponds to the rating of item k_1 by user k_2. Given a subset $\mathcal{K}$ of ratings of items by users, a recommender system finds the items that are expected to be most favored by a target user.

Naive approaches to matrix imputation would be to pad the missing entries with zeros or the average of observed entries. However, such naive methods are not necessarily useful in practice. The idea for imputing the missing element introduced here is to approximate the matrix $\boldsymbol{X}$ with a *low-rank* matrix.

A naive implementation of this idea is to assume that $\boldsymbol{X}$ can be decomposed into the product of $\boldsymbol{U} \in \mathbb{R}^{d_1 \times r}$ and $\boldsymbol{V} \in \mathbb{R}^{r \times d_2}$:

$$\boldsymbol{X} = \boldsymbol{U}\boldsymbol{V} \in \mathbb{R}^{d_1 \times d_2},$$

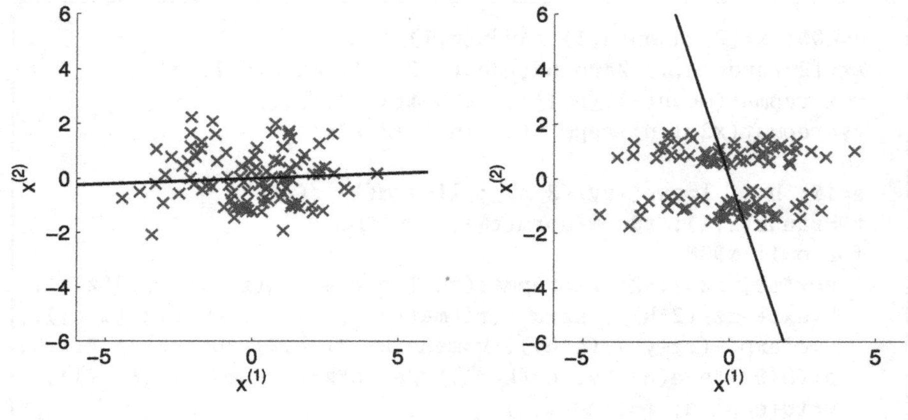

FIGURE 35.20

Example of unsupervised dimensionality reduction based on QMI. The solid line denotes the found subspace to which training samples are projected.

```
d1=20; d2=10; n=100; K=(reshape(randperm(d1*d2),[d1,d2])<=n);
r=3; X=randn(d1,r)*randn(r,d2); T0=randn(d1,d2); e=0.1; l=1;
for o=1:1000
  [U,S,V]=svd(T0-e*((T0.*K)-(X.*K)),'econ');
  S=diag(max(0,diag(S)-e*l)); T=U*S*V';
  if norm(T-T0)<0.001, break, end
  T0=T;
end
figure(1); imagesc(X.*K); figure(2); imagesc(T);
```

FIGURE 35.21

MATLAB code for unsupervised matrix imputation.

where r is the rank of $\boldsymbol{X}$. Then the factorized matrices $\boldsymbol{U}$ and $\boldsymbol{V}$ are estimated from the observed elements, i.e. for $\boldsymbol{\Theta} = \boldsymbol{UV}$, the following optimization problem is solved:

$$\min_{\boldsymbol{U}\in\mathbb{R}^{d_1\times r},\boldsymbol{V}\in\mathbb{R}^{r\times d_2}} \frac{1}{2} \sum_{(k_1,k_2)\in\mathcal{K}} \left(X_{k_1,k_2} - \Theta_{k_1,k_2}\right)^2.$$

Although this minimization problem may be naively solved by gradient descent, finding the global minimizer is not straightforward in practice due to its nonconvexity.

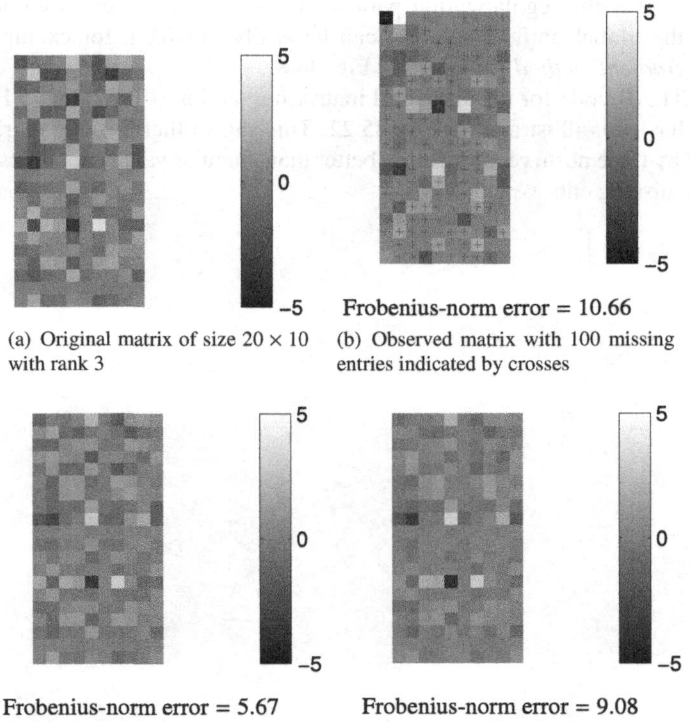

(a) Original matrix of size 20 × 10 with rank 3

(b) Observed matrix with 100 missing entries indicated by crosses

(c) Recovered matrix with rank 3

(d) Zero-filled singular value decomposition with rank 3

FIGURE 35.22

Example of unsupervised matrix imputation. The gray level indicates the value of each entry in [−5, 5].

To overcome the nonconvexity, let us employ regularization with the *trace norm* $\|\boldsymbol{\Theta}\|_{\mathrm{tr}}$ (see Fig. 24.10), which tends to produce a low-rank solution:

$$\|\boldsymbol{\Theta}\|_{\mathrm{tr}} = \sum_{k=1}^{\min(d_1,d_2)} \sigma_k,$$

where σ_k is a *singular value* of $\boldsymbol{\Theta}$ (see Fig. 22.2). Then fitting $\boldsymbol{\Theta}$ to $\boldsymbol{X}$ with trace norm regularization allows us to recover the missing entries:

$$\min_{\boldsymbol{\Theta}\in\mathbb{R}^{d_1\times d_2}} \frac{1}{2} \sum_{(k_1,k_2)\in\mathcal{K}} \left(X_{k_1,k_2} - \Theta_{k_1,k_2}\right)^2 + \lambda\|\boldsymbol{\Theta}\|_{\mathrm{tr}},$$

where $\lambda > 0$ is the regularization parameter. Since this is a convex optimization problem, the global optimal solution can be easily obtained, for example, by the *proximal gradient method* explained in Fig. 34.8.

A MATLAB code for unsupervised matrix imputation is provided in Fig. 35.21, and its behavior is illustrated in Fig. 35.22. This shows that missing entries can be recovered by trace norm regularization better than singular value decomposition after filling the missing entries with zeros.

CHAPTER

NONLINEAR DIMENSIONALITY REDUCTION

36

CHAPTER CONTENTS

In this chapter, supervised and unsupervised methods of *nonlinear dimensionality reduction* are introduced, including approaches based on kernels and neural networks.

36.1 DIMENSIONALITY REDUCTION WITH KERNEL TRICK

The linear dimensionality reduction methods explained in the previous chapter handle training samples only in terms of their inner products. This means that the *kernel trick* (see Section 27.4) is applicable to obtain nonlinear methods. More specifically, training input samples $\{\boldsymbol{x}_i\}_{i=1}^n$ are first transformed by a nonlinear mapping $\boldsymbol{\psi}$ and then ordinary linear dimensionality reduction algorithms are applied in the transformed space, which corresponds to nonlinear dimensionality reduction in the original input space. In this section, kernel-based nonlinear dimensionality reduction methods are introduced.

36.1.1 KERNEL PCA

Let us illustrate the behavior of PCA in a feature space. The original two-dimensional samples $\boldsymbol{x}_i = (x_i^{(1)}, x_i^{(2)})^\top$ are plotted in Fig. 36.1(a). Due to the curved spiral shape,

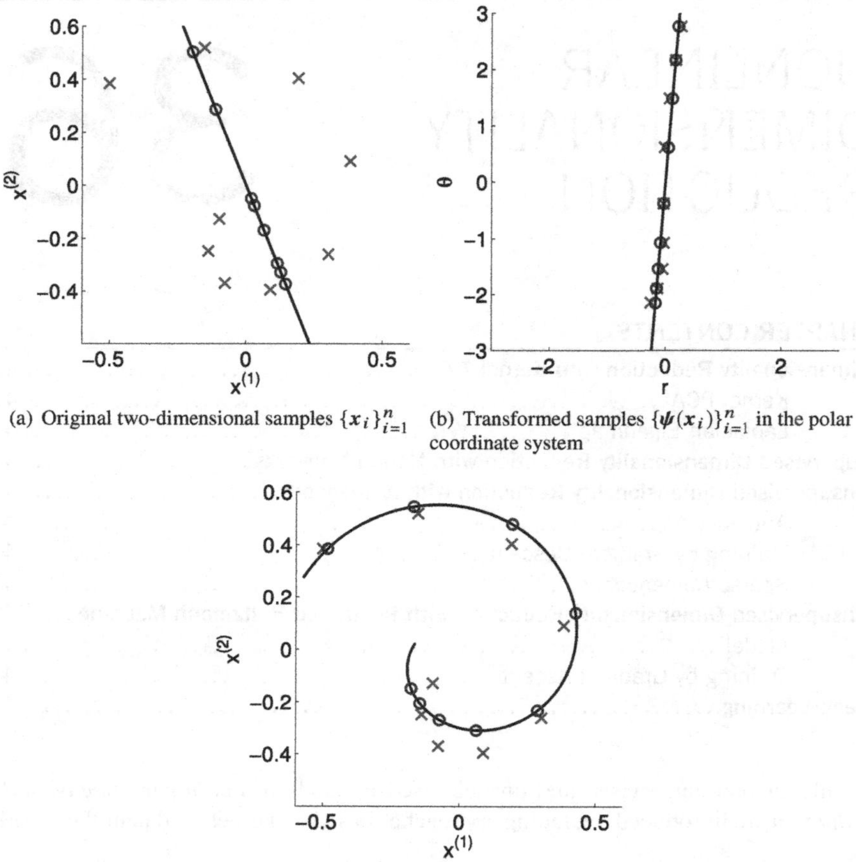

(a) Original two-dimensional samples $\{x_i\}_{i=1}^n$ (b) Transformed samples $\{\psi(x_i)\}_{i=1}^n$ in the polar coordinate system

(c) Result of PCA transformed back to the original input space

FIGURE 36.1

Nonlinear PCA in a feature space. "×" denotes a sample, the solid line denotes the one-dimensional embedding subspace found by PCA, and "○" denotes a projected sample.

linear PCA does not work properly for this data set. If the samples are transformed into the *polar system* (i.e. radius r and angle θ), the spiral shape can be nicely unfolded, as plotted in Fig. 36.1(b). Then linear PCA in the transformed space can well capture the global structure of the unfolded data. Finally, transforming back the projected samples to the original input space gives nice nonlinear projection of the original spiral-shape samples, as plotted in Fig. 36.1(c).

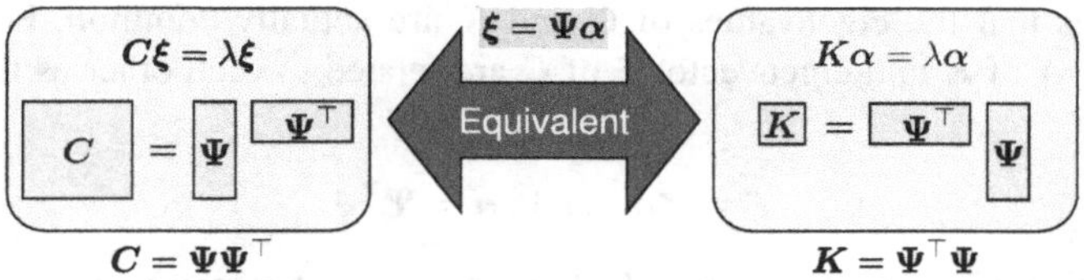

FIGURE 36.2

Eigenvalue problems for PCA. Appropriately choosing the expression of eigenvalue problem depending on whether matrix $\boldsymbol{\Psi}$ is fat or skinny allows us to reduce the computational costs.

If the dimensionality of the feature space is not high, directly performing PCA as explained above is fine. That is, the eigenvalue problem for the total scatter matrix $\boldsymbol{C}$ in the feature space is solved:

$$\boldsymbol{C}\boldsymbol{\xi} = \lambda\boldsymbol{\xi},$$

where

$$\boldsymbol{C} = \sum_{i=1}^{n} \boldsymbol{\psi}(\boldsymbol{x}_i)\boldsymbol{\psi}(\boldsymbol{x}_i)^\top.$$

However, if the dimensionality of the feature space is high, the computational complexity grows significantly. More extremely, if the Gaussian kernel,

$$K(\boldsymbol{x},\boldsymbol{x}') = \exp\left(-\frac{\|\boldsymbol{x}-\boldsymbol{x}'\|^2}{2h^2}\right), \tag{36.1}$$

is used as a nonlinear mapping, the dimensionality of the feature space is actually infinite, and thus explicitly performing PCA in the feature space is not possible.

To cope with this problem, let us consider the eigenvalue problem for the *kernel matrix* $\boldsymbol{K}$ [88]:

$$\boldsymbol{K}\boldsymbol{\alpha} = \lambda\boldsymbol{\alpha},$$

where the (i,i')th element of $\boldsymbol{K}$ is defined as

$$K_{i,i'} = K(\boldsymbol{x}_i,\boldsymbol{x}_{i'}) = \langle\boldsymbol{\psi}(\boldsymbol{x}_i),\boldsymbol{\psi}(\boldsymbol{x}_{i'})\rangle.$$

The matrices $\boldsymbol{C}$ and $\boldsymbol{K}$ can be expressed by using the *design matrix*

$$\boldsymbol{\Psi} = (\boldsymbol{\psi}(\boldsymbol{x}_1),\ldots,\boldsymbol{\psi}(\boldsymbol{x}_n))$$

as

$$\boldsymbol{C} = \boldsymbol{\Psi}\boldsymbol{\Psi}^\top \quad \text{and} \quad \boldsymbol{K} = \boldsymbol{\Psi}^\top\boldsymbol{\Psi}.$$

This implies that the eigenvalues of $\boldsymbol{C}$ and $\boldsymbol{K}$ are actually common. Furthermore, eigenvector $\boldsymbol{\alpha}$ of $\boldsymbol{K}$ and eigenvector $\boldsymbol{\xi}$ of $\boldsymbol{C}$ are related to each other as follows (see Fig. 36.2):

$$\boldsymbol{\xi} = \boldsymbol{\Psi}\boldsymbol{\alpha} \quad \text{and} \quad \boldsymbol{\alpha} = \boldsymbol{\Psi}^\top \boldsymbol{\xi}.$$

While the size of covariance matrix $\boldsymbol{C}$ depends on the dimensionality of the feature space, the size of kernel matrix $\boldsymbol{K}$ depends only on the number of samples and is independent of the dimensionality of the feature space. Thus, if the dimensionality of the feature space is larger than the number of samples, solving the eigenvalue problem with $\boldsymbol{K}$ is computationally more efficient.

As explained in Section 35.2.1, PCA requires centering of samples. However, when the eigenvalue problem with $\boldsymbol{K}$ is solved, feature vectors $\{\boldsymbol{\psi}(\boldsymbol{x}_i)\}_{i=1}^n$ are not explicitly handled and thus feature vectors cannot be centralized directly. In kernel PCA, the kernel matrix $\boldsymbol{K}$ is *implicitly* centralized as

$$\boldsymbol{K} \longleftarrow \boldsymbol{H}\boldsymbol{K}\boldsymbol{H},$$

where

$$\boldsymbol{H} = \boldsymbol{I}_n - \frac{1}{n}\boldsymbol{1}_{n\times n}$$

is the *centering matrix*, $\boldsymbol{I}_n$ denotes the n-dimensional identity matrix, and $\boldsymbol{1}_{n\times n}$ is the $n \times n$ matrix with all ones.

Another issue to be considered is that kernel PCA requires eigenvectors to be normalized as $\|\boldsymbol{\xi}_j\| = 1$, but solving the eigenvalue problem with $\boldsymbol{K}$ usually gives eigenvectors such that $\|\boldsymbol{\alpha}_j\| = 1$. Therefore, normalization of eigenvectors should be carried out explicitly as

$$\boldsymbol{\alpha}_j \longleftarrow \frac{1}{\sqrt{\lambda_j}}\boldsymbol{\alpha}_j \quad \text{for } j = 1, \ldots, m,$$

which comes from

$$\|\boldsymbol{\xi}_j\| = \sqrt{\|\boldsymbol{\xi}_j\|^2} = \sqrt{\|\boldsymbol{\Psi}\boldsymbol{\alpha}_j\|^2} = \sqrt{\langle \boldsymbol{\Psi}^\top\boldsymbol{\Psi}\boldsymbol{\alpha}_j, \boldsymbol{\alpha}_j\rangle} = \sqrt{\langle \boldsymbol{K}\boldsymbol{\alpha}_j, \boldsymbol{\alpha}_j\rangle} = \sqrt{\lambda_j}.$$

Summarizing the above discussions, final embedding solutions of samples $\{\boldsymbol{x}_i\}_{i=1}^n$ by kernel PCA are given by

$$(\boldsymbol{z}_1, \ldots, \boldsymbol{z}_n) = \left(\frac{1}{\sqrt{\lambda_1}}\boldsymbol{\alpha}_1, \ldots, \frac{1}{\sqrt{\lambda_m}}\boldsymbol{\alpha}_m\right)^\top \boldsymbol{H}\boldsymbol{K}\boldsymbol{H},$$

where $\boldsymbol{\alpha}_1, \ldots, \boldsymbol{\alpha}_m$ are eigenvectors of $\boldsymbol{H}\boldsymbol{K}\boldsymbol{H}$ corresponding to the m largest eigenvalues. Similarly, an embedding solution of a new sample $\boldsymbol{x}$ by kernel PCA is given by

$$\boldsymbol{z} = \left(\frac{1}{\sqrt{\lambda_1}}\boldsymbol{\alpha}_1, \ldots, \frac{1}{\sqrt{\lambda_m}}\boldsymbol{\alpha}_m\right)^\top \boldsymbol{H}\left(\boldsymbol{k} - \frac{1}{n}\boldsymbol{K}\boldsymbol{1}_n\right),$$

```
n=200; a=linspace(0,pi,n/2);
u=[a.*cos(a) (a+pi).*cos(a)]';
v=[a.*sin(a) (a+pi).*sin(a)]';
x=[u v]; y=[ones(1,n/2) 2*ones(1,n/2)]';

x2=sum(x.^2,2); hh=2*2^2; H=eye(n)-ones(n,n)/n;
K=exp(-(repmat(x2,1,n)+repmat(x2',n,1)-2*x*x')/hh);
G=H*K*H; [A,L]=eigs(G,2); z=(diag(diag(L).^(-1/2))*A'*G)';

figure(1); clf; hold on;
plot(z(y==1,1),z(y==1,2),'bo');
plot(z(y==2,1),z(y==2,2),'rx');
```

FIGURE 36.3

MATLAB code for kernel PCA with Gaussian kernels.

where $\mathbf{1}_n$ denotes the n-dimensional vectors with all ones and

$$\boldsymbol{k} = (K(\boldsymbol{x},\boldsymbol{x}_1),\ldots,K(\boldsymbol{x},\boldsymbol{x}_n))^\top.$$

A MATLAB code of PCA for Gaussian kernel (36.1) is provided in Fig. 36.3, and its behavior is illustrated in Fig. 36.4. This shows that, while the original two-dimensional samples are not linearly separable, PCA for Gaussian kernel with width $h = 2$ gives linearly separable embedding in the feature space.

36.1.2 LAPLACIAN EIGENMAP

A kernelized version of locality preserving projection is called the *Laplacian eigenmap* [13]. In the generalized eigenvalue problem of locality preserving projection,

$$\boldsymbol{X}\boldsymbol{L}\boldsymbol{X}^\top\boldsymbol{\xi} = \lambda\boldsymbol{X}\boldsymbol{D}\boldsymbol{X}^\top\boldsymbol{\xi},$$

multiplying $\boldsymbol{X}^\top$ from the left-hand side and letting $\boldsymbol{\xi} = \boldsymbol{X}\boldsymbol{\beta}$ yield

$$\boldsymbol{X}^\top\boldsymbol{X}\boldsymbol{L}\boldsymbol{X}^\top\boldsymbol{X}\boldsymbol{\beta} = \lambda\boldsymbol{X}^\top\boldsymbol{X}\boldsymbol{D}\boldsymbol{X}^\top\boldsymbol{X}\boldsymbol{\beta}.$$

If $\boldsymbol{X}^\top\boldsymbol{X}$ is replaced with kernel matrix $\boldsymbol{K}$, a kernelized version of locality preserving projection is obtained as

$$\boldsymbol{K}\boldsymbol{L}\boldsymbol{K}\boldsymbol{\beta} = \lambda\boldsymbol{K}\boldsymbol{D}\boldsymbol{K}\boldsymbol{\beta}.$$

If $\boldsymbol{K}$ is invertible, multiplying $\boldsymbol{K}^{-1}$ from the left-hand side and letting $\boldsymbol{\alpha} = \boldsymbol{K}\boldsymbol{\beta}$ yield

$$\boldsymbol{L}\boldsymbol{\alpha} = \lambda\boldsymbol{D}\boldsymbol{\alpha}, \tag{36.2}$$

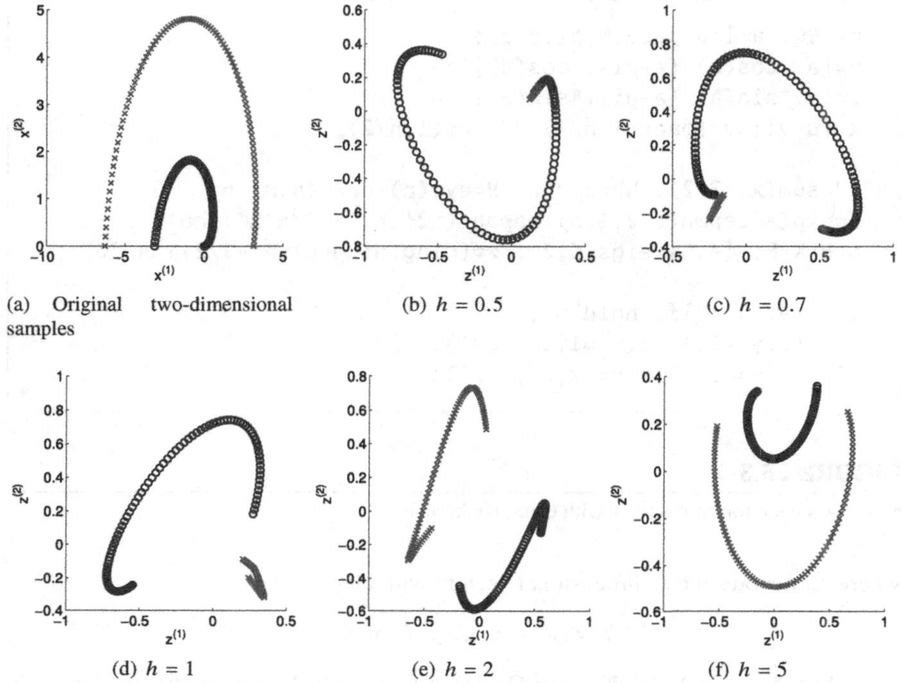

(a) Original two-dimensional samples (b) $h = 0.5$ (c) $h = 0.7$

(d) $h = 1$ (e) $h = 2$ (f) $h = 5$

FIGURE 36.4

Examples of kernel PCA with Gaussian kernels. Original two-dimensional samples are transformed to infinite-dimensional feature space by Gaussian kernels with width h, and then PCA is applied to reduce the dimensionality to two.

which is the eigenvalue problem that the Laplacian eigenmap solves. Let $\lambda_1 \geq \cdots \geq \lambda_n \geq 0$ and $\boldsymbol{\alpha}_1, \ldots, \boldsymbol{\alpha}_n$ be the generalized eigenvalues and generalized eigenvectors of Eq. (36.2). Then the definition of graph Laplacian matrix $\boldsymbol{L}$, Eq. (35.2), yields $\boldsymbol{L}\boldsymbol{1}_n = \boldsymbol{0}_n$. This means that $\lambda_n = 0$ and $\boldsymbol{\alpha}_n = \boldsymbol{1}_n$ hold, which is trivial. In the Laplacian eigenmap, this trivial eigenvector is removed and the final embedding solution is obtained as

$$(z_1, \ldots, z_n) = (\boldsymbol{\alpha}_{n-1}, \boldsymbol{\alpha}_{n-2}, \ldots, \boldsymbol{\alpha}_{n-m})^\top.$$

If the similarity matrix $\boldsymbol{W}$ is *sparse* (see Fig. 35.8), $\boldsymbol{L} = \boldsymbol{D} - \boldsymbol{W}$ is also sparse and thus eigenvalue problem (36.2) can be solved efficiently even if n is large.

A MATLAB code of the Laplacian eigenmap for 10-nearest neighbor similarity (see Fig. 35.8) is provided in Fig. 36.5, and its behavior is illustrated in Fig. 36.6. This shows that the "swiss roll" structure can be nicely unfolded.

```
n=1000; k=10; a=3*pi*rand(n,1);
x=[a.*cos(a) 30*rand(n,1) a.*sin(a)];
x=x-repmat(mean(x),[n,1]); x2=sum(x.^2,2);
d=repmat(x2,1,n)+repmat(x2',n,1)-2*x*x'; [p,i]=sort(d);
W=sparse(d<=ones(n,1)*p(k+1,:)); W=(W+W'~=0);
D=diag(sum(W,2)); L=D-W; [z,v]=eigs(L,D,3,'sm');
figure(1); clf; hold on; view([15 10]);
scatter3(x(:,1),x(:,2),x(:,3),40,a,'o');
figure(2); clf; hold on; scatter(z(:,2),z(:,1),40,a,'o');
```

FIGURE 36.5

MATLAB code of Laplacian eigenmap for 10-nearest neighbor similarity.

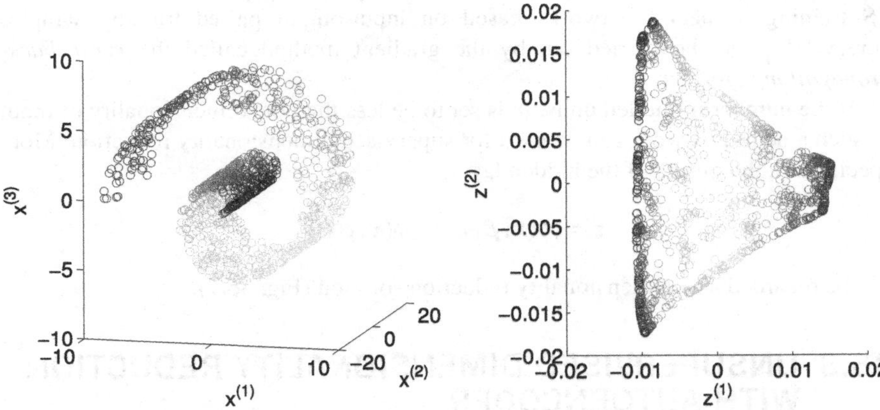

(a) Original three-dimensional samples $\{x_i\}_{i=1}^n$ (b) Embedded two-dimensional samples $\{z_i\}_{i=1}^n$

FIGURE 36.6

Example of Laplacian eigenmap for 10-nearest neighbor similarity.

36.2 SUPERVISED DIMENSIONALITY REDUCTION WITH NEURAL NETWORKS

A *neural network* is a nonlinear model having hierarchical structure. If the number of layers is three (i.e. input, hidden, and output layers), its function is expressed as follows (see Section 21.3):

$$f_\theta(x) = \sum_{j=1}^{b} \alpha_j \phi(x; \beta_j),$$

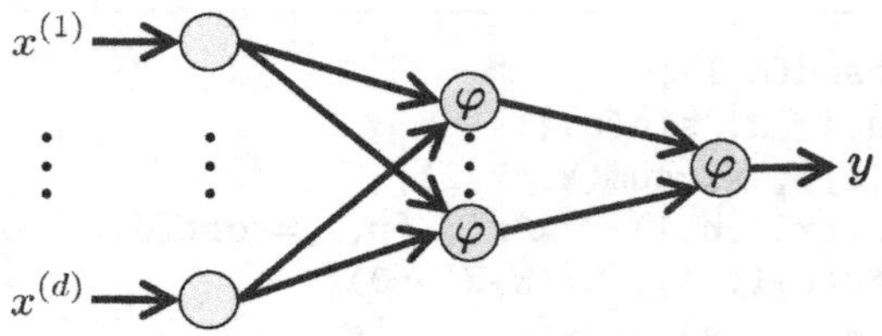

FIGURE 36.7

Dimensionality reduction by neural network. The number of hidden nodes is smaller than the number of input (and output) nodes.

where $\phi(\boldsymbol{x};\boldsymbol{\beta})$ is a basis function parameterized by $\boldsymbol{\beta}$. As explained in Section 22.5, LS training of neural networks based on input-output paired training samples $\{(\boldsymbol{x}_i, y_i)\}_{i=1}^n$ can be carried out by the gradient method called the *error back-propagation* algorithm.

If the number of hidden units, b, is set to be less than the dimensionality of input $\boldsymbol{x}$, such a neural network can be used for supervised dimensionality reduction. More specifically, the output of the hidden layer,

$$\boldsymbol{z} = (\phi(\boldsymbol{x};\boldsymbol{\beta}_1), \ldots, \phi(\boldsymbol{x};\boldsymbol{\beta}_b))^\top,$$

can be regarded as a dimensionality reduction solution (Fig. 36.7).

36.3 UNSUPERVISED DIMENSIONALITY REDUCTION WITH AUTOENCODER

In this section, the back-propagation approach is applied to unsupervised dimensionality reduction.

36.3.1 AUTOENCODER

The *autoencoder* [14, 116] is a three-layer neural network given by

$$\boldsymbol{x} = \varphi\left(\boldsymbol{W}^\top \varphi(\boldsymbol{W}\boldsymbol{x} + \boldsymbol{c}) + \boldsymbol{b}\right),$$

where $\boldsymbol{W}$, $\boldsymbol{b}$, and $\boldsymbol{c}$ are parameters. As illustrated in Fig. 36.8, the autoencoder has a bottleneck layer, and the connection weights between the first and second layers and the connection weights between the second and third layers are shared, i.e. $\boldsymbol{W}$ and $\boldsymbol{W}^\top$ are used. The connection weights are learned so that output is as close to input as possible, by which a compressed expression of input can be obtained in the bottleneck layer. Let us use the *sigmoidal activation function* (see Fig. 21.6) for the

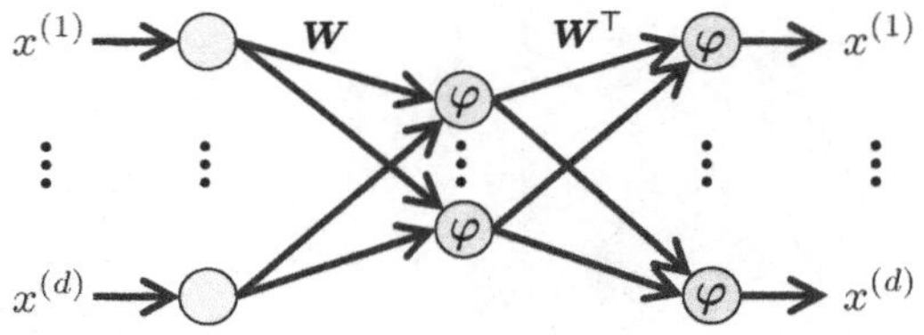

FIGURE 36.8

Autoencoder. Input and output are the same and the number of hidden nodes is smaller than the number of input nodes.

second and third layers:

$$\varphi(\boldsymbol{x}) = \frac{1}{1 + \exp(-\boldsymbol{x}^\top \boldsymbol{w} - a)}.$$

36.3.2 TRAINING BY GRADIENT DESCENT

For d-dimensional input $\boldsymbol{x} = (x^{(1)}, \ldots, x^{(d)})^\top$ and d-dimensional output

$$\boldsymbol{y} = (y^{(1)}, \ldots, y^{(d)})^\top = \boldsymbol{\varphi}\left(\boldsymbol{W}^\top \boldsymbol{\varphi}(\boldsymbol{W}\boldsymbol{x} + \boldsymbol{c}) + \boldsymbol{b}\right),$$

the parameters $\boldsymbol{W}$, $\boldsymbol{b}$, and $\boldsymbol{c}$ are learned by the *stochastic gradient* algorithm introduced in Section 15.3:

$$\begin{aligned}
\boldsymbol{W} &\longleftarrow \boldsymbol{W} - \varepsilon \nabla_{\boldsymbol{W}} J(\boldsymbol{W}, \boldsymbol{b}, \boldsymbol{c}),\\
\boldsymbol{b} &\longleftarrow \boldsymbol{b} - \varepsilon \nabla_{\boldsymbol{b}} J(\boldsymbol{W}, \boldsymbol{b}, \boldsymbol{c}),\\
\boldsymbol{c} &\longleftarrow \boldsymbol{c} - \varepsilon \nabla_{\boldsymbol{c}} J(\boldsymbol{W}, \boldsymbol{b}, \boldsymbol{c}),
\end{aligned}$$

where the *squared loss*,

$$J(\boldsymbol{W}, \boldsymbol{b}, \boldsymbol{c}) = \frac{1}{2} \sum_{k=1}^{d} \left(x^{(k)} - y^{(k)}\right)^2, \tag{36.3}$$

may be used for regression (this corresponds to the log-likelihood of Gaussian distributions), and the *cross entropy loss*,

$$J(\boldsymbol{W}, \boldsymbol{b}, \boldsymbol{c}) = -\sum_{k=1}^{d} \left(x^{(k)} \log y^{(k)} + (1 - x^{(k)}) \log(1 - y^{(k)})\right), \tag{36.4}$$

may be used for binary classification (this corresponds to the log-likelihood of binomial distributions).

For

$$\boldsymbol{y} = \boldsymbol{\varphi}(\boldsymbol{h}), \;\; \boldsymbol{h} = \boldsymbol{W}^\top \boldsymbol{z} + \boldsymbol{b}, \;\; \boldsymbol{z} = \boldsymbol{\varphi}(\boldsymbol{g}), \;\; \text{and} \;\; \boldsymbol{g} = \boldsymbol{W}\boldsymbol{x} + \boldsymbol{c}, \tag{36.5}$$

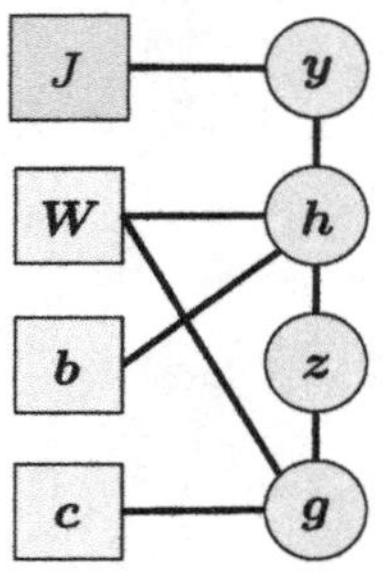

FIGURE 36.9

Chain rule for autoencoder.

the *chain rule* of the derivative (Fig. 36.9) yields

$$\frac{\partial J}{\partial W_{j,k}} = \frac{\partial J}{\partial y^{(k)}}\frac{\partial y^{(k)}}{\partial h_k}\frac{\partial h_k}{\partial W_{j,k}} + \frac{\partial J}{\partial y^{(k)}}\frac{\partial y^{(k)}}{\partial h_k}\frac{\partial h_k}{\partial z^{(j)}}\frac{\partial z^{(j)}}{\partial g_j}\frac{\partial g_j}{\partial W_{j,k}},$$

$$\frac{\partial J}{\partial b_k} = \frac{\partial J}{\partial y^{(k)}}\frac{\partial y^{(k)}}{\partial h_k}\frac{\partial h_k}{\partial b_k},$$

$$\frac{\partial J}{\partial c_j} = \sum_{k=1}^{d}\frac{\partial J}{\partial y^{(k)}}\frac{\partial y^{(k)}}{\partial h_k}\frac{\partial h_k}{\partial z^{(j)}}\frac{\partial z^{(j)}}{\partial g_j}\frac{\partial g_j}{\partial c_j}.$$

Then the gradients are given by

$$\nabla_{\boldsymbol{W}} J = \boldsymbol{z}(\nabla_{\boldsymbol{b}} J)^\top + (\nabla_{\boldsymbol{c}} J)\boldsymbol{x}^\top,$$

$$\nabla_{\boldsymbol{b}} J = \begin{cases} (\boldsymbol{y}-\boldsymbol{x}) * \boldsymbol{y} * (\boldsymbol{1}-\boldsymbol{y}) & \text{(squared loss)}, \\ \boldsymbol{y}-\boldsymbol{x} & \text{(cross entropy loss)}, \end{cases}$$

$$\nabla_{\boldsymbol{c}} J = \boldsymbol{W}\nabla_{\boldsymbol{b}} J * \boldsymbol{z} * (1-\boldsymbol{z}),$$

where "$*$" denotes the elementwise product.

A MATLAB code for dimensionality reduction by the autoencoder is provided in Fig. 36.10, and its behavior is illustrated in Fig. 36.11. Here, the autoencoder is trained with 100 noiseless images of hand-written digit "2" (see Section 12.5 for the details of the hand-written digit data set). Then a test image illustrated in Fig. 36.11(b), which is not included in the training data set, is input to the trained autoencoder. Since the autoencoder is trained only with noiseless images, noise components may be eliminated in the second layer. Then a denoised image illustrated in Fig. 36.11(c) is obtained as output, which looks much better than the output of linear PCA illustrated in Fig. 36.11(d).

```
load digit.mat; x=X(:,1:100,2); [d,n]=size(x); g=min(x,[],2);
x=(x-repmat(g,1,n))./repmat(max(x,[],2)-g,1,n); x=(x>0.5);
m=10; e=0.01/n; W0=randn(m,d); b0=randn(d,1); c0=rand(m,1);
for o=1:100000
  Z=1./(1+exp(-W0*x-repmat(c0,[1 n])));
  Y=1./(1+exp(-W0'*Z-repmat(b0,[1 n])));
  nb=Y-x; % Cross entropy loss
%  nb=(Y-x).*Y.*(1-Y); % Squared loss
  nc=(W0*nb).*Z.*(1-Z); W=W0-e*(Z*nb'+nc*x');
  b=b0-e*sum(nb,2); c=c0-e*sum(nc,2);
  if norm(W-W0)+norm(b-b0)+norm(c-c0)<0.003, break, end
  W0=W; b0=b; c0=c;
end
t=T(:,1,2)>0.5; u=t; u(rand(d,1)>0.9)=1; u(rand(d,1)>0.9)=0;
z=1./(1+exp(-W*t-c)); y=1./(1+exp(-W'*z-b));

figure(1); clf; hold on; colormap gray
subplot(1,3,1); imagesc(reshape(t,[16 16])')
subplot(1,3,2); imagesc(reshape(u,[16 16])')
subplot(1,3,3); imagesc(reshape(y>0.5,[16 16])')
```

FIGURE 36.10

MATLAB code for denoising autoencoder. See Section 12.5 for details of hand-written digit data set "`digit.mat`."

36.3.3 SPARSE AUTOENCODER

Let us consider a nonbottleneck neural network, i.e. the number of hidden nodes, m, is larger than the number of input (and output) nodes, d. If this architecture is used as an autoencoder, just the identity mapping may be learned as a trivial solution. However, if the activations in the hidden layer, $\boldsymbol{z} = (z^{(1)}, \ldots, z^{(m)})$ (see Eq. (36.5)), are enforced to be *sparse*, meaningful features may be extracted.

More specifically, in addition to the loss function $J(\boldsymbol{W}, \boldsymbol{b}, \boldsymbol{c})$ (see Eqs (36.3) and (36.4)), a sparsity-inducing regularization term is minimized at the same time:

$$\min_{\boldsymbol{W},\boldsymbol{b},\boldsymbol{c}} J(\boldsymbol{W}, \boldsymbol{b}, \boldsymbol{c}) + \lambda \sum_{j=1}^{m} R(z^{(j)}),$$

where $\lambda > 0$ is the regularization parameter and $R(z)$ is the regularization functional, e.g. the ℓ_1-norm $R(z) = |z|$ (see Chapter 24), or the KL divergence (see Section 14.2)

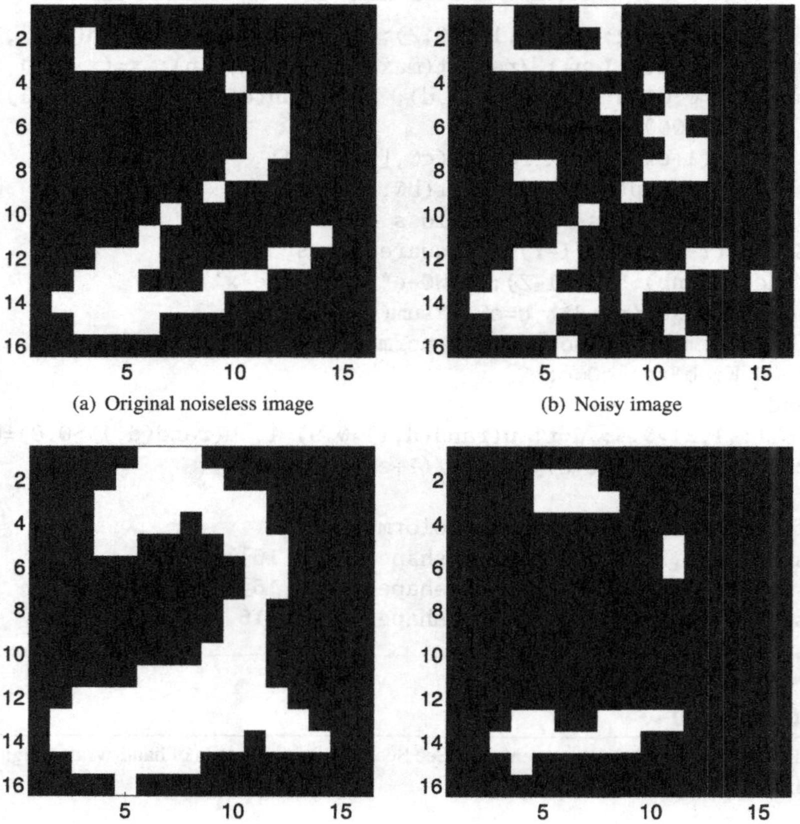

(a) Original noiseless image

(b) Noisy image

(c) Denoised image by autoencoder (10 hidden nodes)

(d) Denoised image by linear PCA (reduced to 10-dimensional subspace)

FIGURE 36.11

Example of denoising autoencoder.

from a constant $\rho \in (0,1)$:

$$R(z) = \rho \log \frac{\rho}{z/m} + (1-\rho) \log \frac{(1-\rho)}{1-z/m}.$$

36.4 UNSUPERVISED DIMENSIONALITY REDUCTION WITH RESTRICTED BOLTZMANN MACHINE

Although the autoencoder is an unsupervised neural network, it is trained in a supervised way by regarding output as input. In this section, another unsupervised

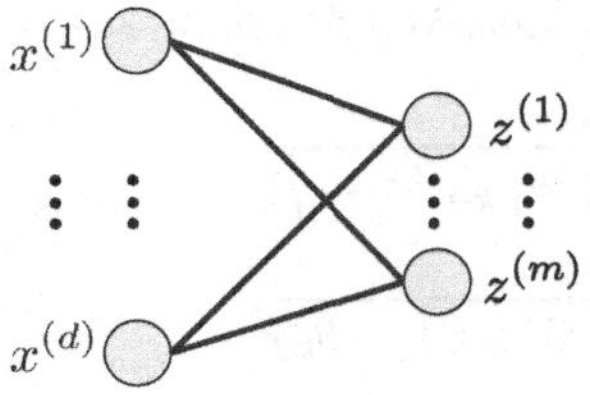

FIGURE 36.12

Restricted Boltzmann machine.

neural network called the *restricted Boltzmann machine* [55, 96] is introduced, which is trained in an unsupervised way.

36.4.1 MODEL

The restricted Boltzmann machine is a two-layered neural network illustrated in Fig. 36.12, where the left layer for d-dimensional input $\boldsymbol{x} = (x^{(1)}, \ldots, x^{(d)})^\top$ is called the *visible layer* and the right layer for m-dimensional hidden variable

$$\boldsymbol{z} = (z^{(1)}, \ldots, z^{(m)})^\top \in \{0, 1\}^m$$

is called the *hidden layer*. Fig. 36.12 shows that any unit in the input layer and any unit in the hidden layer are completely connected. On the other hand, units in the input layer are not connected to each other and units in the hidden layer are not connected to each other. For the moment, let us assume that input is binary: $\boldsymbol{x} \in \{0, 1\}^d$.

The restricted Boltzmann machine is a model of joint probability $q(\boldsymbol{x}, \boldsymbol{z})$ given by

$$q(\boldsymbol{x}, \boldsymbol{z}) = \frac{e^{-E(\boldsymbol{x}, \boldsymbol{z})}}{\sum_{\boldsymbol{x}' \in \{0,1\}^d} \sum_{\boldsymbol{z}' \in \{0,1\}^m} e^{-E(\boldsymbol{x}', \boldsymbol{z}')}}.$$

Here, $E(\boldsymbol{x}, \boldsymbol{z})$ is called the *energy function* defined as

$$E(\boldsymbol{x}, \boldsymbol{z}) = -\boldsymbol{b}^\top \boldsymbol{x} - \boldsymbol{c}^\top \boldsymbol{z} - \boldsymbol{z}^\top \boldsymbol{W} \boldsymbol{x},$$

where $\boldsymbol{W}$, $\boldsymbol{b}$, and $\boldsymbol{c}$ are parameters. Because there is no connection between the input and hidden units, the conditional probabilities can be factorized as

$$q(\boldsymbol{z}|\boldsymbol{x}) = \prod_{j=1}^{m} q(z^{(j)}|\boldsymbol{x}) \quad \text{and} \quad q(\boldsymbol{x}|\boldsymbol{z}) = \prod_{k=1}^{d} q(x^{(k)}|\boldsymbol{z}), \tag{36.6}$$

where each conditional probability is given by the *sigmoidal function*:

$$q(z^{(j)} = 1|\boldsymbol{x}) = \frac{1}{1 + \exp(-\sum_{k=1}^{d} W_{j,k}x^{(k)} - c_j)},$$

$$q(x^{(k)} = 1|\boldsymbol{z}) = \frac{1}{1 + \exp(-\sum_{j=1}^{m} W_{j,k}z^{(j)} - b_k)}.$$

36.4.2 TRAINING BY GRADIENT ASCENT

For training the restricted Boltzmann machine, let us employ MLE (see Chapter 12). However, since hidden variable $\boldsymbol{z}$ is not observable, MLE cannot be performed directly. Here, let us consider the *marginal* model,

$$q(\boldsymbol{x}) = \sum_{\boldsymbol{z}\in\{0,1\}^m} q(\boldsymbol{x},\boldsymbol{z}),$$

and perform MLE for this model:

$$\max_{\boldsymbol{W},\boldsymbol{b},\boldsymbol{c}} L(\boldsymbol{W},\boldsymbol{b},\boldsymbol{c}),$$

where

$$L(\boldsymbol{W},\boldsymbol{b},\boldsymbol{c}) = \frac{1}{n}\sum_{i=1}^{n} \log\left(\sum_{\boldsymbol{z}\in\{0,1\}^m} q(\boldsymbol{x}_i,\boldsymbol{z})\right).$$

The gradient of the log-likelihood is given by

$$\frac{\partial L}{\partial \theta} = -\frac{1}{n}\sum_{i=1}^{n}\sum_{\boldsymbol{z}\in\{0,1\}^m} \frac{\partial E}{\partial \theta} q(\boldsymbol{z}|\boldsymbol{x} = \boldsymbol{x}_i) + \sum_{\boldsymbol{x}\in\{0,1\}^d}\sum_{\boldsymbol{z}\in\{0,1\}^m} \frac{\partial E}{\partial \theta} q(\boldsymbol{x},\boldsymbol{z}), \tag{36.7}$$

where θ represents either $\boldsymbol{W}$, $\boldsymbol{b}$, or $\boldsymbol{c}$, and $\frac{\partial E}{\partial \theta}$ is given by

$$\frac{\partial E}{\partial \boldsymbol{W}} = -\boldsymbol{z}^\top \boldsymbol{x}, \quad \frac{\partial E}{\partial \boldsymbol{b}} = -\boldsymbol{x}, \quad \text{and} \quad \frac{\partial E}{\partial \boldsymbol{c}} = -\boldsymbol{z}.$$

Thanks to factorization (36.6), the first term in Eq. (36.7) can be computed naively as

$$-\frac{1}{n}\sum_{i=1}^{n}\sum_{\boldsymbol{z}\in\{0,1\}^m} \frac{\partial E}{\partial \theta} q(\boldsymbol{z}|\boldsymbol{x} = \boldsymbol{x}_i) = -\frac{1}{n}\sum_{i=1}^{n} \frac{\partial E}{\partial \theta}\prod_{j=1}^{m}\sum_{z^{(j)}\in\{0,1\}} q(z^{(j)}|\boldsymbol{x} = \boldsymbol{x}_i).$$

On the other hand, the second term in Eq. (36.7) can be approximated by the empirical distribution of samples $\{\boldsymbol{x}_i\}_{i=1}^n$ as

$$\begin{aligned}\sum_{\boldsymbol{x}\in\{0,1\}^d}\sum_{\boldsymbol{z}\in\{0,1\}^m} \frac{\partial E}{\partial \theta} q(\boldsymbol{x},\boldsymbol{z}) &= \sum_{\boldsymbol{x}\in\{0,1\}^d}\sum_{\boldsymbol{z}\in\{0,1\}^m} \frac{\partial E}{\partial \theta} q(\boldsymbol{z}|\boldsymbol{x})p(\boldsymbol{x}) \\ &\approx \frac{1}{n}\sum_{i=1}^{n}\sum_{\boldsymbol{z}\in\{0,1\}^m} \frac{\partial E}{\partial \theta} q(\boldsymbol{z}|\boldsymbol{x}_i).\end{aligned}$$

1. Let $\widehat{x}_i = x_i$ for $i = 1, \ldots, n$.
2. Generate $\{\widehat{z}_i\}_{i=1}^n$ from $\{\widehat{x}_i\}_{i=1}^n$ following $q(z|x = \widehat{x}_i)$.
3. Generate $\{\widehat{x}_i\}_{i=1}^n$ from $\{\widehat{z}_i\}_{i=1}^n$ following $q(x|z = \widehat{z}_i)$.
4. Iterate 2–3 until convergence.

FIGURE 36.13

Contrastive divergence algorithm for restricted Boltzmann machine. Note that $q(z|x = \widehat{x}_i)$ and $q(x|z = \widehat{z}_i)$ can be factorized as Eq. (36.6), which allows efficient computation.

```
load digit.mat; x=X(:,1:100,2); [d,n]=size(x); g=min(x,[],2);
x=(x-repmat(g,1,n))./repmat(max(x,[],2)-g,1,n); x=(x>0.5);
m=10; e=0.01/n; W0=randn(m,d); b0=randn(d,1); c0=rand(m,1);
for o=1:100000
  pZ=1./(1+exp(-W0*x-repmat(c0,1,n))); Z=1*(rand(m,n)<pZ);
  pY=1./(1+exp(-W0'*Z-repmat(b0,1,n))); Y=1*(rand(d,n)<pY);
  pX=1./(1+exp(-W0*Y-repmat(c0,1,n))); W=W0+e*(pZ*x'-pX*Y');
  b=b0+e*(x*prod(pZ)'-Y*prod(pX)');
  c=c0+e*(sum(pZ,2)-sum(pX,2));
  if norm(W-W0)+norm(b-b0)+norm(c-c0)<0.007, break, end
  W0=W; b0=b; c0=c;
end
t=T(:,1,2)>0.5; u=t; u(rand(d,1)>0.9)=1; u(rand(d,1)>0.9)=0;
z=1./(1+exp(-W*t-c)); y=1./(1+exp(-W'*z-b));

figure(1); clf; hold on; colormap gray
subplot(1,3,1); imagesc(reshape(t,[16 16])')
subplot(1,3,2); imagesc(reshape(u,[16 16])')
subplot(1,3,3); imagesc(reshape(y>0.5,[16 16])')
```

FIGURE 36.14

MATLAB code for denoising restricted Boltzmann machine. See Section 12.5 for details of hand-written digit data set "`digit.mat`."

However, since this is exactly the same as the negative of the first term, the gradient approximately computed in this way always vanishes and thus is useless in practice. To cope with this problem, let us consider another set of samples $\{\widehat{x}_i\}_{i=1}^n$ generated by the *contrastive divergence algorithm* [54], described in Fig. 36.13, which is based on *Gibbs sampling* (see Section 19.3.3).

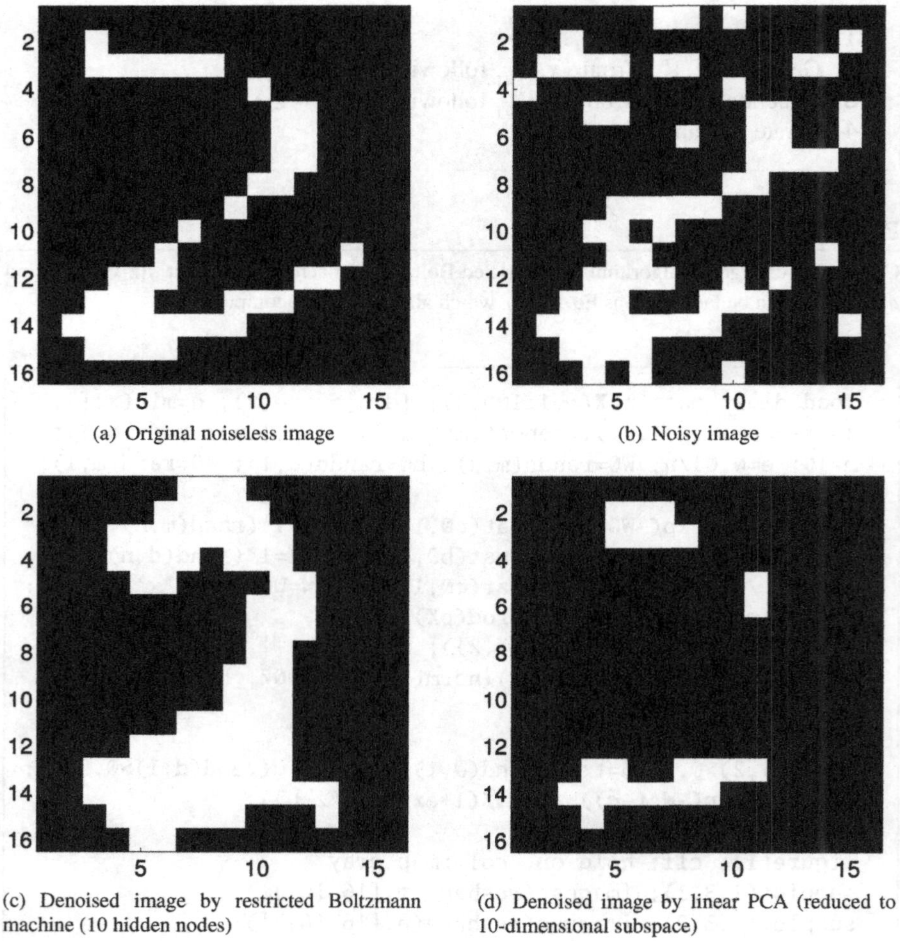

(a) Original noiseless image

(b) Noisy image

(c) Denoised image by restricted Boltzmann machine (10 hidden nodes)

(d) Denoised image by linear PCA (reduced to 10-dimensional subspace)

FIGURE 36.15

Example of denoising restricted Boltzmann machine.

Finally, the gradient of the log-likelihood is approximated as

$$\frac{\partial L}{\partial \theta} \approx \frac{1}{n} \sum_{i=1}^{n} \frac{\partial E}{\partial \theta} \prod_{j=1}^{m} \sum_{z^{(j)} \in \{0,1\}} \Big(- q(z^{(j)} | \boldsymbol{x} = \boldsymbol{x}_i) + q(z^{(j)} | \boldsymbol{x} = \widehat{\boldsymbol{x}}_i) \Big).$$

Based on this approximated gradient, the parameters $\boldsymbol{W}$, $\boldsymbol{b}$, and $\boldsymbol{c}$ can be learned by *gradient ascent* (see Section 15.3).

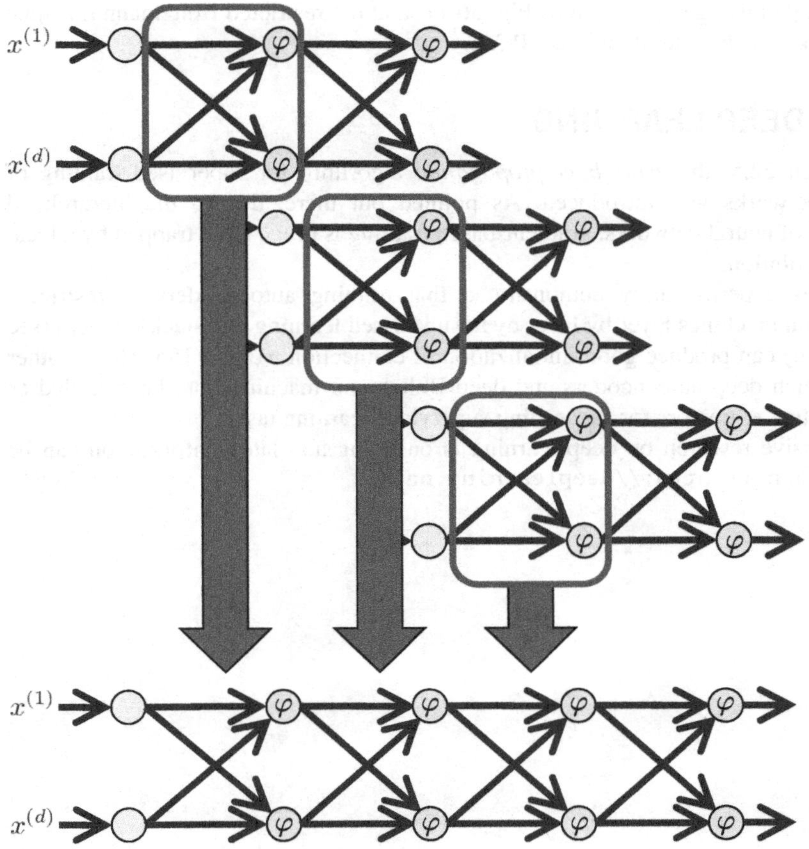

FIGURE 36.16

Construction of deep neural network by stacking.

When input $\boldsymbol{x}$ is continuous, the energy function may be replaced with

$$E(\boldsymbol{x}, \boldsymbol{z}) = \frac{1}{2}\boldsymbol{x}^\top \boldsymbol{x} - \boldsymbol{b}^\top \boldsymbol{x} - \boldsymbol{c}^\top \boldsymbol{z} - \boldsymbol{z}^\top \boldsymbol{W} \boldsymbol{x},$$

which results in the Gaussian conditional probability:

$$q(x^{(k)} = 1|\boldsymbol{z}) = \frac{1}{\sqrt{2\pi}} \exp\left(-\frac{1}{2}\left(x^{(k)} - b_k - \sum_{j=1}^{m} z^{(j)} W_{j,k}\right)^2\right).$$

A MATLAB code for dimensionality reduction by the restricted Boltzmann machine is provided in Fig. 36.14, and its behavior is illustrated in Fig. 36.15. This is the same denoising experiment as Fig. 36.11, and the restricted Boltzmann machine also works much better than linear PCA.

36.5 DEEP LEARNING

In Section 22.5, the *error back-propagation* algorithm for supervised training of neural networks was introduced. As pointed out there, due to the hierarchical structure of neural networks, gradient-based training is prone to be trapped by a local optimal solution.

It was experimentally demonstrated that training autoencoders or restricted Boltzmann machines layer by layer by unsupervised learning and stacking them (see Fig. 36.16) can produce good initialization of connection weights [55, 86]. In other words, such deep autoencoders and deep Boltzmann machines can be regarded as good feature extractors for succeeding supervised learning tasks.

Extensive research on deep learning is on going and latest information can be found, e.g. from "`http://deeplearning.net/`."

CHAPTER

CLUSTERING

37

CHAPTER CONTENTS

In this chapter, the problem of unsupervised classification of input-only samples $\{\boldsymbol{x}_i\}_{i=1}^n$ called *clustering* is discussed, which is aimed at grouping data samples based on their similarity.

37.1 k-MEANS CLUSTERING

The method of *k-means clustering* is one of the most fundamental algorithms for clustering, which aims at finding cluster labels,

$$\{y_i \mid y_i \in \{1,\ldots,c\}\}_{i=1}^n,$$

so that the sum of *within-cluster scatters* is minimized:

$$\min_{y_1,\ldots,y_n \in \{1,\ldots,c\}} \sum_{y=1}^{c} \sum_{i:y_i=y} \|\boldsymbol{x}_i - \boldsymbol{\mu}_y\|^2,$$

where $\sum_{i:y_i=y}$ denotes the sum over i such that $y_i = y$,

$$\boldsymbol{\mu}_y = \frac{1}{n_y} \sum_{i:y_i=y} \boldsymbol{x}_i$$

denotes the center of center y, and n_y denotes the number of samples in cluster y.

However, solving this optimization problem will take exponential computation time with respect to the number of samples n [3], and thus is computationally infeasible if n is large. In practice, a local optimal solution is found by iteratively assigning a single sample to the cluster with the nearest center in a greedy manner:

$$y_i \longleftarrow \underset{y\in\{1,\ldots,c\}}{\operatorname{argmin}} \|\boldsymbol{x}_i - \boldsymbol{\mu}_y\|^2, \quad i = 1,\ldots,n. \tag{37.1}$$

The algorithm of k-means clustering is summarized in Fig. 37.1.

1. Initialize cluster centers $\boldsymbol{\mu}_1,\ldots,\boldsymbol{\mu}_c$.
2. Update cluster labels $y_1,\ldots,y_n$ for samples $\boldsymbol{x}_1,\ldots,\boldsymbol{x}_n$:

$$y_i \longleftarrow \underset{y\in\{1,\ldots,c\}}{\operatorname{argmin}} \|\boldsymbol{x}_i - \boldsymbol{\mu}_y\|^2, \quad i = 1,\ldots,n.$$

3. Update cluster centers $\boldsymbol{\mu}_1,\ldots,\boldsymbol{\mu}_c$:

$$\boldsymbol{\mu}_y \longleftarrow \frac{1}{n_y}\sum_{i:y_i=y} \boldsymbol{x}_i, \quad y = 1,\ldots,c,$$

where n_y denotes the number of samples in cluster y.
4. Iterate 2 and 3 until convergence.

FIGURE 37.1

k-means clustering algorithm.

A MATLAB code for k-means clustering is provided in Fig. 37.2, and its behavior is illustrated in Fig. 37.3. This shows that the k-means clustering algorithm gives a reasonable solution for this data set.

37.2 KERNEL k-MEANS CLUSTERING

Since k-means clustering uses the Euclidean distance $\|\boldsymbol{x}-\boldsymbol{\mu}_y\|$ for determining cluster assignment, it can only produce *linearly separable* clusters, as illustrated in Fig. 37.3. Here, let us apply the *kernel trick* introduced in Section 27.4 to obtain a nonlinear version of k-means clustering, called *kernel k-means clustering* [48].

More specifically, let us express the squared Euclidean distance $\|\boldsymbol{x} - \boldsymbol{\mu}_y\|^2$ in Eq. (37.1) by using the inner product $\langle \boldsymbol{x},\boldsymbol{x}'\rangle$ as

$$\begin{aligned}\|\boldsymbol{x} - \boldsymbol{\mu}_y\|^2 &= \left\|\boldsymbol{x} - \frac{1}{n_y}\sum_{i:y_i=y}\boldsymbol{x}_i\right\|^2\\ &= \langle\boldsymbol{x},\boldsymbol{x}\rangle - \frac{2}{n_y}\sum_{i:y_i=y}\langle\boldsymbol{x},\boldsymbol{x}_i\rangle + \frac{1}{n_y^2}\sum_{i,i':y_i=y_{i'}=y}\langle\boldsymbol{x}_i,\boldsymbol{x}_{i'}\rangle.\end{aligned}$$

Then, replacing the inner product with the kernel function $K(\boldsymbol{x},\boldsymbol{x}')$ immediately gives the kernel k-means clustering algorithm:

$$y \longleftarrow \underset{y\in\{1,\ldots,c\}}{\operatorname{argmin}} \left[-\frac{2}{n_y}\sum_{i:y_i=y}K(\boldsymbol{x},\boldsymbol{x}_i) + \frac{1}{n_y^2}\sum_{i,i':y_i=y_{i'}=y}K(\boldsymbol{x}_i,\boldsymbol{x}_{i'})\right],$$

where irrelevant constant $\langle\boldsymbol{x},\boldsymbol{x}\rangle = K(\boldsymbol{x},\boldsymbol{x})$ is ignored.

```
n=300; c=3; t=randperm(n);
x=[randn(1,n/3)-2 randn(1,n/3) randn(1,n/3)+2;
   randn(1,n/3) randn(1,n/3)+4 randn(1,n/3)]';
m=x(t(1:c),:); x2=sum(x.^2,2); s0(1:c,1)=inf;

for o=1:1000
  m2=sum(m.^2,2);
  [d,y]=min(repmat(m2,1,n)+repmat(x2',c,1)-2*m*x');
  for t=1:c
    m(t,:)=mean(x(y==t,:)); s(t,1)=mean(d(y==t));
  end
  if norm(s-s0)<0.001, break, end
  s0=s;
end

figure(1); clf; hold on;
plot(x(y==1,1),x(y==1,2),'bo');
plot(x(y==2,1),x(y==2,2),'rx');
plot(x(y==3,1),x(y==3,2),'gv');
```

FIGURE 37.2

MATLAB code for k-means clustering.

Since ordinary k-means clustering in the kernel feature space corresponds to nonlinear clustering in the original input space, it can give clustering solutions with nonlinear boundaries. However, due to high nonlinearity brought by the kernel function, optimization of kernel k-means clustering is much harder then that of ordinary k-means clustering and the solution depends even strongly on initialization. For this reason, the use of kernel k-means clustering is not straightforward in practice.

37.3 SPECTRAL CLUSTERING

Strong dependency of kernel k-means clustering on the initial solution is more prominent if the dimensionality of the feature space is higher. *Spectral clustering* addresses this issue by combining clustering with *dimensionality reduction.*

More specifically, after transforming samples to a feature space by a kernel function, spectral clustering applies *locality preserving projection* introduced in Section 35.2.2 to reduce the dimensionality, which has the property that cluster structure of data tends to be preserved. Note that locality preserving projection in the feature space is equivalent to the *Laplacian eigenmap* introduced in Section 36.1.2.

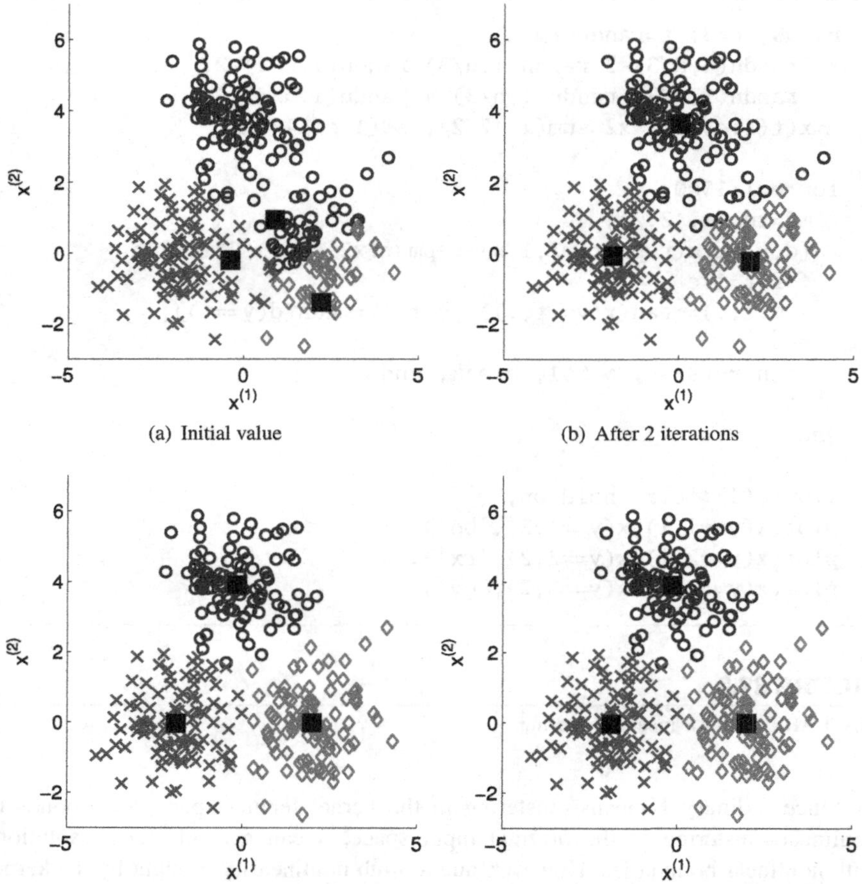

(a) Initial value (b) After 2 iterations

(c) After 4 iterations (d) Final solution (after 6 iterations)

FIGURE 37.3

Example of k-means clustering. A filled square denotes a cluster center.

Then, after dimensionality reduction, spectral clustering employs *ordinary* (not kernel) k-means clustering to obtain cluster labels. The algorithm of spectral clustering is summarized in Fig. 37.4.

A MATLAB code for spectral clustering is provided in Fig. 37.5, and its behavior is illustrated in Fig. 37.6. The original two-dimensional samples $\{\boldsymbol{x}_i\}_{i=1}^n$ in Fig. 37.6(a) are projected onto a one-dimensional subspace in the feature space, which gives embedded samples $\{z_i\}_{i=1}^n$ degenerated into two points as illustrated in Fig. 37.6(b). Then applying ordinary k-means clustering to the degenerated samples

1. Apply Laplacian eigenmap to samples $\{\boldsymbol{x}_i\}_{i=1}^n$ and obtain $(c-1)$-dimensional expressions $\{z_i\}_{i=1}^n$, where c is the number of clusters.
2. Apply(non-kernelized) k-means clustering to the embedded samples $\{z_i\}_{i=1}^n$ and obtain cluster labels $\{y_i\}_{i=1}^n$.

FIGURE 37.4

Algorithm of spectral clustering.

```
n=500; c=2; k=10; t=randperm(n); a=linspace(0,2*pi,n/2)';
x=[a.*cos(a) a.*sin(a); (a+pi).*cos(a) (a+pi).*sin(a)];
x=x+rand(n,2); x=x-repmat(mean(x),[n,1]); x2=sum(x.^2,2);
d=repmat(x2,1,n)+repmat(x2',n,1)-2*x*x'; [p,i]=sort(d);
W=sparse(d<=ones(n,1)*p(k+1,:)); W=(W+W'~=0);
D=diag(sum(W,2)); L=D-W; [z,v]=eigs(L,D,c-1,'sm');

m=z(t(1:c),:); s0(1:c,1)=inf; z2=sum(z.^2,2);
for o=1:1000
  m2=sum(m.^2,2);
  [u,y]=min(repmat(m2,1,n)+repmat(z2',c,1)-2*m*z');
  for t=1:c
    m(t,:)=mean(z(y==t,:)); s(t,1)=mean(d(y==t));
  end
  if norm(s-s0)<0.001, break, end
  s0=s;
end

figure(1); clf; hold on; axis([-10 10 -10 10]);
plot(x(y==1,1),x(y==1,2),'bo');
plot(x(y==2,1),x(y==2,2),'rx');
```

FIGURE 37.5

MATLAB code for spectral clustering.

$\{z_i\}_{i=1}^n$ gives cluster labels $\{y_i\}_{i=1}^n$ illustrated in Fig. 37.6(c). This clustering solution actually corresponds to clustering two spirals in the original input space, as illustrated in Fig. 37.6(d).

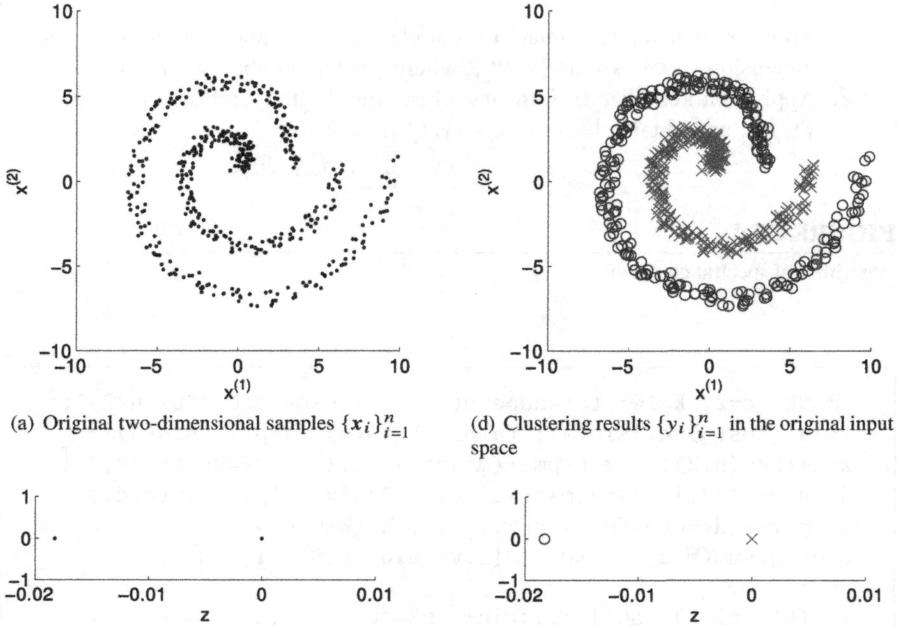

(a) Original two-dimensional samples $\{\boldsymbol{x}_i\}_{i=1}^n$

(d) Clustering results $\{y_i\}_{i=1}^n$ in the original input space

(b) One-dimensional projection $\{z_i\}_{i=1}^n$ by Laplacian eigenmap

(c) Clustering results $\{y_i\}_{i=1}^n$ obtained by ordinary k-means clustering in the feature

FIGURE 37.6

Example of spectral clustering.

37.4 TUNING PARAMETER SELECTION

Clustering solutions obtained by kernel k-means and spectral clustering depend on the choice of kernels such as the Gaussian bandwidth. In this section, the problem of tuning parameter selection is addressed.

Clustering can be regarded as *compressing* d-dimensional samples $\{\boldsymbol{x}_i\}_{i=1}^n$ into c-valued labels $\{y_i\}_{i=1}^n$ (Fig. 37.7). Based on this data compression view, let us choose tuning parameters so that the amount of *information* cluster labels $\{y_i\}_{i=1}^n$ contain on input samples $\{\boldsymbol{x}_i\}_{i=1}^n$ is maximized.

MI [92] allows us to measure the amount of information cluster labels $\{y_i\}_{i=1}^n$ contain on input samples $\{\boldsymbol{x}_i\}_{i=1}^n$:

$$\mathrm{MI} = \int \sum_{y=1}^{c} p(\boldsymbol{x}, y) \log \frac{p(\boldsymbol{x}, y)}{p(\boldsymbol{x})p(y)} \mathrm{d}\boldsymbol{x}.$$

MI is actually the *KL divergence* (see Section 14.2) from the joint probability $p(\boldsymbol{x}, y)$ to the product of marginals $p(\boldsymbol{x})p(y)$. Therefore, it is always non-negative and is

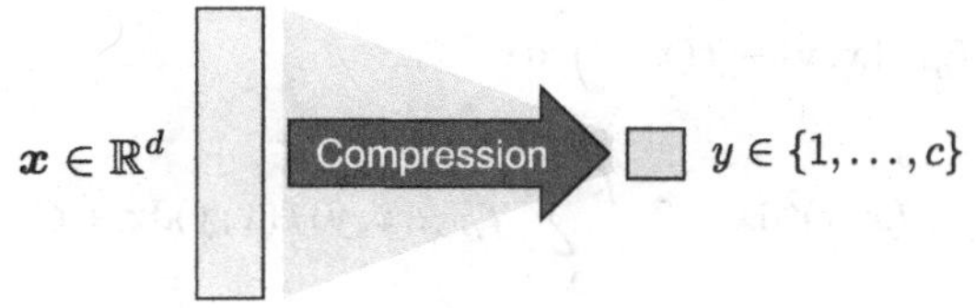

FIGURE 37.7

Clustering can be regarded as compressing d-dimensional vector $\boldsymbol{x}$ into c-valued scalar y.

zero if and only if $p(\boldsymbol{x}, y) = p(\boldsymbol{x})p(y)$, i.e. $\boldsymbol{x}$ and y are *statistically independent* (see Section 5.6). MI allows us to measure the dependency of $\boldsymbol{x}$ and y, i.e. the amount of information y contains on $\boldsymbol{x}$.

The value of MI can be approximated by the *KL density ratio estimator* described in Section 38.3. However, because the log function and the ratio of probability densities are included in MI, it tends to be sensitive to outliers. Here, let us use a variant of MI based on the *L_2-distance* called the QMI [113]:

$$\mathrm{QMI} = \int \sum_{y=1}^{c} \big(p(\boldsymbol{x}, y) - p(\boldsymbol{x})p(y)\big)^2 \mathrm{d}\boldsymbol{x},$$

which is also always non-negative and takes zero if and only if $\boldsymbol{x}$ and y are statistically independent, i.e. $p(\boldsymbol{x}, y) = p(\boldsymbol{x})p(y)$.

Below, an estimator of QMI from $\{(\boldsymbol{x}_i, y_i)\}_{i=1}^n$ called *LS QMI estimation* is introduced, which is an application of *LS density difference estimation* introduced in Section 39.1.3 to QMI. More specifically, LS QMI estimation does not involve estimation of $p(\boldsymbol{x}, y)$, $p(\boldsymbol{x})$, and $p(y)$, but it directly estimates the density difference function:

$$f(\boldsymbol{x}, y) = p(\boldsymbol{x}, y) - p(\boldsymbol{x})p(y).$$

Let us employ the Gaussian kernel model for approximating the density difference $f(\boldsymbol{x}, y)$:

$$f_{\boldsymbol{\alpha}^{(y)}}(\boldsymbol{x}, y) = \sum_{j=1}^{n_y} \alpha_j^{(y)} \exp\left(-\frac{\|\boldsymbol{x} - \boldsymbol{x}_j^{(y)}\|^2}{2h^2}\right),$$

where $\boldsymbol{\alpha}^{(y)} = (\alpha_1, \ldots, \alpha_{n_y})^\top$ is a parameter vector for cluster y, n_y denotes the number of samples in cluster y, and $\{\boldsymbol{x}_i^{(y)}\}_{i=1}^{n_y}$ denote samples in cluster y among $\{\boldsymbol{x}_i\}_{i=1}^n$. The parameter vector $\{\boldsymbol{\alpha}^{(y)}\}_{y=1}^c$ is learned so that the following squared error is minimized:

$$J(\boldsymbol{\alpha}^{(1)},\ldots,\boldsymbol{\alpha}^{(c)}) = \int \sum_{y=1}^{c} \left(f_{\boldsymbol{\alpha}^{(y)}}(\boldsymbol{x},y) - f(\boldsymbol{x},y)\right)^2 \mathrm{d}\boldsymbol{x}$$
$$= \int \sum_{y=1}^{c} f_{\boldsymbol{\alpha}^{(y)}}(\boldsymbol{x},y)^2 \mathrm{d}\boldsymbol{x} - 2\int \sum_{y=1}^{c} f_{\boldsymbol{\alpha}^{(y)}}(\boldsymbol{x},y) f(\boldsymbol{x},y)\mathrm{d}\boldsymbol{x} + C,$$

where $C = \int \sum_{y=1}^{c} f(\boldsymbol{x},y)^2 \mathrm{d}\boldsymbol{x}$ is a constant independent of parameter $\boldsymbol{\alpha}$ and thus is ignored. The expectation in the second term may be approximated by the sample average and $p(y) \approx n_y/n$ as

$$\int f_{\boldsymbol{\alpha}^{(y)}}(\boldsymbol{x},y) f(\boldsymbol{x},y)\mathrm{d}\boldsymbol{x}$$
$$= \int f_{\boldsymbol{\alpha}^{(y)}}(\boldsymbol{x},y) p(\boldsymbol{x}|y)p(y)\mathrm{d}\boldsymbol{x} - \int f_{\boldsymbol{\alpha}^{(y)}}(\boldsymbol{x},y) p(\boldsymbol{x})p(y)\mathrm{d}\boldsymbol{x}$$
$$\approx \frac{1}{n_y}\sum_{i=1}^{n_y} f_{\boldsymbol{\alpha}^{(y)}}(\boldsymbol{x}_i^{(y)},y)\frac{n_y}{n} - \frac{1}{n}\sum_{i=1}^{n} f_{\boldsymbol{\alpha}^{(y)}}(\boldsymbol{x}_i,y)\frac{n_y}{n}$$
$$= \frac{1}{n}\sum_{i=1}^{n_y} f_{\boldsymbol{\alpha}^{(y)}}(\boldsymbol{x}_i^{(y)},y) - \frac{n_y}{n^2}\sum_{i=1}^{n} f_{\boldsymbol{\alpha}^{(y)}}(\boldsymbol{x}_i,y).$$

Further adding the ℓ_2-regularizer yields the following criterion:

$$\min_{\boldsymbol{\alpha}^{(1)},\ldots,\boldsymbol{\alpha}^{(c)}} \sum_{y=1}^{c} \left[\boldsymbol{\alpha}^{(y)\top}\boldsymbol{U}^{(y)}\boldsymbol{\alpha}^{(y)} - 2\boldsymbol{\alpha}^{(y)\top}\widehat{\boldsymbol{v}}^{(y)} + \lambda\|\boldsymbol{\alpha}^{(y)}\|^2\right],$$

where $\lambda \geq 0$ is the regularization parameter and $\boldsymbol{U}^{(y)}$ and $\widehat{\boldsymbol{v}}^{(y)}$ are the $n_y \times n_y$ matrix and the n_y-dimensional vector defined as

$$U_{j,j'}^{(y)} = (\sqrt{\pi}h)^d \exp\left(-\frac{\|\boldsymbol{x}_j^{(y)} - \boldsymbol{x}_{j'}^{(y)}\|^2}{4h^2}\right),$$
$$\widehat{v}_j^{(y)} = \frac{1}{n}\sum_{i=1}^{n_y}\exp\left(-\frac{\|\boldsymbol{x}_i^{(y)} - \boldsymbol{x}_j^{(y)}\|^2}{2h^2}\right) - \frac{n_y}{n^2}\sum_{i=1}^{n}\exp\left(-\frac{\|\boldsymbol{x}_i - \boldsymbol{x}_j^{(y)}\|^2}{2h^2}\right).$$

This optimization problem can be solved analytically and separately for each $y = 1,\ldots,c$ as

$$\widehat{\boldsymbol{\alpha}}^{(y)} = \left(\boldsymbol{U}^{(y)} + \lambda\boldsymbol{I}\right)^{-1}\widehat{\boldsymbol{v}}^{(y)}.$$

Then QMI can be approximated as

$$\widehat{\mathrm{QMI}} = \sum_{y=1}^{c}\left(\frac{1}{n}\sum_{i=1}^{n_y} f_{\widehat{\boldsymbol{\alpha}}^{(y)}}(\boldsymbol{x}_i^{(y)},y) - \frac{n_y}{n^2}\sum_{i=1}^{n} f_{\widehat{\boldsymbol{\alpha}}^{(y)}}(\boldsymbol{x}_i,y)\right)$$
$$= \sum_{y=1}^{c}\widehat{\boldsymbol{\alpha}}^{(y)\top}\widehat{\boldsymbol{v}}^{(y)},$$

```
n=500; a=linspace(0,2*pi,n/2)'; y=[ones(1,n/2) zeros(1,n/2)];
x=[a.*cos(a) a.*sin(a); (a+pi).*cos(a) (a+pi).*sin(a)];
x=x+rand(n,2); x=x-repmat(mean(x),[n,1]); x2=sum(x.^2,2);
d=repmat(x2,1,n)+repmat(x2',n,1)-2*x*x';

hs=[0.5 1 2].^2; ls=10.^[-5 -4 -3]; m=2; r=size(x,2)/2;
u=mod(randperm(n),m)+1; hhs=2*[0.5 1 2].^2;
g=zeros(length(hhs),length(ls),m);

for hk=1:length(hs)
  h=hs(hk); k=exp(-d/(2*h));
  for j=unique(y)
    t=(y==j); U=(pi*h)^r*exp(-d(t,t)/(4*h));
    for i=1:m
      ai=(u~=i); ni=sum(a); aj=(u==i); nj=sum(aj);
      vi=sum(k(t,ai&t),2)/ni-sum(k(t,ai),2)*sum(ai&t)/(ni^2);
      vj=sum(k(t,aj&t),2)/nj-sum(k(t,aj),2)*sum(aj&t)/(nj^2);
      for lk=1:length(ls)
        l=ls(lk); a=(U+l*eye(sum(t)))\vi;
        g(hk,lk,i)=g(hk,lk,i)+a'*U*a-2*vj'*a;
end, end, end, end
g=mean(g,3); [gl,ggl]=min(g,[],2); [ghl,gghl]=min(gl);
L=ls(ggl(gghl)); H=hs(gghl); s=0;
for j=unique(y)
  t=(y==j); ny=sum(t); U=(pi*H)^r*exp(-d(t,t)/(4*H));
  k=exp(-d(t,:)/(2*H)); v=sum(k(:,t),2)/n-sum(k,2)*ny/(n^2);
  a=(U+L*eye(ny))\v; s=s+v'*a;
end
disp(sprintf('Information=%g',s));
```

FIGURE 37.8

MATLAB code for LS QMI estimation.

which comes from the following expression of QMI:

$$\mathrm{QMI} = \int \sum_{y=1}^{c} f(\boldsymbol{x}, y)\big(p(\boldsymbol{x}, y) - p(\boldsymbol{x})p(y)\big)\mathrm{d}\boldsymbol{x}.$$

All tuning parameters such as the regularization parameter λ and the Gaussian bandwidth h can be optimized by cross validation with respect to the squared error J. This is a strong advantage in unsupervised clustering.

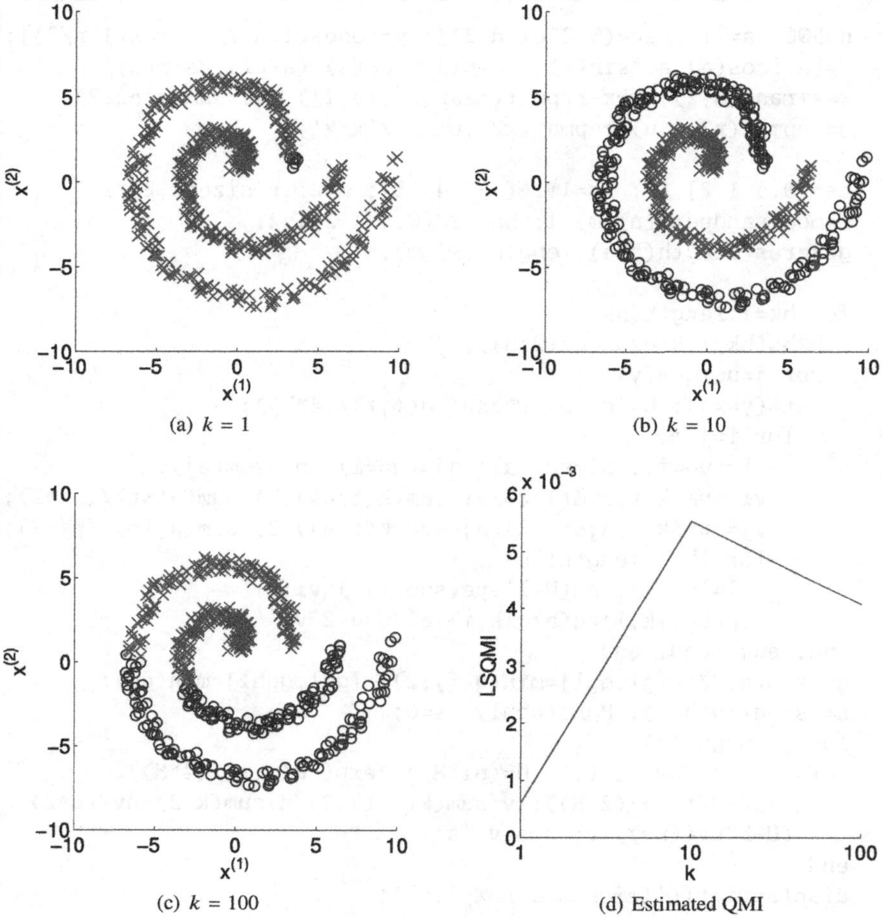

FIGURE 37.9

Example of LS QMI estimation.

A MATLAB code for LS QMI estimation is provided in Fig. 37.8, and its behavior is illustrated in Fig. 37.9. This shows that setting the parameter k included in the nearest neighbor similarity at a too small value (Fig. 37.9(a)) or too large value (Fig. 37.9(c)) does not produce a desirable clustering solution. On the other hand, if the parameter k is chosen so that the LS QMI estimator is maximized, an intermediate value is selected (Fig. 37.9(d)) and this choice works well for this data set (Fig. 37.9(b)).

CHAPTER

OUTLIER DETECTION 38

CHAPTER CONTENTS

The objective of *outlier detection* is to find outlying samples in input-only training samples $\{x_i\}_{i=1}^n$. If the labels of inliers and outliers are available for $\{x_i\}_{i=1}^n$, outlier detection can be formulated as supervised classification. However, since outliers may be highly diverse and their tendency may change over time, learning a stable decision boundary between inliers and outliers is often difficult in practice. In this chapter, various unsupervised outlier detection methods are introduced.

In Part 4, supervised learning methods that are *robust* against outliers were introduced. When the number of outliers is expected not to be too large, robust learning from a data set contaminated with outliers may be useful. On the other hand, if outliers are expected to be more abundant in the data set, removing them in advance by an outlier detection method would be more appropriate.

38.1 DENSITY ESTIMATION AND LOCAL OUTLIER FACTOR

A naive outlier detection method is based on *density estimation*. More specifically, the probability density $p(x)$ of samples $\{x_i\}_{i=1}^n$ is estimated using, e.g. one of the density estimators described in Part 3, and then samples having low probability densities are regarded as outliers. However, since estimating the probability density in low-density regions is difficult due to the shortage of samples, such a density estimation approach may be unreliable for outlier detection purposes. The *local outlier factor* [23] is a stabilized variant of the density estimation approach that finds samples isolated from other samples.

Let us define the *reachability distance* (RD) from x to x' as

$$\mathrm{RD}_k(x, x') = \max\left(\|x - x^{(k)}\|, \|x - x'\|\right),$$

where $x^{(k)}$ is the kth nearest neighbor of x in $\{x_i\}_{i=1}^n$. The RD can be regarded as a stabilized variant of the Euclidean distance $\|x - x'\|$ so that the distance is not less than $\|x - x^{(k)}\|$. Based on the RD, the *local RD* of x is defined as

$$\mathrm{LRD}_k(x) = \left(\frac{1}{k}\sum_{i=1}^{k} \mathrm{RD}_k(x^{(i)}, x)\right)^{-1},$$

which is the inverse of the average RDs from $x^{(i)}$ to x. When x is isolated from surrounding samples, the local RD takes a small value.

The *local outlier factor* of x is defined as

$$\mathrm{LOF}_k(x) = \frac{\frac{1}{k}\sum_{i=1}^{k} \mathrm{LRD}_k(x^{(i)})}{\mathrm{LRD}_k(x)},$$

and x is regarded as an outlier if $\mathrm{LOF}_k(x)$ takes a large value. $\mathrm{LOF}_k(x)$ is the ratio of the average local RD of $x^{(i)}$ and the local RD of x, and x is regarded as an outlier if $x^{(i)}$ is in a high-density region and x is in a low-density region. Conversely, if $x^{(i)}$ is in a low-density region and x is in a high-density region, $\mathrm{LOF}_k(x)$ takes a small value and thus x is regarded as an inlier.

A MATLAB code for computing the local outlier factor is provided in Fig. 38.2, and its behavior is illustrated in Fig. 38.1. This shows that isolated samples tend to have large outlier scores. However, the behavior of the local outlier factor depends on the choice of nearest neighbors, k, and it is not straightforward to optimize k in practice.

38.2 SUPPORT VECTOR DATA DESCRIPTION

Support vector data description [109] is an unsupervised outlier detection algorithm that does not involve explicit density estimation.

Let us consider a *hypersphere* with center c and radius $\sqrt{b}$ on $\mathbb{R}^d$, and learn c and b to include all training samples $\{x_i\}_{i=1}^n$ (i.e., the *minimum enclosing ball* is found):

$$\min_{c,b} b \text{ subject to } \|x_i - c\|^2 \le b \quad \text{for } i = 1, \ldots, n. \tag{38.1}$$

Note that, not the radius $\sqrt{b}$, but the squared radius b is optimized for the convexity of the optimization problem [26].

Support vector data description is a relaxed variant of the minimum enclosing ball problem which finds a hypersphere that contains *most* of the training samples $\{x_i\}_{i=1}^n$ (Fig. 38.3):

```
n=100; x=[(rand(n/2,2)-0.5)*20; randn(n/2,2)]; x(n,1)=14;
k=3; x2=sum(x.^2,2);
[s,t]=sort(sqrt(repmat(x2,1,n)+repmat(x2',n,1)-2*x*x'),2);

for i=1:k+1
  for j=1:k
    RD(:,j)=max(s(t(t(:,i),j+1),k),s(t(:,i),j+1));
  end
  LRD(:,i)=1./mean(RD,2);
end
LOF=mean(LRD(:,2:k+1),2)./LRD(:,1);

figure(1); clf; hold on
plot(x(:,1),x(:,2),'rx');
for i=1:n
  plot(x(i,1),x(i,2),'bo','MarkerSize',LOF(i)*10);
end
```

FIGURE 38.1

MATLAB code for local outlier factor.

$$\begin{aligned}
&\min_{c,b,\xi}\left[b + C\sum_{i=1}^{n}\xi_i\right]\\
&\text{subject to } \|x_i - c\|^2 \leq b + \xi_i \quad \text{for } i = 1,\ldots,n,\\
&\qquad \xi_i \geq 0 \quad \text{for } i = 1,\ldots,n,\\
&\qquad b \geq 0,
\end{aligned} \tag{38.2}$$

where $C > 0$ controls the number of training samples included in the hypersphere and ξ_i is the *margin error* for $\boldsymbol{x}_i$. Samples outside the hypersphere are regarded as outliers, i.e., a test sample $\boldsymbol{x}$ is regarded as an outlier if

$$\|\boldsymbol{x} - \widehat{\boldsymbol{c}}\|^2 > \widehat{b}, \tag{38.3}$$

where $\widehat{\boldsymbol{c}}$ and $\widehat{b}$ are the solutions of optimization problem (38.2).

Optimization problem (38.2) (and also minimum enclosing ball problem (38.1)) has quadratic constraints, and directly handling them in optimization can be computationally expensive. Below, let us consider its *Lagrange dual* problem (Fig. 23.5).

Given the fact that the constraint $b \geq 0$ can be dropped without changing the solution when $C > 1/n$ [27], the Lagrange dual of optimization problem (38.2) is given by

$$\max_{\boldsymbol{\alpha},\boldsymbol{\beta}} \inf_{\boldsymbol{c},b,\boldsymbol{\xi}} L(\boldsymbol{c},b,\boldsymbol{\xi},\boldsymbol{\alpha},\boldsymbol{\beta}) \text{ subject to } \boldsymbol{\alpha} \geq \boldsymbol{0},\ \boldsymbol{\beta} \geq \boldsymbol{0},$$

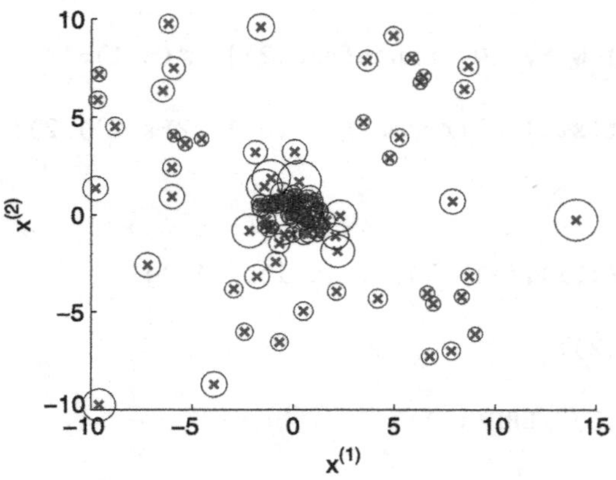

FIGURE 38.2

Example of outlier detection by local outlier factor. The diameter of circles around samples is proportional to the value of local outlier factor.

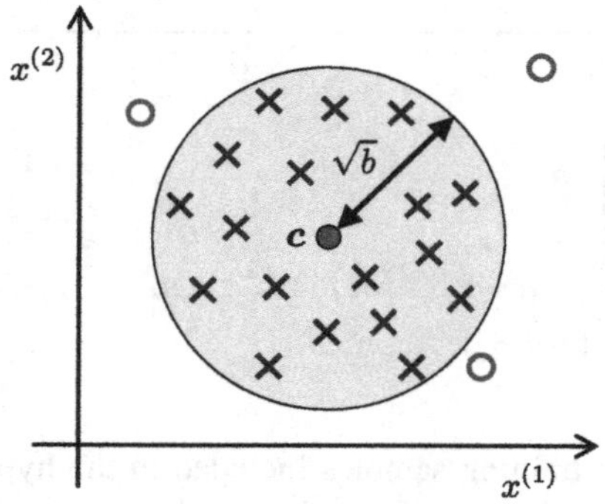

FIGURE 38.3

Support vector data description. A hypersphere that contains *most* of the training samples is found. Samples outside the hypersphere are regarded as outliers.

where α and β are the Lagrange multipliers and $L(c, b, \xi, \alpha, \beta)$ is the Lagrange function defined by

$$L(c, b, \xi, \alpha, \beta) = b + C \sum_{i=1}^{n} \xi_i - \sum_{i=1}^{n} \alpha_i \left(b + \xi_i - \|x_i - c\|^2\right) - \sum_{i=1}^{n} \beta_i \xi_i.$$

The first-order optimality conditions of $\inf_{\boldsymbol{c},b,\boldsymbol{\xi}} L(\boldsymbol{c},b,\boldsymbol{\xi},\boldsymbol{\alpha},\boldsymbol{\beta})$ yield

$$
\begin{aligned}
\frac{\partial L}{\partial b} = 0 &\implies \sum_{i=1}^{n} \alpha_i = 1, \\
\frac{\partial L}{\partial \boldsymbol{c}} = 0 &\implies \boldsymbol{c} = \frac{\sum_{i=1}^{n} \alpha_i \boldsymbol{x}_i}{\sum_{i=1}^{n} \alpha_i} = \sum_{i=1}^{n} \alpha_i \boldsymbol{x}_i, \\
\frac{\partial L}{\partial \xi_i} = 0 &\implies \alpha_i + \beta_i = C, \quad \forall i = 1,\ldots,n,
\end{aligned}
\tag{38.4}
$$

and thus the Lagrange dual problem can be expressed as

$$
\begin{aligned}
&\max_{\boldsymbol{\alpha}} \left[\sum_{i=1}^{n} \alpha_i Q_{i,i} - \sum_{i,j=1}^{n} \alpha_i \alpha_j Q_{i,j} \right] \\
&\text{subject to } 0 \leq \alpha_i \leq C \quad \text{for } i = 1,\ldots,n, \\
&\qquad \sum_{i=1}^{n} \alpha_i = 1,
\end{aligned}
\tag{38.5}
$$

where

$$
Q_{i,j} = \boldsymbol{x}_i^\top \boldsymbol{x}_j.
$$

This is a *quadratic programming* problem (Fig. 27.5), which can be efficiently solved by standard optimization software. However, note that the above quadratic programming problem is convex only when the $n \times n$ matrix $\boldsymbol{Q}$ is nonsingular. When it is singular, $\boldsymbol{Q}$ may be ℓ_2-regularized by adding a small positive constant to the diagonal elements (see Section 23.2) in practice.

Note that, if $C > 1$, optimization problem (38.2) is reduced to minimum enclosing ball problem (38.1), meaning that the solution does not depend on C [27]. This resembles the relation between *hard margin* support vector classification and *soft margin* support vector classification introduced in Chapter 27. On the other hand, when $0 < C \leq 1/n$, $b = 0$ is actually the solution [27], which is not useful for outlier detection because all samples are regarded as outliers. Thus, support vector data description is useful only when

$$
\frac{1}{n} < C \leq 1.
$$

From the KKT conditions (Fig. 27.7) of dual optimization problem (38.5), the following properties hold in the same way as support vector classification (Chapter 27):

- $\alpha_i = 0$ implies $\|\boldsymbol{x}_i - \boldsymbol{c}\|^2 \le b$.
- $0 < \alpha_i < C$ implies $\|\boldsymbol{x}_i - \boldsymbol{c}\|^2 = b$.
- $\alpha_i = C$ implies $\|\boldsymbol{x}_i - \boldsymbol{c}\|^2 \ge b$.
- $\|\boldsymbol{x}_i - \boldsymbol{c}\|^2 < b$ implies $\alpha_i = 0$.
- $\|\boldsymbol{x}_i - \boldsymbol{c}\|^2 > b$ implies $\alpha_i = C$.

Thus, $\boldsymbol{x}_i$ is on the surface or in the interior of the hypersphere when $\alpha_i = 0$, $\boldsymbol{x}_i$ lies on the surface when $0 < \alpha_i < C$, and $\boldsymbol{x}_i$ is on the surface or is in the exterior of the hypersphere when $\alpha_i = C$. On the other hand, $\alpha_i = 0$ when $\boldsymbol{x}_i$ is in the strict interior of the hypersphere and $\alpha_i = C$ when $\boldsymbol{x}_i$ is in the strict exterior of the hypersphere.

Similarly to the case of support vector classification, sample $\boldsymbol{x}_i$ such that $\widehat{\alpha}_i > 0$ is called a *support vector*. From Eq. (38.4), the solution $\widehat{\boldsymbol{c}}$ is given by

$$\widehat{\boldsymbol{c}} = \sum_{i:\widehat{\alpha}_i > 0} \widehat{\alpha}_i \boldsymbol{x}_i.$$

Since $\|\boldsymbol{x}_i - \boldsymbol{c}\|^2 = b$ holds for $\boldsymbol{x}_i$ such that $0 < \alpha_i < C$, the solution $\widehat{b}$ is given by

$$\widehat{b} = \|\boldsymbol{x}_i - \widehat{\boldsymbol{c}}\|^2.$$

As explained in Eq. (38.3), with the solutions $\widehat{\boldsymbol{c}}$ and $\widehat{b}$, a test sample $\boldsymbol{x}$ is regarded as an outlier if, for i such that $0 < \widehat{\alpha}_i < C$,

$$\begin{aligned}\|\boldsymbol{x} - \widehat{\boldsymbol{c}}\|^2 - \widehat{b} &= \|\boldsymbol{x} - \widehat{\boldsymbol{c}}\|^2 - \|\boldsymbol{x}_i - \widehat{\boldsymbol{c}}\|^2 \\ &= \boldsymbol{x}^\top \boldsymbol{x} - 2\sum_{j=1}^{n} \widehat{\alpha}_j \boldsymbol{x}^\top \boldsymbol{x}_j - a_i > 0,\end{aligned} \tag{38.6}$$

where

$$a_i = \boldsymbol{x}_i^\top \boldsymbol{x}_i - 2\sum_{j=1}^{n} \widehat{\alpha}_j \boldsymbol{x}_i^\top \boldsymbol{x}_j.$$

Note that a_i can be computed in advance independent of the test sample $\boldsymbol{x}$.

In the same way as support vector classification, support vector data description can be nonlinearized by the *kernel trick* (Section 27.4). More specifically, for kernel function $K(\boldsymbol{x}, \boldsymbol{x}')$, Lagrange dual problem (38.5) becomes

$$\begin{aligned}&\max_{\boldsymbol{\alpha}} \left[\sum_{i=1}^{n} \alpha_i K(\boldsymbol{x}_i, \boldsymbol{x}_i) - \sum_{i,j=1}^{n} \alpha_i \alpha_j K(\boldsymbol{x}_i, \boldsymbol{x}_j)\right] \\ &\text{subject to } 0 \le \alpha_i \le C \quad \text{for } i = 1, \ldots, n, \\ &\sum_{i=1}^{n} \alpha_i = 1,\end{aligned}$$

```
n=50; x=randn(n,2); x(:,2)=x(:,2)*4; x(1:20,1)=x(1:20,1)*3;
C=0.04; h=[C*ones(n,1); zeros(n,1); 1; -1];
H=[eye(n); -eye(n); ones(1,n); -ones(1,n)]; x2=sum(x.^2,2);
K=exp(-(repmat(x2,1,n)+repmat(x2',n,1)-2*x*x'));
a=quadprog(K,zeros(n,1),H,h); s=ones(n,1)-2*K*a;
s=s-mean(s(find((0<a)&(a<C)))); u=(s>0.001);

figure(1); clf; hold on; axis equal;
plot(x(:,1),x(:,2),'rx'); plot(x(u,1),x(u,2),'bo');
```

FIGURE 38.4

MATLAB code of support vector data description for Gaussian kernel. quadprog.m included in Optimization Toolbox is required. Free alternatives to quadprog.m are available, e.g. from http://www.mathworks.com/matlabcentral/fileexchange/.

and outlier criterion (38.6) becomes

$$K(\boldsymbol{x},\boldsymbol{x}) - 2\sum_{j=1}^{n} \widehat{\alpha}_j K(\boldsymbol{x},\boldsymbol{x}_j) - a_i > 0,$$

where, for i such that $0 < \widehat{\alpha}_i < C$,

$$a_i = K(\boldsymbol{x}_i,\boldsymbol{x}_i) - 2\sum_{j=1}^{n} \widehat{\alpha}_j K(\boldsymbol{x}_i,\boldsymbol{x}_j).$$

When $K(\boldsymbol{x}_i,\boldsymbol{x}_i)$ is constant for all $i = 1,\ldots,n$, which is satisfied, e.g. by the Gaussian kernel,

$$K(\boldsymbol{x},\boldsymbol{x}') = \exp\left(-\frac{\|\boldsymbol{x}-\boldsymbol{x}'\|^2}{2h^2}\right),$$

the above optimization problem is simplified as

$$\min_{\alpha} \sum_{i,j=1}^{n} \alpha_i\alpha_j K(\boldsymbol{x}_i,\boldsymbol{x}_j)$$
$$\text{subject to } 0 \le \alpha_i \le C \quad \text{for } i = 1,\ldots,n,$$
$$\sum_{i=1}^{n} \alpha_i = 1.$$

A MATLAB code of support vector data description for the Gaussian kernel is provided in Fig. 38.4, and its behavior is illustrated in Fig. 38.5. This shows that

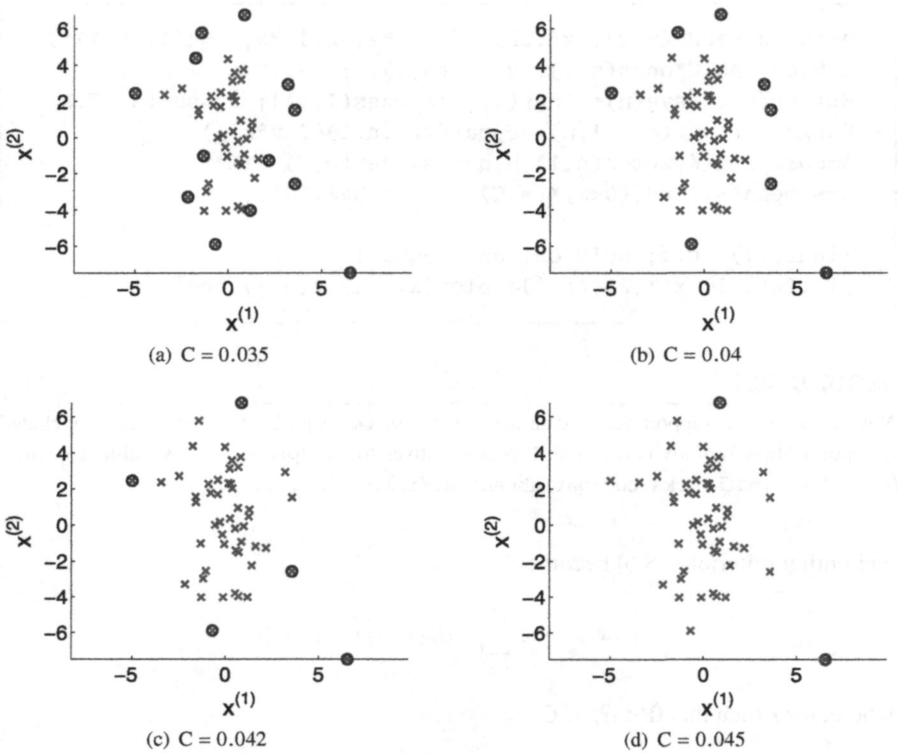

FIGURE 38.5

Examples of support vector data description for Gaussian kernel. Circled samples are regarded as outliers.

results of outlier detection depend on the choice of the trade-off parameter C (and the Gaussian bandwidth), and appropriately determining these tuning parameters is not straightforward in practice because of the unsupervised nature of outlier detection.

38.3 INLIER-BASED OUTLIER DETECTION

Since outliers tend to be highly diverse and their tendency may change over time, it is not easy to directly define outliers. On the other hand, inliers are often stable and thus indirectly defining outliers as samples that are different from inliers would be promising. In this section, such *inlier-based outlier detection* is discussed, under the assumption that, in addition to the test data set $\{\boldsymbol{x}_i\}_{i=1}^{n}$ from which outliers are detected, another data set $\{\boldsymbol{x}'_{i'}\}_{i'=1}^{n'}$ that only contains inliers is available.

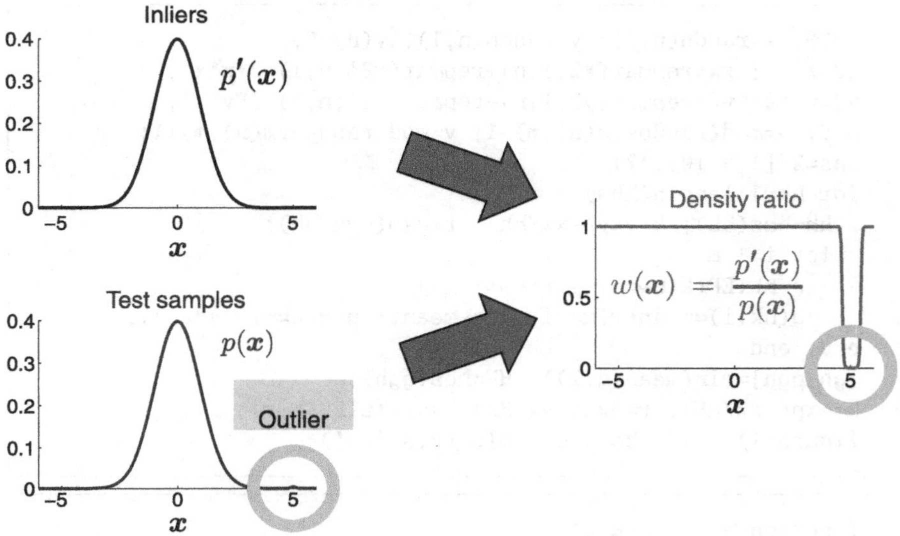

FIGURE 38.6

Inlier-based outlier detection by density ratio estimation. For inlier density $p'(\boldsymbol{x})$ and test sample density $p(\boldsymbol{x})$, the density ratio $w(\boldsymbol{x}) = p'(\boldsymbol{x})/p(\boldsymbol{x})$ is close to one when $\boldsymbol{x}$ is an inlier and it is close to zero when $\boldsymbol{x}$ is an outlier.

A naive approach to inlier-based outlier detection is to estimate the probability density function $p'(\boldsymbol{x})$ of inliers $\{\boldsymbol{x}'_{i'}\}_{i'=1}^{n'}$, and a test sample $\boldsymbol{x}_i$ having a low estimated probability density $\widehat{p}'(\boldsymbol{x}_i)$ is regarded as an outlier. However, this naive approach suffers the same drawback as the density estimation approach introduced in Section 38.1, i.e., estimating the probability density in low-density regions is difficult.

Here, let us consider the ratio of inlier density $p'(\boldsymbol{x})$ and test sample density $p(\boldsymbol{x})$,

$$w(\boldsymbol{x}) = \frac{p'(\boldsymbol{x})}{p(\boldsymbol{x})},$$

which is close to one when $\boldsymbol{x}$ is an inlier and it is close to zero when $\boldsymbol{x}$ is an outlier (Fig. 38.6). Thus, the *density ratio* would be a suitable inlier score.

The density ratio $w(\boldsymbol{x}) = p'(\boldsymbol{x})/p(\boldsymbol{x})$ can be estimated by estimating $p(\boldsymbol{x})$ and $p'(\boldsymbol{x})$ separately from $\{\boldsymbol{x}_i\}_{i=1}^{n}$ and $\{\boldsymbol{x}'_{i'}\}_{i'=1}^{n'}$ and taking their ratio. However, division by an estimated density magnifies the estimation error and thus is not a reliable approach. Here, a method to directly estimate the density ratio $w(\boldsymbol{x}) = p'(\boldsymbol{x})/p(\boldsymbol{x})$ *without* density estimation of $p(\boldsymbol{x})$ and $p'(\boldsymbol{x})$, called *KL density ratio estimation* [76, 104], is introduced.

```
n=50; x=randn(n,1); y=randn(n,1); y(n)=5;
x2=x.^2; xx=repmat(x2,1,n)+repmat(x2',n,1)-2*x*x';
y2=y.^2; yx=repmat(y2,1,n)+repmat(x2',n,1)-2*y*x';
m=5; u=mod(randperm(n),m)+1; v=mod(randperm(n),m)+1;
hhs=2*[1 5 10].^2;
for hk=1:length(hhs)
  hh=hhs(hk); k=exp(-xx/hh); r=exp(-yx/hh);
  for i=1:m
    a=KLIEP(k(u~=i,:),r(v~=i,:));
    g(hk,i)=mean(r(u==i,:)*a-mean(log(k(u==i,:)*a)));
end, end
[gh,ggh]=min(mean(g,2)); HH=hhs(ggh);
k=exp(-xx/HH); r=exp(-yx/HH); s=r*KLIEP(k,r);
figure(1); clf; hold on; plot(y,s,'rx');
```

```
function a=KLIEP(k,r)
a0=rand(size(k,2),1); b=mean(r)'; n=size(k,1);
for o=1:10000
  a=a0-0.001*(b-k'*(1./(k*a0))/n); %a=max(0,a);
  if norm(a-a0)<0.001, break, end
  a0=a;
end
```

FIGURE 38.7

MATLAB code of KL density ratio estimation for Gaussian kernel model with Gaussian bandwidth chosen by cross validation. The bottom function should be saved as "KLIEP.m."

More specifically, let us use a *linear-in-parameter model* for approximating the density ratio $w(\boldsymbol{x}) = p'(\boldsymbol{x})/p(\boldsymbol{x})$:

$$w_{\alpha}(\boldsymbol{x}) = \sum_{j=1}^{b} \alpha_j \psi_j(\boldsymbol{x}) = \boldsymbol{\alpha}^\top \boldsymbol{\psi}(\boldsymbol{x}),$$

where the basis functions $\{\psi_j(\boldsymbol{x})\}_{j=1}^{b}$ are assumed to be non-negative. Since $w_{\alpha}(\boldsymbol{x})p(\boldsymbol{x})$ can be regarded as a model of $p'(\boldsymbol{x})$, the parameter $\boldsymbol{\alpha}$ is learned so that $w_{\alpha}(\boldsymbol{x})p(\boldsymbol{x})$ is as close to $p'(\boldsymbol{x})$ as possible.

For this model matching, let us employ the *generalized KL divergence* for non-negative functions f and g that are not necessarily integrated to 1:

$$\mathrm{gKL}(f\|g) = \int f(\boldsymbol{x}) \log \frac{f(\boldsymbol{x})}{g(\boldsymbol{x})} \mathrm{d}\boldsymbol{x} - \int f(\boldsymbol{x}) \mathrm{d}\boldsymbol{x} + \int g(\boldsymbol{x}) \mathrm{d}\boldsymbol{x}.$$

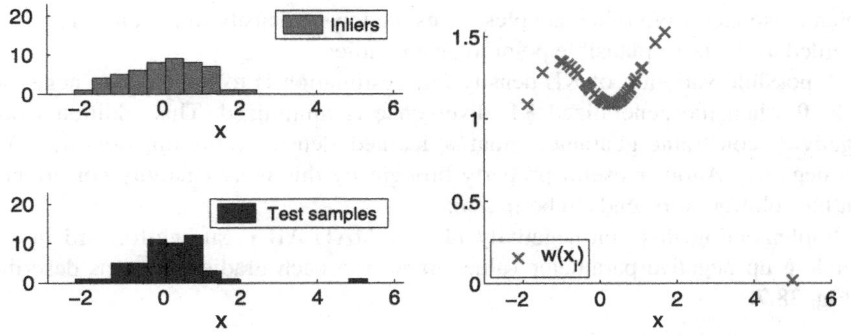

(a) Inliers and test samples

(b) Estimated density ratio

FIGURE 38.8

Example of KL density ratio estimation for Gaussian kernel model.

When f and g are normalized, the above generalized KL divergence is reduced to the ordinary KL divergence since the second and third terms vanish. Under the generalized KL divergence, the parameter α is learned to minimize

$$\mathrm{gKL}(p'\|w_{\alpha}p) = \int p'(\boldsymbol{x}) \log \frac{p'(\boldsymbol{x})}{w_{\alpha}(\boldsymbol{x})p(\boldsymbol{x})} \mathrm{d}\boldsymbol{x} - 1 + \int w_{\alpha}(\boldsymbol{x})p(\boldsymbol{x})\mathrm{d}\boldsymbol{x}.$$

Approximating the expectations by the sample averages and ignoring irrelevant constants yield the following optimization problem:

$$\min_{\alpha} \left[\frac{1}{n} \sum_{i=1}^{n} w_{\alpha}(\boldsymbol{x}_i) - \frac{1}{n'} \sum_{i'=1}^{n'} \log w_{\alpha}(\boldsymbol{x}'_{i'}) \right].$$

This is a convex optimization problem and thus the global optimal solution can be easily obtained, e.g. by a gradient method.

A critical drawback of the *local outlier factor* and *support vector data description* explained in the previous sections is that there is no objective model selection method. Thus, tuning parameters should be selected subjectively based on some prior knowledge. On the other hand, KL density ratio estimation allows objective model selection by *cross validation* (see Section 14.4 and Section 16.4.2) in terms of the KL divergence. This is practically a significant advantage in outlier detection.

A MATLAB code of KL density ratio estimation for the Gaussian kernel model,

$$w_{\alpha}(\boldsymbol{x}) = \sum_{j=1}^{n'} \alpha_j \exp\left(-\frac{\|\boldsymbol{x} - \boldsymbol{x}'_j\|^2}{2h^2} \right), \tag{38.7}$$

is provided in Fig. 38.7, where the Gaussian bandwidth h is chosen by cross validation. Its behavior is illustrated in Fig. 38.8, showing that a sample at $x = 5$, which is isolated from other samples, takes the lowest density ratio value. Thus, it is regarded as the most plausible point to be an outlier.

A possible variation of KL density ratio estimation is to impose non-negativity $\alpha \geq \mathbf{0}$ when the generalized KL divergence is minimized. This additional non-negativity constraint guarantees that a learned density ratio function $w_{\alpha}(\boldsymbol{x})$ is non-negative. Another useful property brought by this non-negativity constraint is that the solution of α tends to be *sparse*.

Implementing this non-negativity idea in MATLAB is straightforward just by rounding up negative parameter values to zero in each gradient step, as described in Fig. 38.7.

CHAPTER

CHANGE DETECTION 39

CHAPTER CONTENTS

The objective of *change detection* is to investigate whether change exists between two data sets $\{\boldsymbol{x}_i\}_{i=1}^n$ and $\{\boldsymbol{x}'_{i'}\}_{i'=1}^{n'}$. In this chapter, two statistical approaches to change detection, *distributional change detection* and *structural change detection*, are explored. Below, $\{\boldsymbol{x}_i\}_{i=1}^n$ and $\{\boldsymbol{x}'_{i'}\}_{i'=1}^{n'}$ are assumed to be drawn independently from the probability distributions with densities $p(\boldsymbol{x})$ and $p'(\boldsymbol{x})$, respectively.

39.1 DISTRIBUTIONAL CHANGE DETECTION

Distributional change detection is aimed at identifying change in probability distributions behind $\{\boldsymbol{x}_i\}_{i=1}^n$ and $\{\boldsymbol{x}'_{i'}\}_{i'=1}^{n'}$. This can be achieved by estimating a *distance* or a *divergence* between $p(\boldsymbol{x})$ and $p'(\boldsymbol{x})$. As explained in Section 14.2, a distance satisfies four conditions: *non-negativity*, *symmetry*, *identity*, and the *triangle inequality*. On the other hand, a divergence is a pseudodistance that still acts like a distance, but it may violate some of the above conditions. In this section, divergence and distance measures between probability distributions that are useful for change detection and their approximators from samples $\{\boldsymbol{x}_i\}_{i=1}^n$ and $\{\boldsymbol{x}'_{i'}\}_{i'=1}^{n'}$ are introduced.

39.1.1 KL DIVERGENCE

The most popular divergence measure in statistics and machine learning would be the KL divergence (see Section 14.2), because of its compatible with MLE (see Chapter 12):

$$\mathrm{KL}(p\|p') = \int p(\boldsymbol{x}) \log \frac{p(\boldsymbol{x})}{p'(\boldsymbol{x})} \mathrm{d}\boldsymbol{x}.$$

Due to the log function which is sharp near zero, the KL divergence may allow sensitive detection of small change in probability distributions. Another advantage of the KL divergence is that it is *invariant* under input metric change, i.e., the value of the KL divergence does not change even if $\boldsymbol{x}$ is transformed to $\widetilde{\boldsymbol{x}}$ by *any* mapping [5].

The KL divergence can be approximated from samples $\{\boldsymbol{x}_i\}_{i=1}^{n}$ and $\{\boldsymbol{x}'_{i'}\}_{i'=1}^{n'}$ by the *direct density ratio estimator* introduced in Section 38.3. More specifically, an estimator $\widehat{w}(\boldsymbol{x})$ of the density ratio function,

$$w(\boldsymbol{x}) = \frac{p(\boldsymbol{x})}{p'(\boldsymbol{x})},$$

can be obtained by *KL density ratio estimation*. Then, the KL divergence can be immediately approximated as

$$\widehat{\mathrm{KL}} = \frac{1}{n} \sum_{i=1}^{n} \log \widehat{w}(\boldsymbol{x}_i).$$

However, due to the log function, estimation of the KL divergence is prone to be sensitive to *outliers*.

39.1.2 PEARSON DIVERGENCE

Ali-Silvey-Csiszár divergences [2, 35], which is also known as *f-divergences*, are generalization of the KL divergence defined as

$$F(p\|p') = \int p'(\boldsymbol{x}) f\left(\frac{p(\boldsymbol{x})}{p'(\boldsymbol{x})}\right) \mathrm{d}\boldsymbol{x},$$

where $f(t)$ is a *convex* function (see Fig. 8.3) such that $f(1) = 0$. It can be easily confirmed that, with $f(t) = t \log t$, the Ali-Silvey-Csiszár divergence is reduced to the KL divergence. Note that all Ali-Silvey-Csiszár divergences are invariant under input metric change.

The *Pearson divergence* [79] is a squared-loss variant of the KL divergence defined as the Ali-Silvey-Csiszár divergence with the squared function $f(t) = (t-1)^2$:

$$\mathrm{PE}(p\|p') = \int p'(\boldsymbol{x}) \left(\frac{p(\boldsymbol{x})}{p'(\boldsymbol{x})} - 1\right)^2 \mathrm{d}\boldsymbol{x}.$$

Since the Pearson divergence does not include the log function, it would be more robust against outliers. However, it still includes the density ratio function $p(\boldsymbol{x})/p'(\boldsymbol{x})$, which tends to be a sharp function and is possibly unbounded. Therefore, its accurate approximation is not straightforward in practice. As discussed in Section 33.2.2, this problem can be mitigated by considering the *relative density ratio* for $\beta \in [0,1]$:

$$w_\beta(\boldsymbol{x}) = \frac{p(\boldsymbol{x})}{\beta p(\boldsymbol{x}) + (1-\beta)p'(\boldsymbol{x})}.$$

The Pearson divergence extended using the relative density ratio is called the *relative Pearson divergence* [121]:

$$\begin{aligned}\mathrm{rPE}(p\|p') &= \mathrm{PE}(p\|\beta p + (1-\beta)p') \\ &= \int \Big(\beta p(\boldsymbol{x}) + (1-\beta)p'(\boldsymbol{x})\Big)\Big(w_\beta(\boldsymbol{x}) - 1\Big)^2 \mathrm{d}\boldsymbol{x}.\end{aligned}$$

The relative Pearson divergence can be approximated from samples $\{\boldsymbol{x}_i\}_{i=1}^n$ and $\{\boldsymbol{x}'_{i'}\}_{i'=1}^{n'}$ by the *direct relative density ratio estimator* introduced in Section 33.2.4. More specifically, an estimator $\widehat{w}_\beta(\boldsymbol{x})$ of the relative density ratio function $w_\beta(\boldsymbol{x})$ can be obtained by *LS relative density ratio estimation*. Then, the relative Pearson divergence can be approximated as

$$\widehat{\mathrm{rPE}} = \frac{1}{n}\sum_{i=1}^{n} \widehat{w}_\beta(\boldsymbol{x}_i) - 1,$$

which comes from the following expression of the relative Pearson divergence:

$$\mathrm{rPE}(p\|p') = \int p(\boldsymbol{x}) w_\beta(\boldsymbol{x}) \mathrm{d}\boldsymbol{x} - 1.$$

The tuning parameter $\beta \in [0,1]$ controls the trade-off between sensitivity and robustness of the divergence measure, which should be appropriately chosen in practice.

39.1.3 L_2-DISTANCE

The *L_2-distance* is another standard distance measure between probability distributions:

$$L_2(p,p') = \int f(\boldsymbol{x})^2 \mathrm{d}\boldsymbol{x},$$

where

$$f(\boldsymbol{x}) = p(\boldsymbol{x}) - p'(\boldsymbol{x}).$$

The L_2-distance is a proper distance measure, and thus it is symmetric and satisfies the triangle inequality unlike the KL divergence and the (relative) Pearson divergence. Furthermore, the density difference $f(\boldsymbol{x})$ is always bounded as long as each density is bounded. Therefore, the L_2-distance is stable, without the need of tuning any control parameter such as β in the relative Pearson divergence.

The L_2-distance can be approximated from samples $\{\boldsymbol{x}_i\}_{i=1}^{n}$ and $\{\boldsymbol{x}'_{i'}\}_{i'=1}^{n'}$ by *LS density difference estimation* [103] that directly estimates the density difference function $f(\boldsymbol{x})$ without estimating $p(\boldsymbol{x})$ and $p'(\boldsymbol{x})$ (see Section 37.4 for direct estimation of $p(\boldsymbol{x}, y) - p(\boldsymbol{x})p(y)$). More specifically, let us consider the following Gaussian density difference model:

$$f_{\boldsymbol{\alpha}}(\boldsymbol{x}) = \sum_{j=1}^{n+n'} \alpha_j \exp\left(-\frac{\|\boldsymbol{x} - \boldsymbol{c}_j\|^2}{2h^2}\right),$$

where $h > 0$ denotes the Gaussian bandwidth and

$$(\boldsymbol{c}_1, \ldots, \boldsymbol{c}_n, \boldsymbol{c}_{n+1}, \ldots, \boldsymbol{c}_{n+n'}) = (\boldsymbol{x}_1, \ldots, \boldsymbol{x}_n, \boldsymbol{x}'_1, \ldots, \boldsymbol{x}'_{n'})$$

are the Gaussian centers. The parameters $\boldsymbol{\alpha} = (\alpha_1, \ldots, \alpha_{n+n'})^\top$ are estimated by LS fitting to the true density difference function $f(\boldsymbol{x})$:

$$\min_{\boldsymbol{\alpha}} \int \left(f_{\boldsymbol{\alpha}}(\boldsymbol{x}) - f(\boldsymbol{x})\right)^2 \mathrm{d}\boldsymbol{x}.$$

Its empirical criterion where an irrelevant constant is ignored and the expectation is approximated by the sample average is given by

$$\min_{\boldsymbol{\alpha}} \left[\boldsymbol{\alpha}^\top \boldsymbol{U} \boldsymbol{\alpha} - 2\boldsymbol{\alpha}^\top \widehat{\boldsymbol{v}} + \lambda \|\boldsymbol{\alpha}\|^2\right],$$

where the ℓ_2-regularizer $\lambda\|\boldsymbol{\alpha}\|^2$ is included. $\boldsymbol{U}$ is the $(n + n') \times (n + n')$ matrix with the (j, j')th element defined by

$$\begin{aligned} U_{j,j'} &= \int \exp\left(-\frac{\|\boldsymbol{x} - \boldsymbol{c}_j\|^2}{2h^2}\right) \exp\left(-\frac{\|\boldsymbol{x} - \boldsymbol{c}_{j'}\|^2}{2h^2}\right) \mathrm{d}\boldsymbol{x} \\ &= (\pi h^2)^{d/2} \exp\left(-\frac{\|\boldsymbol{c}_j - \boldsymbol{c}_{j'}\|^2}{4h^2}\right), \end{aligned}$$

where d denotes the dimensionality of $\boldsymbol{x}$. $\widehat{\boldsymbol{v}}$ is the $(n + n')$-dimensional vector with the jth element defined by

$$\widehat{v}_j = \frac{1}{n} \sum_{i=1}^{n} \exp\left(-\frac{\|\boldsymbol{x}_i - \boldsymbol{c}_j\|^2}{2h^2}\right) - \frac{1}{n'} \sum_{i'=1}^{n'} \exp\left(-\frac{\|\boldsymbol{x}'_{i'} - \boldsymbol{c}_j\|^2}{2h^2}\right).$$

This is a convex optimization problem, and the global optimal solution $\widehat{\boldsymbol{\alpha}}$ can be obtained *analytically* as

$$\widehat{\boldsymbol{\alpha}} = (\boldsymbol{U} + \lambda \boldsymbol{I})^{-1} \widehat{\boldsymbol{v}}.$$

```
n=200; x=randn(n,1); y=randn(n,1)+1;
hhs=2*[0.5 1 3].^2; ls=10.^[-2 -1 0]; m=5;
x2=x.^2; xx=repmat(x2,1,n)+repmat(x2',n,1)-2*x*x';
y2=y.^2; yx=repmat(y2,1,n)+repmat(x2',n,1)-2*y*x';
u=mod(randperm(n),m)+1; v=mod(randperm(n),m)+1;

for hk=1:length(hhs)
  hh=hhs(hk); k=exp(-xx/hh); r=exp(-yx/hh);
  U=(pi*hh/2)^(1/2)*exp(-xx/(2*hh));
  for i=1:m
    vh=mean(k(u~=i,:))'-mean(r(v~=i,:))';
    z=mean(k(u==i,:))-mean(r(v==i,:));
    for lk=1:length(ls)
      l=ls(lk); a=(U+l*eye(n))\vh; g(hk,lk,i)=a'*U*a-2*z*a;
end, end, end
[gl,ggl]=min(mean(g,3),[],2); [ghl,gghl]=min(gl);
L=ls(ggl(gghl)); HH=hhs(gghl);
k=exp(-xx/HH); r=exp(-yx/HH); vh=mean(k)'-mean(r)';
U=(pi*HH/2)^(1/2)*exp(-xx/(2*HH));
a=(U+L*eye(n))\vh; s=[k;r]*a; L2=a'*vh;
figure(1); clf; hold on; plot([x;y],s,'rx');
```

FIGURE 39.1

MATLAB code for LS density difference estimation.

The Gaussian width h and the regularization parameter λ may be optimized by cross validation with respect to the squared error criterion.

Finally, the L_2-distance can be approximated by

$$\widehat{L}^2 = \frac{1}{n}\sum_{i=1}^{n} f_{\widehat{\alpha}}(\boldsymbol{x}_i) - \frac{1}{n'}\sum_{i'=1}^{n'} f_{\widehat{\alpha}}(\boldsymbol{x}'_{i'}) = \widehat{\boldsymbol{v}}^\top \widehat{\boldsymbol{\alpha}}, \tag{39.1}$$

which comes from the following expression of the L_2-distance:

$$L_2(p,p') = \int \big(p(\boldsymbol{x}) - p'(\boldsymbol{x})\big) f(\boldsymbol{x})\mathrm{d}\boldsymbol{x}.$$

A MATLAB code for LS density difference estimation is provided in Fig. 39.1, and its behavior is illustrated in Fig. 39.2. This shows that the density difference function can be accurately estimated.

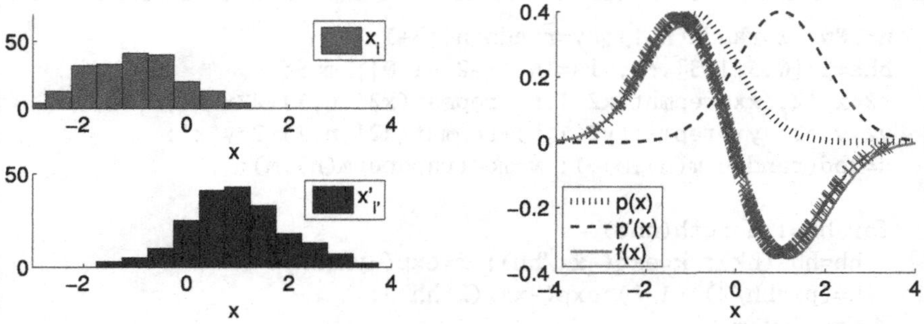

(a) Samples (b) Densities and density differences

FIGURE 39.2

Example of LS density difference estimation. ×'s in the right plot show estimated density difference values at $\{x_i\}_{i=1}^{n}$ and $\{x'_{i'}\}_{i'=1}^{n'}$.

39.1.4 L_1-DISTANCE

The L_2-distance can be generalized to the class of L_s-distances for $s > 0$:

$$L_t(p,p') = \int \left|p(x) - p'(x)\right|^s \mathrm{d}x.$$

Among the class of L_s-distances, the L_1-*distance* is also a member of the class of the Ali-Silvey-Csiszár divergences with the absolute function $f(t) = |t - 1|$ (see Section 39.1.2):

$$L_1(p,p') = \int \left|p(x) - p'(x)\right| \mathrm{d}x = \int p'(x) \left|\frac{p(x)}{p'(x)} - 1\right| \mathrm{d}x.$$

This implies that the L_1-distance is also invariant under input metric change.

Approximation of the L_1-distance from samples $\{x_i\}_{i=1}^{n}$ and $\{x'_{i'}\}_{i'=1}^{n'}$ can actually be performed by the (weighted) *support vector machine* introduced in Chapter 27 [75]. More specifically, since $|t| = \mathrm{sign}(t)\, t$, the L_1-distance can be expressed as

$$L_1(p,p') = \int \mathrm{sign}\big(p(x) - p'(x)\big)\big(p(x) - p'(x)\big) \mathrm{d}x,$$

where sign(t) denotes the sign function:

$$\mathrm{sign}(t) = \begin{cases} 1 & (t > 0), \\ 0 & (t = 0), \\ -1 & (t < 0). \end{cases}$$

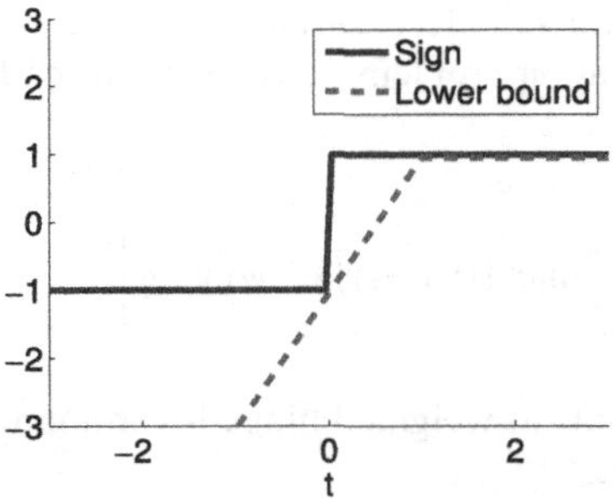

FIGURE 39.3

Lower bound of sign (t) by $-2\max(0, 1-t)+1$.

For a density difference model $f_{\alpha}(\boldsymbol{x})$ with parameter α, the L_1-distance $L_1(p, p')$ can be lower-bounded as

$$\begin{aligned} L_1(p,p') &\geq \max_{\alpha} \int \operatorname{sign}\big(f_{\alpha}(\boldsymbol{x})\big)\big(p(\boldsymbol{x}) - p'(\boldsymbol{x})\big)\mathrm{d}\boldsymbol{x} \\ &= \max_{\alpha} \left[\int \operatorname{sign}\big(f_{\alpha}(\boldsymbol{x})\big)p(\boldsymbol{x})\mathrm{d}\boldsymbol{x} + \int \operatorname{sign}\big(-f_{\alpha}(\boldsymbol{x})\big)p'(\boldsymbol{x})\mathrm{d}\boldsymbol{x} \right], \end{aligned}$$

where the last term is due to $\operatorname{sign}(t) = -\operatorname{sign}(-t)$. As plotted in Fig. 39.3, the sign function can be lower-bounded by

$$-2\max(0, 1-t) + 1 = \begin{cases} 1 & (t > 1), \\ 2t - 1 & (t \leq 1). \end{cases}$$

Based on this, the L_1-distance $L_1(p, p')$ can be further lower-bounded as

$$\begin{aligned} L_1(p,p') \geq 2 - 2\min_{\alpha} \Bigg[&\int \max\big(0, 1 - f_{\alpha}(\boldsymbol{x})\big)p(\boldsymbol{x})\mathrm{d}\boldsymbol{x} \\ &+ \int \max\big(0, 1 + f_{\alpha}(\boldsymbol{x})\big)p'(\boldsymbol{x})\mathrm{d}\boldsymbol{x} \Bigg]. \end{aligned}$$

Let us employ a *linear-in-parameter* density difference model:

$$f_{\alpha}(\boldsymbol{x}) = \sum_{j=1}^{b} \alpha_j \psi_j(\boldsymbol{x}) = \boldsymbol{\alpha}^{\top}\boldsymbol{\psi}(\boldsymbol{x}).$$

Then the empirical version of the above maximization problem (without irrelevant multiplicative and additive constants) is given by

$$\min_{\alpha} \left[\frac{1}{n}\sum_{i=1}^{n} \max\big(0, 1 - \boldsymbol{\alpha}^{\top}\boldsymbol{\psi}(\boldsymbol{x}_i)\big) + \frac{1}{n'}\sum_{i'=1}^{n'} \max\big(0, 1 + \boldsymbol{\alpha}^{\top}\boldsymbol{\psi}(\boldsymbol{x}'_{i'})\big) \right].$$

Let us assign *class labels* $y_i = +1$ to $\boldsymbol{x}_i$ for $i = 1,\ldots,n$ and $y'_{i'} = -1$ to $\boldsymbol{x}'_{i'}$ for $i' = 1,\ldots,n'$. If $n = n'$, the above optimization problem agrees with *hinge loss minimization* (see Section 27.6):

$$\min_{\boldsymbol{\alpha}} \left[\sum_{i=1}^{n} \max\left(0, 1 - y_i \boldsymbol{\alpha}^\top \boldsymbol{\psi}(\boldsymbol{x}_i)\right) + \sum_{i'=1}^{n'} \max\left(0, 1 - y'_{i'} \boldsymbol{\alpha}^\top \boldsymbol{\psi}(\boldsymbol{x}'_{i'})\right) \right].$$

If $n \neq n'$, L_1-distance approximation corresponds to weighted hinge loss minimization with weight $1/n$ for $\{\boldsymbol{x}_i\}_{i=1}^{n}$ and $1/n'$ for $\{\boldsymbol{x}'_{i'}\}_{i'=1}^{n'}$.

The above formulation shows that the support vector machine is actually approximating the sign of the density difference. More specifically, let $p_+(\boldsymbol{x})$ and $p_-(\boldsymbol{x})$ be the probability density functions of samples in the positive class and negative class, respectively. Then, the support vector machine approximates

$$\operatorname{sign}\left(p_+(\boldsymbol{x}) - p_-(\boldsymbol{x})\right),$$

which is the optimal decision function. Thus, support vector classification can be interpreted as directly approximating the optimal decision function without estimating the densities $p_+(\boldsymbol{x})$ and $p_-(\boldsymbol{x})$.

39.1.5 MAXIMUM MEAN DISCREPANCY (MMD)

MMD [17] measures the distance between embeddings of probability distributions in a *reproducing kernel Hilbert space* [9].

More specifically, the MMD between p and p' is defined as

$$\begin{aligned}\mathrm{MMD}(p,p') = \mathbb{E}_{\boldsymbol{x},\widetilde{\boldsymbol{x}}\sim p}[K(\boldsymbol{x},\widetilde{\boldsymbol{x}})] + \mathbb{E}_{\boldsymbol{x}',\widetilde{\boldsymbol{x}}'\sim p'}[K(\boldsymbol{x}',\widetilde{\boldsymbol{x}}')] \\ - 2\mathbb{E}_{\boldsymbol{x}\sim p,\boldsymbol{x}'\sim p'}[K(\boldsymbol{x},\boldsymbol{x}')],\end{aligned}$$

where $K(\boldsymbol{x},\boldsymbol{x}')$ is a reproducing kernel, $\mathbb{E}_{\boldsymbol{x}\sim p}$ denotes the expectation with respect to $\boldsymbol{x}$ following density p. $\mathrm{MMD}(p,p')$ is always non-negative, and $\mathrm{MMD}(p,p') = 0$ if and only if $p = p'$ when $K(\boldsymbol{x},\boldsymbol{x}')$ is a *characteristic kernel* [45] such as the Gaussian kernel.

An advantage of MMD is that it can be directly approximated using samples as

$$\frac{1}{n^2}\sum_{i,\widetilde{i}=1}^{n} K(\boldsymbol{x}_i,\boldsymbol{x}_{\widetilde{i}}) + \frac{1}{n'^2}\sum_{i',\widetilde{i}'=1}^{n'} K(\boldsymbol{x}'_{i'},\boldsymbol{x}'_{\widetilde{i}'}) - \frac{2}{nn'}\sum_{i=1}^{n}\sum_{i'=1}^{n'} K(\boldsymbol{x}_i,\boldsymbol{x}'_{i'}).$$

Thus, no estimation is involved when approximating MMD from samples. However, it is not clear how to choose kernel functions in practice. Using the Gaussian kernel with bandwidth set at the median distance between samples is a popular heuristic [49], but this does not always work well in practice [50].

39.1.6 ENERGY DISTANCE

Another useful distance measure is the *energy distance* [106] introduced in Section 33.3.2:

$$D_{\mathrm{E}}(p,p') = \int_{\mathbb{R}^d} \|\varphi_p(t) - \varphi_{p'}(t)\|^2 \left(\frac{\pi^{\frac{d+1}{2}}}{\Gamma\left(\frac{d+1}{2}\right)} \|t\|^{d+1} \right)^{-1} \mathrm{d}t,$$

where $\|\cdot\|$ denotes the Euclidean norm, φ_p denotes the *characteristic function* (see Section 2.4.3) of p, $\Gamma(\cdot)$ is the *gamma function* (see Section 4.3), and d denotes the dimensionality of $\boldsymbol{x}$.

An important property of the energy distance is that it can be expressed as

$$D_{\mathrm{E}}(p,p') = 2\mathbb{E}_{\boldsymbol{x}\sim p,\boldsymbol{x}'\sim p'}\|\boldsymbol{x} - \boldsymbol{x}'\| - \mathbb{E}_{\boldsymbol{x},\widetilde{\boldsymbol{x}}\sim p}\|\boldsymbol{x} - \widetilde{\boldsymbol{x}}\| - \mathbb{E}_{\boldsymbol{x}',\widetilde{\boldsymbol{x}}'\sim p'}\|\boldsymbol{x}' - \widetilde{\boldsymbol{x}}'\|,$$

where $\mathbb{E}_{\boldsymbol{x}\sim p}$ denotes the expectation with respect to $\boldsymbol{x}$ following density p. This can be directly approximated using samples as

$$\frac{2}{nn'}\sum_{i=1}^{n}\sum_{i'=1}^{n'}\|\boldsymbol{x}_i - \boldsymbol{x}'_{i'}\| - \frac{1}{n^2}\sum_{i,\widetilde{i}=1}^{n}\|\boldsymbol{x}_i - \boldsymbol{x}_{\widetilde{i}}\| - \frac{1}{n'^2}\sum_{i',\widetilde{i}'=1}^{n'}\|\boldsymbol{x}'_{i'} - \boldsymbol{x}'_{\widetilde{i}'}\|.$$

Thus, no estimation and no tuning parameters are involved when approximating the energy distance from samples, which is a useful properties in practice.

Actually, it was shown [91] that the energy distance is a special case of MMD. Indeed, MMD with kernel function defined as

$$K(\boldsymbol{x},\boldsymbol{x}') = -\|\boldsymbol{x} - \boldsymbol{x}'\| + \|\boldsymbol{x}\| + \|\boldsymbol{x}'\|$$

agrees with the energy distance.

39.1.7 APPLICATION TO CHANGE DETECTION IN TIME SERIES

Let us consider the problem of *change detection in time series* (Fig. 39.4(a)). More specifically, given time series samples $\{y_i\}_{i=1}^N$, the objective is to identify whether change in probability distributions exists between y_t and y_{t+1} for some t. This problem can be tackled by estimating a distance (or a divergence) between the probability distributions of $\{y_i\}_{i=t-n+1}^{t}$ and $\{y_i\}_{i=t+1}^{t+n}$.

A challenge in change detection in time series is that samples $\{y_i\}_{i=1}^N$ are often dependent over time, which violates the presumption in this chapter. A practical approach to mitigate this problem is to *vectorize* data [60], as illustrated in Fig. 39.4(b). That is, instead of handling time series sample y_i as it is, its vectorization with k consecutive samples $\boldsymbol{x}_i = (y_i, \ldots, y_{i+k-1})^\top$ is considered, and a distance (or a divergence) is estimated from $\mathcal{D} = \{\boldsymbol{x}_i\}_{i=t-n+1}^{t}$ and $\mathcal{D}' = \{\boldsymbol{x}_i\}_{i=t+1}^{t+n}$.

A MATLAB code for change detection in time series based on the energy distance is provided in Fig. 39.5, and its behavior is illustrated in Fig. 39.6. This shows that the energy distance well captures the distributional change in time series.

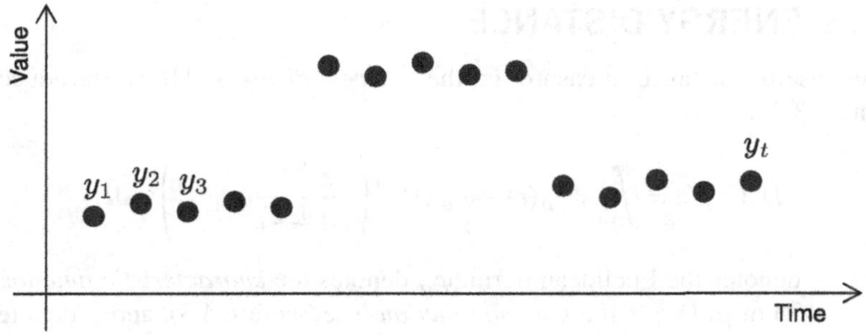

(a) Time series samples

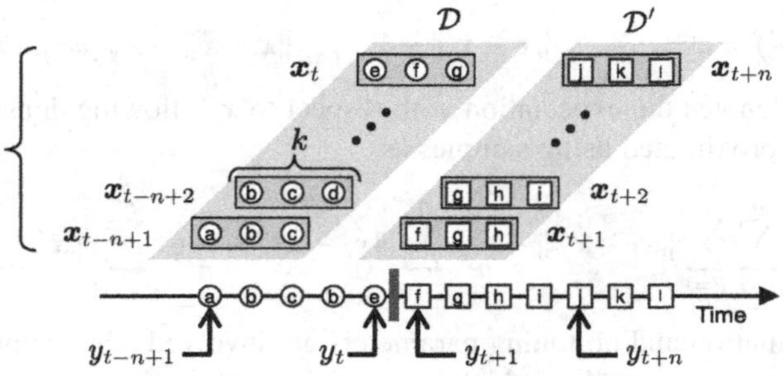

(b) Vectorization to reduce time dependency

FIGURE 39.4

Change detection in time series.

39.2 STRUCTURAL CHANGE DETECTION

Distributional change detection introduced in the previous section focused on investigating whether change exists in probability distributions. The aim of *structural change detection* introduced in this section is to analyze change in the *dependency structure* between elements of d-dimensional variable $\boldsymbol{x} = (x^{(1)}, \ldots, x^{(d)})^\top$.

39.2.1 SPARSE MLE

Let us consider a *Gaussian Markov network*, which is a d-dimensional Gaussian model with expectation zero (Section 6.2):

$$q(\boldsymbol{x}; \boldsymbol{\Theta}) = \frac{\det(\boldsymbol{\Theta})^{1/2}}{(2\pi)^{d/2}} \exp\left(-\frac{1}{2}\boldsymbol{x}^\top \boldsymbol{\Theta} \boldsymbol{x}\right),$$

where not the variance-covariance matrix, but its inverse called the *precision matrix* is parameterized by $\boldsymbol{\Theta}$. If $\boldsymbol{\Theta}$ is regarded as an *adjacency matrix*, the Gaussian Markov network can be visualized as a *graph* (see Fig. 39.7). An advantage of this precision-based parameterization is that the connectivity governs conditional independence.

```
N=300; k=5; n=10; m=N-k+1; E=nan(1,N);
y=zeros(1,N); y(101:200)=3; y=y+randn(1,N);
%y=sin([1:N]/2); y(101:200)=sin([101:200]);
x=toeplitz(y); x=x(1:k,1:m); x2=sum(x.^2);
D=sqrt(repmat(x2',1,m)+repmat(x2,m,1)-2*x'*x);
for t=n:N-n-k+1
  a=[t-n+1:t]; b=[t+1:t+n];
  E(t)=2*mean(mean(D(a,b)))-mean(mean(D(a,a))) ...
       -mean(mean(D(b,b)));
end

figure(1); clf; hold on; plot(y,'b-'); plot(E,'r--');
legend('Time series','Energy distance')
```

FIGURE 39.5

MATLAB code for change detection in time series based on the energy distance.

For example, in the Gaussian Markov network illustrated in the left-hand side of Fig. 39.7, $x^{(1)}$ and $x^{(2)}$ are connected via $x^{(3)}$. This means that $x^{(1)}$ and $x^{(2)}$ are conditionally independent given $x^{(3)}$.

Suppose that $\{\boldsymbol{x}_i\}_{i=1}^{n}$ and $\{\boldsymbol{x}'_{i'}\}_{i'=1}^{n'}$ are drawn independently from the Gaussian Markov networks with precision matrices $\boldsymbol{\Theta}$ and $\boldsymbol{\Theta}'$, respectively. Then analyzing $\boldsymbol{\Theta} - \boldsymbol{\Theta}'$ allows us to identify the change in Markov network structure (see Fig. 39.7 again).

A sparse estimate of $\boldsymbol{\Theta}$ may be obtained by MLE with the ℓ_1-*constraint* (see Chapter 24):

$$\max_{\boldsymbol{\Theta}} \sum_{i=1}^{n} \log q(\boldsymbol{x}_i; \boldsymbol{\Theta}) \text{ subject to } \|\boldsymbol{\Theta}\|_1 \leq R^2,$$

where $R \geq 0$ is the radius of the ℓ_1-ball. This method is also referred to as the *graphical lasso* [44].

The derivative of $\log q(\boldsymbol{x}; \boldsymbol{\Theta})$ with respect to $\boldsymbol{\Theta}$ is given by

$$\frac{\partial \log q(\boldsymbol{x}; \boldsymbol{\Theta})}{\partial \boldsymbol{\Theta}} = \frac{1}{2}\boldsymbol{\Theta}^{-1} - \frac{1}{2}\boldsymbol{x}\boldsymbol{x}^\top,$$

where the following formulas are used for its derivation:

$$\frac{\partial \log \det(\boldsymbol{\Theta})}{\partial \boldsymbol{\Theta}} = \boldsymbol{\Theta}^{-1} \quad \text{and} \quad \frac{\partial \boldsymbol{x}^\top \boldsymbol{\Theta} \boldsymbol{x}}{\partial \boldsymbol{\Theta}} = \boldsymbol{x}\boldsymbol{x}^\top.$$

A MATLAB code of a gradient-projection algorithm of ℓ_1-constraint MLE for Gaussian Markov networks is given in Fig. 39.8, where projection onto the ℓ_1-ball is computed by the method developed in [39].

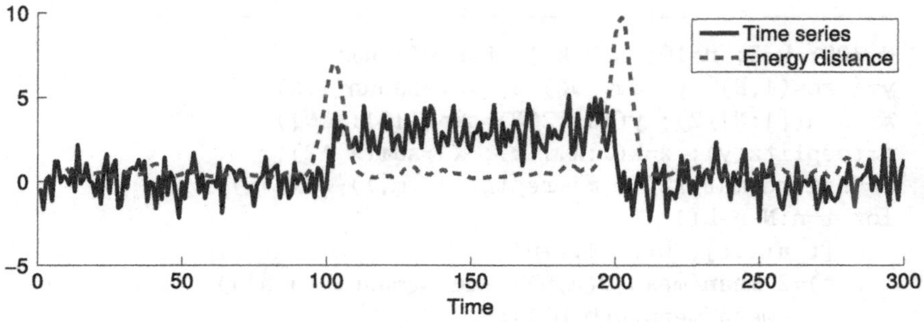

(a) Bias change

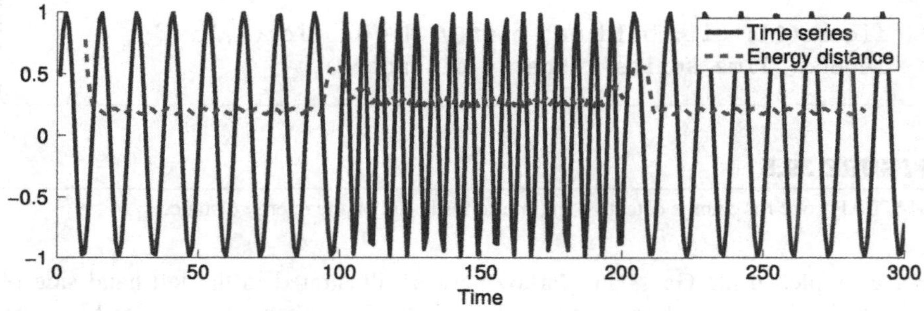

(b) Frequency change

FIGURE 39.6

Examples of change detection in time series based on the energy distance.

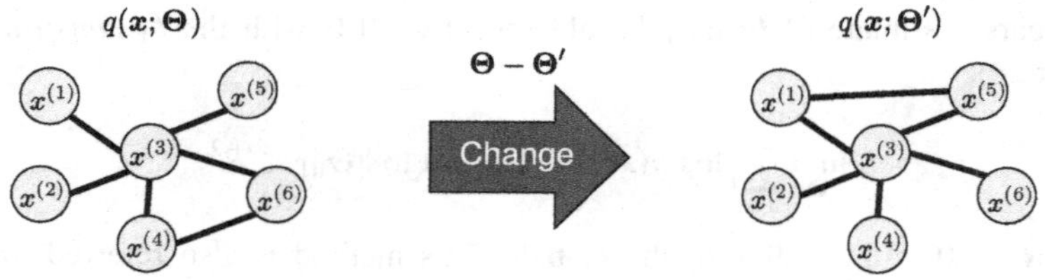

FIGURE 39.7

Structural change in Gaussian Markov networks.

For the true precision matrices

$$\boldsymbol{\Theta} = \begin{pmatrix} 2 & 0 & 1 \\ 0 & 2 & 0 \\ 1 & 0 & 2 \end{pmatrix} \quad \text{and} \quad \boldsymbol{\Theta}' = \begin{pmatrix} 2 & 0 & 0 \\ 0 & 2 & 0 \\ 0 & 0 & 2 \end{pmatrix},$$

```
TT=[2 0 1; 0 2 0; 1 0 2];
%TT=[2 0 0; 0 2 0; 0 0 2];
%TT=[2 1 0; 1 2 1; 0 1 2];
%TT=[2 0 1; 0 2 1; 1 1 2];
d=3; n=50; x=TT^(-1/2)*randn(d,n); S=x*x'/n;
T0=eye(d); C=5; e=0.1;
for o=1:100000
  T=T0+e*(inv(T0)-S);
  T(:)=L1BallProjection(T(:),C);
  if norm(T-T0)<0.00000001, break, end
  T0=T;
end
T, TT
```

```
function w=L1BallProjection(x,C)

u=sort(abs(x),'descend'); s=cumsum(u);
r=find(u>(s-C)./(1:length(u))',1,'last');
w=sign(x).*max(0,abs(x)-max(0,(s(r)-C)/r));
```

FIGURE 39.8

MATLAB code of a gradient-projection algorithm of ℓ_1-constraint MLE for Gaussian Markov networks. The bottom function should be saved as "`L1BallProjection.m`."

sparse MLE gives

$$\widehat{\boldsymbol{\Theta}} = \begin{pmatrix} 1.382 & 0 & 0.201 \\ 0 & 1.788 & 0 \\ 0.201 & 0 & 1.428 \end{pmatrix} \quad \text{and} \quad \widehat{\boldsymbol{\Theta}}' = \begin{pmatrix} 1.617 & 0 & 0 \\ 0 & 1.711 & 0 \\ 0 & 0 & 1.672 \end{pmatrix}.$$

Thus, the true sparsity patterns of $\boldsymbol{\Theta}$ and $\boldsymbol{\Theta}'$ (in off-diagonal elements) can be successfully recovered. Since

$$\boldsymbol{\Theta} - \boldsymbol{\Theta}' = \begin{pmatrix} 0 & 0 & 1 \\ 0 & 0 & 0 \\ 1 & 0 & 0 \end{pmatrix} \quad \text{and} \quad \widehat{\boldsymbol{\Theta}} - \widehat{\boldsymbol{\Theta}}' = \begin{pmatrix} -0.235 & 0 & 0.201 \\ 0 & 0.077 & 0 \\ 0.201 & 0 & -0.244 \end{pmatrix},$$

change in sparsity patterns (in off-diagonal elements) can be correctly identified.

On the other hand, when the true precision matrices are

$$\boldsymbol{\Theta} = \begin{pmatrix} 2 & 1 & 0 \\ 1 & 2 & 1 \\ 0 & 1 & 2 \end{pmatrix} \quad \text{and} \quad \boldsymbol{\Theta}' = \begin{pmatrix} 2 & 0 & 1 \\ 0 & 2 & 1 \\ 1 & 1 & 2 \end{pmatrix},$$

sparse MLE gives

$$\widehat{\boldsymbol{\Theta}} = \begin{pmatrix} 1.303 & 0.348 & 0 \\ 0.348 & 1.157 & 0.240 \\ 0 & 0.240 & 1.365 \end{pmatrix} \quad \text{and} \quad \widehat{\boldsymbol{\Theta}}' = \begin{pmatrix} 1.343 & 0 & 0.297 \\ 0 & 1.435 & 0.236 \\ 0.297 & 0.236 & 1.156 \end{pmatrix}.$$

Thus, the true sparsity patterns of $\boldsymbol{\Theta}$ and $\boldsymbol{\Theta}'$ can still be successfully recovered. However, since

$$\boldsymbol{\Theta} - \boldsymbol{\Theta}' = \begin{pmatrix} 0 & 1 & -1 \\ 1 & 0 & 0 \\ -1 & 0 & 0 \end{pmatrix} \quad \text{and} \quad \widehat{\boldsymbol{\Theta}} - \widehat{\boldsymbol{\Theta}}' = \begin{pmatrix} -0.040 & 0.348 & -0.297 \\ 0.348 & -0.278 & 0.004 \\ -0.297 & 0.004 & 0.209 \end{pmatrix},$$

change in sparsity patterns was not correctly identified (although 0.004 is reasonably close to zero). This shows that, when a nonzero unchanged edge exists, say $\Theta_{k,k'} = \Theta'_{k,k'} > 0$ for some k and k', it is difficult to identify this unchanged edge because $\widehat{\Theta}_{k,k'} \approx \widehat{\Theta}'_{k,k'}$ does not necessarily hold by separate sparse MLE from $\{\boldsymbol{x}_i\}_{i=1}^{n}$ and $\{\boldsymbol{x}'_{i'}\}_{i'=1}^{n'}$.

39.2.2 SPARSE DENSITY RATIO ESTIMATION

As illustrated above, sparse MLE can perform poorly in structural change detection. Another limitation of sparse MLE is the Gaussian assumption. A Gaussian Markov network can be extended to a non-Gaussian model as

$$q(\boldsymbol{x};\boldsymbol{\theta}) = \frac{\overline{q}(\boldsymbol{x};\boldsymbol{\theta})}{\int \overline{q}(\boldsymbol{x};\boldsymbol{\theta})\mathrm{d}\boldsymbol{x}},$$

where, for a *feature vector* $\boldsymbol{f}(x,x')$,

$$\overline{q}(\boldsymbol{x};\boldsymbol{\theta}) = \exp\left(\sum_{k\geq k'} \boldsymbol{\theta}_{k,k'}^{\top} \boldsymbol{f}(x^{(k)},x^{(k')})\right).$$

This model is reduced to the Gaussian Markov network if

$$\boldsymbol{f}(x,x') = -\frac{1}{2}xx',$$

while higher-order correlations can be captured by considering higher-order terms in the feature vector. However, applying sparse MLE to non-Gaussian Markov networks is not straightforward in practice because the normalization term $\int \overline{q}(\boldsymbol{x};\boldsymbol{\theta})\mathrm{d}\boldsymbol{x}$ is often computationally intractable.

To cope with these limitations, let us handle the change in parameters, $\boldsymbol{\theta}_{k,k'} - \boldsymbol{\theta}'_{k,k'}$, directly via the following density ratio function:

$$\frac{q(\boldsymbol{x};\boldsymbol{\theta})}{q(\boldsymbol{x};\boldsymbol{\theta}')} \propto \exp\left(\sum_{k\geq k'} (\boldsymbol{\theta}_{k,k'} - \boldsymbol{\theta}'_{k,k'})^\top \boldsymbol{f}(x^{(k)}, x^{(k')})\right).$$

Based on this expression, let us consider the following density ratio model:

$$r(\boldsymbol{x};\boldsymbol{\alpha}) = \frac{\exp\left(\sum_{k\geq k'} \boldsymbol{\alpha}_{k,k'}^\top \boldsymbol{f}(x^{(k)}, x^{(k')})\right)}{\int p'(\boldsymbol{x})\exp\left(\sum_{k\geq k'} \boldsymbol{\alpha}_{k,k'}^\top \boldsymbol{f}(x^{(k)}, x^{(k')})\right)\mathrm{d}\boldsymbol{x}}, \tag{39.2}$$

where $\boldsymbol{\alpha}_{k,k'}$ is the difference of parameters:

$$\boldsymbol{\alpha}_{k,k'} = \boldsymbol{\theta}_{k,k'} - \boldsymbol{\theta}'_{k,k'}.$$

$p'(\boldsymbol{x})$ in the denominator of Eq. (39.2) comes from the fact that $r(\boldsymbol{x};\boldsymbol{\alpha})$ approximates $p(\boldsymbol{x})/p'(\boldsymbol{x})$ and thus the normalization constraint,

$$\int r(\boldsymbol{x};\boldsymbol{\alpha})p'(\boldsymbol{x})\mathrm{d}\boldsymbol{x} = 1,$$

is imposed.

Let us learn the parameters $\{\boldsymbol{\alpha}_{k,k'}\}_{k\geq k'}$ by a *group-sparse* variant (see Section 24.4.4) of *KL density ratio estimation* explained in Section 38.3 [69]:

$$\begin{aligned}
\min_{\{\boldsymbol{\alpha}_{k,k'}\}_{k\geq k'}} \quad & \log \frac{1}{n'}\sum_{i'=1}^{n'} \exp\left(\sum_{k\geq k'} \boldsymbol{\alpha}_{k,k'}^\top \boldsymbol{f}(x_{i'}^{\prime(k)}, x_{i'}^{\prime(k')})\right) \\
& - \frac{1}{n}\sum_{i=1}^{n}\sum_{k\geq k'} \boldsymbol{\alpha}_{k,k'}^\top \boldsymbol{f}(x_i^{(k)}, x_i^{(k')}) \\
& \text{subject to } \sum_{k\geq k'} \|\boldsymbol{\alpha}_{k,k'}\| \leq R^2,
\end{aligned}$$

where $R \geq 0$ controls the sparseness of the solution.

A MATLAB code of a gradient-projection algorithm of sparse KL density ratio estimation for Gaussian Markov networks is given in Fig. 39.9. For the true precision matrices

$$\boldsymbol{\Theta} - \boldsymbol{\Theta}' = \begin{pmatrix} 2 & 0 & 1 \\ 0 & 2 & 0 \\ 1 & 0 & 2 \end{pmatrix} - \begin{pmatrix} 2 & 0 & 0 \\ 0 & 2 & 0 \\ 0 & 0 & 2 \end{pmatrix} = \begin{pmatrix} 0 & 0 & 1 \\ 0 & 0 & 0 \\ 1 & 0 & 0 \end{pmatrix},$$

```
Tp=[2 0 1; 0 2 0; 1 0 2]; Tq=[2 0 0; 0 2 0; 0 0 2];
Tp=[2 1 0; 1 2 1; 0 1 2]; Tq=[2 0 1; 0 2 1; 1 1 2];
d=3; n=50; xp=Tp^(-1/2)*randn(d,n); Sp=xp*xp'/n;
xq=Tq^(-1/2)*randn(d,n); A0=eye(d); C=1; e=0.1;
for o=1:1000000
  U=exp(sum((A0*xq).*xq));
  A=A0-e*((repmat(U,[d 1]).*xq)*xq'/sum(U)-Sp);
  A(:)=L1BallProjection(A(:),C);
  if norm(A-A0)<0.00000001, break, end
  A0=A;
end
-2*A, Tp-Tq
```

FIGURE 39.9

MATLAB code of a gradient-projection algorithm of ℓ_1-constraint KL density ratio estimation for Gaussian Markov networks. "`L1BallProjection.m`" is given in Fig. 39.8.

sparse KL density ratio estimation gives

$$\begin{pmatrix} 0 & 0 & 1.000 \\ 0 & 0 & 0 \\ 1.000 & 0 & 0 \end{pmatrix}.$$

This implies that the change in sparsity patterns can be correctly identified.

Even when nonzero unchanged edges exist as

$$\boldsymbol{\Theta} - \boldsymbol{\Theta}' = \begin{pmatrix} 2 & 1 & 0 \\ 1 & 2 & 1 \\ 0 & 1 & 2 \end{pmatrix} - \begin{pmatrix} 2 & 0 & 1 \\ 0 & 2 & 1 \\ 1 & 1 & 2 \end{pmatrix} = \begin{pmatrix} 0 & 1 & -1 \\ 1 & 0 & 0 \\ -1 & 0 & 0 \end{pmatrix},$$

sparse KL density ratio estimation gives

$$\begin{pmatrix} 0 & 0.707 & -0.293 \\ 0.707 & 0 & 0 \\ -0.293 & 0 & 0 \end{pmatrix}.$$

Thus, the change in Markov network structure can still be correctly identified.

Bibliography

REFERENCES

1. Akaike H. A new look at the statistical model identification. IEEE Trans Automat Control 1974;AC-19(6):716–23.
2. Ali SM, Silvey SD. A general class of coefficients of divergence of one distribution from another. J Roy Statist Soc Ser B 1966;28(1):131–42.
3. Aloise D, Deshpande A, Hansen P, Popat P. NP-hardness of Euclidean sum-of-squares clustering. Mach Learn 2009;75(2):245–9.
4. Amari S. Theory of adaptive pattern classifiers. IEEE Trans Electron Comput 1967;EC-16(3):299–307.
5. Amari S, Nagaoka H. Methods of information geometry. Providence (RI, USA): Oxford University Press; 2000.
6. Amit Y, Fink M, Srebro N, Ullman S. Uncovering shared structures in multiclass classification. In: Ghahramani Z, editor. Proceedings of the 24th annual international conference on machine learning. Omnipress; 2007. p. 17–24.
7. Anderson TW. An introduction to multivariate statistical analysis. 2nd ed. New York (NY, USA): Wiley; 1984.
8. Argyriou A, Evgeniou T, Pontil M. Convex multi-task feature learning. Mach Learn 2008;73(3):243–72.
9. Aronszajn N. Theory of reproducing kernels. Trans Amer Math Soc 1950;68:337–404.
10. Auer P. Using confidence bounds for exploitation-exploration trade-offs. J Mach Learn Res 2002;3:397–422.
11. Bach FR, Lanckriet GRG, Jordan MI. Multiple kernel learning, conic duality, and the SMO algorithm. In: Proceedings of the twenty-first international conference on machine learning. New York (NY, USA): ACM Press; 2004. p. 6–13.
12. Bartlett P, Pereira FCN, Burges CJC, Bottou L, Weinberger KQ, editors. Practical Bayesian optimization of machine learning algorithms; 2012.
13. Belkin M, Niyogi P. Laplacian eigenmaps for dimensionality reduction and data representation. Neural Comput 2003;15(6):1373–96.
14. Bengio Y. Learning deep architectures for AI. Found Trends Mach Learn 2009;1(2):1–127.
15. Bishop CM. Pattern recognition and machine learning. New York (NY, USA): Springer; 2006.
16. Blei DM, Ng AY, Jordan MI. Latent Dirichlet allocation. J Mach Learn Res 2003;3:993–1022.
17. Borgwardt KM, Gretton A, Rasch MJ, Kriegel H-P, Schölkopf B, Smola AJ. Integrating structured biological data by kernel maximum mean discrepancy. Bioinformatics 2006;22(14):e49–57.
18. Boser BE, Guyon IM, Vapnik VN. A training algorithm for optimal margin classifiers. In: Haussler D, editor. Proceedings of the fifth annual ACM workshop on computational learning theory. ACM Press; 1992. p. 144–52.
19. Boucheron S, Lugosi G, Bousquet O. Concentration inequalities. In: Bousquet O, von Luxburg U, Rätsch G, editors. Advanced lectures on machine learning. Lecture notes in computer science, vol. 3176. Berlin (Heidelberg): Springer; 2004. p. 208–40.
20. Boyd S, Parikh N, Chu E, Peleato B, Eckstein J. Distributed optimization and statistical learning via the alternating direction method of multipliers. Found Trends Mach Learn 2011;3(1):1–122.
21. Breiman L. Bagging predictors. Mach Learn 1996;26(2):123–40.
22. Breiman L. Random forests. Mach Learn 2001;45(1):5–32.

23. Breunig MM, Kriegel H-P, Ng RT, Sander J. LOF: identifying density-based local outliers. In: Chen W, Naughton JF, Bernstein PA, editors. Proceedings of the ACM SIGMOD international conference on management of data; 2000. p. 93–104.
24. Caruana R. Multitask learning. Mach Learn 1997;28:41–75.
25. Chang CC, Lin CJ. LIBSVM: a library for support vector machines. Technical report, Department of Computer Science, National Taiwan University, `http://www.csie.ntu.edu.tw/~cjlin/libsvm/`; 2001.
26. Chang C-C, Tsai H-C, Lee Y-J. A minimum enclosing balls labeling method for support vector clustering. Technical report, National Taiwan University of Science and Technology, 2007.
27. Chang W-C, Lee C-P, Lin C-J. A revisit to support vector data description. Technical report, National Taiwan University, 2013.
28. Chapelle O, Schölkopf B, Zien A, editors. Semi-supervised learning. Cambridge (MA, USA): MIT Press; 2006.
29. Chung FRK. Spectral graph theory. Providence (RI, USA): American Mathematical Society; 1997.
30. Cortes C, Vapnik V. Support-vector networks. Mach Learn 1995;20:273–97.
31. Cramér H. Mathematical methods of statistics. Princeton (NJ, USA): Princeton University Press; 1946.
32. Crammer K, Dekel O, Keshet J, Shalev-Shwartz S, Singer Y. Online passive-aggressive algorithms. J Mach Learn Res 2006;7(March):551–85.
33. Crammer K, Kulesza A, Dredze M. Adaptive regularization of weight vectors. In: Bengio Y, Schuurmans D, Lafferty J, Williams CKI, Culotta A, editors. Advances in neural information processing systems, vol. 22. 2009. p. 414–22.
34. Crammer K, Singer Y. On the algorithmic implementation of multiclass kernel-based vector machines. J Mach Learn Res 2001;2:265–92.
35. Csiszár I. Information-type measures of difference of probability distributions and indirect observation. Studia Sci Math Hungar 1967;2:229–318.
36. Dempster AP, Laird NM, Rubin DB. Maximum likelihood from incomplete data via the EM algorithm. J Roy Statist Soc Ser B 1977;39(1):1–38.
37. Domingo C, Watanabe O. MadaBoost: a modification of AdaBoost. In: Proceedings of the thirteenth annual conference on computational learning theory; 2000. p. 180–9.
38. du Plessis MC, Sugiyama M. Semi-supervised learning of class balance under class-prior change by distribution matching. Neural Netw 2014;50:110–9.
39. Duchi J, Shalev-Shwartz S, Singer Y, Chandra T. Efficient projections onto the ℓ_1-ball for learning in high dimensions. In: McCallum A, Roweis S, editors. Proceedings of the 25th annual international conference on machine learning. Omnipress; 2008. p. 272–9.
40. Evgeniou T, Pontil M. Regularized multi-task learning. In: Proceedings of the tenth ACM SIGKDD international conference on knowledge discovery and data mining. ACM; 2004. p. 109–17.
41. Fisher RA. The use of multiple measurements in taxonomic problems. Ann Eugenics 1936;7(2):179–88.
42. Fisher RA. The use of multiple measurements in taxonomic problems. Ann Eugenics 1936;7:179–88.
43. Friedman J, Hastie T, Tibshirani R. Additive logistic regression: a statistical view of boosting. Ann Statist 2000;28(2):337–407.
44. Friedman J, Hastie T, Tibshirani R. Sparse inverse covariance estimation with the graphical lasso. Biostatistics 2008;9(3):432–41.
45. Fukumizu K, Sriperumbudur BK, Gretton A, Schölkopf B. Characteristic kernels on groups and semigroups. In: Koller D, Schuurmans D, Bengio Y, Bottou L, editors. Advances in neural information processing systems, vol. 21. 2009. p. 473–80.
46. Gärtner T. Kernels for structured data. Singapore: World Scientific; 2008.

47. Geman S, Geman D. Stochastic relaxation, Gibbs distributions and the Bayesian restoration of images. IEEE Trans Pattern Anal Mach Intell 1984;6:721–41.
48. Girolami M. Mercer kernel-based clustering in feature space. IEEE Trans Neural Netw 2002;13(3):780–4.
49. Gretton A, Borgwardt KM, Rasch M, Schölkopf B, Smola AJ. A kernel method for the two-sample-problem. In: Schölkopf B, Platt J, Hoffman T, editors. Advances in neural information processing systems, vol. 19. Cambridge (MA, USA): MIT Press; 2007. p. 513–20.
50. Gretton A, Sriperumbudur B, Sejdinovic D, Strathmann H, Balakrishnan S, Pontil M, et al. Optimal kernel choice for large-scale two-sample tests. In: Bartlett P, Pereira FCN, Burges CJC, Bottou L, Weinberger KQ, editors. Advances in neural information processing systems, vol. 25. 2012. p. 1214–22.
51. Griffiths TL, Steyvers M. Finding scientific topics. Proc Natl Acad Sci USA 2004;101:5228–35.
52. Hastings WK. Monte Carlo sampling methods using Markov chains and their applications. Biometrika 1970;57(1):97–109.
53. He X, Niyogi P. Locality preserving projections. In: Thrun S, Saul L, Schölkopf B, editors. Advances in neural information processing systems, vol. 16. Cambridge (MA, USA): MIT Press; 2004. p. 153–60.
54. Hinton GE. Training products of experts by minimizing contrastive divergence. Neural Comput 2002;14(8):1771–800.
55. Hinton GE, Salakhutdinov RR. Reducing the dimensionality of data with neural networks. Science 2006;313(5786):504–7.
56. Hoerl AE, Kennard RW. Ridge regression: biased estimation for nonorthogonal problems. Technometrics 1970;12(3):55–67.
57. Holland PW, Welsch RE. Robust regression using iteratively reweighted least-squares. Comm Statist Theory Methods 1978;6(9):813–27.
58. Huber PJ. Robust statistics. New York (NY, USA): Wiley; 1981.
59. Jolliffe IT. Principal component analysis. New York (NY, USA): Springer-Verlag; 1986.
60. Kawahara Y, Sugiyama M. Sequential change-point detection based on direct density-ratio estimation. Stat Anal Data Min 2012;5(2):114–27.
61. Kawakubo H, du Plessis M, Sugiyama MC. Coping with class balance change in classification: class-prior estimation with energy distance. Technical report IBISML2014-71, IEICE, 2014.
62. Knuth DE. Seminumerical algorithms. The art of computer programming. vol. 2 Reading (MA, USA): Addison-Wesley; 1998.
63. Konishi S, Kitagawa G. Generalized information criteria in model selection. Biometrika 1996;83(4):875–90.
64. Kullback S, Leibler RA. On information and sufficiency. Ann Math Stat 1951;22:79–86.
65. Lafferty J, Mccallum A, Pereira F. Conditional random fields: probabilistic models for segmenting and labeling sequence data. In: Proceedings of the 18th international conference on machine learning; 2001. p. 282–9.
66. Langford J, Li L, Zhang T. Sparse online learning via truncated gradient. J Mach Learn Res 2009;10:777–801.
67. Li K. Sliced inverse regression for dimension reduction. J Amer Statist Assoc 1991;86(414):316–42.
68. Liu JS. The collapsed Gibbs sampler in Bayesian computations with applications to a gene regulation problem. J Amer Statist Assoc 1994;89(427):958–66.
69. Liu S, Quinn J, Gutmann MU, Sugiyama M. Direct learning of sparse changes in Markov networks by density ratio estimation. Neural Comput 2014;26(6):1169–97.
70. Loftsgaarden DO, QuesenBerry CP. A nonparametric estimate of a multivariate density function. Ann Math Stat 1965;36(3):1049–51.

71. Mackay DJC. Information theory, inference, and learning algorithms. Cambridge (UK): Cambridge University Press; 2003.
72. Metropolis N, Rosenbluth AW, Rosenbluth MN, Teller AH, Teller E. Equations of state calculations by fast computing machines. J Chem Phys 1953;21(6):1087–92.
73. Mosteller F, Tukey JW, editors. Data analysis and regression. Reading (MA, USA): Addison-Wesley; 1977.
74. Murphy KP. Machine learning: a probabilistic perspective. Cambridge (Massachusetts, USA): MIT Press; 2012.
75. Nguyen X, Wainwright MJ, Jordan MI. On surrogate loss functions and f-divergences. Ann Statist 2009;37(2):876–904.
76. Nguyen X, Wainwright MJ, Jordan MI. Estimating divergence functionals and the likelihood ratio by convex risk minimization. IEEE Trans Inform Theory 2010;56(11):5847–61.
77. Orr MJL. Introduction to radial basis function networks. Technical report, Center for Cognitive Science, University of Edinburgh, 1996.
78. Parikh N, Boyd S. Proximal algorithms. Found Trends Optim 2013;1(3):123–231.
79. Pearson K. On the criterion that a given system of deviations from the probable in the case of a correlated system of variables is such that it can be reasonably supposed to have arisen from random sampling. Phil Mag 5 1900;50(302):157–75.
80. Petersen KB, Pedersen MS. The matrix cookbook. Technical report, Technical University of Denmark, 2012.
81. Quiñonero-Candela J, Sugiyama M, Schwaighofer A, Lawrence N, editors. Dataset shift in machine learning. Cambridge (Massachusetts, USA): MIT Press; 2009.
82. Rao C. Information and the accuracy attainable in the estimation of statistical parameters. Bull Calcutta Math Soc 1945;37:81–9.
83. Ricci F, Rokach L, Shapira B, Kantor PB, editors. Recommender systems handbook. New York (NY, USA): Springer; 2010.
84. Rissanen J. Modeling by shortest data description. Automatica 1978;14(5):465–71.
85. Rumelhart DE, Hinton GE, Williams RJ. Learning representations by back-propagating errors. Nature 1986;323:533–6.
86. Salakhutdinov RR, Hinton GE. Deep Boltzmann machines. In: van Dyk D, Welling M, editors. Proceedings of twelfth international conference on artificial intelligence and statistics. JMLR workshop and conference proceedings, vol. 5. Beach (FL, USA): Clearwater; 2009. p. 448–55.
87. Schapire RE, Freund Y. Boosting: foundations and algorithms. Cambridge (Massachusetts, USA): MIT Press; 2012.
88. Schölkopf B, Smola A, Müller K-R. Nonlinear component analysis as a kernel eigenvalue problem. Neural Comput 1998;10(5):1299–319.
89. Schölkopf B, Smola AJ. Learning with kernels. Cambridge (MA, USA): MIT Press; 2002.
90. Scott DW. Multivariate density estimation: theory, practice and visualization. New York (NY, USA): Wiley; 1992.
91. Sejdinovic D, Sriperumbudur B, Gretton A, Fukumizu K. Equivalence of distance-based and RKHS-based statistics in hypothesis testing. Ann Statist 2013;41(5):2263–91.
92. Shannon C. A mathematical theory of communication. Bell Syst Tech J 1948;27:379–423.
93. Silverman BW. Density estimation for statistics and data analysis. London (UK): Chapman and Hall; 1986.
94. Sima V. Algorithms for linear-quadratic optimization. New York (NY, USA): Marcel Dekker; 1996.
95. Smith AFM, Roberts GO. Bayesian computation via the Gibbs sampler and related Markov chain Monte Carlo methods. J Roy Statist Soc Ser B 1993;55:3–24.

96. Smolensky P. Information processing in dynamical systems: foundations of harmony theory. In: Rumelhart DE, McClelland JL, editors. Parallel distributed processing: explorations in the microstructure of cognition, vol. 1. Cambridge (MA, USA): MIT Press; 1986. p. 194–281.
97. Stone M. An asymptotic equivalence of choice of model by cross-validation and Akaike's criterion. J Roy Statist Soc Ser B 1977;39:44–7.
98. Sugiyama M. Dimensionality reduction of multimodal labeled data by local Fisher discriminant analysis. J Mach Learn Res 2007;8(May):1027–61.
99. Sugiyama M. Statistical reinforcement learning: modern machine learning approaches. Boca Raton (Florida, USA): Chapman and Hall, CRC; 2015.
100. Sugiyama M, Idé T, Nakajima S, Sese J. Semi-supervised local Fisher discriminant analysis for dimensionality reduction. Mach Learn 2010;78(1–2):35–61.
101. Sugiyama M, Kawanabe M. Machine learning in non-stationary environments: introduction to covariate shift adaptation. Cambridge (Massachusetts, USA): MIT Press; 2012.
102. Sugiyama M, Krauledat M, Müller K-R. Covariate shift adaptation by importance weighted cross validation. J Mach Learn Res 2007;8(May):985–1005.
103. Sugiyama M, Suzuki T, Kanamori T, du Plessis MC, Liu S, Takeuchi I. Density-difference estimation. Neural Comput 2013;25(10):2734–75.
104. Sugiyama M, Suzuki T, Nakajima S, Kashima H, von Bünau P, Kawanabe M. Direct importance estimation for covariate shift adaptation. Ann Inst Statist Math 2008;60(4):699–746.
105. Sutton RS, Barto GA. Reinforcement learning: an introduction. Cambridge (MA, USA): MIT Press; 1998.
106. Székely GJ, Rizzo ML. Energy statistics: a class of statistics based on distances. J Statist Plann Inference 2013;143(8):1249–72.
107. Takeuchi K. Distribution of information statistics and validity criteria of models. Math Sci 1976;153:12–8 [in Japanese].
108. Tangkaratt V, Sasaki H, Sugiyama M. Direct estimation of the derivative of quadratic mutual information with application in supervised dimension reduction. Technical report 1508.01019, arXiv; 2015.
109. Tax DMJ, Duin RPW. Support vector data description. Mach Learn 2004;54(1):45–66.
110. Tibshirani R. Regression shrinkage and subset selection with the lasso. J Roy Statist Soc Ser B 1996;58(1):267–88.
111. Tibshirani R, Saunders M, Rosset S, Zhu J, Knight K. Sparsity and smoothness via the fused lasso. J Roy Statist Soc Ser B 2005;67:91–108.
112. Tomioka R, Aihara K. Classifying matrices with a spectral regularization. In: Ghahramani Z, editor. Proceedings of the 24th annual international conference on machine learning. Omnipress; 2007. p. 895–902.
113. Torkkola K. Feature extraction by non-parametric mutual information maximization. J Mach Learn Res 2003;3:1415–38.
114. Tsochantaridis I, Joachims T, Hofmann T, Altun Y. Large margin methods for structured and interdependent output variables. J Mach Learn Res 2005;6:1453–84.
115. Vapnik VN. Statistical learning theory. New York (NY, USA): Wiley; 1998.
116. Vincent P, Larochelle H, Bengio Y, Manzagol PA. Extracting and composing robust features with denoising autoencoders. In: Proceedings of 25th annual international conference on machine learning; 2008. p. 1096–103.
117. von Neumann J. Various techniques used in connection with random digits. In: Householder AS, Forsythe GE, Germond HH, editors. Monte Carlo methods; National bureau of standards applied mathematics series, vol. 12. 1951. p. 36–8.

118. Wahba G. Spline models for observational data. Philadelphia (PA, USA): Society for Industrial and Applied Mathematics; 1990.
119. Watanabe S. Algebraic geometry and statistical learning theory. Cambridge (UK): Cambridge University Press; 2009.
120. Wu CFJ. On the convergence properties of the EM algorithm. Ann Statist 1983;11:95–103.
121. Yamada M, Suzuki T, Kanamori T, Hachiya H, Sugiyama M. Relative density-ratio estimation for robust distribution comparison. Neural Comput 2013;25(5):1324–70.
122. Yuan M, Lin Y. Model selection and estimation in regression with grouped variables. J Roy Statist Soc Ser B 2006;68(1):49–67.
123. Zelnik-Manor L, Perona P. Self-tuning spectral clustering. In: Saul LK, Weiss Y, Bottou L, editors. Advances in neural information processing systems, vol. 17. Cambridge (MA, USA): MIT Press; 2005. p. 1601–8.
124. Zou H, Hastie T. Regularization and variable selection via the elastic net. J Roy Statist Soc Ser B 2005;67(2):301–20.

Index

SYMBOLS

A

B

C

D

E

F

G

H

I

J

K

L

M

N

O

P

Q

R

S

T

U

Printed in the United States
By Bookmasters